SFL
GE45 .M37 2002
Environmental modelling and
prediction

G. Peng   L.M. Leslie   Y. Shao

Environmental Modelling
and Prediction

Springer
Berlin
Heidelberg
New York
Barcelona
Hong Kong
London
Milan
Paris
Tokyo

Gongbing Peng   Lance M. Leslie
Yaping Shao   (Eds.)

Environmental Modelling and Prediction

With 168 Figures

PROFESSOR GONGBING PENG
PROFESSOR LANCE M. LESLIE

University of New South Wales
Sydney 2052
Australia

DR. YAPING SHAO
Department of Physics and Material Science
City University of Hong Kong
Kowloon
Hong Kong

ISBN 3-540-67422-5
Springer-Verlag Berlin Heidelberg New York

Library of Congress Cataloging-in-Publication Data

Environmental modelling and prediction / Gongbing Peng, Lance Leslie,
Yaping Shao (eds.). p. cm. Includes bibliographical references and index.
ISBN 3-540-67422-5
1. Environmental sciences--Mathematical models. I. Peng, Gongbing. II. Leslie,
Lance, 1945-III. Shao, Yaping.

This work is subject to copyright. All rights are reserved, whether the whole or part of the material is concerned, specifically the rights of translation, reprinting, reuse of illustrations, recitation, broadcasting, reproduction on microfilm or in any other way, and storage in data banks. Duplication of this publication or parts thereof is permitted only under the provisions of the German Copyright Law of September 9, 1965, in its current version, and permission for use must always be obtained from Springer-Verlag. Violations are liable for prosecution under the German Copyright Law.

Springer-Verlag Berlin Heidelberg New York
a member of BertelsmannSpringer Science+Business Media GmbH

http//:www.springer.de

© Springer-Verlag Berlin Heidelberg 2002
Printed in Germany

The use of general descriptive names, registered names, trademarks, etc. in this publication does not imply, even in the absence of a specific statement, that such names are exempt from the relevant protective laws and regulations and therefore free for general use.

Product liability: The publishers cannot guarantee the accuracy of any information about the application of operative techniques and medications contained in this book. In every individual case the user must check such information by consulting the relevant literature.

Production: camera-ready by authors
Cover design: E. Kirchner, Heidelberg
Printed on acid-free paper SPIN 10685941 32/3130/as 5 4 3 2 1 0

Preface

The environment is the aggregate of the physical, geographical and biochemical conditions under which we live. It significantly influences all aspects of human society, including agriculture, industry, human health, resources, etc. In recent history, we have been challenged not only by the impact of natural environmental variability, but also by human induced environmental changes, such as deforestation, destruction of the ozone layer, increased emission of greenhouse gases and pollution of air, water and soil. It is now increasingly understood that to better protect the environment is an essential task of human society. Environmental modelling and prediction, i.e., the modelling of environmental processes and the prediction of environmental changes, form a critical component of this task.

The environment is a complex interactive system comprising five major components and numerous sub-components. The major components are the atmosphere, the hydrosphere, the cryosphere, the lithosphere and the biosphere. Environmental modelling and prediction is concerned with both the behaviour of the individual components and their interactions, through the mathematical and numerical representation of the the physical, chemical and biological processes taking place within the environmental system. It is also concerned with the variability, stability and predictability of the environment as a dynamic system on a wide range of temporal and spatial scales.

There is an increasing requirement throughout the world for a better understanding of environmental processes and an enhanced capability for predicting environmental changes resulting from natural variations or human interferences. Numerous studies dealing with the modelling and prediction of individual components have been carried out in the past with considerable success. In recent years, environmental research is more focused on integrated approaches in which the environment is treated as a cohesive dynamic system. However, so far there have been very few books dealing exclusively with the latter subject. Our aim in this book is to bring together the collective efforts of scientists from several disciplines, to summarize recent developments in theories, methodologies and applications of integrated environmental modelling and prediction. Our effort in achieving this aim is reflected in Chaps. 1, 2 and 11. In Chap. 1, the subject area is explored and the scope of the book is defined. In Chap. 2, we have attempted to establish a theoretical framework for modelling and predicting the environment as a dynamic system, while in Chap. 11, examples of integrated environmental modelling and prediction are presented.

For environmental modelling and prediction, the modelling and prediction of the weather and climate components are of great importance for two main reasons. First, the atmosphere is the most active part of the environmental system on short time scales (up to a year). Second, the methods used in atmospheric modelling and prediction are arguably the most highly developed and also have the most complete data sets for verification. Chap. 3 of this book is devoted to the modelling and prediction of the atmosphere on a wide range of time and space scales. The next most important component of the environment is the ocean which dominates the behaviour of the environment on time scales from annual to decadal and beyond. Recent developments in ocean modelling and prediction are presented in Chap. 4.

A detailed treatment of the other environmental components is also presented because, although they interact with the atmosphere and the ocean, they have their own distinct dynamic features. It is important for success in environmental modelling and prediction that these individual components and their interactions are also well understood. We summarize the theories, methodologies and achievements in the modelling and prediction of the land-surface, the continental hydrosphere, the cryosphere and the continental biosphere (Chaps. 5, 6, 7 and 8, respectively). The interactions between the atmosphere, the ocean and these environmental components are also discussed. Some aspects of biological modelling and air quality modelling are discussed in Chaps. 9 and 10.

The book also explores recent developments in the theory and applications of environmental prediction, including systems analysis, information theory and natural cybernetics. The further development and application of these theories are very important for environmental modelling and prediction.

This book is a collective effort of many people. In particular, we wish to thank Drs. R. Morison, M. England, P. Oke, P. Irannejad, S. Liu, C. Ciret, Y. Tan, H. Duc, S. Zhao, M. Speer and L. Qi for their contributions to the various chapters. The assistance of Dr. S. Xia in the preparation of the manuscript and graphs is gratefully acknowledged.

Sydney, May 2001

Gongbing Peng
Lance M. Leslie
Yaping Shao

Contents

1. **Environmental Science** 1
   Gongbing Peng, Yaping Shao and Lance M. Leslie
   1.1 The Environment and Environmental Challenges 1
   1.2 Environmental Modelling 8
   1.3 Environmental Prediction 10
   1.4 Recent Developments in Environmental Sciences 12
       1.4.1 Earth Systematics 12
       1.4.2 Natural Cybernetics 14
       1.4.3 Geographic Information System 16
       1.4.4 Non-Linear Dynamical System Analysis and Predictability 17
       1.4.5 Links of Environmental Science to Related Subjects .. 19

2. **The Environmental Dynamic System** 21
   Yaping Shao, Gongbing Peng and Lance M. Leslie
   2.1 The Environment as a Dynamic System 21
   2.2 Mathematical Representations and Simplifications 30
       2.2.1 Governing Equations for Atmosphere 31
       2.2.2 Governing Equations for the Oceans 33
       2.2.3 Basic Equations for Land Surface 34
       2.2.4 Basic Equations for Ground Water 35
       2.2.5 Basic Equations for Ice 36
   2.3 Predictability of the Environmental System 37
   2.4 Methods for Environmental Prediction 45
   2.5 Integrated Environmental Modelling Systems 46
       2.5.1 Atmosphere–Land Surface Interaction Scheme 48
       2.5.2 Atmosphere–Ocean Interaction Scheme 50
   2.6 Ensemble Predictions 52
       2.6.1 Choosing the Initial Perturbations 53
       2.6.2 Cell Mapping 54
   2.7 Dynamic-Stochastic Models 56
       2.7.1 Dissipative Structure 56
       2.7.2 Synergetics 58
       2.7.3 Deterministic Chaos Theory 59

　　　　　2.7.4　Interactive Chaos................................. 62
　　2.8　Natural Cybernetics 66

3. **Atmospheric Modelling and Prediction** 75
   Lance M. Leslie, Russel P. Morison, Milton S. Speer and Lixin Qi
   3.1　Introduction 75
   　　　3.1.1　Impact on Society of Severe Weather and Climate Events　76
   　　　3.1.2　Historical Background 78
   　　　3.1.3　The Role of the Atmosphere in the Environmental System... 79
   　　　3.1.4　Numerical Weather Prediction and Modelling 80
   3.2　Equations Governing Atmospheric Motion 82
   　　　3.2.1　Some Commonly Used NWP Models................ 83
   　　　3.2.2　The UNSW HIRES Model......................... 84
   3.3　Solving the Governing Equations 85
   　　　3.3.1　The Shallow Water Equations...................... 86
   　　　3.3.2　The Finite-difference Techniques 87
   　　　3.3.3　Stability Analyses 88
   　　　3.3.4　Defining the Initial State for the Model Integration ... 91
   3.4　Applications 93
   　　　3.4.1　Very Short Range Forecasting (VSRF) 93
   　　　3.4.2　Short Range Forecasting (SRF) 96
   　　　3.4.3　Medium Range Forecasting (MRF) 99
   　　　3.4.4　Modelling the General Circulation of the Atmosphere . 101
   3.5　Climate Modelling...................................... 104
   　　　3.5.1　The GCM Model Numerics 105
   　　　3.5.2　The Model Physics 106
   　　　3.5.3　Ocean Model Coupling 107
   3.6　Applications and Results 107
   　　　3.6.1　Medium range forecasting experiments 107
   　　　3.6.2　Brief climatology of the model 108
   　　　3.6.3　Regional simulations 109
   3.7　Statistical Models 112
   　　　3.7.1　Deterministic and Statistical Models 113
   　　　3.7.2　Analogue Retrieval Techniques 113
   　　　3.7.3　Combining Deterministic and Statistical Models...... 114
   　　　3.7.4　Multiple Regression Markov Model 115
   3.8　Future Directions...................................... 116
   　　　3.8.1　Impact of Cloud Microphysics...................... 117
   　　　3.8.2　Ensemble Forecasting 117
   　　　3.8.3　Variational Data Assimilation 121

4. Ocean Modelling and Prediction 125
Matthew H. England and Peter R. Oke
- 4.1 Introduction 125
  - 4.1.1 What Is an Ocean Model? 125
  - 4.1.2 Mean Large-scale Ocean Circulation 126
  - 4.1.3 Oceanic Variability 126
  - 4.1.4 The Oceans, Climate, and Forcing 129
- 4.2 A Brief History of Ocean Modelling 131
- 4.3 Anatomy of Ocean Models 132
  - 4.3.1 Governing Physics and Equations 132
  - 4.3.2 Model Choice of Vertical Coordinate 135
  - 4.3.3 Subgrid-scale Processes and Dissipation 136
  - 4.3.4 Boundary Conditions and Surface Forcing 141
- 4.4 Some Commonly Used Ocean Models 145
  - 4.4.1 The GFDL Modular Ocean Model 145
  - 4.4.2 The Princeton Ocean Model 147
  - 4.4.3 The Miami Isopycnic Coordinate Model (MICOM) ... 149
  - 4.4.4 The DieCast Model 149
- 4.5 Ocean Model Applications 150
  - 4.5.1 A Global Coarse Resolution Model 150
  - 4.5.2 Global Eddy-permitting Simulations 153
  - 4.5.3 Regional Simulations in the North Atlantic Ocean 154
  - 4.5.4 ENSO Modelling 156
  - 4.5.5 A Regional Model of the Southern Ocean 157
  - 4.5.6 A Coastal Ocean Model off Eastern Australia 158
  - 4.5.7 A Coastal Model of a River Plume 159
- 4.6 Exploiting Ocean Observations 160
  - 4.6.1 Model Assessment 160
  - 4.6.2 Inverse Methods and Data Assimilation 165
  - 4.6.3 Applications of Data Assimilation to Coastal Ocean Models 168
  - 4.6.4 Application of Data Assimilation to Large-scale Models 169
  - 4.6.5 Variational Data Assimilation, Example 169
- 4.7 Concluding Remarks 171

5. Land Surface Processes 173
Parviz Irannejad and Yaping Shao
- 5.1 The Importance of Land-surface Processes 173
- 5.2 Land-surface Parameterization Schemes 176
- 5.3 Modelling Soil Hydrology in Land-surface Schemes 179
  - 5.3.1 Bucket Model 179
  - 5.3.2 Force-restore Model 180
  - 5.3.3 Multi-layer Soil Model 181
- 5.4 Soil Temperature 185
- 5.5 Surface Energy Balance 187

	5.5.1 Surface Fluxes 188
	5.5.2 Evaporation from Unsaturated Soil 190
	5.5.3 Transpiration 191
	5.5.4 Bulk Transfer Coefficients 192
	5.5.5 Canopy (Vegetation) Resistance 194
	5.5.6 Canopy Water Storage 195
	5.5.7 Surface Albedo 196
5.6	Carbon-dioxide Budget 198
5.7	Verification of Land-surface Schemes 198
5.8	Land Surface Modelling over Heterogeneous Surfaces 202
	5.8.1 Models of Dynamic Effects 204
	5.8.2 Models of Aggregation Effects 206
5.9	Land-surface Parameters 209
5.10	The Impact of Land Surface on Climate and Weather: Some Case Studies 210

6. Hydrological Modelling and Forecasting 215
Suxia Liu

6.1	The Hydrosphere 215
6.2	Mathematical Representation and Simplification 219
	6.2.1 Water Vapour 219
	6.2.2 Precipitation 222
	6.2.3 Soil Water 224
	6.2.4 Infiltration 226
	6.2.5 Depression Storage 229
	6.2.6 Vegetation Water 230
	6.2.7 Interception 232
	6.2.8 Evapotranspiration 233
	6.2.9 Channel Flow 235
	6.2.10 Overland Flow 236
	6.2.11 Groundwater 237
	6.2.12 Water Quality 239
6.3	Hydrological Modelling 244
	6.3.1 Xinanjiang Model 245
	6.3.2 TOPOMODEL 248
	6.3.3 Integrated Hydrological Modelling System 250
6.4	Hydrological Forecasting 254
	6.4.1 Short-term Hydrological Forecasting 255
	6.4.2 Mid- and Long-term Hydrological Forecasting 262
	6.4.3 Water Quality Forecasting 264
	6.4.4 Water Supply and Water Demand Forecasting 267

7. Simulation and Prediction of Ice-snow Cover 275
Gongbing Peng
- 7.1 Basic Characteristics of the Global Ice-snow Cover 276
  - 7.1.1 Sea-ice Cover.. 276
  - 7.1.2 Snow Cover 277
- 7.2 Ice-snow Interactions with Other Environmental Components 281
  - 7.2.1 Influences of Ice-snow Cover on the Atmosphere 281
  - 7.2.2 Effects of the General Circulation of the Atmosphere and Climate on Ice-snow Cover 283
  - 7.2.3 Main Physical Processes of Interactions among Ice-Snow Cover, Atmospheric Circulation and Water Cycle 284
- 7.3 Numerical Simulation 285
  - 7.3.1 Energy Balance Models 286
  - 7.3.2 Dynamic-Thermodynamic Models 291
  - 7.3.3 The Holland Model 293
  - 7.3.4 Treatment of Ice-snow Cover in GCMs 294
  - 7.3.5 Other Models 299
- 7.4 Prediction of Sea-ice and Related Components 299
  - 7.4.1 Numerical Forecasting of Sea-ice 301
  - 7.4.2 Meteorological and Hydrological Predictions using Ice-snow Cover as an Indicator 312

8. Modelling and Prediction of the Terrestrial Biosphere 317
Catherine Ciret
- 8.1 Introduction and Background 317
- 8.2 Terrestrial Ecosystem Models 319
  - 8.2.1 Model Overview................................... 319
  - 8.2.2 Model Validation 324
  - 8.2.3 Model Limitations and Research Needs.............. 327
- 8.3 Modelling Biosphere–Atmosphere Interactions.............. 329
  - 8.3.1 Biosphere–Atmosphere Feedbacks: Overview of Numerical Experiments 329
  - 8.3.2 Different Approaches Toward the Development of an Interactive Biosphere Model 332
- 8.4 Simulation of an Interactive Vegetation Canopy in a Climate Model: Example of the Plant Production and Phenology (PPP) Model 336
  - 8.4.1 Description of the Plant Production and Phenology Model 336
  - 8.4.2 Application of the PPP Model 340
- 8.5 Summary... 342

9. Theoretical Basis of Biological Models in Environmental Simulation 345
Yunhu Tan
- 9.1 Introduction 345
- 9.2 Single Reactant or Substrate 347
- 9.3 Multiple Reactants or Substrates 350
- 9.4 Substrate Inhibition 353
- 9.5 pH Effects 354
- 9.6 Comparison of Models with Observations 357
- 9.7 Discussion and Conclusions 358

10. Modelling of Photochemical Smog 361
Hiep Duc, Vo Anh and Merched Azzi
- 10.1 Air Chemistry of Smog Formation 361
  - 10.1.1 Chemistry of Ground Level Ozone Formation 361
  - 10.1.2 Generic Reaction Set (GRS) Mechanism 363
  - 10.1.3 Integrated Empirical Rate (EIR) Model 364
  - 10.1.4 Application 368
- 10.2 Air Chemistry in Air Quality Modelling 371
  - 10.2.1 Review of Air Quality Modelling 371
  - 10.2.2 Air Pollution Forecasting Using GRS and Reactive State-space Models 373
  - 10.2.3 The Extended Space-time Model 374
  - 10.2.4 The GRS Mechanism 374
  - 10.2.5 The State-space Form 375
  - 10.2.6 Application 377

11. Applications of Integrated Environmental Modelling 383
Yaping Shao and Sixiong Zhao
- 11.1 Recent Developments in Integrated Environmental Modelling 383
- 11.2 Atmosphere-Ocean Interactions 386
- 11.3 Atmosphere-Land Interactions 393
  - 11.3.1 Soil Moisture Simulation 394
  - 11.3.2 Influence of Land-surface Processes on Weather and Climate 399
- 11.4 Atmosphere-Ocean-Ice Interactions 402
- 11.5 Prediction of Environmental Cycles 407
  - 11.5.1 Energy and Water Cycle 407
  - 11.5.2 Dust Cycle 410
- 11.6 Specific Events: Air Quality 418
- 11.7 Performance of Integrated Environmental Modelling Systems 423

References 425

Index 469

List of Contributors

G. Peng (Chaps. 1, 2, 7)	School of Mathematics, University of New South Wales, Sydney 2052, Australia
L. Leslie (Chaps. 1, 2, 3)	As above
Y. Shao (Chaps. 1, 2, 5, 11)	Department of Physics and Material Science, City University of Hong Kong, Kowloon, Hong Kong
R. Morison (Chap. 3)	School of Mathematics, University of New South Wales, Sydney 2052, Australia
M. England (Chap. 4)	As above
P. Oke (Chap. 4)	College of Oceanic and Atmospheric Sciences, Oregon State University, 104 Ocean Admin. Bldg. Corvallis, Oregon
P. Irannejad (Chap. 5)	Environment Division, Australian Nuclear Science & Technology Organisation, PMB 1 Menai, NSW 2234, Australia
S. Liu (Chap. 6)	School of Mathematics, Univrsity of New South Wales, Sydney 2052, Australia
C. Ciret (Chap. 8)	As above
Y. Tan (Chap. 9)	ET Group International, GPO Box 159 Canberra, ACT 2601, Auslralia
H. Duc (Chap. 10)	Environment Protection Authority, NSW, PO Box 29, Lidcombe, NSW 2141, Australia
S. Zhao (Chap. 11)	Institute of Atmospheric Physics, Chinese Academy of Sciences, Beijing 100080, China
M. Speer (Chap. 3)	Bureau of Meteorology, Sydney, NSW 2000, Australia
L. Qi (Chap. 3)	School of Mathematics, University of New South Wales, Sydney 2052, Australia

1. Environmental Science

Gongbing Peng, Yaping Shao and Lance M. Leslie

1.1 The Environment and Environmental Challenges

It is difficult to define precisely the environment , but in general, it embraces the social, geographical, physical and biochemical conditions under which we live. Conceptually, it can be divided into the social and the natural environment. The former is created by human activities, including industrial complexes, cities, villages etc., while the latter is a system comprising the atmosphere, hydrosphere, lithosphere, cryosphere and biosphere. The components of the natural environment undergo changes on a range of temporal and spatial scales, which impact upon the social and economic activities of human society. Figure 1.1 is a schematic illustration of the environmental components and the interactions between them.

The content of this book is comprised mainly with the study of the natural environment, which we refer to simply as the environment. Environmental science is concerned with the scientific laws governing the evolution of the environment and its impact on the human society. Further, it focuses on environmental assessment, prediction, management and control. It is a multidisciplinary subject, embracing physics, chemistry, mathematics, biology, ecology, geography, hydrology, computational science and astronomy, among many others. One of the cornerstones of environmental science is environmental modelling and prediction.

A profound feature of the environment is that its processes take place over a wide range of time scales, from micro-seconds to millions of years, and even billions of years if geological processes are also considered. The spatial domain of the environment depends upon the time scales we are interested in. In general, it extends from the upper levels of the stratosphere, about 50 km from the Earth's surface, to the upper level of the lithosphere, about 5 to 6 km over land in average and 4 km over the bottom of ocean. According to this definition, the time scales of the environmental processes, including some geological processes, extend over millions of years. The most obvious and intensive interactions between the environment and human society take place on time scales less than several hundred years. From this perspective,

1. Environmental Science

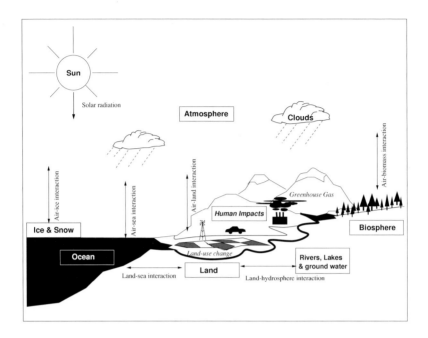

Fig. 1.1. A schematic illustration of the environmental components and the interactions among them (modified from Trenberth et al., 1996).

the deep lithosphere can be excluded from consideration. In this book, the environment is defined to comprise the upper stratosphere down to the top part of the lithosphere, about several hundred meters deep. The geological processes taking place in the deep lithosphere are not considered. While these processes are not unimportant, they can be considered to be either stationary or sporadic (e.g. volcanic eruptions and earthquakes) on the time scales of concern to this book.

In the new millennium, human society faces unprecedented challenges arising from environmental changes, brought about by both natural and human-induced processes. Some of the major challenges are described below.

Population. The world population has been rising rapidly from 2.5 billion in 1950, 4 billion in 1975 to over 6 billion in 2000 and will rise further to about 9 billion in 2050. The present rate of population growth is about 80 million per year. Over the next 50 years, global economic productivity will be several times its current level. These increases will exert serious pressures of global dimensions on natural resources. Every year about 5 to 7 million hectares of agricultural land is lost, giving a total of over 2 billion hectares (or 20 million sq kilometres) since records became available. There is now a serious shortage of cultivated land due to the combined effects of land degradation and urbanization. The forested areas of the world have also been falling sharply. At the beginning of the 18th century, they occupied 34% of the total land surface, in contrast to 24% at the end of the 20th century (Qu, 1987; Zhang, 1999). Presently, they are shrinking at a rate of about 11 million hectares per year. Because forests play an important role in the cycles of energy, water, CO_2 and aerosols, their destruction disturbs the balance between the different components of the environment, influencing atmospheric temperature and precipitation, and resulting in the degeneration of the entire ecosystem.

Global Warming. The burning of fossil fuels, industrial activities and deforestation significantly affect the atmospheric composition. Increases in CO_2 and other trace gases have led to global warming. The global surface temperature has increased by 0.3–0.6°C since the late 19th century, with the greatest warming over the region between 40 and 70°N. The most rapid increase appears to have occurred during the period 1955–1994 (Fig. 1.2). The global surface temperature appears to have been higher in the late 20th century than any other similar period in the past 600 years. In some regions, the 20th century has been the warmest for some thousands of years (Nicholls et al., 1996). The 1990s have been the hottest period on record. In 1998, the global mean surface temperature reached 16.86°C, which is higher than the mean value for 1931–1990 by 0.56°C. The high temperatures in 1998 caused many serious environmental problems. Corresponding increases in ocean and soil temperatures have been observed. Closely related to global warming of the atmosphere and oceans is rising sea-levels. In the 20th century, the global mean sea-level has risen about 10 to 20 cm, at a rate of 2.35 mm per year (Zhang, 1999), associated with an estimated ocean temperature increase of more then 0.1 °C.

Desertification. Desertification is one of the most serious environmental challenges. The drought affected areas on Earth total almost 48.8 million km^2 (about 1/3 of the total land surface) and 3/4 of this area is experiencing desertification, affecting about 1 billion people. The land surfaces in these areas are often subject to severe wind and water erosion. About 21 million hectares of farmland are affected each year by erosion activities. For example, dust storms frequently occurred in northern China in the 20 century. During the 1990s, dust storms occurred almost every year and were most severe in 1998 and 2000. Figure 1.3 shows the dust clouds associated with an intense wind erosion episode in the Takla Makan desert. The desert areas in northwestern China have been rising sharply, extending by 1560 km^2 per year in the 1970s, 2100 km^2 in the 1980s to 2400 km^2 in the 1990s (Chen, 1998).

Water Resources. Water covers about 71% of the Earth's surface, but water resources in many regions of the world are limited. Fresh water from lakes and rivers suitable for drinking constitutes less than 1% of the total amount of water. According to the Food and Agriculture Organization of the United Nations, by 2000, 51% of the available water will be consumed by agricultural

4   1. Environmental Science

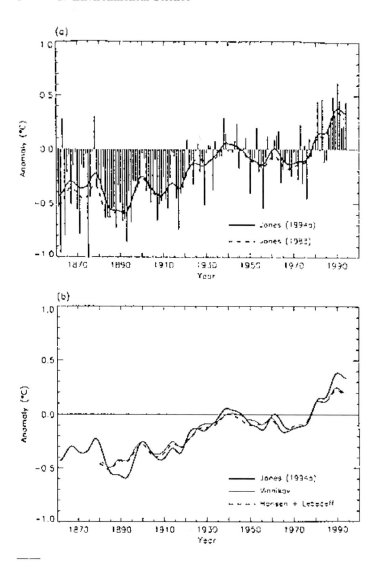

Fig. 1.2. (a) Annual averages of surface air temperature anomalies (°C) for land areas, 1981 to 1994, relative to 1961 to 1990. Bars and solid curve are from Jones (1994); dashed curve from Jones (1988). The smoothed curves are created using a 21-point binomial filter. (b) As (a) but updated from Hansen and Lebedeff (1988) – dashed line; Vinnikov et al. (1990) – thin line. Thick solid line is from Jones (1994), as in (a) (from Nicholls et al., 1996).

1.1 The Environment and Environmental Challenges   5

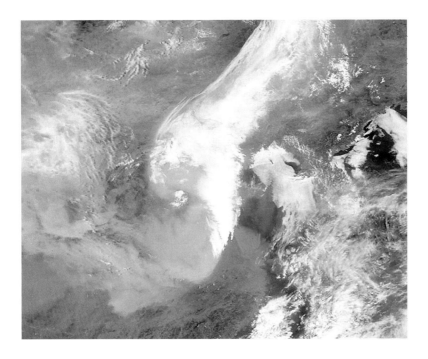

Fig. 1.3. Satellite image of the dust storm originating in the Takla Makan desert of northwest China on 14 April 1998. The image, from 16 April 1998, shows the dust clouds behind the cold front and near the centre of the storm (SeaWiFS image produced by Norman Kuring, NASA GSFC, with acknowledgment).

activities, 40% by industrial activities and 9% by other living purposes. The industrial use of water will increase further in the near future. The number of countries in the world with water shortages has been increasing. It was estimated at 18 in 1990, but will increase to about 30 by 2050. The situation of "cutoff of the river water flow" has happened in several countries, including China, India, Egypt and the USA (Zhang, 1999)and has serious consequences. For instance, in the lower reaches of the Yellow River, during the 26 year period 1972–1997, the flow cutoff occurred on 20 of those years. In 1997, the duration of the cutoff reached 226 days, stretching along a river length of up to 704 km (Ren, 1999).

Ozone Layer Destruction. Smith and Warr (1991) have reported that measurements from the British Antarctic Survey Station Halley Bay showed the presence of an ozone "hole" since 1985. In some cases, the amount of ozone decreased by 60%. NASA's analysis of 1986 showed that ozone decreased 2.3% to 6.2% between 30°N and 60°N for the winter months (December to March), although no significant trend was found for the summer months. The measurements indicate that the global ozone concentration decreased about

3.4–3.6% in average during 1978–1987. Further analyses with more recent data suggest that the decreasing may be accelerating. According to Zhang (1999), the area of the Antarctic ozone hole was 1.5 million km^2 in 1982, which has increased to 25 million km^2 in 1998. The decrease in ozone is probably related to the emissions of NO$_X$ from airplanes flying in the stratosphere and chlorofluoro carbons (CF$_2$Cl$_2$ or CFCl$_3$) arising from the use of refrigerating equipment. The change in NO$_2$ has a strong impact on ozone concentration and distribution (Turco, 1992). The ozone layer in the stratosphere absorbs 99% of solar ultraviolet radiation. The destruction of the ozone layer has disastrous consequences to human health. For example, it is known to lead to increased skin cancer as people are exposed to stronger ultraviolet solar radiation.

Air, Water and Soil Pollution. Watson et al. (1990) have presented air pollution data which show the rapid increase of air pollution in the recent history (Table 1.1). Turco (1992), Houghton et al. (1992) and Prather et al. (1995) have also presented evidence which supports the findings of Watson et al. (1990). Associated with air pollution is acid rain, caused by the emissions of SO, SO$_2$ and NO$_X$ from the consumption of coal and petroleum. Acid rain first appeared in Northern European countries (e.g. Norway and Sweden), then observed in many places around the world. In the last 30 years, the acidity of rainfall has increased about 10% every year. Acid rain affects more than 1 million km^2 in China, which is 11.4% of the total area of that country (Chen, 1998). Similar observations can be made in the pollution of water and soil. For example, increases in the fluxes of trace gases such as CO$_2$, CH$_4$ and NO$_2$ from the undisturbed forest soils in central Germany and agricultural areas in North China Plain have been observed by Dong et al. (1998, 2000).

Table 1.1. Air pollution in the recent history (Watson et al., 1990).

Concentrations	CO$_2$ [ppmv]	CH$_4$ [ppmv]	CFC-11 [pptv]	CF-12 [pptv]	N$_2$O [ppbv]
Before industrialization (1750–1800)	280	0.8	0	0	288
Recent Concentrations (1990)	353	1.72	280	484	310
Annual rate of increase	0.5%	0.9%	4%	4%	0.25%

Natural Disasters. According to the frequencies of their occurrence worldwide in the last three decades, the most serious natural disasters are the following: floods, tropical cyclones, tornados and whirlwinds, earthquakes, thunderstorms, snow-storms and blizzards, heat waves, cold spells, volcanic eruptions, landslides and avalanches, tidal waves, tsunamis, frosts, droughts and dust storms. Figure 1.4 shows the satellite image of a Typhoon Maggie approaching Taiwan (June 5, 1999). Just one tropical cyclone of the annual global average of over 50 can devistate a country or region, especialy in developing countries.

Fig. 1.4. Visible satellite image of the cloud system of Typhoon Maggie over the tropical Pacific Ocean to the southeast of Taiwan, which is in the top left hand corner of the figure. The spiral cloud bands, the eye of the typhoon and the strong convective region embedded in the spiral cloud bands and the diverging cirrus can be clearly seen (courtesy of National Satellite Meteorological Center, China Meteorological Administration).

Most of the above listed disasters are associated with severe weather events and/or other intense short-term oceanic and land-surface processes. These disasters often occur in localized areas in a relatively short period of time, but can pose major threat to human life and cause severe damage to property. According to the statistical analysis of Clarke and Munasinghe (1994), during the period of 20 years from the 1970s to 1990s, earthquakes, volcanic eruptions, landslides, floods, tropical storms, droughts and other disasters have killed 3 million people, inflicted injury, disease, homelessness and social disruption on 1 billion others. As an example, the April 1999 hail storm that struck Sydney, Australia (Fig. 1.5) caused economic damage of more than 1.4 billion Australian dollars within a time period of about one hour. To some extend, human activities may result in quasi-natural disasters. Increased air, water and soil pollution and the destruction of forests are typical examples.

Fig. 1.5. A photo showing the size of hail during the 14 April 1999 Sydney hail storm. Hail size reached up to 11 cm.

1.2 Environmental Modelling

An increasing awareness of environmental issues has been the driving force behind the rapid development of environmental modelling. The key questions are how human society can be better prepared for major environmental challenges, prevent human induced environmental damages and better manage and protect the environment. The demand by society for a better understanding of the environment and an improved capacity for environmental prediction have been ever increasing.

Non-linearity is a key feature of the environment. The evolution of the environment as a whole involves the interactions between its components, e.g., between the ocean and the atmosphere, the ocean and the sea-ice, the atmosphere and the land surface, the biosphere and the atmosphere, the land-surface and the continent hydrosphere, etc. These interactions are essentially nonlinear. That is, the responses of a component (e.g. the atmosphere) to the forces exerted by another (e.g. the ocean) cannot be described using a linear equation system. The non-linear nature of the interactions is responsible for many of the fundamental difficulties in environmental modelling and prediction. Complexity is another key feature of the environment. The interactions among the environmental components occur through processes of different nature, either physical and chemical or biological. Moreover, these processes are subject to external forcing, such as solar radiation. Further, the environmental processes take place on a wide range of scales. The temporal scales range from micro-seconds to thousands of years, whereas the spatial scales range from micro-metres to thousands of kilometers.

As a consequence, the environment is a large dynamic system involving many processes and interactions which generate a wide range of temporal and spatial fluctuations of a large number of variables. How can such a vast and complicated system be understood and predicted? This is the central problem of environmental modelling and prediction.

The past several decades have seen the developments of many computational environmental models. In these models, environmental processes are represented by mathematical equations, often comprising sets of coupled non-linear partial differential equations, derived from first principles, or equations derived empirically from observations. These equations are solved using numerical algorithms on powerful computers. Environmental modelling is achieved through the numerical solution of these equations for given boundary and initial conditions, by means of advanced computing techniques and machines.

Environmental modelling has become an essential tool for environmental studies, complementary to observation. While measurements provide our knowledge base for the environment, they are inevitably limited to certain aspects of the complex system. In contrast, environmental modelling allows exploration from an integrated perspective. A well-tested model can be a good representation of the environment as a whole, its dynamics and its responses to possible external changes. They can be used as virtual laboratories in which environmental phenomena can be reproduced, examined and controlled through numerical experiments. Environmental models also provide the framework for integrating the knowledge, evaluating the progress in understanding and creating new scientific concepts. Most importantly, environmental modelling provides the foundation for environmental prediction. Environmental models are useful for testing hypotheses, designing field experiments and developing scenarios.

Environmental modelling can be classified into two basic categories. One is the modelling of individual environmental components, e.g., the atmosphere and the ocean. The other is the simulation of the entire environment as a dynamical system, especially through the modelling of the component interactions and the feedback consequences.

An example of the first category are numerical weather prediction models, under development since the 1920s but with the first successful prediction not appearing until the late 1940s. These models are based on the approximation of the partial differential equations governing the thermo-dynamical processes of the atmosphere. These equations are, in essence, the conservation laws for momentum, mass and energy plus the equation of state. Finite difference and spectral methods (e.g. Krishnamurti et al., 1998; Durran, 1999; Jacobson, 1999) are most popular methodologies for solving them. These models have enjoyed considerable success in weather forecasting and climate simulations.

An example of the second category are general circulation models, also called global climate models (GCMs) which are essentially coupled systems of modules for the atmosphere, ocean, land and ice-snow. GCMs can be considered as the first versions of future environmental models. In contrast to current GCMs, which focus on the atmosphere, environmental models will increasingly and systematically consider the interactions among the environmental components and the behaviour of the environment as a whole. To this end, new sets of variables, including climate variables, will be required.

Until now, most environmental models are developed either for individual environmental components, or for a subset of the environment with two or three components, e.g., the atmosphere and ocean coupled subset and the atmosphere, ocean and ice coupled subset. None of the existing models fully comprises all environmental processes. In some GCMs, physical, chemical and biological process are considered, but most models only deal with one or two types of processes. The task of developing models capable of representing the entire environment is the ultimate challenge and one which will require decades of further research effort.

1.3 Environmental Prediction

Environmental prediction is the quantification of the future environment states, given its current and/or past conditions. It provides the scientific guidance for environmental protection and management. Three classes of environmental predictions can be identified, namely,

- Key variables which quantify the states of individual components, e.g., wind, air temperature, humidity and precipitation for the atmosphere; flow speed, temperature and salinity for the ocean; mass and area distribution of ice and snow, etc.
- Exchanges of mass, momentum and energy between the components, in association with major cycles in the environmental system, e.g., water and energy cycles; and

- Environmental phenomena, such as, air quality, ozone concentration, dust storms, desertification, etc.

From a time-scale perspective, environmental prediction can be short range (up to several days), middle range (weeks to months) and long range (annual to decadal). For short and middle range predictions, physical processes are highlighted, while long-range predictions must involve not only physical, but also chemical and biological processes.

Environmental prediction employs three main types of models, namely, dynamic, statistical and dynamic-stochastic. Dynamic models, for instance, are widely used for short-range atmospheric predictions. Tese are based on the understanding that the processes under consideration are governed by laws which can be expressed as deterministic mathematical equations. For given boundary and initial conditions, approximate solutions for these equations are obtained through numerical integrations. These approximate solutions constitute the predictions for the future states of the atmosphere. Dynamic models have the advantage of clear physical representations of the processes involved. However, the range of this type of prediction is limited because environmental processes are often deterministic only for a short time period. For instance, recent studies using AGCMs (Atmospheric Global Circulation Models) have shown that for middle-latitude synoptic systems, the predictability time scale using a dynamic model is about two weeks, and for planetary scale systems, it is about four weeks (Li, 1995). The main reason for the limitation in predictability lies in dynamic instability and non-linear interactions within the atmosphere, but variations in external factors may also play a significant role.

Statistical methods are developed on the basis of probability theory. These methods have been used widely for long-range weather and climate forecasts. Well-established statistical methods used in these areas include correlation matrix, regression equations, spectral analysis, and experimental orthogonal transformation. More recently, new statistical methods using singular-spectrum analysis, neural network, entropy analysis and combination of phase-space components have been developed. The statistical methods have been shown to be effective for long-range weather forecasts, but their disadvantage is that they often do not have a solid physical basis.

Dynamic-stochastic models are a combination of the two approaches described above. This is a promising method for long-range environmental predictions, as it retains the advantages of both the dynamic and statistical methods. Some discussion of this method is given in Chap. 2.

Whichever model is used, environmental modelling and prediction must be based on a sound understanding of the key processes and factors responsible for environmental changes. These processes and factors can be both interactions within the environmental system and external forcing (Lau, 1985; Walsh and Ross, 1988; Manabe et al., 1991 and Cane, 1992). In numerical models, this understanding is reflected in the conceptual model which provides the basis for constructing more sophisticated predictive models. In statistical models, this understanding is reflected in the choice of factors which represent the physical interpretation and causality of the statistical behaviour of the environment. Suppose the statistical (e.g., regression) model we construct can be written simply as

$$Y = F(X_i)$$

where Y represents the variable to be predicted, X_i the factors. It is often the case that the choice of X_i is more important than the statistical method used, which is reflected in the possible functional form of F. For example, as long-range climate predictions are concerned, attention should be paid to factors such as El-Nino (Dickinson et al., 1996; Wang and Eltahir, 1999), global ice-snow coverage (Peng et al., 1987, 1992, 1996; Li et al., 1996), concentrations of greenhouse gases (Manabe and Wetherald, 1987), soil moisture and temperature (Dickinson, 1992), vegetation cover (Aber, 1992), as well as factors such as solar activities (Reid, 1999), the Earth pole displacement and the Earth-rotation velocity (Peng and Lu, 1982, 1983; Peng and Si, 1983). The impact of some processes on the environment can be very significant. Again, the El-Nino phenomenon arising from the atmosphere-ocean interactions in the tropical Pacific is a good example. The occurrence of El-Nino is accompanied by a sea-surface temperature increase in the eastern Pacific equatorial ocean, which results in changes in the position and intensity of the Walker and the Hadley circulations. These eventually lead to changes in the patterns of air temperature and rainfall, forcing further adjustment in the continental hydrosphere and biosphere. Figure 1.6 shows the difference in the atmospheric circulation between normal and El-Nino conditions.

1.4 Recent Developments in Environmental Sciences

Recent years have seen substantial progress in environmental sciences. This is reflected in the emergence of new theories and new models as well as in the implementation of new technologies in environmental observations. A brief summary of these progresses is given below.

1.4.1 Earth Systematics

Earth systematics is an aggregation of many related sciences with cybernetics as its core. Since the 1960s, with the development of computer technology, great progress has been made in this area. The general system theory of Von Bertalanffy (1968), the thermodynamic theory of structure (Glansdorff and Prigogine, 1971) and synergetics (Haken, 1983) are examples. Qian, a pioneer of engineering cybernetics (e.g. Tsien, 1954), suggested applying cybernetics and system theory to the environmental system (e.g. Qian, 1994).

1.4 Recent Developments in Environmental Sciences

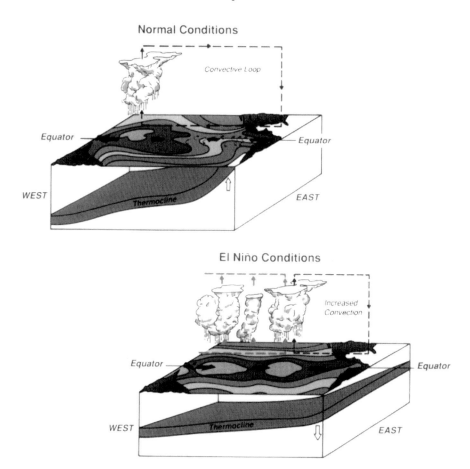

Fig. 1.6. Schematic illustration of the differences between normal and El-Nino conditions (after Dickinson et al., 1996).

The concepts and strategies for studying the environment as a system are outlined in the NASA document entitled "Earth System Science" (1988). In that document, the Earth is considered as an integrated system of interacting components, whose study must transcend disciplinary boundaries.

The creation of earth systematics is a logical extension of the advancement in satellite-based remote sensing of the Earth and the maturation of the individual disciplines related to the environmental components, including oceanography, meteorology, cryology, geography, geology, biology and geophysics. It also reflects the increasing impacts of human activities on the global environment.

Earth systematics is a framework for studying the Earth scaiences, which are considered as a set of interacting components, rather than a collection of individual components that are more or less independent. The emphasis of the theory is to describe and understand the evolution of the earth system on planetary scales. This is achieved through a combination of global observational techniques and sophisticated (both conceptual and numerical) models.

The evolution of the earth system has five distinctly identifiable time scales, namely, seconds to hours, days to seasons, decades to centuries, millennia and millions to billions of years. The characteristic spatial scales and the respective representative processes are shown in Fig. 2.2. The first two of the five time scales are scales are related to atmospheric, biological and oceanic processes. The intermediate timescale (decades to centuries) are of particular importance to the strategic response of the human society to environmental changes. On this time scale, the interactions between the natural environment and human society are strongest. The central themes for this time scale are climate change and the associated impact on the hydrosphere and biosphere. The principal processes for the interactions between the atmosphere and the biosphere lie in the time frame between decades and centuries. The very long time scales (millions to billions of years) include geophysical, geological and geochemical processes.

The earth systematics, also referred to as earth system science, approach has four steps: 1) the acquisition of observations; 2) the analysis and interpretation of the observed data; 3) the construction of conceptual and numerical models; and 4) the verification of the models together with their use to furnish statistical predictions of future trends. Such a framework enormously increases the range and quantity of observations to be accommodated, the scope of data analysis and interpretation, and the complexity of the models. The core of the approach is the establishment of the earth-information system. Such a system is used for receiving, processing and archiving a large quantity of observed data; for promoting data analysis and interpretation; for supporting the development of numerical models; for extrapolating results as well as identifying and simulating future trends. Mathematical and numerical models are used to carry the analysis further, incorporating numerical algorithms for processes that permit quantitative links with other processes and hence the incorporation of interactions among them. Verification of the predictions completes the final step of the cyclical research approach.

1.4.2 Natural Cybernetics

Natural cybernetics is a new branch of environmental studies. The expression, "Natural Cybernetics", was introduced by Zeng (1995). Its central concern is the optimization of human induced environmental control under the constraints of natural environmental processes and social-economic requirements.

1.4 Recent Developments in Environmental Sciences

Natural cybernetics is a multi-disciplinary subject. In particular, it embraces various subjects in environmental sciences, cybernetics, mathematics and engineering. The essence of natural cybernetics can be represented by several general equations. Suppose $X(p,t)$ denotes the ensemble of environmental variables and $Y(p,t)$ denotes the ensemble of variables related to human activities, where p is space and t is time. Y affects X directly or indirectly, such that X is determined both by itself and Y. The evolution of both X and Y is expressed by mathematical equations (usually partial differential equations). These equations can be solved with given initial and boundary conditions. In natural cybernetics, the most suitable Y is sought to achieve a predefined optimization (see Chap. 2 for more details).

Several problems associated with natural cybernetics have been studied by Zeng (1995, 1996) and Zhu et al. (1999). They have shown how the theory can be applied to optimal engineering of silt sedimentation in navigation channels, hydraulic design, pollution control, control of climate, artificial weather modification and others. Among the examples mentioned above, the study of the control of silt sedimentation is the most complete, as outlined in Chap. 2. Here, we use the control of the global mean surface temperature as an example to illustrate the basis of natural cybernetics.

Suppose the models for climate and the cycle of CO_2 are sufficiently accurate. For different scenarios of CO_2 emission reduction, the increase of the global mean surface temperature, T_s, in year 2050 can be estimated. Suppose the global mean emission rate of CO_2 is $\dot{c}(t)$. Then the increase in T_s, $\Delta T_s = T_s(t) - T_s^0$, is a function of $\dot{c}(t)$ with T_s^0 being the current value of T_s. It is required that

$$\overline{\Delta T_s} = \left[\int_{t_1}^{t_2} \int_S [T_s(s,t) - T_s^0(s,t)]dsdt\right] \frac{1}{S(t_2 - t_1)} \leq \delta \tag{1.1}$$

where δ is the allowed change of temperature in the time interval $(t_2 - t_1)$ and S is the area of land surface. The problem is then to find the maximum $\dot{c}(t)$ under the constraint of (1.1), i.e.

$$\int_{t_1}^{t_2} \dot{c}(t)dt = \max \tag{1.2}$$

The developments of different economies and the "fairness" for different countries also need to be considered. This is reflected in a weighting function $R_2(s,t)$. It follows that (1.2) should be replaced by

$$\int_{t_1}^{t_2} \int\int_S R_2(s,t)\dot{c}(s,t)dsdt = \max \tag{1.3}$$

While (1.1) is an obvious over-simplification, considering the problems involved in the reduction of CO_2, it is sufficient here for the purpose of illustrating the basics of natural cybernetics.

Natural cybernetics has several specific characteristics, as summarized below. 1) The system to be controlled is an open and complicated one of nonlinear nature in the natural environment. 2) The evolution of the system is basically deterministic, but is complex. Uncertainties arise from internal processes, initial and boundary influences and other random factors. The control is applied to the long-term behaviour of the system. 3) The control is only applied to localized and sensitive parts of the system. The results of the control are small changes in the evolution of the system, but they are most concerned and important to the human society. 4) The evolution of the system is described by unsteady partial differential equations with inclusion of some ordinary differential equations. 5) The related control problems are usually complicated and difficult and, therefore, they often require intensive computation.

The implementation of natural cybernetics proceeds as follows.

- Carry out observations of the processes involved to obtain sufficient information;
- Solve the governing equations of the system to understand its structure and behaviour and to identify the control mechanisms;
- Combine observations with the dynamic equations to derive unknown parameters and initial and boundary conditions;
- Design the optimization scheme which determine the control of system evolution;
- Examine the sensitivity and stability of the direct and the optimal control problems. It is desired that the controlled parts are the most sensitive in the system; and
- Study more generalized feedback-correction methods to correct the control at the next stage and studying the correspondent control techniques.

1.4.3 Geographic Information System

Geographic Information System (GIS) has been under development since the middle of the 1960s. By the early 1990s, it has become an essential tool for environmental studies. GIS provides a powerful framework for the integration of large amounts of many different types of spatial data obtained from field, airborne and satellite observations of the environment. On the technical level, it is a combination of computer-based databases and a wide range of application softwares. A typical GIS comprises a number of subsystems involving data input, storage, retrieval, manipulation, analysis and reporting.

The construction of a GIS involves the establishment of databases, data management, acquisition and retrieval functions. It requires the conversion of source data into information that is easy to use. Many countries have now developed environmental GIS for the purposes of model development, knowledge library and decision making (Chen et al., 1999).

The development of a GIS is closely related to remote sensing, air survey technology, the global positioning system (GPS) and the Internet. While a large proportion of current GIS data are obtained by digitizing the existing conventional maps based on aerial photographing, remotely sensed data, with increased availability and improved quality, are rapidly becoming the major data sources of GIS. In the past 20 years, both GIS and remote sensing technology have reached high levels of development. GIS has led to the extensive use of remotely sensed data, while remote sensing has in turn greatly improved the data quality and types of data required by modern GIS (Peuquet and Marble, 1990). In recent years, remote sensing has been increasingly applied to mineral resource exploration and to the monitoring and assessment of environmental events on scales from local to global, such as flood, drought, fire and land-surface changes. In the 1990s, geoinformation science has been developed on the basis of GIS and the earth systematics. The latest development in Digitzing Earth and Cyberspace has promoted GIS to a new stage.

GIS combined with remote sensing can be used not only in environmental monitoring, but also in environmental prediction. While GIS and environmental prediction using dynamic models have been developed more or less in parallel, there is an emerging trend to couple GIS and mathematical models, for integrated environmental modelling and prediction. This coupling utilizes the strengths of GIS in representing spatial data and that of dynamic models in representing temporal processes. The technique has been shown to be effective in the simulation and prediction of complex environmental processes which have not been possible before, such as the prediction of soil moisture and dust storms (e.g. Shao et al., 1997; Shao and Leslie, 1997). Cheng (1999) has also shown that the accuracy of flood forecasts can be improved using an integrated analysis of satellite data, radar measurement and remote measuring system of rainfall, using a digital elevation model and a databse of land-surface characteristics.

1.4.4 Non-Linear Dynamical System Analysis and Predictability

In studying weather predictability, Lorenz (1963) developed the theory of deterministic chaos. Based on the study by Saltzman (1962), Lorenz obtained a simple system of thee coupled non-linear equations for Bernoulli convection

$$\frac{\mathrm{d}X}{\mathrm{d}t} = -\sigma(X - Y) \tag{1.4a}$$

$$\frac{\mathrm{d}Y}{\mathrm{d}t} = rX - Y - XZ \tag{1.4b}$$

$$\frac{\mathrm{d}Z}{\mathrm{d}t} = XY - bZ \tag{1.4c}$$

where X, Y and Z are the intensity of convective motion, the temperature difference between the ascending and descending currents and the distortion of the vertical temperature profile from linearity, respectively; σ is the Prandtl number, r is the Rayleigh number and b is a constant.

The stability of the equilibrium states of the Lorenz system depends critically on the values of r. The equilibrium state is unstable when $r \geq 24.74$ (with $\sigma = 10$ and $b = 8/3$). In these cases, the trajectories of the system in the phase space move ceaselessly and irregularly, covering a region which appears like a butterfly (Fig. 1.7). This is now known as the Lorenz butterfly attractor. The Lorenz system is a simple example of chaos.

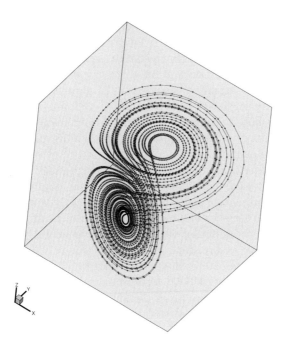

Fig. 1.7. The Lorenz butterfly attractor.

Chaos is the behaviour of bounded solutions of certain nonlinear dynamic systems which are sensitive to initial conditions. Non-periodic solutions are ordinarily unstable with respect to small modifications, so that slightly differing initial states can evolve into profoundly different ones. The existence of chaos creates the intrinsic unpredictable behavior caused by the nonlinear terms. In studying the Hadley circulation of the atmosphere, Lorenz (1984) used a low-order model, defined by the three ordinary differential equations:

$$\frac{dX}{dt} = -Y^2 - Z^2 - aX + aF \tag{1.5a}$$

$$\frac{dY}{dt} = XY - bXZ - Y + G \tag{1.5b}$$

$$\frac{dZ}{dt} = bXY + XY - Z \tag{1.5c}$$

where X represents the intensity of the symmetric globe encircling westerly wind (related to the pole-ward temperature gradient), and Y and Z are the cosine and sine phases of a chain of superposed large-scale eddies. XY and XZ indicate the amplification of the eddies. The terms $-bXZ$ and bXY are displacements of the eddies by the westerly current. The coefficients a and b influence the development of the westerly current and the displacement for the eddies and the terms aF and G express symmetric and asymmetric thermal forcing.

Lorenz used this model to study the behaviour of the Hadley Cell, discussing the conditions for the existence of chaos and of bifurcation processes. His analysis showed that the time-dependent solution was sensitive with respect to small disturbances. In particular, the solution of nonlinear equations is extremely sensitive to initial values.

Based on chaos theory, Chou (1995) and Guo et al. (1996) have developed a quasi-dynamic-stochastic method for long-term climatic predictions. They suggested that long-term climate prediction can be understood as the estimation of the state probability distribution on the attractors. Predicability is a measure of the degree of chaos. The work of the latter authors is an interesting attempt to apply the chaos theory to long-range environmental predictions.

1.4.5 Links of Environmental Science to Related Subjects

Earth systematics and natural cybernetics are newly developed subjects with distinct theoretical frameworks and methodologies. Their objectives are to increase our knowledge of the environment for its prediction and control. These two research fields are directly related to the environmental science and they share common characteristics. All three research fields are concerned with the changes of the environment caused not only by physical, chemical and biological processes, but also by human activities, most obviously in the geobiological cycle. As they deal with the same system, they share the same theoretical basis which involves almost all traditional subjects, including mathematics, physics, chemistry, astronomy, geophysics and biology. In terms of methodology, they all require the progress in contemporary cross subjects which include system theory, cybernetics, information theory, synergetics, non-linear dynamics, optimization theory and mutation theory.

The emphases of the three research fields overlap but also differ somewhat. Earth systematics is more a basic research field, combining geophysics, geology, geography etc. It has a focus on the relationships between the five main spheres (i.e., the atmosphere, hydrosphere, lithosphere, cryosphere and biosphere) and their interactions with the human society. While it is concerned with the physical, chemical and biological processes which result in global system variations, its emphasis is largely on global variations on time scales from decades to centuries.

Natural cybernetics is a cross-branch of environmental science and engineering. It is concerned with the control mechanism of the environment and with the theory and method of artificial control. Much attention is being paid to the stability and sensitivity of the natural processes and their non-linear variations.

Environmental science is both fundamental and applied, with an emphasis on environmental monitoring, forecast, management and control. Similar to earth systematics, this science is concerned with the formation and evolution of the environment, including the five main spheres and the interactions between the physical, chemical and biological processes. However, it is mostly concerned with the Earth's surface, while earth systematics also involves deeper layers of the earth. From the perspective of time scales, the environmental science is concerned not only with environmental changes on time scales of decades to centuries, but also those on much shorter time scales, including diurnal, seasonal and annual.

2. The Environmental Dynamic System

Yaping Shao, Gongbing Peng and Lance M. Leslie

2.1 The Environment as a Dynamic System

The environment is a dynamic system which comprises the atmosphere, the ocean, the land surface, the cryosphere, the continental hydrosphere and the terrestrial biosphere. Here, we will focus on that part of the atmosphere consisting mainly the troposphere and the lower part of the stratosphere, and on the land surface including the soil layer down to depth of some 50 to 100 m. This system has several fundamental features. First, a wide range of physical, chemical and biological processes take place on a spectrum of temporal and spatial scales. Second, while the individual components of the system may follow different laws, they are inter-dependent and their evolution cannot be understood fully when considered in isolation. Third, the system is open. The openness lies primarily in the fact that the Earth receives its incoming energy from the solar radiation, 90% of which is in the wavelength range of 0.3 to 3 μm, but which releases energy in a much longer wavelength radiation (90% in 3 to 30 μm). This openness results in the continuous adjustment and complex interactions among the components within the system. Finally, the interactions between the motions on different scales and those between the different components are mostly non-linear. Figure 2.1 is a schematic illustration of the environment and the major interactions between its key components. These include the interactions between the atmosphere and ocean, the atmosphere and sea-ice, the atmosphere and land surface, the land surface and hydrological system and the ocean and sea-ice. It remains an enormous challenge to the scientific community to determine how such a complex system can be approached.

Table 2.1 gives a summary of some relevant physical properties of the components.

The interactions between the environmental components are manifested as exchanges of energy, mass and momentum at the interfaces between them, the consequent dynamic adjustments within the components involved and the complex (often non-linear) feedback processes. For the individual components, such as the atmosphere and the ocean, there are well established

22   2. The Environmental Dynamic System

Table 2.1. Summary of the physical properties of the major components of the Earth's environment.

	Ocean	Atmosphere	Land Surface	Biosphere	Cryosphere
Total mass [Mt]	1.4×10^{12}	5.1×10^9	3.0×10^9		2.5×10^{10}
Total area [km^2]	3.6×10^8	5.1×10^8	1.5×10^8		1.6×10^7
Specific heat [J kg^{-1}K^{-1}]	4218	1004	1500		2106
Velocity [ms^{-1}]	1	10	10^{-4}	10^{-4}	10^{-4}

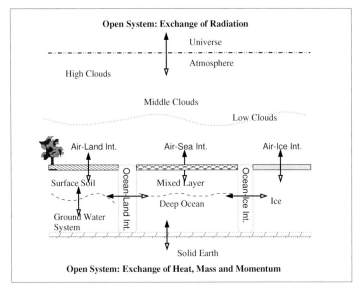

Fig. 2.1. A schematic illustration of the Earth's environment as an open system, the major components and the important interactions.

frameworks for research. However, for the environment as a whole, such a framework is yet to be fully developed.

Environmental modelling and prediction is concerned with the quantification of the current and future states of the environment. The traditional mainstream effort has been devoted to the development of theories and models for individual components and to the simulation and prediction of them. It is common, for instance, to consider the atmosphere as a thermal-dynamic system which obeys a set of partial differential equations, and to quantify its state using variables such as temperature, flow speed, pressure, etc. In modelling the atmosphere, its interactions with the other components of the environment were specified through the boundary conditions for the partial differential equations. In contrast, the environmental system as a whole is far larger and more complex than every individual component. Here, a range of inter-dependent processes of different nature, either physical, chemical or biological, take place on spatial scales ranging from several micro-metres to thousands of kilometres, and on temporal scales ranging from seconds to millions of years. Before we can consider the modelling and simulation of the environmental system, it is natural to ask first how such a complex system can be adequately quantified and how an adequate set of state variables can be defined. Indeed, it is virtually impossible to fully define the state of the environment using a set of variables, as such a set would be too largeto solve, unless we focus on a particular subset of problems.

It cannot be over-emphasised that the fundamental challenges of environmental modelling and prediction lie in the interactions within the environmental system. It is plausible to consider the interactions to be of two different kinds, namely, the interactions between the processes on different temporal and spatial scales and those between different components of the system. In order to simplify the environment to a dynamic system that is mathematically and computationally managable, we must focus our attention on the environmental phenomena occurring on particular temporal and spatial scales.

The typical spatial and temporal scales of the environmental phenomena are tentatively summarised in Fig. 2.2. The time scale of 10^6 to 10^9 years is an astronomical one which corresponds to the formation of the Earth, the evolution of life and the chemical composition of the atmosphere. While important and interesting, the processes on this time scale are not of concern within the context of this book. The emphasis of environmental modelling and prediction is placed upon shorter time scales. Five distinguishable time scales can be identified, as listed below:

- Geological Time Scale (10^3 to 10^6 years): Within this time frame, ice and non-ice age periods occur, developments of soils take place and there are ecological evolutions and redistributions. These environmental phenomena are associated with certain astronomical processes, such as the changes in the rotation of the Earth around the sun and solar activities. Geological processes also play important roles, the drift of the continental plates and changes in surface topography can force significant shifts in climate patterns and associated with them, changes in the ecosystem.
- Decadal Time Scale (10 to 10^3 years): Significant changes in the atmospheric composition occur, e.g. in the concentration of greenhouse gases such as carbon dioxide and ozone. Within this time frame, global climate variations can be significant and human induced environmental changes are most obvious, such as desertification related to deforestation and acid rain related to the emission of sulfur dioxide gases. In turn, the environmental processes on this time scale also impact significantly on human activities. In this time frame, the physical climatological and the biochemical processes are dominant. The physical climatological processes include those

related to atmospheric physics, atmosphere and ocean dynamics, land surface and the energy balance between the components of the environment. The bio-chemical processes include ones related to ocean biology, continental ecological system and chemical reactions in the stratosphere.
- Seasonal Time Scale (1 Year): Environmental processes on this time scale are fundamentally associated with the seasonal variations of the distribution of solar radiation. This is also the time scale for circulations in the oceans, fluctuations in sea ice coverage, hydrological circles in rivers and lakes and the period of plant growth.
- Sub-seasonal Time Scale (days to weeks): On this time scale, synoptic events occur. Individual atmospheric high- and low-pressure systems, fronts and hurricanes develop. Some of these systems lead to severe environmental events, such as flood and air-pollution episodes, that significantly impact on social and economic activities of the human society. This is also the time scale of large ocean eddies which last several weeks.
- Diurnal Time Scale (several hours to 1 day): This time scale is determined by the rotation of the earth, accompanied by the heating and cooling of the surface caused by the variation of solar radiation. The exchanges of energy, momentum and mass between the atmosphere, the land surface and the ocean are controlled by the physical and biological processes on this time scale. Between the atmosphere and the land surface, these exchanges, for instance, are determined by the development of turbulence that is closely associated with the heating and cooling of the surface. A range of phenomena with a diurnal variation can be identified, such as the variations of surface temperature, surface soil moisture and the photo-synthetic activities of plants.

It is natural to introduce a set of different temporal and spatial scales, (T_i, L_i), to characterise the ith subset of environmental processes. Suppose we are concerned mainly with the environment of a temporal scale T and a spatial scale L. Then the processes with $T_i \gg T$ and $L_i \gg L$ are slow processes and those with $T_i \ll T$ and $L_i \ll L$ are fast ones. In environmental modelling and prediction, it is often the case that our primary interests are on the behaviour on particular temporal and spatial scales. In that case, it is then reasonable to assume the slow processes to be stationary and the fast processes to be stochastic. For the latter ones, statistical tools can be implemented to represent them in the modelling and prediction of the environment under consideration.

Having defined the scales of the problems, it is then important to identify the major interactions within the environmental dynamic system and to quantify the associated feedback processes. Clearly, this identification is related to the environmental problems we are addressing and to their characteristic temporal and spatial scales. In the following, we consider global climate changes on seasonal to decadal scales and give an outline of the interactions as an illustration.

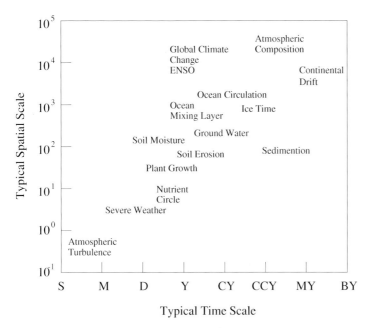

Fig. 2.2. Spatial and temporal scales of some representative processes taking place in the Earth's environment.

Atmosphere and Ocean Interactions. Atmosphere-ocean interactions are of primary importance to the environmental system. In general, the ocean has a longer response time than the other components of the environment, because it has a larger mass and a larger reservoir of energy which is available for exchange. Haken (1983) pointed out that in complex systems such as the environment, long-lived systems usually act as "masters" which "slave" the short-lived systems. The ocean is the dominant component in the behaviour of the environmental system on time scales similar to that of the ocean. One possible scenario is that the environment as a dynamic system should adjust toward an equilibrium that is dictated by the ocean. However, such an equilibrium may never be achieved, because during the adjustment of the environment to a given state of the ocean, this state may change due to external forces (e.g. astronomical), and more importantly, as a result of the interactions.

The atmosphere comes into direct contact with all other sub-components except the ground-water system and is involved in the interactions with all of them. The velocity scale of atmospheric motion, around $10\,\mathrm{ms}^{-1}$, is by far the largest for any of the components. The velocity scale of ocean flow is around $1\,\mathrm{ms}^{-1}$ and the movement of substances in soil is around $10^{-4}\,\mathrm{ms}^{-1}$. The atmosphere is a component which redistributes energy, mass and momentum throughout the entire system. As the atmosphere-and-ocean interactions are generally non-linear, the disturbances in the atmosphere and the ocean on certain length scales offen result in disturbances on other length scales.

As a consequence, the atmosphere-ocean interaction is the most important one in the environmental dynamic system. Recent studies have demonstrated that the ocean-circulation pattern and sea-surface temperature have a profound impact on the behaviour of the atmosphere, such as the well know El-Nino and La-Nina phenomena in the Pacific ocean, resulting in large fluctuations in precipitation around the global and hence further influencing the behaviour of the biosphere and the continental hydrosphere.

The El-Nino and Southern Oscillation, or ENSO, is a well known example of atmosphere-ocean interactions (see Chap. 4). It has major consequences on the behaviour of the environment as a whole. The ENSO phenomenon takes place in the equatorial Pacific. El-Nino is now accepted as one underlying mechanism responsible for extensive drought in Australia and abnormal weather events, such as the extreme floods in China in 1998. During the El-Nino years, anonamously warm water in the eastern Pacific, notably along the tropical coast of South America, occurs. The El-Nino phenomenon is closely associated with the phenomenon of Southern Oscillation in the tropical atmosphere, which represents the variations of the Walker circulation. El-Nino and Southern Oscillation are two closely related processes, one taking place in the Pacific Ocean and the other in the tropical atmosphere.

The Walker circulation (Fig. 2.3) is a longitudual circulation taking place in the low latitudes. It arises from the nonuniform distribution of land and sea and the variations of sea-surface temperature across the ocean, which introduce zonal assymetries in heating and the east-west circulation. Air rises at longitudes of relative heating and sinks at other longitudes of relative cooling. Under normal conditions of the ocean and atmosphere across the Pacific Ocean, in the tropical latitudes off the west coast of South America, the south or south-east trade winds generate a surface current of cold water, and this is deflected towards the west due to the Coriolis effect. The surface water drifting away is replaced by up-welling of colder deep ocean water. Consequently, a strong temperature gradient in sea-surface temperature extends from this area across the Pacific to the warm waters of the Indonesian archipelago. This temperature contrast drives an east-west circulation cell in the atmosphere, known as the Walker circulation. Over the Indonesian region the air is warm and light, air pressure is low and the air rises producing much cloud and rainfall. Over the eastern Pacific, the relatively cool air sinks, atmospheric pressure is high and rainfall is scanty by tropical standards. To complete the cell, there are easterly winds across the Pacific at the surface and westerlies in the upper atmosphere.

On occasion, the easterlies of the tropical Pacific weaken with resultant reduced wind stress. Warm water flows eastward across the Pacific displacing the cold water off the South American coast. With the appearance of warm surface waters in the central to eastern Pacific, the difference in temperature

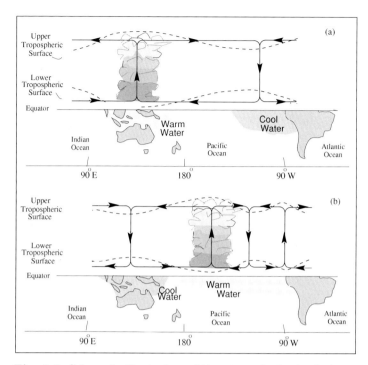

Fig. 2.3. Schematic illustration of the atmospheric circulations over the tropical Pacific. (a) The Walker circulation in normal conditions associated with warm water and high rainfall in eastern Australia, and cold surface water in the tropical east Pacific. (b) Typical circulation pattern during El-Nino events with warm waters in the tropical east Pacific. Areas consistently influenced by the ENSO phenomenon include Australia, South East Asian countries, the eastern part of China, the western part of the United States and some areas in South America and Africa.

across the Pacific is reduced. The Walker circulation becomes weaker, or can break down altogether and a more complicated circulation cell structure can develop in its place. As a result, cloud formation and rainfall is reduced in the Australian/Indonesian region but increased in the central to eastern Pacific. There is also evidence that these warm waters in the central Pacific can induce a series of atmospheric waves propagating into the Northern Hemisphere. For some El-Nino events these atmospheric disturbances can be the cause of severe weather events over the entire globe. Areas consistently influenced by the El-Nino phenomena include Australia, South East Asian countries, the eastern part of China and the western part of the United States and sections in South America and Africa.

Atmosphere and Land Interactions. Land surface occupies less than 30% of the Earth's surface. The total exchange of energy and mass (in terms of water vapour, carbon and other elements) between the atmosphere and the land surface is less than that between the atmosphere and ocean. However, atmosphere-land interactions remain of primary importance for several reasons. First, compared with the ocean, the land surface has a much larger aerodynamic roughness. The topographic properties of the surface produce strong disturbances in the atmosphere which impact on other components of the system. Second, the land surface has a much smaller heat capacity than the ocean and allows strong temperature contrasts to form between land and ocean on diurnal and longer, e.g., seasonal time scales. The small heat capacity and small heat conductivity of the land surface lead to strong fluctuations in land-surface temperature on diurnal to annual scales, resulting in fluctuations in the atmosphere. The combined dynamic and thermal effect of the land surface on the atmosphere can be clearly seen in major circulation systems such as the Indian and East Asian Monsoon, which have an enormous effect on the entire environment. Third, the land surface is much more heterogeneous in space and time than the ocean, which again leads to disturbances in the atmosphere. Finally, most human activities are concentrated on land and these activities leads to strong modifications of the land surface properties, such as deforestation, urban development and agriculture. Human activities have a significant influence on the dynamic behaviour of the entire environment, but most obviously on the immediate environment in which humans live.

Atmosphere and Ice Interactions. Snow and ice are located mainly in the polar regions. Compared to the land surface, snow and ice cover a much smaller area. As the solar radiation in the polar region is small, the energy exchange between the atmosphere and the ice surface over the polar regions must also be smaller in general than over the land surface. The impact of ice on the environment has three aspects, namely, 1) the presence of ice increases surface albedo and enhances the reflection of short-wave solar radiation back into space; 2) ice absorbs heat from the atmosphere and/or ocean during melting and hence modifies the variations in the atmospheric and ocean temperatures; 3) ice is a source of fresh water, melted ice reduces the salt concentration in ocean currents and affects the circulation of ocean.

As described in Chap. 7, the coverage of snow and ice over the entire globe can undergo significant seasonal and annual variations. Associated with these variations are the variations of the overall albedo of the globe. The albedo of ice surface is larger than all other surfaces in the natural environment. Albedo of sea ice with a surface temperature smaller than 269 K is around 0.8, and for ice surface with temperature at 271 K, albedo is still around 0.5. This is in contrast to the value of 0.2 for most vegetated and bare-soil surfaces and 0.07 for the ocean. Hence, the global coverage of ice significantly influences the energy balance of the entire environment, by reflecting short-wave solar radiation directly back to space.

The ice-albedo feedback hypothesis is an excellent illustration of the role ice plays in the environmental system. This is a positive feedback with a range of consequences. If there is a sufficiently strong disturbance in the environmental system that leads to increased ice coverage, the albedo of the surface will increase and, hence, more short-wave solar radiation will be reflected back into the space, leading to further cooling of the atmosphere and growing ice cover. On the other hand, if there is a disturbance which leads to decreased ice cover, the albedo of the surface also decreases, which results in increased warming of the environment and further melting of ice. Both feedbacks are positive.

In general, fluctuations in the total amount of ice is very slow in comparison with those taking place in the atmosphere and ocean. Hence, the variations in ice is a good indicator of the long-term climate variations.

Atmosphere–Biosphere Interaction. While the biosphere plays a passive role in environmental dynamics, the annual to long term variations in plant growth lead to changes in the physical properties of the land surface, such as surface albedo and aerodynamic roughness length, which in turn influence the atmosphere–land interactions. The transpiration of plants also plays an important role in the water circulation. The photo-synthetic effect of plants influences the ambient concentration of greenhouse gases, such as carbon-dioxide and methane, which further impacts upon the atmospheric radiative processes. Human activities may cause changes to the biosphere, such as large scale deforestation, which can impact on the global climate. These interactions are extremely important, as they produce the background setting for many other processes, e.g., plant growth, dry land salinity, soil erosion and water supply.

In studying the dynamics of deserts and drought in the Sahel, Charney (1975) proposed a self-induction effect for desert development through the enhancement of albedo, which occurs when a desert has formed or is forming. The lack of rainfall in desert areas results in a lack of vegetation. As a consequence, the surface albedo increases, as bare surfaces have a higher albedo than vegetated ones. Hence, the desert region reflects more short-wave solar radiation into space than the surroundings. In addition, desert surfaces are hotter than the surrounding regions and the air above them less cloudy. Hence, deserts emit more terrestrial radiation to the space. The combined effect is that the desert is a radiative sink of heat relative to the surroundings. In order to maintain thermal equilibrium, the air must descend and compress adiabatically, reducing the relative humidity in the atmosphere and enhances the dryness of the desert. This is a positive feedback process.

Hierarchy of Interactions. The hierarchy of interactions is reflected in the reactions of the other components of the system to a certain dynamic process in the environmental system. We consider again the ice-albedo feedback mentioned earlier. This feedback process is constrained by the reactions of the ocean and the atmosphere. An increasing ice coverage inevitably decreases the open surface of the ocean, leading to reduced evaporation and reduced cloud cover in the atmosphere. As a consequence, an increased amount of short-wave solar radiation will penetrate the atmosphere without being reflected back to space by clouds. With increased solar radiation reaching the surface, the temperature of the environment will increase, compensating the effect created directly by the ice-albedo feedback.

There are many examples of a hierarchy of interactions. Suppose that an ENSO event produces rainfall deficits in certain areas of the world, as is now almost certainly known to be the case. These deficits lead to reduced surface-vegetation cover, reduced soil moisture and hence evapotranspiration, but increased sensible heat flux. Such changes in the land-surface properties will lead to the adjustment of the atmosphere on the local scale. Reduced rainfall and vegetation cover then lead to increased potential for soil erosion, reduced fresh-water reserve in rivers and lakes, causing further a wide range of environmental problems.

The hierarchy of interactions is also reflected in the interactions between the different scales of the individual components. This type of interactions represent the typical path of energy flow: for instance, tropical convection and global circulation occur on two completely different scales, yet, they are closely related.

In the context of this book, environmental modelling and prediction has an implicit meaning of studying environmental phenomena on time scales below that of, or equal to, the decadal time scale. We are mainly concerned with four different scales, namely, decadal, seasonal, sub-seasonal and diurnal time scales. As attention changes from one time scale to the other, the principal interactions and the processes involved become different. It is on this philosophy that the simplifications and the mathematical representations of the system are based.

2.2 Mathematical Representations and Simplifications

The processes taking place in the environmental dynamic system can be classified as being geo-fluid dynamic, physical and bio-chemical. Each component of the system may have its own distinctly different dynamic, physical and bio-chemical processes that need to be represented using a set of mathematical equations. For the atmosphere and the ocean geo-fluid dynamic processes, hydrological processes and flow in porous media, the equations are well established. They are mostly represented by a set of partial differential equations. The basics of many physical and chemical processes in the environment are also well understood, for example, radiation in the atmosphere. However, a range of physical and bio-chemical processes, most of which are involved in the interactions among the environmental components, are not well understood. Often, mathematical expressions for these processes cannot be derived from first principles. Rather, we must resolve to techniques such as parameterisations. The governing set of equations for the environment as a whole is a collective set of equations. Inevitably, such a set of equations is very large

2.2.1 Governing Equations for Atmosphere

Depending on the coordinate system being used, the governing equations for the atmosphere obtain somewhat different forms. To facilitate discussion here, we present the so called primitive equations. To satisfy the needs of weather and climate predictions, the thermodynamic state of the atmosphere is represented using flow velocity, air density, temperature, humidity and pressure. The evolution of the atmospheric thermodynamic state is thus fully quantified using a set of seven equations, including three equations of motion, the conservation equations for air mass, enthalpy and water vapour plus the equation of the state. In a local catersian coordinate system, (x, y, z) with z pointing in the upward direction (to facilitate the use of the cartesian-tensor indicical notation, we shall also set $x \equiv x_1$, $y \equiv x_2$ and $z \equiv x_3$), these equations can be written, in tensor notion, as

$$\frac{\partial u_i}{\partial t} + u_j \frac{\partial u_i}{\partial x_j} = -\delta_{i3} g - 2\epsilon_{ijk} \Omega_j u_k - \frac{1}{\rho}\frac{\partial p}{\partial x_i} + \frac{1}{\rho}\frac{\partial \tau_{ij}}{\partial x_j} \tag{2.1a}$$

$$\frac{\partial \rho}{\partial t} + \frac{\partial \rho u_j}{\partial x_j} = 0 \tag{2.1b}$$

$$\frac{\partial c_p T}{\partial t} + u_j \frac{\partial c_p T}{\partial x_j} = k_T \frac{\partial^2 c_p T}{\partial x_j^2} + s_T \tag{2.1c}$$

$$\frac{\partial q}{\partial t} + u_j \frac{\partial q}{\partial x_j} = k_q \frac{\partial^2 q}{\partial x_j^2} + s_q \tag{2.1d}$$

$$p = \rho R_a T \tag{2.1e}$$

In (2.1a), which constitutes the equations of motion (for velocity components u, v and w with $u \equiv u_1$, $v \equiv u_2$, $w \equiv u_3$), t is time, g is acceleration due to gravity, ρ is air density and p is pressure. The Kronecker and Alternating tensors are respectively defined by:

$$\delta_{ij} = \begin{cases} 1 & \text{for } i = j \\ 0 & \text{for } i \neq j \end{cases} \tag{2.2}$$

and

$$\epsilon_{ijk} = \begin{cases} 1 & \text{for cases 123, 231, 312} \\ -1 & \text{for cases 321, 213, 132} \\ 0 & \text{otherwise} \end{cases} \tag{2.3}$$

Ω_j is the jth component of the Earth's rotation vector $\Omega = [0, \omega \cos \varphi, \omega \sin \varphi]$, with $\omega = 7.27 \times 10^{-5}$ s^{-1} being the approximate angular speed of the Earth's rotation and φ the latitude. The terms τ_{ij} represent the viscous shear stress tensor which causes the fluid to deform due to intermolecular forces.

Equation (2.1b) is known as the continuity equation. It represents the conservation of mass (as well as ensuring that the atmosphere behaves as a physical continuum). Equation (2.1c) is the enthalpy ($c_p T$) conservation equation, where c_p is the specific heat of air at constant pressure and k_T is the molecular diffusivity for heat. The source term, s_T, arises from the effects of radiation (convergence or divergence of radiative flux) and phase changes of atmospheric water vapor (e.g. evaporation, condensation etc.). Equation (2.1e) is the equation of state, which describes the relationship between pressure, temperature and density. R_a is the specific gas constant of air. For dry air, $R_a = R/M_d = 287\,\mathrm{J\,kg^{-1}K^{-1}}$, with R the universal gas constant and M_d the molecular weight of dry air. For humid air, $R_a = (1 + 0.61q)R/M_d$, where q is specific humidity. The behaviour of q obeys (2.1d), where s_q is a source/sink term related to the phase changes of water pater (from vapor to liquid etc.).

The Equations (2.1a) to (2.1e) form a closed system, which describes the spatial and temporal dependence of the 7 dependent variables, namely, u, v, w, p, ρ, T and q. These equations are usually manipulated and may attain different forms in different coordinate systems after various simplifications. For example, if atmospheric aerosols are of interest, an aerosol conservation equation can be written as

$$\frac{\partial c}{\partial t} + u_{pj}\frac{\partial c}{\partial x_j} = k_p \frac{\partial^2 c}{\partial x_j^2} + s_c \tag{2.4}$$

where k_p is molecular diffusivity for aerosols, s_c is an aerosol source/sink term and u_{pj} are advection velocities for aerosols.

For global atmospheric circulations, and indeed for atmospheric phenomena in a limited area down to of several hundreds of square kilometres (mesoscale), the governing equations are often written in the σ-coordinate with σ being defined as

$$\sigma = \frac{p}{p_s}$$

where p_s is the pressure at the surface (Simmons and Bengtsson, 1984). In this system, the scalar form of the equations is

$$\frac{\partial}{\partial t}(p_s u) = -\frac{\partial}{\partial x}(p_s uu) - \frac{\partial}{\partial y}(p_s uv) - \frac{\partial}{\partial \sigma}(p_s u\dot\sigma)$$
$$+fp_s v - p_s \frac{\partial \phi}{\partial x} - RT\frac{\partial p_s}{\partial x} + F_u + D_u \tag{2.5a}$$

$$\frac{\partial}{\partial t}(p_s v) = -\frac{\partial}{\partial x}(p_s uv) - \frac{\partial}{\partial y}(p_s vv) - \frac{\partial}{\partial \sigma}(p_s \dot\sigma v)$$
$$-fp_s u - p_s \frac{\partial \phi}{\partial y} - RT\frac{\partial p_s}{\partial y} + F_v + D_v \tag{2.5b}$$

$$\frac{\partial \phi}{\partial \sigma} = -\frac{p_s}{\rho} \tag{2.5c}$$

$$\frac{\partial p_s}{\partial t} = -\frac{\partial}{\partial x}(up_s) - \frac{\partial}{\partial y}(vp_s) - \frac{\partial}{\partial \sigma}(\dot{\sigma}p_s) \tag{2.5d}$$

$$c_p \frac{\partial T}{\partial t} = -c_p \left\{ u\frac{\partial T}{\partial x} + v\frac{\partial T}{\partial y} + \dot{\sigma}\frac{\partial T}{\partial \sigma} \right\} + RT \left\{ \frac{\dot{\sigma}}{\sigma} + \frac{d\ln p_s}{dt} \right\}$$
$$+ Q_T + F_T + D_T \tag{2.5e}$$

$$\frac{\partial q}{\partial t} = -u\frac{\partial q}{\partial x} - v\frac{\partial q}{\partial y} - \dot{\sigma}\frac{\partial q}{\partial \sigma} + Q_q + F_q + D_q \tag{2.5f}$$

In the above equations, $\dot{\sigma}$ is the vertical "velocity" in the σ-coordinate system, $\phi = gz$ is the geopotential and $f = 2\Omega \sin \varphi$ is the Coriolis parameter. F_u and F_v are the horizontal frictional forces, and F_T and F_q are the heating and moisture changes arising from subgrid scale turbulent exchanges. Q_T and Q_q are the energy and moisture source/sink terms, and D_u, D_v, D_T and D_q are the lateral diffusions of momentum, heat and moisture, respectively. The hydrostatic assumption is implicitly embedded in the equations of motion and the mass conservation equation, which is valid for modelling large-scale atmospheric systems and when the spatial resolution is larger than 5 km. For higher resolution, the effects of surface topography and convection may generate strong vertical accelerations which violate the hydrostatic assumption. For modelling some sub-synoptic atmospheric systems and intense weather events, such as squall lines, non-hydrostatic models are necessary.

The equation system for the atmosphere can be solved numerically, stepping forward in time, using various computational schemes (discussed further in Chap. 3), provided that the initial and boundary conditions can be adequately specified and the source/sink terms (the Q terms) as well as the subgrid flux terms (the F and D terms) can also be specified. It is in the boundary conditions, the source/sink terms and the subgrid flux terms, that the interactions between the atmosphere and the other components of the environment mostly take place.

2.2.2 Governing Equations for the Oceans

In a manner similar to the atmosphere, the thermodynamic state of the ocean is represented using flow velocity, density, temperature, salt content and pressure. The evolution of the ocean thermodynamic state is also described using a set of seven equations, including three equations of motion, the conservation equations for water mass, enthalpy and salt content plus the equation of the state. The governing equations for the ocean are very similar to those for the atmosphere, except for the salt-concentration equation, the equation of state, and the nature of the forcings.

$$\frac{\partial u_i}{\partial t} + u_j \frac{\partial u_i}{\partial x_j} = -\delta_{i3}g - 2\epsilon_{ijk}\Omega_j u_k - \frac{1}{\rho_o}\frac{\partial p}{\partial x_i} + \frac{1}{\rho_o}\frac{\partial \tau_{ij}}{\partial x_j} \tag{2.6a}$$

$$\frac{\partial \rho_o}{\partial t} + \frac{\partial \rho_o u_j}{\partial x_j} = 0 \qquad (2.6b)$$

$$\frac{\partial c_p T}{\partial t} + u_j \frac{\partial c_p T}{\partial x_j} = k_T \frac{\partial^2 c_p T}{\partial x_j^2} + s_T \qquad (2.6c)$$

$$\frac{\partial c_s}{\partial t} + u_j \frac{\partial c_s}{\partial x_j} = k_{cs} \frac{\partial^2 c_s}{\partial x_j^2} + s_{cs} \qquad (2.6d)$$

where ρ_o is sea water density, c_s is specific salt content in sea water, which has dimensions of [M M^{-1}]. All other variables have similar interpretations as their counterparts for the atmosphere. One important feature of the sea is that the density of sea water is determined both by temperature, pressure and the concentration of sea-salt. Thus, the density of sea water is typically a function of the form

$$\rho_o = \rho_o(p, c_s, T) \qquad (2.7)$$

While the equation of state for gas, like air, can be derived analytically from statistical mechanics. This is not possible for liquids. Instead, empirical equations for estimating ρ_o are used (e.g. Kraus and Businger, 1994).

As for the atmosphere, simplifications for the above equation system can be carried out. For instance, it is accurate to assume that the seawater is an incompressible fluid and hence the continuity equation reduces to

$$\frac{\partial u_j}{\partial x_j} = 0 \qquad (2.8)$$

It is also reasonable to assume that the motion is hydrostatic and hence the equation of vertical motion reduces to

$$\rho_o g + \frac{\partial p}{\partial z} = 0 \qquad (2.9)$$

2.2.3 Basic Equations for Land Surface

The land surface consists of mainly unsaturated soil, vegetation and surface water. The physical and biochemical processes taking place at the land surface are extremely complex, as it encompasses plant growth of various species, water movements in rivers and porous media, sediment movements and emission of greenhouse gases, such as CO_2 and methane, to name just a few. These processes are also subject to human interference. At the atmosphere–land interface, exchanges of energy, momentum and mass take place, similar to that occurring at the atmosphere–ocean interface. For atmospheric processes, the fluxes of energy, mass and momentum arising from these exchanges serve as the lower boundary conditions. At the same time, these fluxes serve as the upper boundary conditions to, for instance, surface soil hydrological processes.

For many land-surface processes, precise and explicit forms of the governing equations are not readily available. However, for two of the most important processes, namely, the thermodynamic and hydrodynamic ones, the conservation equations for heat and moisture can be written as

$$\frac{\partial T}{\partial t} = -\frac{1}{C}\frac{\partial G_j}{\partial x_j} + s_T \tag{2.10}$$

$$\frac{\partial W}{\partial t} = -\frac{1}{\rho_w}\frac{\partial Q_j}{\partial x_j} + s_W \tag{2.11}$$

where T is soil temperature, C is volumetric soil heat capacity, W is volumetric soil moisture, ρ_w is liquid-water density and s_T and s_W are source terms. The definition of soil moisture is as follows: pores of various sizes occur in soil and a proportion of them is filled with liquid water; the relative volume of this water in a unit volume of soil is the volumetric soil moisture. The jth component of heat flux through the soil, G_j, obeys a simple flux-gradient relationship

$$G_j = -K_{sj}\frac{\partial T}{\partial x_j} \tag{2.12}$$

where K_{sj} is the jth thermal conductivity. The jth water flux through the soil, Q_j, obeys Darcy's law

$$Q_j = -K_{wj}\rho_w\frac{\partial(\psi + z)}{\partial x_j} \tag{2.13}$$

where K_{wj} is the jth component of hydraulic conductivity and ψ is the matrix potential (or the pressure head) due to capillary forces, which is normally negative (see Domenico and Schwartz, 1997). A substitution of (2.12) and (2.13) into (2.10) and (2.11) gives advection-diffusion type prognostic equations for soil temperature and soil moisture

$$\frac{\partial T}{\partial t} = \frac{1}{C}\frac{\partial}{\partial x_j}K_{sj}\frac{\partial T}{\partial x_j} + s_T \tag{2.14}$$

$$\frac{\partial W}{\partial t} = \frac{\partial}{\partial x_j}K_{wj}\frac{\partial(\psi + z)}{\partial x_j} + s_W \tag{2.15}$$

2.2.4 Basic Equations for Ground Water

One of the central tasks of hydro-geological studies is to model the movement of water in the phreatic zone and the vadose zone, and the associated movement of other matter. The vadose zone is the zone between the surface of the land and the water table and can be conveniently considered as part of the land surface. The governing equation for water movement in the phreatic zone is the equation for the hydraulic head, H_g,

$$r_s \frac{\partial H_g}{\partial t} = \frac{\partial}{\partial x_j} \left(K_{wj} \frac{\partial H_g}{\partial x_j} \right) + s_H \qquad (2.16)$$

where r_s is specific storage with the dimensions of $[L^{-1}]$; s_H is a general source term that defines the volume of inflow to the system per unit volume of aquifer per unit of time. The water fluxes in the x_j directions, q_j, are given by $-K_{wj} \frac{\partial H_g}{\partial x_j}$. Boundary conditions are required for applying the equation to the simulation of the ground-water system in a given domain, especially the rate of water exchange between the vadose and the phreatic zones, in addition to s_H. The upper-boundary condition for ground-water modelling becomes the lower-boundary condition for the land-surface model.

2.2.5 Basic Equations for Ice

The governing equations for ice also consist of the equations of motion, mass conservation and energy conservation (Hibler and Flato, 1992). However, these equations take somewhat different forms for different ice types. We shall concentrate on those for sea ice. The major tasks of sea-ice modelling include the modelling of ice movement, the distribution of ice thickness and its growth rate. Sea ice moves in response to wind and water currents and the internal stresses. The equation of ice motion can be written as

$$m_c \frac{\partial \mathbf{v}_{\text{ice}}}{\partial t} + (\mathbf{v}_{\text{ice}} \cdot \nabla) \mathbf{v}_{\text{ice}} = -m_c f \mathbf{k} \times \mathbf{v}_{\text{ice}} + \tau_a + \tau_0 + \\ + \nabla \cdot \sigma - m_c g \nabla h \qquad (2.17)$$

where m_c is ice mass per unit area, $\mathbf{v}_{\text{ice}}$ is ice velocity, σ is the second order internal stress tensor in the sea ice due to ice interaction, τ_a and τ_0 are the forces due to atmospheric and ocean currents and h is the height of the sea surface. From the perspective of ice motion, the interactions between sea ice, the atmosphere and the ocean are reflected in τ_a and τ_0.

Equation (2.17) involves ice mass per unit area, and hence it is important to determine the depth of ice, H_{ice}. The spatial distribution of H_{ice} is best described using the area distribution density, $p_{\text{ice}}(H_{\text{ice}})$. The fraction of surface area covered with ice of depth H_{ice} is given by $p_{\text{ice}}(H_{\text{ice}})dH_{\text{ice}}$. The temporal variation of p_{ice} responds to deformation, advection, growth and decay of ice. The mass conservation equation is then written in terms of p_{ice} as

$$\frac{\partial p_{\text{ice}}}{\partial t} + \nabla \cdot (\mathbf{v}_{\text{ice}} p_{\text{ice}}) + \frac{\partial (f_{\text{ice}} p_{\text{ice}})}{\partial H_{\text{ice}}} = \Psi \qquad (2.18)$$

where f_{ice} is the vertical growth rate of ice thickness and Ψ, a function of p_{ice} and H_{ice}, is a redistribution function that describes the creation of open water and the transfer of ice from one thickness to an other by rafting and ridging.

Finally, the growth rate f_{ice} is determined by the energy balance for the ice layer. Suppose the heat flux from the atmosphere is G_a and the heat flux to the ocean is G_o, then the vertical growth rate of the ice layer is

$$\rho_{ice}\lambda \frac{dH_{ice}}{dt} = G_a - G_o \qquad (2.19)$$

The calculation of G_a can be determined from the surface-energy balance, in a similar fashion as for calculating ground-heat flux in land-surface modelling (see Chap. 5)

$$G_a = R_n - H - \lambda E \qquad (2.20)$$

where R_n is net radiation, H is sensible heat flux, E is latent heat flux and λ is the latent heat coefficient for sublimation. From the energy-conservation perspective, the interactions between sea ice, the atmosphere and the ocean are reflected in G_a and G_o.

2.3 Predictability of the Environmental System

The above discussion shows that there is a large set of coupled, non-linear equations at our disposal with numerous feedback processes. Even more equations can be derived for other physical and bio-chemical processes as required. With the rapid advances in computing power and the increasing availability of data required by the equations, accurate environmental prediction is a real possibility. Indeed, this has been the prevailing conviction of scientists engaged in the study.

However, we must recognize that the environment is a system that differs profoundly from most other systems. A simple conceptual model is illustrated in Fig. 2.4a. An external forcing (input) causes the system to respond and produce an output that does not affect the input at future times (no feedback). Such a conceptual model is useful for some engineering problems, but not so when applied to the environment. For instance, if the entire Earth is considered to be one system, then the input would be the solar radiation and the output is the terrestrial radiation. The system, as illustrated in Fig. 2.4a is trivial as the output on average is approximately equal to the input. To study the behaviour of the environment, especially its components, the input cannot be specified without involving the system itself, as the input must be related to the system state. Consider the atmosphere as an example and, for simplicity, consider only the atmospheric temperature. The input for the atmospheric system includes the radiative fluxes and the heat fluxes at the surfaces of land, ocean and ice. It is readily understood that these fluxes must depend on the system output, i.e. the atmospheric temperature in this case. Hence, the environment has a much more complex behaviour than that depicted in Fig. 2.4a. An illustration is given in Fig. 2.4b. The environment is divided into several components. Here, the total input for the environment is partitioned among the components and this partitioning depends on the state of the environment. In addition, the input for each component depends on the relationships among the components. For such a system, we are mainly interested in the partitioning of the input into different components and their reactions to it.

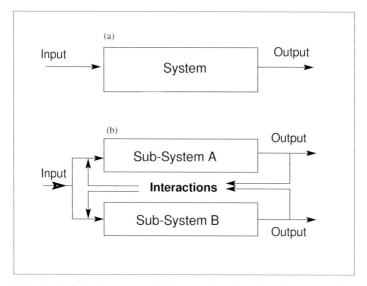

Fig. 2.4. (a) A conceptual illustration of a simple system. An external forcing generates through the system an output. (b) A conceptual illustration of the environmental system. An input must be partitioned between the components (subsystems), which depends on the state of the components and the inputs for the individual components also depend on the interactions among them.

Given the profound effects of the non-linear interactions within each component of the environment (e.g., within the atmosphere) and those between the components (e.g., atmosphere and ocean), is the environment at all predictable? Some understanding of the predictability of the environmental system is warranted before we endeavor to develop sophisticated predictive models. The predictability of the environmental system, especially centered on the behaviour of the atmosphere, has been under active research for some years (e.g. Lorenz, 1963). Used loosely, the term predictability refers to whether some kind of prediction is possible (more formal definitions are given below).

2.3 Predictability of the Environmental System

Hence, in addressing predictability, we must be clear about what exactly is being predicted.

Suppose we use a set of variables, $\chi_i(\mathbf{x}, t)$ ($i = 1, ..., N_\chi$) to quantify the state of the environment, where $\mathbf{x}$ is space and t is time. In a sense, these variables must always be regarded as being averages. It is sufficient here to consider the temporal average, namely,

$$\bar{\chi}_i(\mathbf{x}, t) = \frac{1}{\tau} \int_{t-\tau/2}^{t+\tau/2} \chi_i(\mathbf{x}, t') dt' \qquad (2.21)$$

bearing in mind that, in practice, the average also involves space. Depending on the size of τ, we would obtain different predictions of the future environment. In case $\tau = 0$, we have the instantaneous future state, while in case $\tau \to \infty$, we have the true statistical mean.

Predictions can be divided into the categories of deterministic and statistical predictions. The former describes the precise value of the future state variables $\bar{\chi}(\mathbf{x}, t)$ at a given future time, while the latter describes the probability of $\bar{\chi}(\mathbf{x}, t)$ for a particular realization.

Let us consider a specific example from the view point of fluid dynamics. Suppose χ_i denotes the Lagrangian velocity of an air parcel in the atmosphere. If the Lagrangian velocities of all parcels are determined, then the behaviour of the atmospheric flow is completely specified. The auto-correlation function of the velocities is defined as

$$R(\tau) = <\chi_i(t)\chi_i(t+\tau)> \qquad (2.22)$$

where the operator $<>$ denotes the ensemble average over many parcels. As time progresses, the Lagrangian velocity of an air parcel follow approximately a Markov process and the auto-correlation function is known to be approximately

$$R(\tau) = \exp(-\tau/T_L) \qquad (2.23)$$

where T_L is the Lagrangian time scale over which χ_i is significantly correlated. Observations show that in turbulent flows of the atmospheric boundary layer, T_L is of the order of 100 s. Thus, for a given initial velocity at time t, $\chi_i(t)$, there is a natural limit, i.e., T_L, beyond which the prediction of the instantaneous Lagrangian velocity will not be possible. That is, the instantaneous velocity of an air parcel is predictable only within time T_L. Beyond this time, the memory of the parcel in this initial velocity disappears.

A more rigorous definition of the predictability time scale can be given. Suppose the environment is described by a n-dimensional variable χ_i which obeys a set of non-linear equations

$$\frac{d\chi_i}{dt} = f_i(\chi_1, ..., \chi_n) \qquad i = 1, ..., n \qquad (2.24)$$

Suppose a solution of (2.24) is $\chi_i(t)$. The graphical representation of this solution is a trajectory in the n-dimensional phase space. If the $\chi_i(t)$ is subject to a small perturbation $\delta\chi_i$, then the growth rate of the perturbation can be approximated by a set of linear equations, using a Taylor expansion

$$\frac{d\delta\chi_i}{dt} = \sum_{j=1}^{n} A_{ij}\delta\chi_j \tag{2.25}$$

where $A_{ij} = \partial f_i/\partial \chi_j |_{(\chi_1,...,\chi_n)}$ is the Jacobi matrix of the $f_i(\chi_1, ..., \chi_n)$ function. The eigenvalues of A_{ij}, give the exponential growth rate of $\delta\chi_i$ along the direction of the corresponding eigenvectors, for a given time. The real components of the time-averaged eigenvalues, λ_i, are the Lyapunov exponents, which can be used as a measure of the predictability of the dynamic system. This is because that the sum of all positive λ_i

$$K = \sum \lambda_i \qquad \lambda_i > 0 \tag{2.26}$$

gives the expansion rate of the δx_i in the phase space. K^{-1} is known as the average predictability time scale.

For environmental systems, if we exclude the influences of external forces occurring at the boundaries, then its predictability is determined by the dynamic instability and non-linear interactions within the system. These dynamic instability and non-linear interactions (actually) refer to those in geofluids. Numerical tests using Atmospheric (Global) General Circulation Models (AGCMs) suggest that for middle latitude weather patterns, the predictability limit is around two weeks, while for global circulation patterns it can increase to 4 weeks. These tests imply that the time scale over which a prediction is effective depends on the spatial scale of the phenomenon under consideration. On the basis of existing research in relation to the atmosphere, we can find a rough picture of the relationship between this time and the spatial scale (Fig. 2.5).

As the environment is such an open complex system, its predictability is of course not simply determined by the dynamic instability and non-linear interactions in the fluid motion. Suppose the state of an environmental component is represented by χ, e.g., the flow speed of the ocean. Then we can formally express its future value as

$$\chi(\mathbf{x}, t) = \int_{t_0}^{t} A(\chi, t) dt + \int_{t_0}^{t} F_i(\chi, t, \psi) dt + \int_{t_0}^{t} F_e(\chi, t) dt \tag{2.27}$$

where A represents the internal forces (e.g., the advection and diffusion terms in (2.6a)), F_i represents the forces acting on its boundaries (e.g., the wind shear at the ocean surface) arising from the interactions with other components of the environment and F_e denotes possible forces outside of the environment. This simple consideration suggests that it is useful to distinguish the following three kinds of predictabilities.

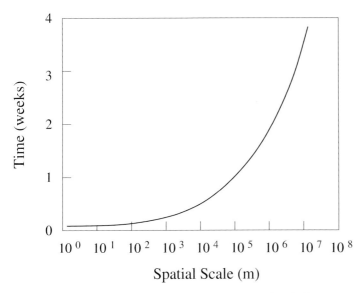

Fig. 2.5. Indicative relationship between time scale of effective prediction and the typical spatial scale of the atmospheric system.

Internal Dynamic Predictability. This kind of predictability refers to the deterministic behaviour of the system. It is closely related to the dynamic instability and non-linear interactions of fluid motions in the environmental system. Since the initial conditions, especially those for the atmosphere and ocean, always differ in some degree from the true state of the environment. Uncertainties in the initial conditions may grow with time and propagate to lower frequencies, owing to dynamic instability and non-linear interactions. As a consequence, the predictions, being based on initial fields which have uncertainties, no matter how insignificant they seem to be, will lead to large uncertainties in the prediction of the future state of the environment. The internal dynamic predictability is formally related to the integral of $\int_{t_0}^{t} A(\chi,t) \mathrm{d}t$ in (2.27). Figure 2.6 shows the numerical experiment of Shukla (1981) using an atmospheric general circulation model (AGCM). This experiment consists of a control run and two sensitivity runs for January 1975 and 1976. In the sensitivity runs, Gaussian random perturbations (with zero mean and $3\,\mathrm{m\,s^{-1}}$ standard deviation) were superposed on the initial horizontal velocity fields of the control run. The temporal evolution of the sea-surface pressure (averaged over 22°N–38°N, 10°W–45°E) and that of the temperature on the 500 hPa level (averaged over 38°N–58°N, 110°E–165°E) show that the uncertainties in the predictions significantly increase with time and the errors in the predictions become unacceptable after approximately two weeks.

Internal dynamic predictability is most likely related to the possible chaotic behaviour of the geofluid dynamic system and, hence, it is associated

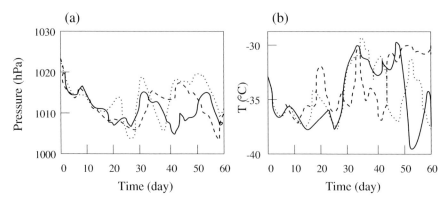

Fig. 2.6. Temporal evolutions of sea-surface pressure and temperature on the 500 hPa level show the limited predictability of middle latitude atmospheric systems (redrawn from Shukla, 1981).

with the flow patterns of the atmosphere and ocean. In general, chaos refers to the intrinsic stochastic and unpredictable behaviour of a system which obeys a deterministic set of equations, such as the Navier-Stokes equations. Deterministic systems are often dynamic systems with a set of governing equations, normally ordinary differential equations, partial differential equations or iterative equations. The parameters of the equation system are all deterministic, and hence in principle, these equations give precise solutions for given initial and boundary conditions. Some systems, under no influence of external stochastic forcing, are extremely sensitive to the initial and boundary conditions for a certain range of control parameters: an infinitely small variation in these conditions or even computer round-off error may lead to dramatic changes in the solution. Such stochastic behaviour is caused by the non-linear interactions within the system, not by stochastic external forces. Lorenz (1963) has shown that a much simplified version of the Navier-Stokes equations, which govern the motions of the atmosphere and the ocean, is a chaotic system. More discussion will be given in Sect. 2.7.

Forcing Predictability. The environment is also subject to external forces related, for example, to solar activity and geological processes, which are stochastic in nature. Solar radiation is the ultimate energy source of all environmental processes, ranging from circulations of the atmosphere and ocean to photo-synthesis of plants. The total energy received by the environment is affected both by solar activity (e.g., solar black spots) and the processes within the environmental system (e.g., variations in albedo for solar radiation). For example, as the total amount of cloud increases, more solar radiation is reflected back into the space and hence, less energy is available for the environment. Satellite measurements of the solar constant fall between $1350\,\mathrm{Wm}^{-2}$ to $1390\,\mathrm{Wm}^{-2}$. The solar radiation is mainly emitted through the wavelength range between 0.25 and 4.0 μm with 99% of the radiation through the wavelength range between 0.275 and 4.67 μm. The sun itself is a dynamic system and inevitably, the activities of the sun influence the solar radiation and the solar constant. The observed time series of the number of solar black spots between 1850 and 1980 (Fig. 2.7) shows an approximate 11-year period of solar black spots, but the number of black spots is stochastic. It is also known that the occurrence of solar black spots are related to variations of the solar constant (Kondratyev and Nikolsky, 1970). The above description shows that the environment as a whole is subject to external forces which are unpredictable in nature. The stochastic nature of the external forces has a significant impact on the predictability of the environment, although this predictability is not yet well understood.

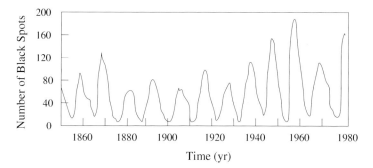

Fig. 2.7. The number of solar black spots between 1850 and 1980 (after Newkirk, 1982).

Interaction Predictability. The behaviour of an individual component of the environment (e.g., the atmosphere) is affected by the forcing exerted on its boundary (e.g., heating of the atmosphere from the ocean surface). The forcing is determined by the interactions involving often two or more components of the environment. Obviously, whether the environment can be predicted greatly depends upon whether these interactions and the associated changes in each components are predictable.

The predictability of ENSO is an excellent example to illustrate interaction predictability. Historical records show that the period of ENSO falls between 2 and 7 years with an average around 3.5 years. The dynamic mechanisms which lead to an ENSO are closely related to the non-linear interactions between the atmosphere and the ocean. There have been attempts (e.g., Fraedrich, 1988) to estimate the predictability of ENSO by reconstructing the phase space geometry of the dynamic system from the time series of certain state variables (e.g., atmospheric pressure measured at Darwin, Australia). Suppose the time series of a state variable is ψ_j

$$\psi_j = \psi(t_0 + j\Delta t) \qquad j = 1, ..., N \qquad (2.28)$$

where t_0 is the starting observation time, Δt is the time interval and N is the length of the time series. A m-dimensional phase space R^m can be reconstructed by introducing a time lag τ

$$\psi_m(t_j) = \{\psi(t_j), \psi(t_j + \tau), ..., \psi[t_j + (m-1)\tau]\} \tag{2.29}$$

If m is sufficiently large then the reconstructed system has the same geometric properties as the original dynamic system.

The predictability of the dynamic system can be estimated by calculating the cumulative distribution function

$$C_m(l) = \frac{1}{N} \sum \Theta[l - r_{ij}(m)] \tag{2.30}$$

where

$$\Theta(x) = \begin{cases} 0 & x \geq 0 \\ 1 & x < 0 \end{cases} \tag{2.31}$$

and $r_{ij}(m) = |\psi_m(t_i) - \psi_m(t_j)|$ is the distance between $\psi_m(t_i)$ and $\psi_m(t_j)$. $C_m(l)$ describes the probability for two points in the phase space R^m to be smaller than l. For $N \to \infty$ and $l \to 0$, $C_m(l)$ is approximately

$$C_m(l) \approx l^{D_2} \exp(-m\tau K_2) \tag{2.32}$$

where D_2 is the correlation dimension of the phase space attractor and K_2 is the Kolmogorov entropy. K_2^{-1} is a description of the predictability time scale. D_2 and K_2 can be determined by

$$D_2 = \frac{\ln C_m(l)}{\ln l} \tag{2.33}$$

$$K_2 = \frac{1}{\tau} \ln \left(\frac{C_m}{C_{m+1}} \right) \tag{2.34}$$

Denote the monthly standard deviation of pressure at Darwin as DP and the running averages of DP over 3, 6 and 9 months respectively as DP3, DP6 and DP9. Using DP3, DP6 and DP9 as time series for reconstructing the dynamics of ENSO, Yang and Chen (1990) found that the predictability of ENSO is approximately 2 years.

The other profound effect of the interactions among the components of the environmental system is that these interactions lead to changes in the environment, which impact on the external forcing. The ice–albedo feedback discussed in Sect. 11.4 is a typical example: the ice–ocean–atmosphere interactions virtually leads to the unpredictable behaviour of the albedo of the environment (here, clouds, ice surface and ocean surface). This in turn leads to the unpredictability of the radiative energy received by the environmental system.

2.4 Methods for Environmental Prediction

While the predictability of the environmental system requires a great deal of further research, the above discussion suggests that long-term (several years to several decades) deterministic predictions of the environment do not seem to be possible. Hence, long-term environmental predictions are essentially statistical predictions, in which the statistical behaviour of the future environment is described. However, on shorter time scales (e.g., from days to several weeks for the atmosphere and up to several years for ocean), deterministic predictions are possible.

There are two basic philosophical approaches to the modelling and prediction of the environment:

- the integrated modelling systems approach; and
- the stochastic-dynamic modelling approach.

In the first approach, the environment is considered as a deterministic system and, for given initial conditions, its future state can be determined by numerically solving the governing equations. This approach has been shown to be effective for the assessment and short-term predictions of the environment. While the nature of this approach is deterministic, it can be used to obtain ensemble predictions as discussed in Sect. 2.6. The latter give a probabilistic description of the future environment on the basis of a number of individual (deterministic) predictions, namely the ensemble. The development of integrated modelling systems is based on the following facts:

1. A number of deterministic models, such as AGCMs, OGCMs, land-surface models, ecological models etc., which have been developed in the past for the individual components of the environment, have matured and have been shown to be very useful in practice. These models are becoming increasingly tightly coupled to form integrated modelling systems;
2. There has been a continuing and dramatic increase in computing power which enables the implementation of integrated modelling systems with increased spatial and temporal resolutions;
3. Increasing availability of geographic information data which provides the input parameters for integrated modelling; and
4. Data assimilation skills have been improved significantly.

Integrated modelling systems provide direct and quantitative predictions of a wide range of environmental problems, ranging from severe weather events, flood, soil water availability, dust storms, salinisation, air quality, sea state etc. Also, they significantly increase our understanding of the physical and biochemical processes governing the environmental system. However, because of the intrinsic possible non-linear interactions among the components, we are currently not in a position to produce accurate long-term predictions.

In the dynamic-stochastic approach, the environment is viewed as a dynamic system which has intrinsic stochastic behaviour and is subject to external stochastic forces. It is thus impossible to predict deterministically the future state of the environment beyond the predictability time limit. Suppose the environment can be quantified using a set of n independent variables and all likely states of the environment occupy certain regions of the n-dimensional phase space, namely, the attractors. The stochastic-dynamic approach attempts to predict the probability distribution of the future environment on these attractors. Based on this philosophy, both ensemble-prediction techniques and dynamic-stochastic models have been developed.

2.5 Integrated Environmental Modelling Systems

In an integrated environmental modelling system, we attempt to represent all major processes and interactions mathematically and obtain approximate numerical solutions to the environment. A logical procedure for developing such a system is to develop models for the individual components of the environment, each using a set of governing equations, control parameters and a different grid optimized according to the specific structure of the spatial domain. Assuming models for the individual components are established, the important tasks for developing integrated modelling systems lie in the identification of feedback processes and in the design of interfaces between the component models. Most GCMs have various degrees of coupling, especially atmosphere–and–ocean coupling and atmosphere–and–land coupling. It is certain that future environmental models will rapidly progress along the lines of system integration.

The physical, chemical and biological processes of the environment often involve the interactions of several components. These processes and interactions need to be represented mathematically and numerically in integrated modelling systems. As they often occur on a wide range of temporal and spatial scales, it is usually not possible and unnecessary to include all processes in a single integrated modelling system. Integrated modelling systems are normally focused on specific time scales of environmental phenomena. Thus, an important issue that arises is how models for individual components can be coupled consistently.

Depending on the time scale we are interested in, the processes that need to be represented in the integrated system are also different. For instance, GCMs are concerned with the variability of the climate on seasonal to decadal time scales. Thus, GCMs must include the entire ocean and its interactions with the atmosphere. For sub-seasonal time scales, deep ocean circulation and ground-water hydrological processes and deep ice, can all be considered as almost stationary. Hence, sub-seasonal environmental models comprise mainly a regional atmospheric model, a slab ocean comprising only the well-mixed layer and a land surface of only about 1 m deep.

The framework of an integrated modelling system is illustrated in Fig. 2.8. Such a system comprises four basic components, namely: 1) computational models for individual components of the environment; 2) interaction schemes; 3) data assimilation procedures; and 4) GIS databases. Computational models for individual components are discussed in Chaps. 3 to 10 of this book. A brief outline of the atmosphere–land and the atmosphere–ocean interfaces is given in the following section.

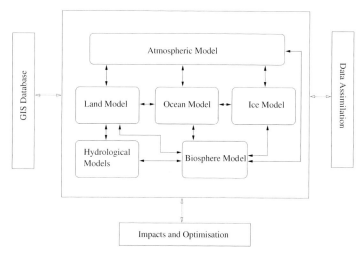

Fig. 2.8. A schematic illustration of the framework of an integrated environmental modelling system.

Integrated models for studying water resources on catchment scales is another example. Such models couple an atmospheric prediction model with a land-surface model and a hydrological model and have the support of remote sensing data and field observations. These models are powerful in studying the energy and water balances on catchment to continental scales and the issues related to water resource, such as precipitation, evapotranspiration and stream flows. Figure 2.9 is a schematic illustration of an integrated water-resource model. This system consists of an atmospheric model with a spatial resolution of about 100 km with a temporal resolution of one week, a precipitation model of around 30 km and a temporal resolution of 1 hr and a hydrological model with a spatial resolution of 5 km and a temporal resolution of 1 hr.

In Table 2.2, T refers to "thermodynamic" and R to dynamic sea ice with rheology. Dr stands for free drift sea ice. H, W, τ, T stand for flux adjustment of heat, fresh water, surface stress and ocean surface temperature, respectively. B refers to a simple bucket; BB refers to modified bucket; C includes canopy processes; r_s denotes inclusion of stomatal resistance. The

Table 2.2. List of coupled climate models and features (from Gates et al., 1995).

Group	Country	AGCM resolution	OGCM resolution	Sea ice	Flux Correction	Land-surface scheme
BMRC	Australia	R21 L17	$3.2° \times 5.6°$ L12	T	none	Chameleon Surface Model
CCC	Canada	T32 L10	$1.8° \times 1.8°$ L29	T	H, W, T	BB
CERFACS	France	T42 L31	$1.0° \times 2.0°$ L20		none	B
COLA	USA	R15 L9	$3.0° \times 3.0°$ L16	T	none	C r_s
CSIRO	Australia	R21 L9	$3.2° \times 5.6°$ L12	T/R	H,W,τ,T	C r_s
GFDL	USA	R30 L14	$2.0° \times 2.0°$ L18	T/Dr	H,W	B
GISS	USA	$4° \times 5°$ L9	$4.0° \times 5.0°$ L16	T	none	C
IAP	China	$4° \times 5°$ L2	$4.0° \times 5.0°$ L20	T	H,W	B
LMD/OPA	France	$3.6° \times 2.4°$ L15	$1.0° \times 2.0°$ L20	T	none	B
MPI	Germany	T21 L19	$2.8° \times 2.8°$ L9	T	H,W,τ,T	B
MRI	Japan	$4° \times 5°$ L15	$.5 - 2° \times 2.5°$ L21	T/Dr	H,W	B
NCAR	USA	R15 L9	$1.0° \times 1.0°$ L20	T/R	none	B
UCLA	USA	$4° \times 5°$ L9	$1.0° \times 1.0°$ L15		none	fixed wetness
UKMO	UK	$2.5° \times 3.8°$ L19	$2.5° \times 3.8°$ L20	T/Dr	H,W	C r_s

information presented in this table applies to 1995 therefore will not be up-to-date, as the models are continuously and rapidly improved.

2.5.1 Atmosphere–Land Surface Interaction Scheme

The energy, mass and momentum exchanges between the atmosphere and land surface significantly influence the cycles of these quantities in the environment. One of the key concerns of atmosphere–land surface interaction schemes, known as land-surface schemes, is to quantify these exchanges. Most current land-surface schemes are basically one-dimensional, but with treatment of spatial issues. The starting point of a land-surface scheme are the conservation equations for temperature and soil moisture (2.10) and (2.11). A full account of land-surface schemes is given in Chap. 5. An atmosphere–ice interaction scheme is very similar to a land-surface scheme as described in Chap. 11.

Consider a thin air layer of depth Δz_a in contact with the surface. The mean temperature of this layer obeys

$$\frac{\partial T_a}{\partial t} = -\frac{1}{\rho_a c_a}(R_n + \lambda E + H - G_0)/\Delta z_a \qquad (2.35)$$

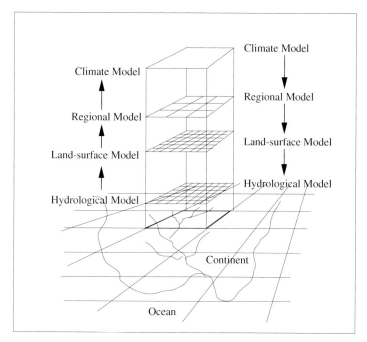

Fig. 2.9. Schematic structure of the coupled water-resource model.

where ρ_a is air density, c_a is the specific heat capacity of air, R_n is net radiation, E is evaporation and H is sensible heat flux, G_0 is ground heat flux. As $\Delta z_a \to 0$, we obtain the surface energy-balance equation

$$R_n + \lambda E + H - G_0 = 0 \qquad (2.36)$$

The net radiation R_n at the surface obeys

$$R_n = R_{is} + R_{es} + R_{il} + R_{el} \qquad (2.37)$$

where R_{is} is short-wave irradiance, R_{es} is exiting short-wave radiation, R_{il} is long-wave irradiance and R_{el} is exiting long-wave radiation. A knowledge of the net radiation at the surface is essential for determining the heat budget.

Similarly, a water-balance model for a thin soil layer of depth Δz_s can be written as

$$\rho_w \Delta z_s \frac{dW}{dt} = P_r - E - R - D_r \qquad (2.38)$$

where W is volumetric soil moisture averaged over Δz_s, P_r is precipitation rate, R is surface runoff and D_r is drainage. As $\Delta z_s \to 0$, we obtain the water-balance equation at the atmosphere–land interface

$$Pr - E - R - I_r = 0 \qquad (2.39)$$

where I_r is infiltration. Equations (2.36) and (2.39) are coupled through E, which appears in both of them. Hence, the energy balance at the surface must be considered in conjunction with surface hydrological processes. The emphasis of a land-surface scheme is to determine all components of (2.36) and (2.39). The processes need to be considered in a land-surface scheme can be broadly divided into three categories: sub-surface thermal and hydraulic processes, bare soil transfer processes, and vegetation processes. In recent years, a large number of schemes has been developed, which treat these processes very differently.

The vertical heat flux G though soil obeys a simple flux-gradient relationship

$$G = -K_s \frac{\partial T}{\partial z} \tag{2.40}$$

where K_s is the thermal conductivity. The vertical soil water flux Q obeys Darcy's law

$$Q = -K_w \rho_w \frac{\partial(\psi + z)}{\partial z} \tag{2.41}$$

where K_w is the hydraulic conductivity and ψ is the matrix potential due to capillary force.

A land-surface scheme is the algorithm required to solve this equation system. The details of the processes involved in the energy fluxes (e.g., H and λE), the water fluxes R and I_r, as well as terms such as S_T, S_W and G_0 are complex in detail and hence, as described in Chap. 5, these terms are mostly parameterized. The parameterization introduces a number of parameters which depend on the properties of the land surface. Thus, land-surface schemes require the support of quality data sets for specifying these parameters.

2.5.2 Atmosphere–Ocean Interaction Scheme

Atmosphere-and-ocean interactions also take place through exchanges of energy, mass and momentum. More specifically, ocean releases sensible and latent heat into the atmosphere influencing the structure of atmosphere and hence atmospheric circulations, while atmosphere releases momentum to the ocean through shear stress at the atmosphere–ocean interface. Detailed descriptions of atmosphere and ocean interactions can be found in Kraus and Businger (1994).

The coupling of the atmosphere and ocean is also realised through an interface parameterization scheme. The essence of the atmosphere–ocean interface, which we shall call an ocean-surface scheme, is similar to that of a land-surface scheme. The energy balance at the ocean surface obeys (2.36), where G_o should be understood as the heat flux into the ocean. Again, a knowledge of the net radiation at the sea surface is essential for determining the ocean heat budget. One method for estimating H and E is the bulk-transfer method, namely,

$$H = -c_p \rho C_H |U_a - U_o|(T_a - T_o) \tag{2.42}$$
$$E = -\rho C_E |U_a - U_o|(q_a - q_o) \tag{2.43}$$

where C_H and C_E are bulk-transfer coefficients for sensible heat and water vapor, respectively, U_a is the atmospheric flow speed at the reference level (commonly set to 10 m), U_o is the flow speed at the ocean surface, T_a and T_o are temperatures of air and the ocean surface and q_a and q_o are specific humidities of air and at the ocean surface, respectively.

The shear stress at the ocean surface is the main driving force for ocean currents. Under steady state conditions, the x and y components of the ocean current, u_o and v_o, obey the equations

$$-fv_o = \frac{1}{\rho_o}\frac{\partial p}{\partial x} + \frac{1}{\rho_o}\frac{\partial \tau_x}{\partial z} \tag{2.44}$$

$$fu_o = \frac{1}{\rho_o}\frac{\partial p}{\partial y} + \frac{1}{\rho_o}\frac{\partial \tau_y}{\partial z} \tag{2.45}$$

Integration of the above two equations over the depth of h leads to

$$fM_y = \int_{-h}^{0} \rho_o f v_o dz = \int_{-h}^{0} \frac{\partial p}{\partial x} dz - \tau_{x0} \tag{2.46}$$

$$fM_x = \int_{-h}^{0} \rho_o f u_o dz = -\int_{-h}^{0} \frac{\partial p}{\partial y} dz + \tau_{y0} \tag{2.47}$$

where M_x and M_y are the depth averaged mass transfer in the east-west direction and in the south-north direction, respectively. The above equations show that these mass transfers are related to the shear stress at the ocean surface, τ_{x0} and τ_{y0}. At the ocean surface, we have

$$\rho_a u_*^2 = \sqrt{\tau_{x0}^2 + \tau_{y0}^2} \tag{2.48}$$

where the friction velocity at the ocean surface is given by

$$u_*^2 = C_D (U_a - U_o)^2 \tag{2.49}$$

with $U_o = \sqrt{u_o^2 + v_o^2}$. C_D is the drag coefficient determined by the properties of the atmospheric turbulence. Detailed descriptions of methods for calculating C_D can be found in Kraus and Businger (1994). The formation of ocean current on the global scale and the well known double Ekman spiral in the atmospheric and ocean boundary layers are typical phenomena arising from the momentum exchange between the atmosphere and ocean.

2.6 Ensemble Predictions

As discussed in Sect. ??, the chaotic behaviour of the environment effectively limits its predictability. A small perturbation in the initial field may lead to growing uncertainties and eventual failure of the prediction. For a complex dynamic system such as the environment, it is in general not clear exactly how uncertainties grow, but their growth rates depend on factors such as scale, external forcing and model domain. Assuming that a particular modelling system sufficiently well represents the true dynamics of the environment, it is possible to obtain information about its inherent predictability by running the model from a number of initial conditions (internal dynamic predictability) and under a number of external forces (forcing predictability), both within a certain range. While each individual forecast is a deterministic one defined uniquely by the given initial and boundary conditions, an ensemble of forecasts from different initial and boundary conditions represent the chaotic nature of the environment. For ensemble predictions, an ensemble of forecasts will be produced rather than a single forecast for the future states, which can be evaluated using statistical tools, for instance, a probability density function of the predicted variables. Ensemble prediction offers the best possible forecast, in a statistical sense. The averaging of the ensemble provides a forecast statistically more reliable than any single forecast (Leith, 1974). In addition, from the variations in the ensemble, the reliability of the predictions can be estimated and, for a sufficiently large ensemble, the predicted quantities can be expressed in terms of probabilities. These probabilities convey all the information available regarding future environmental status. The basic idea of ensemble prediction is as illustrated in Fig. 2.10.

Ensemble prediction is increasingly widely used in weather and climate predictions. Leith (1974) has proposed ensemble prediction using a Monte Carlo approach. Other techniques have subsequently been developed by Hoffman and Kalnay (1983), Toth and Kalnay (1993), Moteni (1993), Houtekamer (1995) and Moteni et al. (1996), among many others. Examples of using ensemble prediction can be found in, for instance, Brooks et al. (1996), using a regional spectral model for short range atmospheric forecasts; in Toth and Kalnay (1993) using a medium-range forecast model on a global domain. Ensemble predictions for climate variations using coupled ocean and atmosphere model have also been carried out.

Of course, it is not certain whether the ensemble truly represents the inherent predictability of the dynamic system or the unreliability of the numerical model, as numerical models are often an imperfect representation of the true dynamics of the environment. It is most likely that the uncertainties in a forecast arise from both the inadequacy of the model and the inherent unpredictability of the real environment. In most current ensemble predictions, the models are assumed to be "perfect" and perturbations are introduced only to the initial conditions and external forcing. However, a

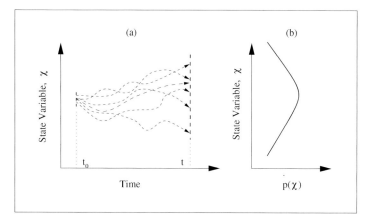

Fig. 2.10. An illustration of ensemble prediction. (**a**) An ensemble of forecasts for the state variable χ at a future time t are generated using perturbed initial conditions at t_0 and perturbed boundary forcing. (**b**) The ensemble forecasts are analysed using statistical methods, e.g., determining the probability density function at time $p(\chi)$.

considerable amount of works has been carried out on the sensitivity to the physical processes in the model.

2.6.1 Choosing the Initial Perturbations

The key to ensemble prediction is how to construct the initial perturbations. The simplest ensemble prediction is the Monte Carlo Forecast (MCF), in which the uncertainties in the initial field are assumed to obey the white noise distribution (i.e., all errors occur with the same probability). In MCF, randomly generated perturbation fields are superposed on a given initial field to produce an ensemble of N initial fields. These initial fields are integrated with time using the dynamic model to produce an ensemble of N forecasts. Suppose F_i denotes the forecast derived from the ith initial field. Then the ensemble average of the forecasts is

$$F = \frac{1}{N} \sum_{i=1}^{N} F_i \qquad (2.50)$$

The computational expense of MCF is N times that for the conventional single forecast, but has been shown in theory and practice to be superior to the latter (Leith, 1974). The main theoretical drawback of MCF is that the randomly generated perturbations fields do not necessarily represent the inherent uncertainties in the initial conditions, because the error distribution in the initial field is far from white noise. Another problem with MCF is that there are too many small perturbations that are damped very quickly by the forecast model. One alternative to MCF is the lagged average forecasting (LAF) (Hoffman and Kalnay, 1983). For LAF, the ensemble of the N initial fields are observed or analysed fields at different times $t = 0, -\tau, -2\tau, ..., -(N-1)\tau$ and the prediction is obtained using (2.50).

In constructing the initial perturbations, we should also consider how to implement ensemble forecasts so that the computation is economic. Note that not all errors in the initial field are likely to grow and it would be a waste of computing time if these errors are included in the initial fields. It is important to identify the initial uncertainties that result in diverging forecasts. This involves the technique of determining the nature of errors in the initial fields, which depends on the dynamic problem. For short-range weather prediction, for example, the initial field is the optimal analysis of observed data and previous forecast at the initial time. The initial fields obtained using optimal analysis retains both forecast errors and observational errors. Errors that grew in the previous forecasts have a larger chance of remaining in the analysis than errors that had decayed. These growing errors will then amplify again in the next forecast. It follows that the analysis contains errors that are dynamically created by the repetitive use of the model to create the initial fields. This is referred to as the Breeding of Growing Modes (BGM). These fast growing errors are above and beyond the traditionally recognised random errors that result from observations. The observation errors generally do not grow rapidly since they are not organized dynamically. It turns out that the growing errors in the analysis are related to the local Lyapunov vectors of the atmosphere (which are mathematical phase space directions that can grow fastest in a sustainable manner). These vectors are estimated by the breeding method (Toth and Kalnay, 1993).

2.6.2 Cell Mapping

Ensemble prediction also produces estimates of transition probabilities between current and future times. Suppose the initial field V consists of n variables, $v_1, v_2, ..., v_n$ and the range of error of V is η, every given field of V represents infinite number of fields within the range of $V \pm \eta/2$. The n dimensional space is now divided into L n-dimensional cubes of size η, which are called cells. The number of cells for the ith component of V is L_i, then $L = \prod_{i=1}^{i=n} L_i$. For a given initial field of V, the true precise value of V is not known. Figure (2.11) shows the situation of a two-dimensional case. There, the space of v_1 and v_2 is divided $L_1 L_2$ cells, denoted by the index (i_1, i_2). To facilitate discussion, the 2-dimensional cells can be denoted using the index j with $j = i_1 + (i_2 - 1)L_1$. For the n dimensional case, we use i_k ($k = 1, 2, ..., n$) to represent the cells and

$$j = i_1 + \sum_{k=1}^{n} \left[(i_k - 1) \prod_{m=1}^{n-k} L_m \right] \qquad (2.51)$$

Suppose the initial field is located in cell j_0. Then the true initial field can be any point within this cell. If the probability for every given point within the cell to be the true value is known (e.g., MCF means that the probability for every given point within the cell is equal, whereas other methods use different probability distributions), then the projection of these points after time τ can be determined using the dynamic model. The probability for the system to take a value in the cell j is $p_{j_0,j}(\tau)$, which is the transition probability for a V in cell j_0 at time $t = 0$ to locate in cell j at time $t = \tau$. The transition probability $p_{j_0,j}$ can be estimated by choosing Γ points in the j_0 cell and applying the dynamic model to these Γ points and integrate with time for the period of τ. In doing so, we obtain Γ projections distributed in the L cells. Suppose Γ_j points fell into cell j, the transition probability $p_{j_0,j}$ can be calculated by using

$$p_{j_0,j} = \frac{\Gamma_j}{\Gamma} \tag{2.52}$$

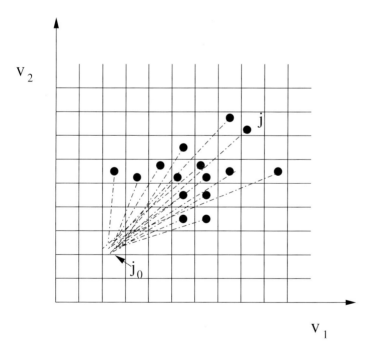

Fig. 2.11. An illustration of a two dimensional cell and the estimate of the transition probability between the cells.

Since non-linear systems are sensitive to the initial conditions, we would obtain different outcomes after time τ for different points from cell j_0. Ensemble prediction allows an estimate of the predictability. To achieve this, we divide the j_0 cell further into K smaller cells. In this case, the transition probability $p_{j_0,j}$ is given by

$$p_{j_0,j} = \frac{1}{K} \sum_{k=1}^{K} p_{j_0^k,j} \tag{2.53}$$

where j_0^k is used to indicate the kth small cell in cell j_0. The standard deviation for $p_{j_0^k,j}$

$$R_k = \left[\frac{1}{L} \sum_{j=1}^{j=L} (p_{j_0^k,j} - p_{j_0,j})^2 \right]^{1/2} \tag{2.54}$$

describes the difference between predictions using j_0^k and those using j_0. Clearly, the larger R_k is, the less predictable is the system.

Ensemble forecasts produce a large amount of information that requires statistical analysis. The simplest statistical parameter is the ensemble mean or a weighted average of the ensemble forecasts. The ensemble variance also can be used to provide information on the uncertainties and/or confidence limits, as it reflects the overall degree of variability amongst the ensemble members. Larger values indicate areas of substantial disagreement and hence less confidence in any individual prediction, including the ensemble mean, and visa versa. The maps of spread thus provide an evolving measure of the relative confidence geographically and with respect to individual weather systems. Another commonly used technique is to group ensemble members that have similar feature into clusters, each of which can be considered as a possible prediction. Ensemble forecasting provides quantitative estimates of probabilities. Probability estimates are the percentage of predictions out of the total ensemble that satisfy certain specified criterion. Some examples of ensemble prediction are given in Chap. 3.

2.7 Dynamic-Stochastic Models

Considerable developments have been made in recent years in the study of systems, especially in the areas of synergetics, systems with dissipative structure, chaos theory and fractals. All these new techniques are related to non-linear complex systems, of which the environment is an example. We shall briefly discuss several aspects of non-linear theories and consider their possible applications to environmental modelling and prediction.

2.7.1 Dissipative Structure

The theory of dissipative structure studies the ordered structure of non-linear systems at states far from equilibriums. What then is a dissipative structure?

Suppose a system consists of a number of subsystems on the microscopic level and the continuous motions of the subsystems construct a macroscopic stable structure which is maintained by external supplies of mass and energy. This differs from the equilibrium state in that, in the latter state, there are no subcomponent interactions on the macroscopic level and the macroscopic structure of the whole system does not evolve.

Systems of dissipative structure have the following features: 1) ordered structure in space; 2) oscillation in time; 3) the temporal and spatial structure of the system depends on the supplies of external energy and mass; 4) minor structural change may lead to major functional changes. A well known example is Bénard convection. The studies on Bénard convection by Saltzman (1962) and Lorenz (1963) led to the historical development of chaos theory.

Dissipative structure occurs under several conditions. First, the system is governed primarily by non-linear interactions. These interactions result in a convoluted structure of the system components which cannot be separated from each other. The non-linear interactions also allow small perturbations to amplify and to create perturbations on other scales of the system motion. The dynamics of system evolution depends on the competition and mutual constraint between the subsystems. The organization and the structure that occur from the self-organization do not depend on the external forces but on the interactions among the subcomponents. The interactions and constraints result in that the system has several possible tendencies for evolution, and in certain conditions, one of these tendencies prevails and creates ordered structure from disorder. Second, the system must be an open system. If the system is closed then it always evolves to further disorder. For open systems, there exists a continuous supply of mass, energy and/or other signals. If this external supply reaches a certain level, self-organization takes place, which leads to the dissipative structure of the system. Third, the state of the system must be sufficiently far from equilibrium. This is because, in the vicinity of equilibrium, non-linear interactions are generally weak and the interactions between the subcomponents are approximately linear. Finally, during the process of self-organization, stochastic fluctuations play an important role. In almost all natural systems, external or internal stochastic fluctuations occur. These fluctuations may trigger interactions between the subcomponents of the system and forces the system to evolve to achieve newer structure. Of course, the effect of the perturbations depends on the stability of the system. For systems at stable equilibration, small perturbations would not change the state of the system. At some critical point, the system becomes sensitive to the perturbations.

The atmosphere and ocean, when observed as systems, satisfy the above conditions. The motions of the atmosphere and the oceans certainly show dissipative structures, e.g., the Bénard convection in the atmospheric boundary layer. The environment as a whole also satisfies the above conditions. However, how the theory of dissipative structure can be applied to environmental modelling and prediction requires further research.

2.7.2 Synergetics

Synergetics (Haken, 1983) studies the mechanisms and behaviour of systems at non-equilibrium. The system is studied from both the macroscopic and microscopic points of view. A system is considered to consist of a large number of subsystems. During the evolution of the system, basic structural components are formed of subsystems, which are called meso-scopic components. Again, Bénard convection is a good example. In this convection, the microscopic components are individual molecules which cause heat conduction and momentum transfer through random motion. When the surface is heated from below and the flow becomes thermally unstable, convective cells, namely the meso-scopic components, would develop, which eventually form large convective rolls and hexagons of the Bénard convection.

Statistical descriptions can be applied to the evolutions both from the microscopic to the meso-scopic level and from the meso-scopic to the macroscopic level. Synergetics theory considers systems as having two basic actions. First, there exist irregular and random motions which lead to disorder of the entire system. In Bénard convection, the random motions of molecules constitute these basic irregular motions on the microscopic level. From the view point of fluid dynamics, this is dissipation (either molecular or small-eddy). Second, there are interactions among the subsystems, which result in the structure of the system, or, order. Different interactions result in different macro-structures of the system.

Slaving theory is the cornerstone of synergetics. The mathematical formation of the system dynamics is represented by a set of partial differential equations, in which the interactions between the subsystems are represented by non-linear terms such as

$$\frac{dq_i}{dt} = \alpha_i q_i + \beta_{ij} q_i q_j + f_i \tag{2.55}$$

where α_i are descriptions of the response characteristics of q_i and β_{ij} represent the intensity of the interactions and depend on external forcing. The equations also contain stochastic terms f_i which are pre-specified in terms of their probability distributions. The system consists of components which respond rapidly to forces exerted on them (large α_i) and components which respond slowly to forces (small α_i). Hence, the q_i can be divided into fast and slow variables. Fast variables are slaved by slow responding variables. This is the slaving theory of synergetics. The mathematical treatment depends on the time scales of the problems under investigation (i.e. $1/\alpha_i$). For short time scale behaviour, the slow variables are almost constant and the evolution of the system is mainly characterised by fast variables. On longer time scales, the evolution of the system is determined by slow variables and the behaviour of the fast variables becomes similar to that of stochastic variables.

From the view point of synergetics, the environment can be considered to be a system which is slaved largely by the ocean while the atmospheric variables are fast ones. The slaving process is relative easy to understand for constant external conditions. However, the environment is such a system affected by an external forcing (when simplified) which is periodic with two distinct periods (diurnal and annual changes of solar radiation). As a consequence, the slaving processes (the large scale ocean processes) must also evolve with time. Unfortunately, we do not yet know how the slaving processes behave under such periodic external forcing.

2.7.3 Deterministic Chaos Theory

Chaos refers to some kind of stochastic behaviour of a dynamic system. Of particular interest is deterministic chaos which specifically refers to unpredictable phenomena governed by deterministic equations, such as (1.5a)–(1.5c). If these equations consist of non-linear terms, their solution may demonstrate irregular characteristics for a certain range of control parameters, e.g., a and b in (1.5a)–(1.5c). The occurrence of deterministic chaos reflects the intrinsic unpredictable behaviour caused by the non-linear terms, not by stochastic external forces.

Deterministic chaos is closely related to strange attractor. Attractors are equilibrium points and/or regions in the phase space, to which the trajectories of the dynamic system in phase space approach as time increases. To illustrate this, we consider a three dimensional dynamic system

$$\dot{x}_1 = f_1(x_1, x_2, x_3) \tag{2.56a}$$
$$\dot{x}_2 = f_2(x_1, x_2, x_3) \tag{2.56b}$$
$$\dot{x}_3 = f_3(x_1, x_2, x_3) \tag{2.56c}$$

The equilibrium points are defined by $\dot{x}_i = 0$. Suppose the system is in equilibrium at (x_{10}, x_{20}, x_{30}). Then we have

$$f_1(x_{10}, x_{20}, x_{30}) = f_2(x_{10}, x_{20}, x_{30}) = f_3(x_{10}, x_{20}, x_{30}) = 0$$

We are particularly interested in the dynamic stability at the equilibrium points and hence, we consider a small disturbance and examine how the disturbance would evolve with time. From the above equations, we obtain

$$\delta \dot{x}_i = A_{ij} |_{x_{10}, x_{20}, x_{30}} \, \delta x_j \tag{2.57}$$

where $A_{ij} = \partial f_i / \partial x_j$ is a component of the Jacobian matrix $\underline{A}$. Denote the eigenvalues of $\underline{A}$ as s_i and the real parts of s_i as λ_i, which are the Lyapunov exponents. Four types of attractors may occur, namely, ordinary, periodic, quasi-periodic and strange attractors, depending on the values of λ_i. In a three dimensional system, the Lyapunov exponents can be ordered as

$$\lambda_1 \leq \lambda_2 \leq \lambda_3$$

The corresponding arrangements of λ_i corresponding to ordinary, periodic, quasi-periodic and strange attractors are as shown in Table 2.3.

Table 2.3. Arrangements of λ_i corresponding to different attractors.

	λ_1	λ_2	λ_3
Ordinary attractor	-	-	-
Periodic attractor	0	-	-
Quasi-Periodic attractor	0	0	-
Strange attractor	+	0	-

As an example for these different types of attractors, the behaviour of the Lorenz system for different values of Rayleigh number is depicted in Fig. 2.12. For small Rayleigh numbers, the Lorenz system has an ordinary attractor, while as the Rayleigh number increases the system has an strange attractor. In the latter case, the solution of the system (the location of the trajectory) at time t is extremely sensitive to the initial condition (the initial location of the trajectory). If the initial condition cannot be exactly specified the precise future state of the system cannot be predicted. Of course, the Lorenz system is relatively simple and describes best some aspects of the Bénard convection in fluids. The importance of the study of Lorenz (1963) lies in that it has demonstrated a concrete example of the existence of chaos. However, such a system remains distant from the real environment.

We now ask how chaos theory can be applied to long-term environmental predictions. Suppose Equations (2.56a)–(2.56c) represent a chaotic system. The trajectory of the system in the phase space for a given initial condition $x_i(t_0)$ is then given by $x_i[t, x_i(t_0)]$. As $t \to \infty$, all trajectories are attracted to a set of points in the phase space (the strange attractor), regardless of the initial conditions. While each individual trajectory is sensitive to its initial condition, the features of the strange attractor are independent of initial conditions. Hence, it is possible to described the strange attractor statistically in terms of, for instance, probability distributions functions of the attractive set of points and the transition probabilities between them. This is the approach taken in dynamic-stochastic models for long-term environmental predictions, in particular for long-term climate predictions.

The framework of the dynamic-stochastic approach is outlined below, which consists of three basic components.

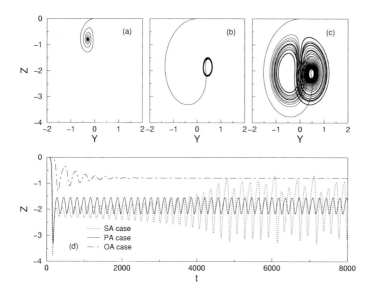

Fig. 2.12. (a) Ordinary, (b) periodic and (c) strange attractor of the Lorenz system for different values of the Rayleigh number. Example time series for the three cases are shown in (d).

1. Strange Attractor: The first component is to quantify the strange attractor which represents all future states of the environment on the basis of a dynamic model. Suppose the environment is correctly represented with a set of n non-linear equations with its solutions approaching a strange attractor as time increases. Then, through the integration of these equations, trajectories of the dynamic system can be determined for initial conditions which are specified within a reasonable range. Of course, these trajectories do not represent the true long-term behaviour of the environment, as they are sensitive to the initial conditions which always contain a degree of error. However, this kind of integrations allows the estimation of the topological features of the strange attractor which is in itself independent of the initial conditions.

   The strange attractor represents all possible future states of the environment. From this view point, long-term environmental prediction is not to make a deterministic forecast of a particular realization at a given time in future, but a statistical statement of the probability (distribution) of a possible future realization on the strange attractor. Ultimately, the strange attractor is a volume in a n-dimensional space, despite the fact that such a space might be fractional and it consists of infinite number of points. In theory, the ensemble environment, namely, the strange attractor, can be estimated by integrating the governing equations from t_0 to $t = \infty$ for all possible initial conditions with known control parameters. If the external conditions (reflected in the control parameters), such as the solar radiation, do not change with time, then the ensemble environment would be independent of time.

2. Transition Probability in Phase Space: The topological features of the strange attractor can be quantified through the estimates of the Lyapunov exponents, fractional dimensions and the analysis of energy spectrum, etc. However, we are most interested in the joint probability density function of $x_i(t=\infty)$ in the phase space, i.e., $p_{sa}(x_1, x_2, ..., x_i, ..., x_{n-1}, x_n)$. In principle, p_{sa} is not difficult to estimate numerically, as illustrated in Fig. 2.13 for a two dimensional case. In this case, the strange attractor is covered by $M \times N$ cells. For cell (i, j) in that figure,

$$p_{sa} = \lim_{t \to \infty} \frac{S_{i,j}}{S_{total} \Delta x \Delta y} \tag{2.58}$$

where $S_{i,j}$ is the length of the trajectory in cell (i, j) and S_{total} is the total trajectory length. The probability for the environment to take the values in cell (i, j) is $p_{sa} \Delta x \Delta y$.

The prediction of the future state of the environment is achieved through the estimate of the transition probability in phase space. Suppose the current environment is located in cell (i, j) at $t = t_0$. Then the probability for the environment in cell (m, n) at $t = t_1$ can be estimated from the transition probability $p_{sa}(m, n, t \mid i, j, t_0)$. The method for estimating the transition probability has been described in Sect. 2.6.2 (see also Fig. 2.11). This component of the dynamic-stochastic approach is based on the statistical evaluation of the strange attractor.

3. Evolution of the Strange Attractor: In the above, we have assumed that the strange attractor does not evolve within the time period of $t_1 - t_0$. However, if (and only if) the external conditions for the environment change, the characteristics of the strange attractor may evolve with time, as illustrated in Fig. 2.14 for a strange attractor similar to the Lorenz attractor. Hence, the third component of the dynamic-stochastic approach is to identify the possible evolution of the strange attractor with time under the influence of external forces.

2.7.4 Interactive Chaos

Based on the philosophy of the dynamic-stochastic model, as outlined above, the basic strategy of long-term environmental prediction is to estimate the probability of the future environment on the strange attractor that embraces all future states of the environment. A crucial step in this strategy is to identify the chaotic behaviour of the environment. Although there are no known concrete examples which prove the environment is a chaotic system, it is generally assumed that it has inherent chaotic behaviour (Lorenz, 1984).

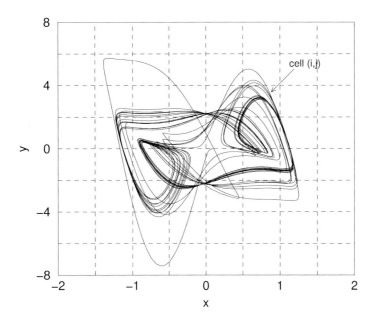

Fig. 2.13. An illustration for estimating joint probability density function, p_{sa}, and the transition probability, $\tilde{p}_{sa}$, on a strange attractor. The strange attractor is covered by $M \times N$ cells and p_{sa} and $\tilde{p}_{sa}$ can be estimated using (2.58) and (2.53), respectively.

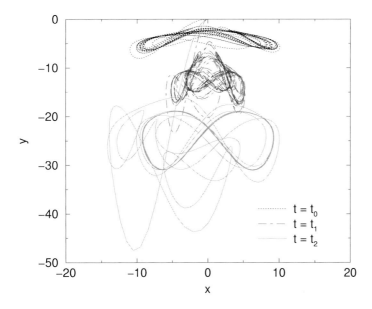

Fig. 2.14. Evolution of a strange attractor with time under the influence of external forcing.

The governing equations given earlier in Chap. 2 provide the foundation for identifying environmental chaos, but these equations need to be simplified to capture the essence of the environmental behaviour. An interesting example of atmosphere-land interactive chaos has been studied by Guo et al. (1996). In contrast to the Lorenz system, the atmosphere-land interactive chaos is not a purely fluid dynamic problem. Rather, it is a coupled system of atmospheric fluid motion and the thermal processes of the land surface. In this sense, the chaotic system studied by Guo et al. (1996) is more closely related to the coupled environment. A brief outline of their model is described below.

For a simplified climatic problem, we have the following equation system in a spherical coordinate system (λ, φ, p), where λ is longitude, φ is latitude and p is pressure.

$$\left(-2\Omega \cos\varphi \nabla^2 + \frac{2\Omega \sin\varphi}{a_0^2} \frac{\partial}{\partial \varphi}\right) \chi = \frac{2\Omega}{a_0^2} \frac{\partial \Psi}{\partial \lambda} \tag{2.59}$$

$$\left(2\Omega \cos\varphi \nabla^2 - \frac{2\Omega \sin\varphi}{a_0^2} \frac{\partial}{\partial \varphi}\right) \Psi = \nabla^2 \phi \tag{2.60}$$

$$\nabla^2 \chi + \frac{\partial \omega}{\partial p} = 0 \tag{2.61}$$

$$\frac{R}{p} T = -\frac{\partial \phi}{\partial p} \tag{2.62}$$

$$\frac{\partial T}{\partial t} + \frac{1}{a_0^2 \sin\varphi}\left(\frac{\partial \Psi}{\partial \varphi}\frac{\partial T}{\partial \lambda} - \frac{\partial T}{\partial \varphi}\frac{\partial \Psi}{\partial \lambda}\right) - \frac{\gamma_d - \bar{\gamma}}{gp} R\bar{T}\omega$$
$$= \frac{\partial}{\partial p}\nu\left(\frac{gp}{RT}\right)^2 \frac{\partial T}{\partial p} + \mu \nabla^2 T \tag{2.63}$$

where χ is velocity potential function and Ψ is stream function; ϕ is the geopotential, T is temperature, a_0 is average radius of the earth γ_d is adiabatic laps rate and $\bar{\gamma}$ is the mean lapse rate of the atmosphere.

We now expand the variables into spherical functions and use finite difference in the p direction. By dividing the atmosphere into 5 levels in the vertical and mathematical manipulation of the finite difference equations, we obtain the following ordinary differential equations for the temperature for the 500 hPa level, T_{500} (see Fig. 2.15)

$$\frac{dT_1}{dt} = a_{11}T_1 + a_{12}T_2 + a_{13}T_3 - \sigma_1(T_2^2 + T_3^2) + b_1 \tag{2.64}$$

$$\frac{dT_2}{dt} = a_{21}T_1 + a_{22}T_2 + a_{23}T_3 + \sigma_1 T_1 T_2 - \sigma_2 T_1 T_3 + b_2 \tag{2.65}$$

$$\frac{dT_3}{dt} = a_{31}T_1 + a_{32}T_2 + a_{33}T_3 + \sigma_2 T_1 T_2 + \sigma_1 T_1 T_3 + b_3 \tag{2.66}$$

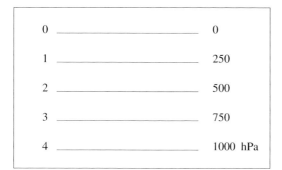

Fig. 2.15. Levels in the atmosphere used by Guo et al. (1996).

where

$$T_{500} = cT_1P_2^0 + T_2\cos 2\lambda P_4^2 + T_3\sin 2\lambda P_4^2 \tag{2.67}$$

with

$$c = \sqrt{\frac{\int_0^\pi (P_4^2)^2 \sin\varphi \, d\varphi}{2\int_0^\pi (P_2^0)^2 \sin\varphi \, d\varphi}} \tag{2.68}$$

and

$$P_1^0 = \cos\varphi \tag{2.69}$$

$$P_2^0 = \frac{1}{2}(3\cos^2\varphi - 1) \tag{2.70}$$

$$P_3^2 = 15(1 - \cos^2\varphi)\cos\varphi \tag{2.71}$$

$$P_4^2 = \frac{15}{2}(-7\cos^4\varphi + 8\cos^2\varphi - 1) \tag{2.72}$$

where a_{ij}, σ_i and b_i are coefficients. These coefficients are functions of temperature for the 1000 hPa level, T_{1000} (Fig. 2.15), which in turn, depends on the atmosphere-land interactions. Similar to T_{500}, we can express T_{1000} as

$$T_{1000} = cT_{1s}P_2^0 + T_{2s}\cos 2\lambda P_4^2 + T_{3s}\sin 2\lambda P_4^2 \tag{2.73}$$

Hence, a simplified coupled system for the prediction of atmospheric temperature is established. The interactions between the atmosphere and the land surface is determined by the bulk transfer coefficient C_H, which determines the transfer of heat and is an indicator of the interactions between the atmosphere and land. Choosing realistic model parameters, Guo et al. (1996) have found chaotic behaviour of both the atmospheric temperature at 500 hPa and the temperature of the land surface, as shown in Figs. 2.16 and 2.17. The structure of this interactive chaos is similar to that studied by Lorenz (1984), but more specifically related to the atmosphere-land interactions.

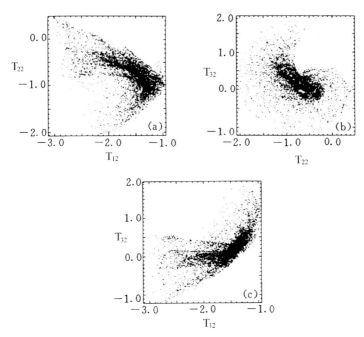

Fig. 2.16. Projections of T_1, T_2, T_3 of the coupled system for a 20 year integration (from Guo et al., 1996).

2.8 Natural Cybernetics

The ultimate purpose of environmental modelling and prediction is to allow us to optimize productive activities within the constrains of the environment. Therefore, research in environmental modelling and prediction has a major interest in the impact of human activities and how natural laws can be utilized in these activities, such as large engineering projects. A emerging new subject is natural cybernetics which combines environmental modelling and prediction with optimization. The theoretical framework of natural cybernetics is first developed by Zeng who has also given concrete examples of its application (e.g. Zeng, 1995).

The basic idea of natural cybernetics is as outlined below. Denote the set of environmental variables we wish to consider as $\mathbf{E}(\mathbf{x},t)$, which is a m-dimensional vector varying in space $\mathbf{x}$ and time t, and correspondingly, denote the set of variables related to human activities as $\mathbf{H}(\mathbf{x},t)$ which is a n-dimensional vector, also varying in space and time. As the changes of the latter set of variables have an impact on the former, we have

$$\frac{\partial E_i}{\partial t} = \Gamma_i(\mathbf{E}, \mathbf{H}, t) \qquad i = 1, 2, ..., m \qquad (2.74)$$

where Γ_i is a function. The evolution of E_i is governed by the above equations with the initial condition $E_i \mid_{t=0} = E_i^0(\mathbf{x})$ and boundary condition

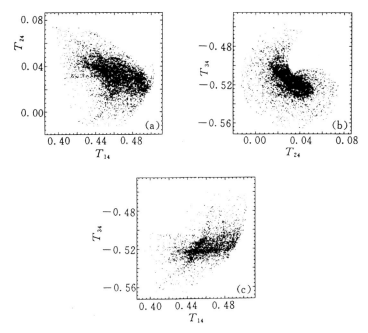

Fig. 2.17. Projections of T_{1s}, T_{2s} and T_{3s} of the coupled system for a 20 year integration (from Guo et al., 1996).

$\Gamma_i(\mathbf{E}, \mathbf{H}, t) |_{\partial \Omega} = \Gamma_i^{\partial \Omega}(t)$, where $E_i^0(\mathbf{x})$ and $\Gamma_i^{\partial \Omega}(t)$ are known functions. The variations of $\mathbf{H}$ are limited to a certain range and thus the modulus of $\mathbf{H}$ obeys

$$\| \mathbf{H} \| \leq C \tag{2.75}$$

where C is a limiting constant. Suppose the ideal environment for human society is $\mathbf{E}_0$. The changes brought about by $\mathbf{H}$ should then obey the following condition

$$\| \mathbf{E} - \mathbf{E}_0 \| \leq D \tag{2.76}$$

The central topic of natural cybernetics is find an optimal $\mathbf{H}$ under the conditions of (2.75) and/or (2.76). Suppose the criteria for optimisation is given by a function $P(\mathbf{E}, \mathbf{H})$. Then, we request that $P(\mathbf{E}, \mathbf{H})$ be optimal, i.e.,

$$\frac{\partial P}{\partial H_j} = 0 \qquad j = 1, 2, ..., n \tag{2.77}$$

Zeng (1995) has presented in detail the formulation of natural cybernetics and an example of its application to the engineering problem of sedimentation in water channels (Fig. 2.18). Silt sedimentation is an environmental process that has a profound impact on the evolution of water channels. The dynamics of silt movement and deposition can be formulated as an initial-boundary values problem consisting of equations of motion, mass conservation, silt transport and river floor topography. The engineering problem is to maintain an optimal navigation channel under the constraint of fixed expenses. This optimisation problem must be solved together with the dynamic model.

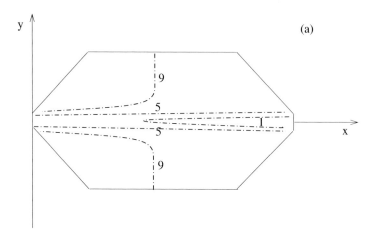

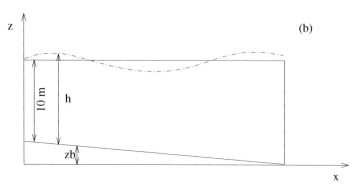

Fig. 2.18. (a) The model area of river and lake with initial distribution of bottom elevation; (b) the initial slope of river channel along the x-axis (redrawn from Zeng, 1995).

Suppose the upper water surface is z_s and the bottom topography is z_b. The water depth is then

$$h = z_s - z_b \tag{2.78}$$

If the silt mixing ratio in water is s, then the density of the water is

$$\rho = \rho_0(1+s) \tag{2.79}$$

where ρ_0 is the density of water free of silt. As the typical time scales involved in the fluid motion are much smaller than those of sedimentation, h and ρ vary rather slowly with time and hence it is useful to define

$$h = \tilde{h} + h'; \rho = \tilde{\rho} + \rho' \tag{2.80}$$

where $\tilde{h}$ and $\tilde{\rho}$ are smoothed values of h and ρ, while h' and ρ' are small within the time scale of fluid motion. The equations for the fast hydrodynamic processes are

$$\frac{\partial \mathbf{v}}{\partial t} + (\mathbf{v} \cdot \nabla)\mathbf{v} = -g\nabla \tilde{z}_s - (2\tilde{\rho})^{-1} g\tilde{h}\nabla\rho + \mathbf{F} + \mathbf{D}_\mathrm{m} \tag{2.81a}$$

$$\frac{\partial \tilde{\rho}\tilde{h}}{\partial t} + \nabla \cdot \tilde{\rho}\tilde{h}\mathbf{v} = 0 \tag{2.81b}$$

$$\frac{\partial \tilde{\rho}}{\partial t} + \mathbf{v} \cdot \nabla \rho = 0 \tag{2.81c}$$

where $\mathbf{v}$ is flow velocity, $\mathbf{F}$ is external force (e.g., wind stress and tidal force) and $\mathbf{D}_\mathrm{m}$ is dissipation. The equations for the slow sedimentation processes are as follows

$$\frac{\partial \rho_0 h}{\partial t} + \nabla \cdot \rho_0 \tilde{h}\mathbf{v} = \rho_0 F_w \tag{2.82a}$$

$$\frac{\partial \rho_0 s \tilde{h}}{\partial t} + \nabla \cdot \rho_0 s \tilde{h}\mathbf{v} = D_i - \rho_0 S_e \tag{2.82b}$$

$$\frac{\partial z_b}{\partial t} = \sigma S_e - \nabla \cdot \mathbf{T}_\mathrm{r} - \dot{z}_{b2} \tag{2.82c}$$

where F_w is net water flux at z_s, D_i is the diffusion term for silt, S_e is the removal rate of silt due to sedimentation, σ is the water to silt density ratio, $\mathbf{T}_\mathrm{r}$ is the transport rate of silt along the bed surface and $\dot{z}_{b2}$ is the change rate of z_b arising from the mechanical removal of silt for the maintenance of navigation channels.

The above set of equations can be solved with pre-specified initial and boundary conditions. The suitable numerical methods are those described by Zeng and Zhang (1982), combined with the time splitting scheme as described by Marchuk (1982). The time splitting scheme consists of solving the following sub-problems

- linear gravity waves (very fast process);
- advection of water flow (medium fast process);
- advection of density and density flow (medium fast process);
- transport of water and silt (medium fast process);
- the dissipative process (slow process);
- the change in bed topography (very slow process).

Figure 2.19 shows the accumulated sediment after 2767 days of integration of an artificial computational domain.

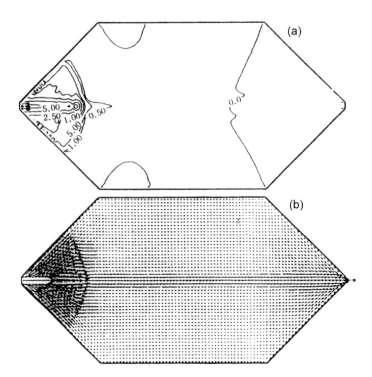

Fig. 2.19. Simulated accumulated sediment after 2767 days of integration for an artificial computational domain (from Zeng, 1995).

We now consider the optimum engineering of the navigation channel. Let the maximum total cost for mechanically removing silt in the time interval T is a fixed constant, c, e.g.,

$$\int_0^T \int_\Omega R_1 \dot{z}_{b2} \mathrm{d}\Omega \mathrm{d}t \leq c \qquad (2.83)$$

where $R_1 = R_1(x, y, t)$ is the cost per unit time and unit area, a known function. The optimization problem is to find $\dot{z}_{b2}(x, y, t)$ $\{(x, y) \in \Omega, t \in [0, T]\}$, such that

$$\int_0^T \int_\Omega R_2 h \mathrm{d}\Omega \mathrm{d}t = \max \qquad (2.84)$$

where $R_2 = R_2(x, y, t)$ is a known weighting function, as normally the channels cover only a fraction of Ω.

The general form of the above described problem can be summarized as

$$\frac{\partial \Phi}{\partial t} = L\Phi + F + \Psi \qquad (x, y) \in \Omega, t \in [0, T] \qquad (2.85\mathrm{a})$$

$$\Lambda \Phi = G \qquad (x, y) \in \partial\Omega \qquad (2.85\mathrm{b})$$

$$\lim_{t \to 0} \Phi = \Phi^{(0)} \qquad (x,y) \in \Omega \qquad (2.85c)$$

$$\int_0^T \int_\Omega K_1 \Psi \mathrm{d}\Omega \mathrm{d}t \leq C \qquad (\Psi \geq 0) \qquad (2.85d)$$

$$\int_0^T \int_\Omega K_2 \Phi \mathrm{d}\Omega \mathrm{d}t = \max \qquad (2.85e)$$

where Φ is a vector (e.g. consisting u, v, h, s and z_b); $F, G, \Phi^{(0)}, K_1$ and K_2 are known vectors; L and Λ are linear or non-linear operators, Ψ is the vector to be determined, $K_1\Psi$ and $K_2\Phi$ are scalar products. The optimization problem can be solved by iterative process, if L and Λ are non-linear operators. Suppose the solution of the n-iteration is

$$\Gamma_{n+1} = \Gamma_n + \delta\Gamma \qquad (\Gamma \text{ being } \Phi, \Psi, L \text{ or } \Lambda) \qquad (2.86)$$

Taking the first variation of (2.85a) to (2.85e), we have the following linear problem

$$\frac{\partial \delta\Phi}{\partial t} = \tilde{L}_n \delta\Phi + \delta\Psi \qquad (2.87)$$

$$\tilde{\Lambda}_n \delta\Phi = 0 \qquad (2.88)$$

$$\delta\Phi^{(0)} = 0 \qquad (2.89)$$

$$\int_0^T \int_\Omega K_1 \delta\Psi \mathrm{d}\Omega \mathrm{d}t \leq \left[c - \int_0^T \int_\Omega K_1 \psi_n \mathrm{d}\Omega \mathrm{d}t \right] \leq c_n \qquad (2.90)$$

$$\int_0^T \int_\Omega K_2 \delta\Phi \mathrm{d}\Omega \mathrm{d}t = \max \qquad (2.91)$$

Here, $\tilde{L}_n = L_n \delta\Phi + (\delta L)\Phi_n$ and $\tilde{\Lambda}_n = \Lambda_n \delta\Phi - (\delta\Lambda)\Phi_n$ are new operators.

The above linear problem can be solved using the method of adjoint operator. Denote the inner product of two functions in Ω as

$$(a, b) \equiv \int_\Omega ab \mathrm{d}\Omega$$

The adjoint problem can be set up as follows

$$\frac{\partial K_1 \Phi^*}{\partial t} = -\tilde{L}_n^* \delta\Phi^* - K_2 \qquad (2.92)$$

$$\tilde{\Lambda}_n^* \delta\Phi^* = 0 \qquad (2.93)$$

$$\delta\Phi^* |_{t=T} = 0 \qquad (2.94)$$

Here, the set of operators $\tilde{L}_n^*$ and $\tilde{\Lambda}_n^*$ is adjoint to the set of operators $\tilde{L}_n$ and $\tilde{\Lambda}_n$ defined by

$$(K_1 \tilde{L}_n A, B) = (A, \tilde{L}_n^* B)$$

72   2. The Environmental Dynamic System

with A and B satisfying the boundary conditions
$$\tilde{\Lambda}_n^* A = 0, \quad \tilde{\Lambda}_n^* B = 0$$

It follows from (2.87)–(2.89) and (2.92)–(2.94) that

$$\int_0^T (K_1 \delta \Phi^*, \delta \Psi) dt = \int_0^T (K_2 \delta \Phi) dt \tag{2.95}$$

Therefore, the problem is transformed in finding $\delta \Phi^*$ which is the solution of the adjoint problem (2.92)–(2.94), satisfying

$$\int_0^T (K_1 \delta \Phi^*, \delta \Psi) dt = \max \tag{2.96}$$

In the silt sedimentation problem, K_1 and Ψ are of separated variable type, e.g.,

$$K_1(x, y, t) = K_1^{(1)}(t) K_1^{(2)}(x, y) \tag{2.97}$$

with

$$\begin{aligned} K_1^{(1)}(t) &> 0 & t &\in [0, T] \\ K_1^{(2)}(x, y) &\geq 0 & [(x, y) &\in \Omega] \end{aligned}$$

Also

$$\delta \Psi = \delta \Psi^{(1)} + \delta \Psi^{(2)} \tag{2.98}$$
$$(K_1, \delta \Psi^{(1)}) = K_1^{(1)}(t)(K_1^{(2)}, \delta \Psi) \tag{2.99}$$
$$(K_1, \delta \Psi^{(2)}) = 0 \tag{2.100}$$

We obtain

$$\delta \Psi^{(1)} = c_n K_1^{(2)}(x, y) / [K_1^{(1)}(t) \cdot |K_1^{(2)}|^2] \tag{2.101}$$

Similar for $K_1 \delta \Phi^*$, we have

$$K_1 \delta \Phi^* = \lambda K_1^{(1)}(t) K_1^{(2)}(x, y) + X(x, y, t) \tag{2.102}$$

where $(X, K_1^{(2)}) = 0$, and

$$\lambda = (K_1 \delta \Phi^*, K_1) / [K_1^{(1)}(t)]^2 \cdot \|K_1^{(2)}\|^2 \tag{2.103}$$

Since $\lambda K_1(x, y, t)$ and X are known,

$$\delta \Psi^{(2)} = \gamma X \tag{2.104}$$

where γ is the maximum value to satisfy

$$\max_\gamma (\gamma X - \delta \Psi^{(1)}) \geq 0 \tag{2.105}$$

While the basic idea of natural cybernetics is not too difficult to understand, its implementation can be very complex and requires much further research. Clearly, the prerequisite for natural cybernetics is environmental modelling and prediction. One of the difficulties we currently face is that, apart from specific areas, environmental modelling and prediction is not yet fully developed. In particular, the research problems related to the long-term are far from solved. Another profound difficulty is the definition of the optimization function. Unless the environmental impact of human activities are fully understood, the optimization function is extremely difficult to define. Even if they are well defined, the optimization problem may not lead to an unique solution of **E**. Furthermore, the social and economical objectives which ultimately define the meaning of optimization are often not inconsistent. Here, we start to bridge the gap between the environmental and social sciences.

3. Atmospheric Modelling and Prediction

Lance M. Leslie, Russel P. Morison, Milton S. Speer and Lixin Qi

3.1 Introduction

This chapter is intended to provide the reader with an introduction to the development and application of models of the Earth's atmospheric flow. It describes a range of past and present methods for solving the equations governing the behaviour of the Earth's atmosphere. These equations can be cast in predictive form and their solution therefore provides forecasts of the future state of the atmosphere. The governing equations are mathematical expressions of the various conservation laws governing the atmosphere, as a fluid in motion. The equations also allow for the effects of a wide range of forcing mechanisms. Here we mention but a few, such as the differential heating of the sun's radiation between the equator and the poles, dissipation of momentum, heat and moisture at the Earth's surface and boundary layer, turbulence in the free atmosphere, the drag on the atmosphere from steep, high or rough terrain, and the profound influence of Earth's rotation. The complexity of the interaction between the Earth's atmosphere and the other components of the Earth's environmental system are illustrated schematically in Fig. 3.1, which shows the coupled atmosphere-ocean-landsurface model, which is the focus of much of this book. Each of the components represented in Fig. 3.1 is a model in its own right and interacts with the atmospheric model in two main ways. The early and most commonly used procedures are the so-called "one-way" interactions, in which the atmospheric model is either the only true dynamical model or, if the other components are complex models, they are forced by the atmospheric model. More recently, "two-way" or coupled models have appeared. They explicitly treat complicated non-linear interactions between the atmospheric model and the other environmental components. Examples include coupled atmospheric-land surface/soil models in which there is a feedback, or exchange, between both the atmospheric model and the land surface/soil model, as well as coupled ocean atmosphere models where there is a two way feedback of moisture, momentum and heat.

76   3. Atmospheric Modelling and Prediction

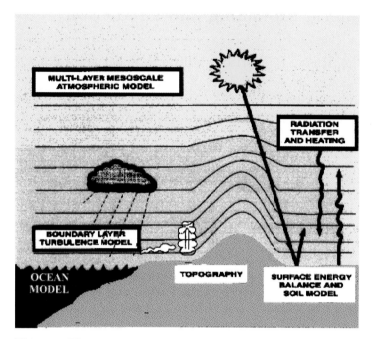

Fig. 3.1. The components of the coupled atmosphere-ocean-landsurface model.

This chapter is divided into the following seven sections. First, there is a brief history of numerical weather prediction (NWP), followed by a discussion of the basic concepts and equations governing the evolution of the atmosphere. Next is the derivation and evaluation of sums of the many available finite-difference numerical methods for solving the atmospheric equations. In this chapter, other types of numerical methods for solving the atmospheric equations are not discussed. However, it should be mentioned that techniques such as spectral models are very popular. The interested reader is referred to other texts, such as the recent book by Krishnamurti (1996). Applications of NWP are given below on a range of significant forecasting problems, over a large spread of time and space scales from nowcasting, through short range and medium range forecasting to extended range predictions and, finally, to general circulation and climate models . Each of these time scales is of great importance in environmental modelling and prediction. Finally, there is a discussion of some of the powerful new techniques developed in recent years for enhanced modelling and prediction of the troposphere.

3.1.1 Impact on Society of Severe Weather and Climate Events

A major motivation for this chapter is the fact that weather and climate affect almost all aspects of human activity. Severe or extreme events, can have devastating impact on the human and natural environments. Moreover,

3.1 Introduction

each year weather events result in an average of hundreds of thousands of human deaths. Table 3.1 below shows a summary of deaths from natural disasters, including weather and weather-related phenomena in the 25 year period 1967 to 1991 (adapted from the World Meteorological Organization, 1998). The figures reveal that approximately half of the human fatalities caused by atmospheric events are due to weather phenomena and the other half to disasters on much longer time scales, particularly seasonal and climate time scales.

Table 3.1. Deaths worldwide from weather disasters, 1967 to 1991 (WMO, 1998).

Type	No. of events	Persons killed
Weather Events		
Hurricanes	894	896,063
Flood	1,358	304,870
Storm	819	54,500
Heat and cold	133	4,926
Drought	430	1,333,728
Weather-related		
Avalanche	29	1,237
Landslide	238	41,992
Fire	729	81,970
Famine	15	605,832
Food shortage	22	0
Insect	68	252
Epidemic	291	124,338

Table 3.1 reveals that the worst causes of deaths from weather and weather-related disasters (such as storm surges in Bangladesh) are tropical cyclones (also called hurricanes or typhoons depending on location) and drought. These phenomena operate over very different space and time scales. Hurricanes typically last no more than a day or two at a given location whereas the effects of drought occur over seasons or even years. Economic losses from severe and extreme weather events have increased enormously over the last thirty years with estimated costs of around US$30 billion per year in the 1990s.

3.1.2 Historical Background

Forecasting of the weather by solving the equations governing the behaviour of the atmosphere, using high-speed electronic computers, is now over 50 years old. The first truly successful prediction was carried out in the late 1940s in the USA by Charney, Fjortoft and von Neumann, who published their one day forecast in 1950 (Charney et al., 1950). Since that time the field has grown dramatically, to the extent that all major weather services around the world depend heavily upon numerical weather prediction (NWP) models to provide forecast guidance. Moreover, NWP models are now very large systems with the individual components such as the representation of cloud processes, radiation, or boundary layer fluxes all being closely coupled together as part of the overall system. This is true also of the generation of the initial state for the NWP model. In modern data assimilation and analysis systems, the observations and the model are closely connected, with the model acting as a formal constraint in some of the most advanced systems. The level of dependence of weather centres on NWP output is especially high for weather events that are intrinsically difficult to forecast and that are potentially hazardous. Examples include such diverse events as severe weather phenomena and high-level pollution days. The NWP centres also have strong research and development programs for producing predictions that are at least several days but usually a week or more in advance. Progress in NWP since the time of Charney and co-workers has been enormous and has been driven by advances in four main areas. These areas are intertwined, and each is vital to improving the quality of NWP models. They include: data collection and analysis; theoretical advances in our understanding of the behaviour of the atmosphere and its translation into the set of equations governing the behaviour of the atmosphere; innovations in the accuracy, efficiency and stability characteristics of the algorithms used to solve the governing equations; and continued geometric growth in the power of electronic computers used to carry out the computations, produce the graphical output and disseminate the forecasts to the various centres.

It has been convenient to classify weather events into four groups according to their time scales. The first scale is commonly referred to as very short range forecasting (VSRF) and occupies the 0–12 hour forecast period. Spatial resolutions are from hundreds of metres to 5 km. It overlaps with so-called nowcasting, which is in the 0–3 hour range or sometimes the 0–6 hour range. The second scale is short range forecasting (SRF), which is conventionally intended to cover the period 0–48 hours, and typical spatial resolutions of between 5 and 35 km, depending on the computer power available at the weather centre. They are concerned with regions that cover significant parts of the globe, such as Europe, North America and the Asia-Pacific region. The next, third, scale is the so-called medium range forecasting (MRF) scale, which varies between centres but is typically between 7 and 15 days. Spatial resolutions are now between about 40 km and 200 km, again depending on the computing facilities available. The fourth weather forecasting scale is the extended range forecast (ERF), which is used to describe forecasts out to one month in advance. Both the MRF and ERF scales require global models. However, instead of forecasting the weather for a particular day, the aim of ERF is to provide statistical averages of the key output variables, such as temperature or precipitation, and to express these in terms of means and deviations from the monthly averages. Beyond the ERF limits the predictions move from sub-climate to seasonal and climate predictions, but the basic requirements are similar to those of ERF simulations, namely deviations from means. The main emphasis in this chapter will be on the shorter time scales but some longer term ERF to climate simulations will also be presented. It also is pointed out that the activities affected by weather and climate are very large in number and are not readily separated from one another. Typically, agricultural activities depend not only on the amount of rain that falls in a given year, but in addition, they depend critically on the amount and timing of the individual weather events that actually produce the rain. The farmers' planting agenda depends as much, perhaps more, on knowing a day or two in advance how much rain will fall as on how much will fall over a calendar year. Similarly, pollution events are usually short term and related to specific weather conditions, although the overall classification of a city as having poor air quality is a climate-related feature. There are many other instances of this kind of interdependence, such as the importance of accurate forecasts of wind and temperature on the demand for power from the energy industry, and potential damage to the transmission lines that carry the energy.

3.1.3 The Role of the Atmosphere in the Environmental System

As mentioned in detail in Chap. 2 of this book, the atmosphere is one of at least five components of the Earth's environmental systems. As such it is an environmental subsystem. The atmosphere is a complex, coupled, forced, non-linear environmental system that has numerous feedback processes. Although the atmosphere is just one component of the entire coupled system it is arguably the most significant, for the following reasons. Without the atmosphere, there would be no human life as we know it, as the human population lives in the thin atmospheric envelope that extends to about 20 km above the Earth's surface. As a result, the atmosphere has always been the most accessible component and therefore the most studied of the five Earth system components. This factor has led to a much greater body of knowledge about the atmosphere than is available for the other components, at this point in time. The atmosphere also is unique in being the only component of the Earth's environmental system that is truly connected to all the other components. It is coupled to the oceans in a two-way manner, drawing moisture and heat from the oceans, while returning energy to the oceans in the form of wind stresses that drive the ocean surface and waves. Many systems from the tropics to the high latitudes draw their energy from the oceans. One of the better known examples is the analogy of typhoons with Carnot cycle engines developed by Emmanuel (see, for example, Lighthill, 1992). The atmosphere also has some unique characteristics. It has very fast response times when compared with the other components such as the oceans, the land-surface, the cryosphere and so on. As a direct consequence of this rapid response time, the transport of materials such as energy, or pollutants, is very fast compared with transport rates in the other systems. Related to this aspect are the vast range of space and time scales that operate in the atmosphere and the very large number of circulation patterns from planetary scales down to micro-scales. In Sect. 3.4 of this chapter, examples of applications drawn from this range of scales will be presented in detail to illustrate the modelling and predictive capacity of NWP modelling systems, as we enter the twenty-first century.

3.1.4 Numerical Weather Prediction and Modelling

The history of routine numerical weather prediction (NWP), or operational NWP as it is most commonly known, dates back to the mid-1950s. Although much has changed in the details of the operational NWP process, the basic structure has remained largely the same and is shown schematically in Fig. 3.2. The steps in Fig. 3.2 are now individually traced. To begin with, a starting analysis, or initial state, is obtained by combining observational data with the latest model forecast. Problems known as model "shock" and model "spin-up" arise from the fact that the analysis is derived from large sets of widely scattered observations, and cannot be an exact solution to the governing equations. As a consequence, spurious gravity wave modes can be, and are, excited. The forecast model typically requires about 3 to 6 hours for geostrophic adjustment and other balances to take place. Moreover, some processes, such as precipitation, do not commence immediately but lag behind the real atmosphere for a similar period of time. The problems of model shock and spin-up have been the subject of much research and are steadily being eliminated in the most advanced NWP systems. The procedure of ameliorating this problem is known as initialization and its aim is to modify the analysis such that it produces a much better set of initial conditions. The initialization process has many forms (for a summary see, for example, Daley, 1991). Finally, the model is run from the initial conditions for the desired time period and the output is stored at pre-specified intervals. When the model run is completed, a post-processor step is executed to obtain those atmospheric variables that the model user, such as the forecaster, requires. There are other steps in the process. Model output statistics (MOS) are statistical modifications of the deterministic NWP model predictions based on regression equations. The regressions are carried out between the model output and a very large number of atmospheric variables such as maximum and minimum temperatures, rainfall amounts, wind speed and direction and sea level pressure values, to name just a small selection. Statistical models are very widely used both as stand-alone techniques and in combination with the NWP and climate models. They will be discussed in greater detail in this chapter in Sect. 3.4, below. In that section, mention will also be made of the very important "downscaling" technique which has as its aim the production of detailed information at high resolution from the much coarser resolution general circulation models (GCMs). The GCMs are now all global models and they also form the basis of climate models. The field of model output visualization, which draws very heavily on adapting mainstream information technology (IT) has advanced extremely rapidly in the past decade or so. The graphical rendering of model forecasts has become one of the most important components of the NWP and GCM systems. Its importance lies largely in the need to make the forecasts readily accessible to both the forecasters and other end users. The raw model output usually is not user-friendly and graphical user interfaces (GUIs) are becoming almost universally available. Visualization also is important in providing three- and four-dimensional images of the atmosphere, animation of the predictions, variable perspectives and most recently, the capacity to present more than one field in a graphical image (five-dimensional visualization). Finally, the visualization process is combined with powerful GIS data bases to produce even more accessible and useful output for the end-users. The last, but not least, important step in the operational forecast process is the dissemination of the forecast products. The standard procedure in a major NWP centre is to produce a forecast for a particular country or region and then to distribute it to a large number of localized regions. This can be achieved in a number of ways. The centre producing the NWP output can transmit the model output to selected groups connected to it by the internet or, conversely, it can make the same variables accessible to the users by establishing a web-site for the purpose. Each approach has its own advantages and disadvantages.

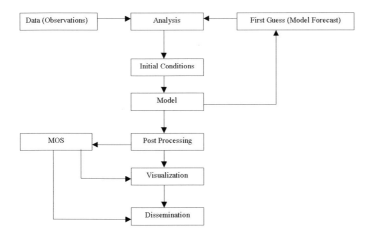

Fig. 3.2. Simplified flow diagram of a numerical weather prediction (NWP) system.

3.2 Equations Governing Atmospheric Motion

The equations governing the dynamical behaviour of the atmosphere are well-known and readily accessible in many textbooks (Haltiner and Williams, 1981; Krishnamurti, 1996). The governing equations are based on conservation laws including the conservation of momentum; the conservation of mass, the first law of thermodynamics and the equation of state for a perfect gas. Conservation equations, such as for water vapour, are part of the system. The outstanding characteristic of the governing equations is that they form a deterministic set of equations for predicting future values of basic atmospheric variables that are known at some earlier time. As such, the system is an initial value problem on the sphere, or an initial-boundary value problem over a smaller region of the globe, depending on which is of interest.

In their simplest form the governing equations are seven in number, for the seven primary variables comprising the three wind components, temperature, pressure, density and moisture. Table 3.2 defines the symbols used in the equations below. In practice, the number of variables to be predicted is much greater, so the number of equations to be solved is larger.

Table 3.2. Definitions of the symbols used in the governing equations

Symble	Meaning
x, y	Horizontal co-ordinate system
p	Atmospheric pressure
p_s	Atmospheric Pressure at the surface
σ	Sigma co-ordinate, $\sigma = p/p_s$
u, v, w	Zonal, meridional and vertical components of velocity
T	Temperature
θ	Potential temperature
R	Gas constant
ρ	Density
ϕ	Geopotential
f	Coriolis
t	Time

The equations for the three wind components come from Newton's Second Law of motion which expresses conservation of momentum:

$$\frac{du}{dt} - fv = -\frac{\partial \phi}{\partial x} - RT\frac{1}{p_s}\frac{\partial p_s}{\partial x} \qquad (3.1)$$

$$\frac{dv}{dt} - fu = -\frac{\partial \phi}{\partial y} + RT\frac{1}{p_s}\frac{\partial p_s}{\partial y} \tag{3.2}$$

$$\frac{\partial \phi}{\partial \sigma} = -RT/\sigma \tag{3.3}$$

The equation for pressure comes from the so-called continuity equation which expresses conservation of mass:

$$\frac{\partial p_s}{\partial t} = -\frac{\partial p_s u}{\partial x} - \frac{\partial p_s v}{\partial y} - p_s\frac{\partial w}{\partial \sigma} \tag{3.4}$$

The temperature equation is obtained from the First Law of Thermodynamics and expresses the conservation of entropy:

$$\frac{d\theta}{dt} = 0 \tag{3.5}$$

where $\theta = T(p_s/p)^K$, $K = R/C_p$ and $\dfrac{d}{dt} = \dfrac{\partial}{\partial t} + u\dfrac{\partial}{\partial x} + v\dfrac{\partial}{\partial y} + w\dfrac{\partial}{\partial \sigma}$

Density is derived from the equation of state:

$$p = \rho RT \tag{3.6}$$

Finally, the moisture field is obtained from the conservation of water vapour:

$$\frac{dq}{dt} = 0 \tag{3.7}$$

The above seven equations (3.1 to 3.7) are used to update the values of the seven variables u, v, p, T, D, ϕ and q.

We note here that for climate models, the problem is quite different, as it is no longer an initial-value problem since details of the forcing are much more important than for NWP. This aspect was considered at length by Lorenz (1973), who described atmospheric modelling as being composed of two very different kinds of predictions. Predictions of the first kind are initial-value problems and as such they refer to NWP. Predictions of the second kind study the changes in the statistical properties of the atmosphere, and refer to extended range out to climate simulations.

3.2.1 Some Commonly Used NWP Models

Since the most meteorological institutions have NWP models there is a large number of them and there is no possibility of mentioning all available models. In addition most of the models are different, sometimes in quite fundamental ways. However, in some countries, notably the USA and Europe, there is a set of NWP models that either belong to an individual institution, or are community models used by many other institutions. A recent summary of the NWP models available in the USA is given in Table 3.3, below. The

84   3. Atmospheric Modelling and Prediction

Table illustrates how widespread is the use of the selection of NWP models developed in the USA. Many of these models are used in other parts of the world.

Table 3.3. Selected examples of real-time/operational numerical weather prediction models in the USA. The table is based on Kuo (2000).

The acronyms used in the table are defined below:
- NCEP: National Centers for Environmental Prediction
- FNMOC: Fleet Numerical Meteorology and Oceanography Center
- UW: University Washington
- CSU: Colorado State University
- NCSU: North Carolina State University
- UO: University of Oklahoma
- KSC: Kennedy Space Center
- ATEC: Army Test and Evaluation Center

Institution	Modelling system	Resolution (km)	Vertical Levels	Domain
NCEP	ETA	32	45	North America
FNMOC	COAMPS	81/27/9/3,45/15/5	Variable	North America
UW	MM5	36/12/4	33	Pacific Northwest
CSU	RAMS	72/18	30	Western US
NCSU	MASS	45/15	25	US/N.C.
UO	ARPS	9/3, 36/18/12	43/50	Southern Great Plains
KSC	RAMS	60/15/1.25	36	Florida
ATEC	MM5	30/10/3.3/1.1	31	Utah

3.2.2 The UNSW HIRES Model

The model used most commonly by the authors has been developed at The University of New South Wales (UNSW), Sydney, Australia over the past 5 years or so. It has a number of optional forms for both the numerical procedures and for the representation of physical processes in the atmosphere. The model is known as HIRES, an acronym for HIgh RESolution model. It can be deployed anywhere on the globe and can be "nested" within itself until the desired resolution is achieved. This multiple nesting facility is shown in Fig. 3.3 which is shown set up for a study over southeastern Australia. The three resolutions are, respectively, 50 km, 15 km and 5 km.

The model characteristics present in the version of late 2000 are summarised in Table 3.4. As can be seen, it is a state of the art NWP system and it continues to undergo a program of research development. The HIRES model forms the basis for most of the applications described in Sect. 3.4 of this chapter.

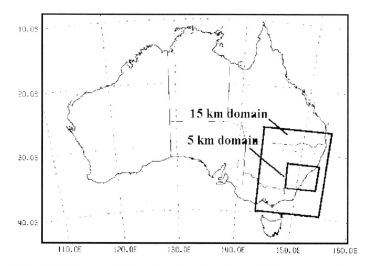

Fig. 3.3. The model domain, showing the multiple nesting of domains. The outer domain has its boundaries over the oceans and the inner two domains are nested within the outer domain. In the example shown here the resolutions of the three domains are 50km, 15km and 5km, respectively.

Table 3.4. The main features of a commonly used version of the UNSW HIRES NWP model

MODEL FEATURE	HIRES
Horizontal resolution	50 km, 15 km and 5 km
Numerical scheme	Split semi-implicit (high order)
No. of vertical levels	25
Assimilation scheme	6-hourly cycling
Initialization	Dynamic
Orography	2 minute resolution
Boundary layer scheme	Mellor-Yamada, Level 2.25
Sea-surface temperatures	5-day average

3.3 Solving the Governing Equations

Solving the primitive equations has taken a number of forms, but in this Chapter we will focus entirely on the method of finite-differences, in which the continuous derivatives are approximated by deriving difference approximations based largely on Taylor series expansions on a uniform or non-uniform three-dimensional grid. There are many textbooks describing the ideas behind the finite-difference method and we refer the reader to the books by Haltiner and Williams (1980) and Krishnamurti and Bounoua (1996) and

Durran (1999). The primitive equations for the atmosphere are given above in equations (3.1) to (3.7). In order to indicate how they are solved in practice it is sufficient to consider a simplified form of the equations, known as the shallow water equations. These equations contain the essence of the full equations in terms of their numerical properties but are far easier to manipulate. Moreover, the main implications that arise from analysing the shallow water equations usually carry over very closely to the full governing equations. As already mentioned, spectral models are a viable alternative to finite-difference schemes and are in widespread usage around the world. They are not discussed in this chapter through lack of space.

3.3.1 The Shallow Water Equations

The linearized shallow water equations make the following assumptions:

- The fluid is homogeneous. This assumption means there is no buoyancy, and filters out acoustic waves.
- The fluid has a free surface at which the pressure is constant.
- The hydrostatic assumption is valid. This means the vertical acceleration is much smaller than the vertical pressure gradient force and the force due to gravity. The depth scale must be considerably less than the horizontal length scale. For this reason, the word "shallow" is used. Note that the hydrostatic assumption does not imply that the vertical velocity is zero.
- The initial motion field is height independent.
- The bottom of the fluid is the surface $z = 0$. This is not a requirement of the shallow water equations, but we shall adopt it here for simplicity.
- There is no friction.
- The displacement of the free surface is very much less than the undisturbed fluid depth.

The shallow water equations then become:

$$\frac{Du}{Dt} = \frac{\partial u}{\partial t} + u\frac{\partial u}{\partial x} + v\frac{\partial u}{\partial y} = fv - g\frac{\partial h}{\partial x} \tag{3.8}$$

$$\frac{Dv}{Dt} = \frac{\partial v}{\partial t} + u\frac{\partial v}{\partial x} + v\frac{\partial v}{\partial y} = fu - g\frac{\partial h}{\partial y} \tag{3.9}$$

$$\frac{Dh}{Dt} = \frac{\partial h}{\partial t} + u\frac{\partial h}{\partial x} + v\frac{\partial h}{\partial y} = -h\left(\frac{\partial u}{\partial x} + \frac{\partial v}{\partial y}\right) \tag{3.10}$$

where u, v, x, y, f, t are defined in Table 3.2, g is gravity and h is fluid depth.

The derivation of equations (3.8) to (3.10) is well-known (e.g. Holton, 1972).

Numerical schemes are chosen on the basis of many factors. Here we focus on the accuracy of a scheme, that is, its ability to reproduce the phase and amplitude of various known numerical solutions, such as a shape that is transported by the advection equation. This is a enormous area of research and needs only to be outlined here and some examples given for a number of popular schemes.

3.3.2 The Finite-difference Techniques

The equations of motion for the atmosphere, such as the shallow water equations (3.8) to (3.10) are predictive in form and a number of techniques exist for solving the equations, that is, Smarching T the solutions forward in time from a known set of initial conditions. In this Chapter only the finite-difference technique will be discussed in detail. The finite-difference approach involves replacing the continuous derivatives in (3.8) to (3.10) by discrete differences at a set of points in space and time. These points are referred to as grid points. Most commonly, the grid points are evenly spaced but variable grids also are quite popular. We mention in passing that in atmospheric prediction, the so-called spectral method is also widely used. Spectral methods replace the dependent variables by expanding them as a finite sum of orthogonal functions known as basis functions. The problem then becomes one of predicting the coefficients of the expansions. The reader is referred to Haltiner and Williams (1980) for an early, but still excellent, discussion of spectral methods.

Returning to the finite-difference approach, the standard approximations to the continuous derivatives is in terms of Taylor series, where it is assumed that the dependent variables are sufficiently differentiable. Immediately, it is clear that finite-difference approximations to the continuous derivatives are not unique. Define a regular finite-difference grid in one spatial and one temporal scale by the following discrete values of x and t: $x = j\Delta x$, $t = n\Delta t$. Use lower case u to denote the continuous velocity and upper case U to denote the finite difference approixmation to u. The superscript denotes time on the grid and the subscript deontes space, so U_j^n is U at time n, and at x position j. Using this, the first derivative of the velocity component, u, may be approximated by at least the following two discrete expressions:

$$\frac{\partial u}{\partial x} = \frac{U_{i+1}^n - U_i^n}{\Delta x} + O(\Delta x) \tag{3.11}$$

or

$$\frac{\partial u}{\partial x} = \frac{U_{i+1}^n - U_{i-1}^n}{2\Delta x} + O(\Delta x^2) \tag{3.12}$$

Similarly, the second derivative may be approximated by

$$\frac{\partial^2 u}{\partial x^2} = \frac{U_{i+1}^n - 2U_i^n + U_{i-1}^n}{(\Delta x)^2} + O(\Delta x^2) \tag{3.13}$$

Finally, the finite-difference approximation to the advection equation can take many forms, for example,

$$\frac{\partial u}{\partial t} + U\frac{\partial u}{\partial x} = \frac{u_i^{n+1} - u_i^{n-1}}{2\Delta t} + U\frac{u_{i+1}^n - u_{i-1}^n}{2\Delta x} + O(\Delta x^2 + \Delta t^2) \quad (3.14)$$

Note that in some cases the first order continuous advection equation has been replaced by a second order discrete approximation. In so doing, a spurious solution known as the computational mode has been introduced. This is just one of the many complications involved in solving the governing equations using the finite-difference technique (see Durran, 1999).

Choosing a Finite-difference Scheme. Finite-difference schemes are chosen on the basis of many factors. Here we focus on the accuracy of a scheme, that is, its ability to reproduce the phase and amplitude of various known numerical solutions, such as a shape that is transported by the advection equation. This is a very large area of research and need only be outlined here and some examples given for a number of popular schemes. There are three important properties of a finite-difference scheme. The first one is how closely the scheme approximates the continuous system that is being solved. If the difference between the two can be made arbitrarily small, then the finite-difference scheme is said to be consistent with the continuous system of equations. The next property of the finite-difference solution is that it can be made arbitrarily close to the exact solution of the continuous system. If this is possible, then the scheme is said to be convergent. Finally, a scheme is said to be stable if the solution does not amplify from one timestep to the next. The three concepts just described are tied together in Lax Rs Equivalence Theorem which states that a consistent, stable scheme converges to the continuous solution. Therefore, given that we will always choose a consistent scheme, our focus in choosing a scheme is on its stability properties. We turn our attention to this aspect in the next section and illustrate the approach using a number of popular schemes.

3.3.3 Stability Analyses

As already mentioned, the selection of a suitable scheme rests on three factors that can vary in order of importance depending upon the purpose of the prediction. We will assume in this section that accuracy is the primary consideration, rather than speed or reliability, which are arguably more important in operational weather centres. In this case, the choice of scheme revolves around the stability of the schemes and the associated characteristics of amplitude and phase errors. Other important factors are the efficiency of the algorithm and its portability between different computing platforms, such as from supercomputers to workstations, or even personal computers.

In this section, a comparison is made between perhaps two schemes that have been most widely used in finite-difference numerical models, during the history of NWP. The first scheme is the second-order Leapfrog scheme (also known as the centred-time and centred-space scheme). The second scheme is the semi-Lagrangian scheme. They are representatives of two broader groups of schemes known as Eulerian schemes and semi-Lagrangian schemes, respectively. Both schemes have their advantages and disadvantages and frequently the choice is a matter of "horses for courses".

The Leapfrog Scheme. Consider the advection equation:

$$\frac{\partial u}{\partial t} + c\frac{\partial u}{\partial x} = 0 \tag{3.15}$$

If we use the grid defined in section 3.3.2 and use lower case u to denote the solution of the PDE (continuous) and upper case U to denote the solution of the finite difference equation (FDE), the second-order Leapfrog scheme becomes:

$$\frac{U_j^{n+1} - U_j^{n-1}}{2\Delta t} + c\frac{U_{j+1}^n - U_{j-1}^n}{2\Delta x} = 0 \tag{3.16}$$

To demonstrate the Von Neumann stability method, define

$$U_j^n = \sum_p Z_p^n e^{ipj} = \sum_p A\rho_p^n e^{ipj} \tag{3.17}$$

Substituting (3.17) into (3.16) gives:

$$\frac{\rho^{n+1} - \rho^{n-1}}{2\Delta t} + c\frac{\rho^n(e^{ip} - e^{-ip})}{2\Delta x} = 0 \tag{3.18}$$

Rearranging gives the quadratic equation $\rho^2 + 2i\mu \sin p\rho - 1 = 0$, where $\mu = c\Delta t/\Delta x$ is the Courant number.

Because we have three, ρ^{n+1}, ρ^n and ρ^{n-1}, we now have a quadratic equation and so there are two solutions for the amplification factor. One solution is the true, or physical mode. The other solution was introduced by the second-order discretisation of the first-order continuous equations and is known as the computational mode. The two solutions are:

$$\rho = (-i\mu \sin p) \pm \sqrt{-\mu^2 \sin^2 p + 1} \tag{3.19}$$

From this equation, the product of the roots is -1; therefore $\sqrt{(-\mu^2 \sin^2 p + 1)}$ must be real, as the roots would be then be purely imagery and one of them would be greater than unity, thereby violating stability criterion. For $\sqrt{(-\mu^2 \sin^2 p + 1)}$ to be real for all p, we must have $\mu^2 \leq 1$. The stability condition for the leapfrog scheme becomes

$$-1 \leq \mu = c\Delta t/\Delta x \leq +1, \text{or} \Delta t \leq \Delta x/c \tag{3.20}$$

It can easily be seen that this stability condition demands a prohibitively small timestep for high resolution predictions such as short-range, mesoscale forecasts. With a maximum wind speed of 100 ms^{-1} and a resolution of 10 km the timestep would need to be less than 100 seconds for the scheme to remain stable.

90   3. Atmospheric Modelling and Prediction

The Semi-Lagrangian Scheme. Another numerical method that has become very popular in NWP since the mid- to late 1980s is the semi-Lagrangian scheme. This scheme was formerly known as the method of characteristics. The equations of motion, as we have seen, can in general be written as conservation equations

$$du/dt = S(u) \tag{3.21}$$

where the left hand side of the equation represents a total time derivative (following and individual parcel) of the vector of dependent variables u. The total time derivative, or Lagrangian time derivative is conserved for a parcel, except for the changes introduced by the source or sink term, S.

In a Lagrangian scheme, one follows individual parcels (by transporting them with the 3-dimensional fluid velocity). However, this is not convenient because one has to keep track of many individual parcels, and with time they may "bunch up" in certain areas of the fluid. In the semi-Lagrangian scheme, we use a regular grid, as in the Eulerian schemes. However, at every new time step we need to find out where the parcel arriving at each grid point (AP) at time $n + 1$ came from in the previous time step, that is, the departure point (DP). For a fluid parcel at a grid point AP, at time level $n + 1$, its value is determined by interpolation from values near the departure point. The interpolation is required as the departure point will, in general, not be a grid point at time n, as is shown in Fig. 3.4.

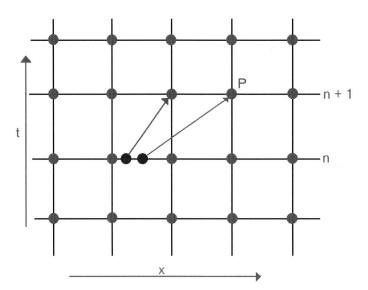

Fig. 3.4. Stencil for the Semi-Lagrangian Scheme. The trajectories of two points are shown, starting ending at gridpoints at time level $n + 1$, but starting between gridpoints at time level n. In general, the path, or trajectory, will not be a straight line.

The stability analysis is quite simple. As mentioned, the value of u along the trajectory is obtained by interpolating the values of the grid points surrounding the departure point. The distance a parcel has travelled in a single time step (Δt) is $U\Delta t$. By defining $\alpha = (U\Delta t)/(\Delta x)$, and $P = [\alpha]$, and $\alpha' = \alpha - p$, the departure point is a fraction α' of a grid spacing from grid point $(i - p)$. We then calculate the advected quantity:

$$u_i^{n+1} = u_*^n = (1 - \alpha')u_{i-p}^n + \alpha' u_{i-p-1}^n \qquad (3.22)$$

using linear interpolation. In practice, at least second order interpolation is used, but linear interpolation will serve to illustrate the properties of the scheme. The stability analysis using the Von-Neumann method shown earlier applied to the Leapfrog scheme is obtained by substituting $u_j^n = \lambda^n C e^{-ikx_j}$ into the equation above and gives

$$\lambda = [1 - \alpha'(1 - e^{-ikx_j})]e^{-ipx\Delta x} \qquad (3.23)$$

The amplitude of λ is the amplification factor. After some simple algebra we obtain the following equation for the amplification factor:

$$|\lambda| = [1 - 2\alpha'(1 - \alpha')(1 - \cos\beta)]^{1/2}, \quad \beta = ikj\Delta x \qquad (3.24)$$

Now since $0 \leq \alpha' \leq 1$ and $-1 \leq \cos\beta \leq 1$, then the amplification factor satisfies the inequality $|\lambda| \leq 1$. Consequently, the semi-Lagrangian scheme is stable for all Courant numbers, and therefore for all time step sizes.

Another commonly used Eulerian numerical scheme is the third order upwinding scheme (see, e.g., Durran, 1999). Odd order upwinding schemes have excellent phase properties and good amplitude properties, provided they are of order 3 or higher. The Australian Bureau of Meteorology has adopted this approach for its new regional NWP model, LAPS (Puri et al., 1998). For constant c, it is readily shown that upwind schemes are formally equivalent to semi-Lagrangian schemes (Dietachmayer, 1990). Figures 3.5 and 3.6 below are comparisons of the above mentioned schemes, with the final graph being a semi-Lagrangian scheme in mass conserving form, Purser and Leslie, (1995).

Excellent schemes are now available in both Eulerian and semi-Lagrangian formulations. Choosing between the two has become a matter of the importance of other factors, notably the computer platforms available.

3.3.4 Defining the Initial State for the Model Integration

There are numerous ways of defining the initial state for the model integration. The most common of these is an analysis scheme, or a data assimilation cycle. Analysis schemes are significant less costly in computer time, but have the disadvantage that they do not generally use the model equations as a constraint on the data. They can be classified into a number of different types including: surface fitting techniques; empirical linear techniques, such

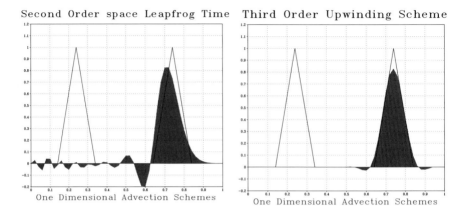

Fig. 3.5. The popular centred in space and time (leapfrog) scheme (left) and the third-order upwind scheme (right) that is becoming more widely used.

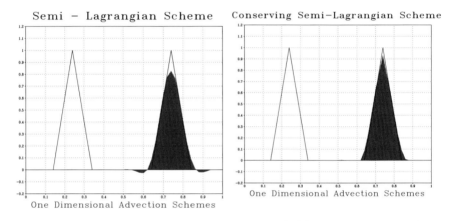

Fig. 3.6. The third-order semi-Lagrangian scheme (left) and the conserving form (right) of the scheme. Note that the conserving scheme is now monotonic and has improved amplitude.

as the popular Cressman and Barnes analysis techniques; statistical interpolation (SI) procedures; and continuous assimilation approaches, such as the variational techniques. Multivariate SI has the main technique used by operational centres. Recently, four-dimensional analyses have become popular and much work is being carried out on schemes such as variational data assimilation and Kalman filters. The reader is referred to Daley (1991) for a more detailed description of data analysis techniques. One of the problems with standard analysis techniques was they created a current picture of the atmosphere which could be out of balance with the operational model. This would lead to model shock when the model was initiated with the analysis.

This problem was solved with the use of a dynamical initialisation scheme at the start of the model integration.

Data assimilation procedures are now the most commonly used methods at operational centres. The big advantage of data assimilation is that at the zero time the model variables will be close to the assimilated data and will also be in closer balance with the model equations and parameterisations. Multi-dimensional variational assimilation (e.g. 3D-var or 4D-var) techniques are still very expensive computationally but are rapidly growing in popularity. Such methods are beyond the scope of this chapter, but they are of sufficient importance that there will be discussion of them in Sect. 3.7, below.

3.4 Applications

The aim of this subsection is to present examples of each of the four prediction scales mentioned in the Introduction. To repeat, these are as follows. Very short range forecasting (VSRF), short range forecasting (SRF), medium range forecasting (MRF), extended range forecasting (ERF) and general circulation modelling (GCM). GCMs also form the basis of the longest time scale simulations, namely, climate modelling. We will address each of these categories in turn and will present examples from each of them. The model used in most of the forecasts was developed over a number of years at the University of New South Wales (UNSW). It is a high-resolution, high-order, mass-conserving, semi-implicit, semi-Lagrangian model with state-of-the-art representation of physical processes and an attached Geographic Information System (GIS) database (Leslie and Skinner, 1994; Leslie and Purser, 1995; Shao and Leslie, 1997).

3.4.1 Very Short Range Forecasting (VSRF)

As has already been mentioned, the spatial and temporal scales involved here are typically from 0 to 12 hours and from 100 m to 25 km over areas ranging from very local areas to regions several hundred kilometres in each direction. The shorter end of the scale, the period 0–3 hours, is referred to as nowcasting, and techniques other than NWP models are important in assisting the forecasters. The prediction of developing thunderstorms or other severe weather approaching a populated area relies heavily on extrapolations from, for instance, radar, profiler and automatic weather station (AWS) observational networks. As might be expected, each scale has its own forecasting techniques and emphasis. For VSRF, the general emphasis is on a high level of local detail in the initial field, being very much an initial value problem, and on very accurate prediction algorithms. Much work has been carried out in recent times at high resolutions, and using ever more accurate techniques for solving the NWP equations. VSRF is an extremely difficult because of the short time scales and the associated very rapid error growth rates (which can have error-doubling time scales of minutes or tens of minutes). Other problems include lack of data at the spatial scales involved and an inadequate understanding of some of the important physical and dynamical processes such as the cloud microphysics. Although there are exceptions at the lower resolution end of the scale, in most centres the models used are not fully "operational", as they are not being used routinely to provide real-time forecasts at major weather centres. Most often they are used in case studies to increase our understanding of the underlying mechanisms, and in the further development of the VSRF systems. This situation is changing rapidly and in the near future VSRF will indeed become an operational activity. Two case studies follow.

Severe weather from a thunderstorm squall line. The development of multicellular organised convection is one of the most damaging forms of severe weather. This case occurred over the Sydney metropolitan area and resulted in flash flooding, with associated disruption of the city for several hours. Such storms regularly result in loss of life and property, and if hail is present, in severe crop damage. In New South Wales alone, the annual damage cost from severe weather events is almost one billion dollars. In this event, the line of storms organised itself in a very short time scale of about 3 hours as it approached the Sydney metropolitan area. It was very slow moving and at its peak remained quasi-stationary for about one hour. Moderate vertical wind shear was present in the lowest 1–2 km and there was little vertical wind shear above the steering level. These are optimal conditions for the development of long-lived squall lines (Thorpe et al., 1985). The results of the numerical simulation of this severe weather event are presented by showing simulations from the model, which was run over an area 250 km × 250 km at a horizontal resolution of 5 km. The initial conditions and boundary conditions were provided from a larger-scale model operating at 50 km resolution. The high-resolution model was then run for 6 hours from the initial time of 00 UTC February 10, 1990, which covered the entire cycle from clear skies over Sydney to the development and passage of the squall line. The accumulated rainfall was about 100 mm, with instantaneous rain rates of over 70 mm per hour. At 3 hours into the forecast the squall line was lying in a south-west to north-east orientation over the city. There was an abundance of low level moisture, the Convective Available Potential Energy (CAPE) level was moderate to high and the convective inhibition was small. There was an area of cyclonic relative vorticity present in the initial state, as could be seen in the 850 hPa fields. A series of cross-sections through the line of thunderstorms was made to examine the evolution of the relative vorticity and vertical velocity fields. In particular, the vertical velocity cross-section showed the development at 6 hours of an intense updraft of almost 25 ms^{-1} and the presence of a downdraft immediately behind. Figures 3.7a and 3.7b show the 3-hour model forecast of the squall line and the observed rainfall over the same 3-hour period, respectively.

Fig. 3.7. The simulated thunderstorm line over the Sydney metropolitan area indicated by **(a)** the 3 hourly accumulated rainfall from HIRES. **(b)** The observed rainfall for the same period as **(a)**.

Atmospheric pollution from forest fires. Smoke pollution from the burning of forest areas is a major problem for populated areas in the Asia-Pacific region, even for cities at a distance from the fires. In this example the city of Perth, in Western Australia, was badly affected by smoke from forest fires several hundred kilometres to the south of the city in November 1995. It was Perth's worst-ever pollution day and was produced by the conjunction of three meteorological factors. First, a strong inversion kept the smoke particulates trapped near the earth's surface, below the 950 hPa level. Second, a strong sea breeze developed in the afternoon and transported the smoke that was heading out to sea back over the land, including the Perth area. Finally, a meso-beta scale low pressure system developed to the east of the city and kept transporting the smoke from the fire area over the city. Examples are shown in Fig. 3.8a of the streamlines at 950 hPa driving the air from the fire areas located to the south of Perth over the city. Trajectories from a three-dimensional dispersion model which was run in conjunction with the NWP model illustrate how the particulates were transported over the city of Perth from large distances (Fig. 3.8b). Clearly, the application of real-time air quality prediction models is a high priority, and success is possible provided the model can capture both the fine details of the weather situation, such as the near-surface inversion, and the larger-scale meteorological evolution.

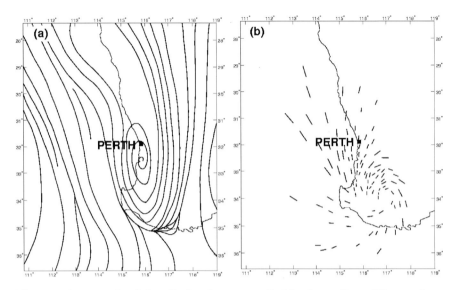

Fig. 3.8. An example of air pollution from controlled fire burnoff over Western Australia. (a) The streamfunction from HIRES valid at 6 am November 15, 1995 and (b) the smoke transport indicated by tracers calculated from forward trajectories computed from the 950 hPa velocity field.

3.4.2 Short Range Forecasting (SRF)

The scales of interest here are generally accepted as being 0–48 hours, covering a geographical area the size of a region such as South-East Asia, the Australian region or the Pacific basin. The size of the region therefore is typically 5000 km^2, with a grid resolution of 10–25 km. Once again the emphasis of the model analysis and prediction system is that of collecting as much data as possible, and in using very highly accurate model numerical algorithms for solving the governing equations. Three very different case studies are presented from the SRF scale. Two are tropical cyclones, one of which undergoes extra-tropical transition, and a cold front rainfall event are presented.

Tropical Cyclone Drena. Tropical Cyclone Drena was a long-lived tropical cyclone that lasted from January 2 to 10, 1997, and moved in basically a north-to-south direction over the south-western Pacific Ocean, crossing three island groups and affecting a fourth. About halfway through its life cycle it underwent transition to an extra-tropical cyclone, and during this transition phase it passed over the eastern edge of Norfolk Island. As it began to approach Norfolk island, a series of additional soundings were made at the Norfolk Island meteorological station and provided a valuable set of 10 radiosonde flights over a 54-hour period. These provide valuable data to compare the model output with actual observations, especially vertical structure.

The best track locations of TC Drena for the period 00 UTC January 6, 1997 to 00 UTC January 11, 1997 are shown in Fig. 3.9a. The model predicted cyclone eye passing over the eastern edge of Norfolk Island at 1800 local time (06 UTC) January 9, 1997 is displayed in Fig. 3.9b. Given the lack of data in the south-western Pacific, the actual passage over the island at 1745 local time (0545 UTC) January 9 as recorded by the observed change in wind speed and direction is quite invaluable. Excellent comparisons were also obtained between the observed and model-predicted vertical temperature and dew point soundings (Buckley and Leslie, 1998). These results, along with other recent work by the authors using high-quality and high solution satellite-derived observations, indicate that forecasting tropical cyclone tracks has substantially increased in accuracy with many centres reporting reductions in errors in recent years.

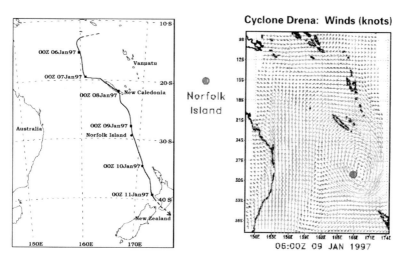

Fig. 3.9. (a) The track of tropical cyclone Drena in the South Pacific Ocean is shown on the left. Note that the track passes over Norfolk Island at about 06 UTC on January 9, 1997. (b) Shown on the right, the 78 hour forecast position of Cyclone Drena, which is very close to the observed position.

Tropical Cyclone Abe. Tropical Cyclone Abe was a typhoon that occurred during a major international field experiment TCM-90 that took place in the north-west Pacific in 1990. The abbreviation TCM stands for Tropical Cyclone Motion. Tropical Cyclone Abe moved north-west over the period 1100 UTC August 28 to 1100 UTC August 31, 1990, making landfall on the People's Republic of China central coast, to the south of Shanghai. Two forecasts were made, one with the conventional data and the other with the additional data obtained during TCM-90. The impact of additional data was large, producing a more accurate prediction of the track and the intensity of

Tropical Cyclone Abe. The reduction in the mean track forecast error for Abe was over 15%, which is consistent with improvements using other cyclones in the Asian region, both during TCM-90 and in subsequent work following the launch of the new Japanese geosynchronous satellite GMS-5, which has provided high-resolution cloud and water vapour derived winds. Figures 3.10a and 3.10b show the 24 and 48 hour forecast positions and wind vectors for Abe.

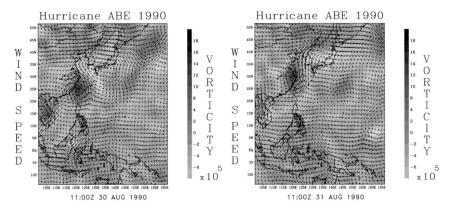

Fig. 3.10. (a) 24 hour forecast for the position of TC Abe, showing the position and the wind vectors indicating speed and direction. The vorticity fields indicate intensity. (b) The 48 hour forecast.

Rainfall prediction for agriculture. On June 21, 1996, a cold front was approaching the New South Wales (NSW) which is one of Australia's main winter grain crop areas. The amount of rainfall expected was critical in planning for the winter crop sowing, especially as the area had been very dry for some time. Provided a substantial area of central NSW received 20 mm or more in the 24-hour period, heavy winter sowing of grain would occur. If the rainfall was less, the entire winter crop was in danger of almost complete failure. This was indeed the case, as the recorded rainfall in the 24-hour observational period ending at 9.00 a.m. on June 23, 1996 showed large parts of central NSW receiving between 20 and 30 mm. A great deal of effort has been put into improving initial moisture fields in the UNSW model (Speer and Leslie, 1998), and the model forecast precipitation has been improved significantly, as is shown in Fig. 3.11. An ensemble procedure was used and compared with the results from an unperturbed forecast. Ensemble procedures are discussed in some detail in Sect. 3.7. The 24-hour rainfall forecast compares very favourably with the observed rainfall and is far superior to the previous operational model, which had much lower resolution (75 km compared with 25 km) and a poorer moisture analysis scheme. Again, the implications of such an improvement in rainfall predictions are very promising for the agricultural sector not only in Australia but elsewhere in the Asia-Pacific region.

Fig. 3.11. The figure on the left is the observed 24 hour rainfall, the middle figure is the routine, unperturbed forecast, and the figure on the right is the ensemble mean forecast. Note that the ensemble mean forecast is markedly superior to the unperturbed prediction.

3.4.3 Medium Range Forecasting (MRF)

Medium range forecasting, like short range forecasting, is presently at the heart of the NWP systems in use at weather services around the world. The definition varies between centres from about 7 days to 15 days, but all centres recognise that the skill level falls away rapidly after about 4 to 5 days. There are exceptions, but this rate of degradation in skill is true in general. The models are global at these time scales, and have spatial resolutions between 50 km and 250 km. With further advances in computing power the resolution is expected to increase further in both the horizontal (30 km or less) and the vertical (at least 50 levels). The medium range models also are readily available on the Global Telecommunication System (GTS), so most centres can compare forecasts from a range of models. Frequently, these predictions are very different, and leave the forecasters in a state of considerable confusion. Moreover, given that the skill level begins to diminish after 5 days or so, there is widespread use of ensemble prediction techniques of various kinds. Given that the forecasting problem is no longer a just an initial value problem, the aim of these techniques is to provide some measure of forecast skill, and to produce error bars around the ensemble mean forecast. Substantial additional useful information can be obtained in this way. If the ensemble members exhibit a wide divergence then the skill of the ensemble mean is doubtful. On the other hand, if the forecasts remain close together, the forecast is regarded as possessing a high level of skill.

The summertime trough-ridge system over northeastern Australia. One of the major forecasting problems over northeastern Australia, especially in the warmer months, is the bias in the location of the trough-ridge system located over the coast and adjacent ranges. An example of the trough-ridge system is shown in Fig. 3.12a. The operational model run by the Australian Bureau of Meteorology has large westward biases as shown in Fig. 3.12b. The reason for these biases was the failure of the model to adequately represent the coastal ranges in both height and steepness and to underestimate the level of surface heating in the interior, away from the coast. When a high resolution version of HIRES was used, together with improved orography and representation of surface heating, there was a dramatic improvement in the model simulations. In fact, the biases were almost entirely removed, as shown in Fig. 3.12c.

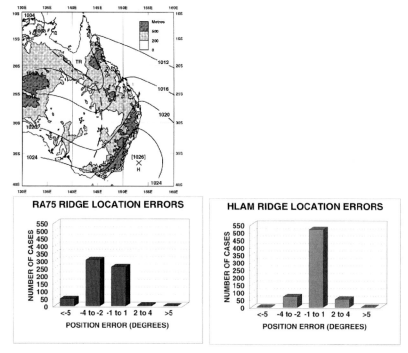

Fig. 3.12. (a) A typical ridge-tough system over northeastern Australia, 00 UTC, Jan 2, 1997. Note the location of the trough exactly on the coast and the well-defined inland trough. (b) The distribution of 24 hour forecast errors in the near-coastal ridge for the operational model during the period November 1993–April 1994. (c) As in (b) except for the HIRES model.

A cold outbreak over south-eastern Australia. On June 21, 1995, Sydney experienced its lowest maximum on record for June. The temperature did not rise above 10.4 °C, compared with an official forecast maximum of 14 °C. Over inland New South Wales the temperatures were much lower, particularly at night when severe frosts occurred over a number of days. These frosts were responsible for the destruction of much of the stone fruit and other frost-sensitive crops for that year. The cold outbreak was caused by an intense anticyclone located over the Southern Ocean, which moved only very slowly eastwards, maintaining a cold southerly airstream over south-eastern Australia for several days. Air reaching Sydney at latitude 34 °S originated from as far as 55 °S. The synoptic situation is not unusual other than in its intensity, with the anticyclone reaching central pressures of 1045 hPa, which occurs relatively infrequently, and this was the most severe example recorded. The predicted maximum temperatures from the NWP model and the official (forecasters) maximum predicted temperatures will also be presented in detail, as they illustrate the natural tendency of forecasters not to predict a record-breaking forecast. The forecasts in Fig. 3.13 are shown from 4 days ahead and are compared with the observed maximum temperatures. The NWP model clearly outperformed the forecasters, showing that extreme events such as this case are predictable using the NWP model.

3.4.4 Modelling the General Circulation of the Atmosphere

The final category of forecasting is the general circulation, or climate, modelling which attempts to predict weather conditions out to months, seasons and years in advance. While climate modelling is normally carried out over decades, there have been simulations over periods of thousands of years. In particular, possible major deviations from the average weather conditions are a major aim of the forecasting technique, as they provide the most information that is of value to the forecasters and to the public. Two examples of shorter range general circulation modelling are discussed below. Then, attention is focussed on climate modelling, which is a subject in its own right and is discussed in Sect. 3.5.

General circulation models (GCMs). General circulation models (GCMs) are models which attempt to portray "average" conditions over the earth, rather than atmospheric evolution from an initial state. As such, they are not purely deterministic models like short and medium range NWP models. In the first instance, GCMs were developed to assist in understanding many aspects of the general circulation of the atmosphere, hence their name. This understanding includes the dynamics of the formation, strength and location of jet streams, the subtropical ridges, the rudiments of the monsoonal circulation and so on. Currently, GCMs are used in many ways, including the generation of scenarios for the future states of the atmosphere.

A major application of GCMs is to the now-established existence of global warming of the Earth's atmosphere. GCMs have been used in scenario mode to explore the underlying nature of, and consequences arising from, this enhanced greenhouse effect. It produces atmospheric warming owing to the increasing accumulation of carbon dioxide (CO_2), methane (CH_4) and other gases with pronounced radiative properties in the atmosphere. Any attempt at a comprehensive survey of GCMs in studying the greenhouse effect is beyond the scope of this note.

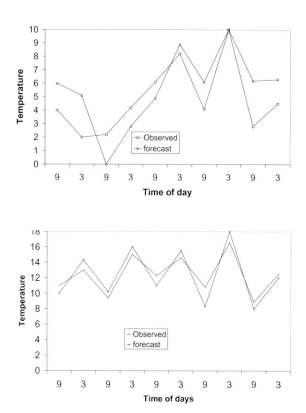

Fig. 3.13. A comparison of the observed and forecast screen temperature errors for two stations near Sydney, Australia from the HIRES atmospheric model. The station on the first is an inland station over high terrain, while the station on the second is near the coast.

GCMs have demonstrated that the mean temperature of the earth is very sensitive to the amount of cloud in the atmosphere. This fact is complicated by the debate about precisely how much cloud is actually present in the atmosphere. This discussion is especially true of cirrus clouds. Initially, GCMs were run with a fixed fraction of the atmosphere covered by cloud. This fraction was set at $1/2$. More recent satellite studies indicate that the level of cloudiness, particularly cirrus, is considerably higher, and about 60% of the globe is covered with cloud at any instant. Most current models allow a variable amount of cloud. High, middle, and low cloud appear to have different effects. Increasing the fraction of low cloud appears to cause a net earth cooling, while an increase in the fraction of high-level cloud appears to be responsible for a mean surface temperature increase. Many climate models show an increase in the high-level cloud in double CO_2 experiments, consistent with the traditional view that increasing CO_2 levels will cause a net warming of the atmosphere. Recent work (since 1993) shows that aerosols, especially sulfate ions emitted from northern hemisphere power stations are a powerful source of reflection of solar radiation. This helps to lower daytime temperatures, but possibly increases night-time temperatures, as they act as condensation nuclei for clouds. It appears that cloudiness is increasing over industrialised areas. Furthermore, there is growing evidence that cloud radiation properties may be different in an atmosphere with greater CO_2 concentrations than one with less CO_2. The most recent studies and observations confirm that mean temperatures are increasing and much of the contribution is due to higher overnight minimum temperatures.

The role of the oceans has to be taken into account in double CO_2 experiments, since the oceans absorb CO_2 and heat. The thermal inertia of the oceans is extremely large and the study of coupled atmosphere/ocean models is necessary for greenhouse modelling. To date, coupled models have examined the impact of the upper layers of the oceans in detail, but the role of the ocean subsurface is still not fully understood, despite rapid growth of knowledge in recent years. Ocean models have become essential tools as the significance of the ocean in the general circulation has become more recognized as paramount. Ocean models are simpler in some ways than the modelling of the atmosphere, but more complex in others. For example, the hydrostatic approximation is of even greater validity in the ocean than in the atmosphere. Also, the upper and lower boundaries of the ocean are fixed whereas the atmosphere's upper boundary is not. On the other hand, the equation of state for the ocean is more difficult than for the atmosphere, and is of the form $\rho_w = f(T, S, p)$ where ρ_w is the density of seawater, and S is its salinity. In general, f comprises a set of polynomials which depend on T, S, and p. The possibility of catastrophic rises in mean sea level has frequently been discussed. Recent results from GCMs indicate that the sea level will rise due to the thermal expansion of surface water and melting about the edges of the Greenland and Antarctic ice sheets. This is partially offset by a growth in the centres of the major ice sheets as more water vapour is advected over them from a moister atmosphere. After some time, it is believed that this effect will dominate the effects of thermal expansion and edge ice melting. At this stage, the sea level will fall. However, if the atmosphere continues to warm, the sea level will again rise.

A major criticism of GCMs has long been that they have been unable to predict the actual temperature changes observed in the atmosphere over the past, say 100 years. However, in 1995, the Hadley Centre in the UK announced a successful prediction of mean atmospheric temperatures since the start of the Industrial Revolution. This simulation contained many of the observed features of the past several hundred years. Subsequently, climate models from other centres have produced similar close matches with observed phenomena. In addition, there has been a recent proliferation of successful GCM simulations of observed changes in the atmosphere and the oceans, most notably ENSO and in global warming. With increased ability to run GCMs at higher resolutions and improvement in model physics many of the questionable procedures used in climate models, such as flux correction, have ben vastly improved. Higher resolution coupled atmospheric/oceanic GCMs, known as AOGCMs, have substantially reduced the problems of climate drift which have plagued GCMs in the past.

Soil moisture and soil erosion. One of the most important applications of GCMs is to the prediction of soil moisture and temperature. If soil moisture levels are low, then the consequences can be devastating. For example, 10% soil moisture levels can result in poor agricultural yields and in severe weather events such as dust storms that produce destructive erosion. On the other hand, excessive soil moisture in areas cleared of vegetation can produce the much-feared problem of saltation and water erosion. Recent work between UNSW, the Chinese Academy of Science Institute of Geography, and the Australian Bureau of Resource Sciences is producing promising results (Shao and Leslie, 1997). This work is one of the first attempts to predict soil moisture over periods of a month or more using a coupled atmospheric model-land surface/soil model. Figure 3.14 shows the predictions of soil moisture deficit as a cold front crossed southern Australia in February 1996. Predictions of soil moisture deficit are now being produced routinely, but to date there has been little work in this area using modelling of this kind. The results obtained have not been published in major international journals after careful comparisons with the few available data sets.

Drought. Although drought is usually considered on scales of seasons or longer, important indications can be obtained from monthly anomalies in various indicators such as precipitation, temperature and pressure. The UNSW model has been run in general circulation mode for this purpose, in both hindcast and forecast mode for cases in which there were severe deficiencies in rainfall. Results of the simulations have been compared with observational anomalies (Leslie and Fraedrich, 1997). Thus far the work is promising but is only qualitatively correct. Much more work needs to be done before confidence can be placed in the forecasts. Some forecasts have proved to be accurate; others have been poor. Drought and failed crops are of extreme importance to most countries in the Asia-Pacific area, as their economics depend heavily on agricultural commodity production for both internal consumption and export earnings.

3.5 Climate Modelling

In this section, a particular general GCM will be discussed in detail to discuss general circulation modelling in general. There are two main aims of

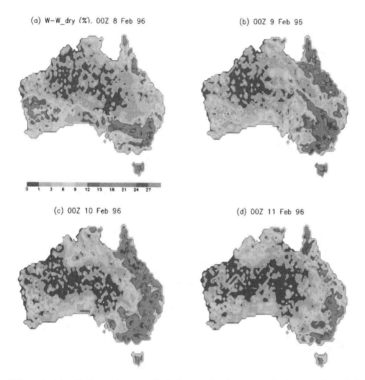

Fig. 3.14. Soil moisture deficit at 24 hour periods predicted by the coupled atmospheric-landsurface/soil modelling system HIRES-ALSIS.

this section. The first aim simply is to describe the salient features of the new model The second aim is to demonstrate the performance of the new model in a range of applications regarded as necessary for a credible GCM. These applications involve assessing the model as: (i) a medium range forecast model providing daily 10-day forecasts using operational global data; (ii) a climate model, integrated over a 10-year period, and then compared with global analyses for the same period; (iii) an atmospheric model driven by a regional SST anomaly field over a twelve month period that was an El-Nino year; and (iv) a down-scaling application in which a new, more efficient procedure requires only one forecast and permits two-way interaction between the GCM and the region of interest.

3.5.1 The GCM Model Numerics

Much of the following is drawn from work by Leslie and Fraedrich (1997) funded by a Max Planck Prize. Much of the wording has been retained.

The model is based on the semi-implicit, semi-Lagrangian regional/global model of Leslie and Purser (1991, 1995). The most advanced form of the model employs 6th-order differencing and forward trajectories. However, the version used in this study is that which has been used most heavily in previous research applications. It is a two-time level, bi-cubic semi-Lagrangian scheme, with third-order accurate, backward trajectory calculations (Leslie and Purser, 1991). The governing equations are in conservation form wherever possible, and the local mass-conserving scheme of Leslie and Purser (1995) also has been incorporated so that the model can be run as a climate model. The novel features of the numerics compared with other GCMs are the high-order accurate differencing and the mass-conserving formulation of the semi-Lagrangian scheme. Together, these features enable very long-term integrations of the model to be carried out with only a small change in the global total mass and total energy. The high-order differencing (greater than or equal to third order) is made possible by the "cascade" interpolation procedure which divides the multi-dimensional interpolation into a sequence of one-dimensional interpolations. The cascade interpolation, which produces enormous computational savings for medium and large grid dimensions, is described in full detail by Purser and Leslie (1991). The cascade interpolation procedure also is used for the exact, conservation of mass and tracers by working with the cumulative distributions of the variable to be conserved (Leslie and Purser, 1995). Unlike the a posteriori semi-Lagrangian conservation schemes (Priestley, 1993; Gravel and Staniforth, 1994), which can allow spurious local corrections to be made to ensure global conservation, the direct conserving scheme employed here ensures that a high level of local accuracy is maintained.

3.5.2 The Model Physics

The representation of physical processes in the model is summarized next. The vertical fluxes of momentum, heat and moisture in the boundary layer are the same as those used by Louis (1979), while shallow convection is the same as that described by Geleyn (1987).

Large-scale precipitation falls if the relative humidity exceeds 95% and cumulus convection is based on the mass flux scheme of Tiedtke (1989). Gravity wave drag is obtained from the formulation of Palmer et al. (1986). The model uses the modified Schwarzkopf and Fels (1985, 1991) scheme for the long-wave radiation and the Lacis and Hansen (1974) scheme for the short wave radiation. Cloud cover diagnosis for the radiation schemes is calculated in three layers: low, middle and high, using a modification of the scheme developed by Slingo (1987). The surface temperatures are calculated using a prognostic equation based on a surface energy budget, with a three-layer soil model (Leslie et al., 1985). Finally, a crude ice model formulated by Parkinson and Washington (1979) is used to compute ice growth.

3.5.3 Ocean Model Coupling

Coupling with an ocean model is still in test mode and at this point the only calculations completed have been those in which SST anomalies were provided for the Pacific ocean. Elsewhere the SST field has been specified from the new Reynolds monthly climatologies, which became available in 1994. The Reynolds climatologies have been combined with the analyzed SST anomalies from the Australian Bureau of Meteorology's operational archives to obtain the actual SST field used in the model simulations.

3.6 Applications and Results

The model was applied to a variety of problems in an assessment of its performance as both a NWP model and a GCM. It has already been shown to have a high level of skill as a short range (1 to 2 days) NWP model in a number of studies (Leslie and Purser, 1995). As a GCM the model is run at a resolution of about 300 km, and has 19 levels in the vertical. A time step of 30 minutes is used in all the forecasts/simulations.

3.6.1 Medium range forecasting experiments

The model was tested as a medium range forecast model by assessing its performance for the entire year of 1994. The measure of skill adopted for the model was the commonly used anomaly correlation coefficient. The model was run out to 10 days, once per day, using archived global analyses at 00 UTC. On the parallel workstation cluster the 10-day forecast took approximately one hour for each 10 day run, at 1.5 degrees resolution. The relatively small CPU time was due in large measure to the large time step size of 30 minutes. For higher resolution runs, even more efficiency is possible as the 30 min step was chosen as the maximum value that could be "risked" in order not to cause difficulties with the representation of physical processes.

At higher resolutions of, say, 0.57 the efficiency gains are greater as the same time step can be maintained in the semi-implicit, semi-Lagrangian framework. The anomaly correlation coefficients for the 500 hPa height fields over the Southern Hemisphere for the 10-day forecasts are shown in Figs. 3.15a and 3.15b. They are comparable with those from other modelling centres around the world. The anomaly correlation is above 0.8 at day 3, and does not drop below 0.6 until after day 5. Beyond day 6 there is still skill that extends even out to day 10 in particular situations. However, techniques such as ensemble forecasting procedures (Toth and Kalnay, 1993) are required to help determine which forecasts have skill at these longer time periods. Over the Northern Hemisphere the anomaly correlation coefficients remain above 0.6 out to day 7, a full day or so longer than for the Southern Hemisphere.

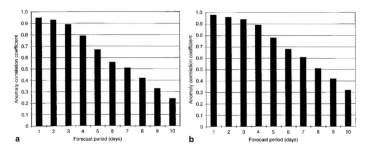

Fig. 3.15. (a) Southern Hemisphere anomaly correlation coefficients for days 1 to 10 of the new GCM, averaged over the calendar year January 1, 1994 to December 31, 1994; (b) as in (a) except for Northern Hemisphere.

3.6.2 Brief climatology of the model

The ECMWF analyses have long been taken as the standard analyses of temperature, wind and SLP, and we use them here. Mean global precipitation was obtained from the widely used climatology developed by Jaeger (1976). The distributions of zonal mean temperature and zonal mean wind components are shown in Fig. 3.16a, b. The corresponding SLP and precipitation zonal means are given in Fig.3.16c, d. Close inspection of the model simulated climatologies reveals that they are very similar in quality to the earlier ECHAM2/ECHAM3 model climatologies as reported by Roeckner et al. (1992). This comparison with an established set of GCM simulations suggests that the new model is performing at a satisfactory skill level as a GCM.

The interest in ENSO is largely in the capability of coupled ocean-atmospheric models to simulate it, and ultimately to predict its structure and occurrence over seasons or even years ahead. The aim here is much more modest, being determined to a large extent by the fact that the authors are still working on the coupled atmospheric-oceanic model. Therefore, since only the atmospheric response is possible, we used an SST comprising the new Reynolds monthly climatological SST field plus an anomaly SST field over the South Pacific Ocean provided by the Australian Bureau of Meteorology. The simulation therefore was a simple one in that the climatological forcing was provided solely by the SST distribution, so that the experiment was similar to that carried out by Smith (1994). The main differences from the Smith (1994) simulations resulted from the fact that the aims were different so that the simulation period was much shorter, being over the El Nino year of 1994, and that analyzed SST anomalies provided by the Australian Bureau of Meteorology were used, rather than just climatological monthly means. The period November–December 1994, which were effectively the 11 and 12 month predictions, was one in which there were large positive pressure anomalies in the SLP and 500 hPa fields over eastern Australia and the adjacent south Pacific Ocean. In this simple example of atmospheric modelling-SST coupling, the predicted and observed anomalies are shown in Fig. 3.17.

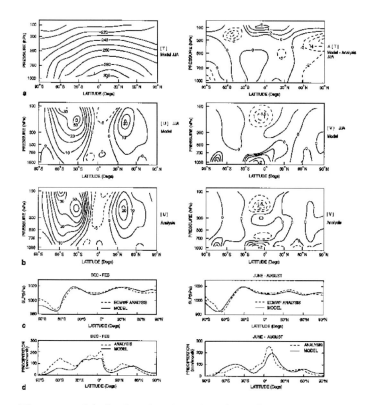

Fig. 3.16. (a) Analyzed and predicted zonal mean temperature for the 10 year period 1981–1990; (b) as in (a) except for u and v components of wind; (c) as in (a) except for SLP; (d) as in (a) except for precipitation analyses.

Figure 3.17a, b shows the observed and predicted SLP mean anomalies (hPa) for November–December 1994, and the results are in good agreement given the constraints of the numerical experiment. The areas of positive anomalies are very similar over eastern Australia and the South Pacific, however the intensity is under forecast. Figure 3.16c, d shows the corresponding observed and simulated 500 hPa height anomalies and the results are again in good agreement.

3.6.3 Regional simulations

An additional bonus, apart from speed, provided by the parallel computer was the ease with which an alternative approach to "down-scaling" could be implemented. Simply defined, down-scaling is the transfer of model information from larger scales (for example, relatively coarse resolution GCM runs) to finer scales, such as specific regions (for example, Australia or Europe). A number of down-scaling schemes have been devised and these may

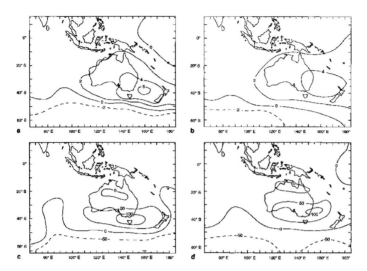

Fig. 3.17. (a) Observed SLP anomaly (hPa) for November–December 1994; (b) predicted SLP anomaly (hPa) for November–December 1994; (c) as in (a) except for 500 hPa geopotential height anomaly (metres); (d) as in (b) except for 500 hPa geopotential height anomaly (metres).

broadly be divided into two main classes: statistical down-scaling and numerical down-scaling. In statistical down-scaling, a search is made for correlations between the large-scale flow and regional variables such as rainfall patterns. This method can be quite powerful when the correlations are large and statistically significant. In the case of numerical down-scaling, the GCM is run for some pre-determined period and the simulations are then used to drive a much higher resolution regional model. There are numerous examples of numerical down-scaling, such as the work of Giorgi and Mearns (1991). Also, there are two main problems. GCMs are typically run at resolutions of around 300 km, thereby imposing completely artificial lower limits on the resolvable length scales. Such an approach greatly reduces the possibility of physically realistic energy fluxes taking place. Moreover, as the GCM is run first, and then used to drive the regional numerical model, the information exchange is only one-way, hence the literal meaning of the term "down-scaling". As a result, there is considerable debate over whether or not additional information is being provided by the regional model (see, e.g. Anthes, 1985). That is, extra detail is not necessarily extra information as the larger scales determine the scales that are resolved by the regional model.

In the alternative procedure employed here, the GCM is run at standard resolutions everywhere except over the region of interest, where a graded mesh is used. The grid spacing is set to be constant over the region of interest and is slowly increased (typically by a factor of 10 per cent per grid point) over an intermediate region surrounding the region of interest until it matches

3.6 Applications and Results   111

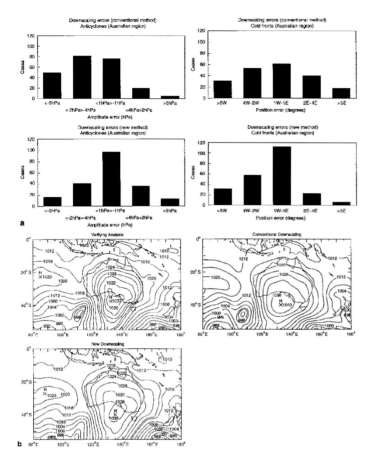

Fig. 3.18. (a) Histograms of the location errors in the standard downscaling procedure for Australian region anticyclones and cold fronts (top two panels) and the new procedure (bottom two panels); (b) verifying analysis valid at 00 UTC August 5, 1994 (left top panel) and the two 72 hr forecasts over the Australian region using standard numerical down-scaling (middle panel) and the new procedure (bottom panel).

the GCM resolution at some suitably chosen distance from the centre of the region of interest. The grid spacing over the region of interest itself is set at the same value (50 km) as in the conventional procedure. There are three advantages in this approach: two-way interaction is allowed between the GCM and the region of interest; the model needs to be run only once; and the graded mesh eliminates any problem of boundary conditions that normally cause difficulties when two models are involved. Finally, the approach fits in very well with parallel architectures.

A simple test of the procedure has been carried out over Australia. The GCM was run for the 2 year period January 1, 1993 to December 31, 1994, the first year being regarded as a "spin-up" period, and the second year used for an assessment of the performance of the model. the positioning of two of the most significant and readily identifiable features of the flow patterns over Southern Australia, namely the subtropical ridge (the quasi-permanent anticyclone centres) and the longwave troughs were examined. The Australian region poleward of about 30 °S is dominated by these systems. The predicted positions of the ridges and troughs forecast by the model were tabulated and compared with the positions of the observed systems. To avoid confusion, it is emphasized that this was not a case of comparing the deterministic forecasts with observations of individual high- and low-pressure cells over the 12 month period. It was comparing the mean forecast locations of systems that were predicted by the model over the 12 month period with the mean observed positions of the subtropical ridge and the long-wave trough. The conventional down-scaling procedure was used with a regional version of the GCM being run at 50 km resolution and the statistics are shown in the two upper panels of Fig. 3.18a.

The corresponding mean position errors for anticyclones and cold fronts using the new regional climate down-scaling procedure are shown in the two bottom panels of Fig. 3.18a. It is clear from the naked eye, and verified at greater than the 99% confidence level using a standard t-test that the new approach has improved the prediction of the location of the systems. A case study illustrating the gains obtained from the new procedure over the Australian region is presented in Fig. 3.18b in which 72 hr forecasts verifying at 00 UTC August 5, 1994 are compared using the conventional and new down-scaling procedures.

The upper panel of Fig. 18b shows the verifying analysis for the 72 hr forecast; the middle panel the regional forecast obtained from the conventional down-scaling; and the lower panel the corresponding forecast from the new two-way down-scaling procedure. Close inspection of the forecasts show that the conventional down-scaling procedure produces obvious phase lags in the systems to the west, southwest and southeast of the Australian continent. There are other gains from the new down-scaling procedure that can be noted by comparing the forecasts with the verifying analysis. In summary, it is obvious that the forecasts from the conventional down-scaling procedure lag those from the new procedure, both in the interior of the domain and in the boundary regions.

3.7 Statistical Models

The use of statistical techniques has long been part of meteorology. Almost all aspects of weather and climate have a large statistical component. This is true whether we are discussing data, the analysis, the models or the verification and assessment procedures.

3.7.1 Deterministic and Statistical Models

The earliest models that were of any value were exclusively statistical in nature. They were based almost entirely on climatological records and were not particularly skillful, with few exceptions. It was only with the advent of electronic computers in the 1940s that deterministic forecasts became available. The first deterministic forecasts were short range, out to 24 hours ahead. Within a decade, the deterministic models had been extended to simulate the general circulation of the atmosphere and the first GCM/climate models appeared. The excitement and success of the deterministic forecasts pushed development of statistical models into the background and it was not until the 1960s and 1970s that researchers again began to point out the merits of statistical approaches to modelling the evolution of the earth's atmosphere. It is also true to say that the dominance of the deterministic approach, or paradigm, was so strong at the major weather centres that it became almost heretical to suggest that statistical models had a role to play in atmospheric simulation.

Fortunately, the situation has now changed and both deterministic and statistical models are recognized as having roles to play, either separately or in combination. As will be discussed in Sect. 3.6.3 below, major successes have been achieved when deterministic and statistical models are used in some kind of optimal linear or non-linear combination.

3.7.2 Analogue Retrieval Techniques

The idea behind analogue retrieval techniques is that a search is made for a "similar", or analogous, weather pattern that occurred in the past. This matching process is accomplished simply by simply accessing computer analyses stored in a data base and computing various differences between them and the present weather chart. The standard measures used for matching the weather patterns typically are correlation coefficients and root-mean-squared errors, or some combination of these measures. Once the best analogues have been selected, the weather patterns for the following days are based on those of the past evolution of the best analogues. The idea, therefore, is one of using the knowledge of what happened to an analogue of the present weather pattern to make some a judgement about the forecast. Commonly, the best three or four analogues are retrieved from the archives. A disappointing feature of this approach is the lack of many good analogues and the fact that even the better analogues deteriorate rapidly after about 12 to 24 hours.

The problem is one of the most difficult, but most important, facing meteorologists at present. Pattern recognition, as the research area is referred to, is so far proving to be more difficult for a computer to carry out with success than for a human. An experienced team of forecasters is often very difficult to outperform by analogue techniques. An example of a recently

114   3. Atmospheric Modelling and Prediction

implemented analogue technique is part of the the Australian Bureau of Meteorology's Australian Integrated Forecast System (AIFS). New techniques are being tested to improve the analogue approach, especially the fact is that there are very few exceptionally good analogues and that these become fewer in number as the domain size increases. This deficiency is largely due to the very short period (in meteorological terms) over which numerical analyses have been archived.

3.7.3 Combining Deterministic and Statistical Models

It was shown by Thompson (1977) and later by Fraedrich and Leslie (1988) that numerical weather predictions and stochastic model forecasts can be combined in an error minimizing manner to yield a prediction with skill greater than each forecast individually.

In theory, independent forecasts can be combined to produce a forecast which, on average (i.e. over a large number of outcomes) is better than either of the individual forecasts. By better, we mean the mean square error is less. To illustrate, we wish to predict some parameter ϕ at some point at various times. There are two independent, unbiased predictions, ϕ_1 and ϕ_2 that correlate equally well with ϕ. (If they did not, one would clearly be superior to the other, and we could simply disregard the inferior one.) By subtracting the mean (climatological) values, over a long period of time, ϕ_1 and ϕ_2 would have zero mean.

Let $\phi^* = A\phi_1 + B\phi_2$ be our prediction for ϕ. We need to find those values of A and B that give the smallest error (best prediction), in a mean-square sense. The mean square error E of ϕ^* is given by

$$E = \frac{1}{N}\sum_N (A\phi_1 + B\phi_2 - \phi)^2 \tag{3.25}$$

The expression for E is minimised by differentiating with respect to A and B, and setting these to zero. We get

$$\frac{\partial E}{\partial A} = 2A\sum_N \phi_1\phi_1 + 2B\sum_N \phi_1\phi_2 - 2\sum_N \phi_1\phi = 0 \tag{3.26}$$

$$\frac{\partial E}{\partial B} = 2B\sum_N \phi_2\phi_2 + 2A\sum_N \phi_1\phi_2 - 2\sum_N \phi_2\phi = 0 \tag{3.27}$$

For the combined estimate to be unbiased $A + B = 1$. Since the covariances are easily calculated from the observations these are just three linear equations in the two the unknowns A and B and are easily solved, we get:

$$A = \frac{(\sum_N \phi_1\phi \sum_N \phi_2\phi_2 - \sum_N \phi_2\phi \sum_N \phi_1\phi_2)}{(\sum_N \phi_1\phi_1 \sum_N \phi_2\phi_2 - \sum_N \phi_1\phi_2 \sum_N \phi_1\phi_2)} \tag{3.28}$$

and

$$B = \frac{(\sum_N \phi_1\phi_1 \sum_N \phi_2\phi - \sum_N \phi_1\phi \sum_N \phi_1\phi_2)}{(\sum_N \phi_1\phi_1 \sum_N \phi_2\phi_2 - \sum_N \phi_1\phi_2 \sum_N \phi_1\phi_2)} \qquad (3.29)$$

The value of ϕ^*, which is a linear combination of ϕ_1 and ϕ_2 is a better estimate ϕ, over a very large number of events, than either ϕ_1 or ϕ_2, and by normalising ϕ_1, ϕ_2 and ϕ both A and B can be expressed in terms of correlation, rather than sums.

Numerous examples exist of the effectiveness of this procedure, including predictions of the main model variables, long range predictions and applications to specific phenomena such as tropical cyclone tracks.

3.7.4 Multiple Regression Markov Model

Miller and Leslie (1984) developed a scheme that is used by the Australian Bureau of Meteorology to provide predictions of the probability of precipitation (POP). The basic idea is based upon the observation that wet (or dry) weather tends to persist. Suppose that the probability (based on statistics) of a wet (i.e. any measurable precipitation) day in Melbourne at this time of the year is 0.4. However, if the preceding day is wet, the probability that the day is wet may increase to 0.45. On the other hand, if the preceding day is dry, the probability the day is wet may decrease to 0.36. The object behind POP is to improve upon climatology by taking this extra information into account.

The principle behind POP forecasting is to assume a meteorological processes model as a Markov process. This means that the probability of a state occurring depends on one or more previous states, but not on all previous states. The example of a Markov process is a two-state, first-order process. Let the two possible states be wet and dry. Let us denote the state "The probability of state A occurring, given that state B has already occurred" by $P_r\{A|B\}$. Then, a two-state, first-order process would mean, for example:

P_r {Friday will be wet | Monday was wet, Tuesday was wet, Wednesday was dry, and Thursday was dry } = P_r {Friday will be wet | Thursday was dry }

Similarly, a second order process would be, using the following example:

P_r { Friday will be wet | Monday was wet, Tuesday was wet, Wednesday was dry, and Thursday was dry } = P_r {Friday will be wet | Wednesday was dry, and Thursday was dry }

So a first order Markov process depends only on the previous state, a second order process on the two previous states, third order on 3, and so on.

The Bureau's scheme for POP (at time of writing) is a second order process. The basic assumption is that the POP in the future depends only upon the present and immediately preceding (i.e. 3 hours earlier) weather states. Four weather states are defined:

- State 1: 0–2 oktas cloud and no rain
- State 2: 3–5 oktas cloud and no rain
- State 3: 6–8 oktas cloud and no rain
- State 4: 0.2 mm rain or more

The state becomes a covariate to be put into an equation. Fifteen other covariates are defined. Thirteen of these depend only upon the present and immediately preceding surface observations. (For example, the surface dew point depression is a covariate). Two further covariates are used to represent the annual cycle. Then, the probability of rain over the next few hours is determined by a formula of the type below:

$$P_r(j,h,t,m) = a(j,h,t,m) + \sum_{k=1}^{14} b_k(j,h) X_k + $$
$$+ b_{15}(j,h) \sin\left(\frac{\pi m}{6}\right) + b_{16} \cos\left(\frac{\pi m}{6}\right) \tag{3.30}$$

In this equation, j represents the weather, h the number of hours of validity of the forecast, t the time of day, and m the month. The X_k are covariates derived from the current observations and those which occurred three hours before. The a and b_k are derived from statistics, using the linear technique. P_r is a value between 0 (no chance of rain) and 1 (certainty of rain).

The skill of the method can be checked by using a half-Brier score B_r. This is defined by

$$B_r = \frac{1}{N} \sum_{j=1}^{N} (\delta_i - P_{ri})^2 \tag{3.31}$$

where N is the number of occasions being verified, $\delta_i = 1$ if rain did occur in the forecast period, and $\delta_i = 0$ if it was dry. P_{ri} are the forecast probabilities given by (3.30) for the occasions being checked.

The lowest possible Brier score is zero, implying a perfect forecast every time, and the worst possible score is one, implying an erroneous score on each occasion. Typical values are of the order 0.16, but there is substantial variation from station to station, and from season to season. As an extreme example, the score for Darwin in the dry season is close to zero, but becomes quite large in the wet season.

3.8 Future Directions

It is hoped that the above presentation has served the intended purposes of describing in some detail, by the use of a wide variety of examples, the present status of numerical weather prediction (NWP). There are a number of applications from different parts of the globe, but with special emphasis on examples drawn from the Asia-Pacific region. Active programs are in place, and many of the problems once regarded as intractable are now being attempted with considerable success. This raises the question of what lies ahead for weather forecasting on the various time scales discussed here. Research and applications of topics such as the incorporation of cloud microphysics, predictability, ensemble forecasting, four dimensional variational data assimilation (4DVAR) in either weak or strong constraint mode, artificial intelligence/machine learning, coupling of the atmospheric model with ocean and land surface models, parallel computing techniques, optimal networks and the role of statistical-dynamical prediction all should become, or remain, active research areas in the coming years. In this section, we will look only at three of these research areas in detail, namely the impact of explicit representation of cloud microphysics on precipitation forecasting, ensemble forecasting techniques and the impact of variational data assimilation on the quality of the initial state and subsequent forecast.

3.8.1 Impact of Cloud Microphysics

In a simple but clean and direct comparison, the widely used Kuo parameterisation of convection scheme was compared with an explicit 6 water-ice phase microphysics scheme for a rain event over southeastern Australia. Figure 3.19a shows the observed rainfall for the 24 hour period 9am September 1, 1997 to September 2, 1997. The rainfall obtained from the Kuo scheme is shown in Fig. 3.19b and compares quite well with the observed rainfall over most of the region except for the middle part of the state of New South Wales where a large area of heavy convective rainfall was not forecast by the Kuo scheme. By contrast, the precipitation predicted from the explicit scheme (Fig. 3.19c) is uniformly excellent and captures the heavy rainfall area not predicted by the Kuo scheme. Extensive testing of the explicit scheme against other precipitation parameterisations is required, but these early results are very encouraging.

3.8.2 Ensemble Forecasting

The idea behind ensemble forecasting is to create an initial sample of equally likely initial states by adding other members to the original initial state. As shown in Fig. 3.20, a "cloud" of points is created in phase space and describes the uncertainty in the initial state. This cloud is the ensemble of initial conditions, with each representing a feasible initial state of the atmosphere, consistent with the observational/analysis uncertainties. The evolution of the cloud, or probability density function (PDF) is meant to simulate the change in the PDF during the forecast period. It can be shown that following the

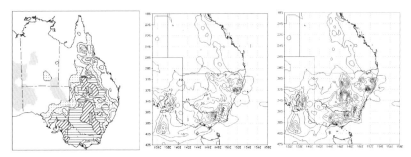

Fig. 3.19. (a) The observed 24 hour rainfall for the period 9 am September 1, 1997 to September 2, 1997; (b) Forecast precipitation from the Kuo scheme; (c) as in (b) except for the explicit scheme.

path of the mean initial state provides the best estimate of the state of the atmosphere, noting that all members of the evolving PDF remain equally likely states of the atmosphere.

Choosing the ensemble members is one of the two major areas of research interest. The other area is determining how well the spread of the ensemble correlates with the mean forecast error. If the correlation is high, then the ensemble spread can provide very valuable information about the confidence in the forecast. If the correlation is not high then the spread is of little value. In studies of extra-tropical cyclones, Stensrud et al. (1999) and Leslie and Speer (1998) found that the correlation was too low to be of value and further research is needed.

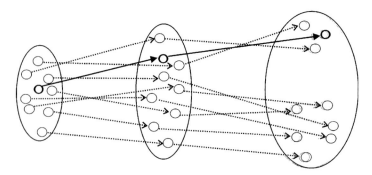

Fig. 3.20. Evolution of the ensemble of possible initial states of the atmosphere (left) through two forecast periods (middle and right). The heavy line indicates the evolution of the "best" initial state, while the dashed lines indicate the evolution of the remaining ensemble members.

Presently, there are three main methods in common use for the creation of ensemble members. The National Centers for Environmental Prediction

(NCEP) uses the breeding method (Toth and Kalnay, 1993) and the European Centre for Medium-Range Weather Forecasts (ECMWF) uses the singular vector decomposition approach (Molteni et al., 1996). We have chosen a third approach, the Monte Carlo procedure, which has also been used successfully (see, Mullen and Baumhefner 1989, 1994; hereafter referred to as MB89 and MB94) but has suffered in the past from the prohibitive computational cost of generating a sufficiently large ensemble. Questions have rightly been raised about the relative merits of allocating resources to better single forecast approaches or using degraded resolution models in the SREF approach (Brooks et al., 1995). As will be explained later, this choice was not particularly important here. However, the continued exponential growth in computing power means that it is feasible now to perform large numbers of runs of a high-resolution limited-area model in real time. For example, this has been achieved over an area the size of southeastern Australia using the boundary conditions from the operational model forecast to nest the limited area model developed jointly by Leslie and Purser (1995). Details of the model are summarized in Table 3.4. The 15-km horizontal resolution domain is shown in Fig. 3.3. Initial and boundary conditions were provided from LAPS forecast for the coarse mesh run (75-km). The 15-km HIRES runs were produced by nesting within the 75-km LAPS model and without varying the boundary conditions for any of the ensemble members.

Monte Carlo method. The general procedure for carrying out a Monte Carlo forecast involves the generation of a finite sample of equally likely initial states, advancing each of those states using model equations and the determination of an estimate of the true state at any later time using the sample mean. The uncertainty of the initial state is given by covariances generated by the BoM's operational analysis scheme. As the sample size approaches infinity, the forecast based on the sample mean becomes "best" in the sense that it is unbiased and possesses minimum variance. From earlier work on turbulence flows (Leith, 1974), relative mean square velocity errors were reduced significantly up to a sample as small as 8 or 10 initial states. Larger numbers of initial states are now used in operations, for example, at NCEP and at ECMWF. Although similar to MB89 and MB94, the perturbation methodology adopted has significant differences. Other studies performed by the first author (LML) with tropical cyclones involved perturbing the wind field only, whereas MB89 (p.2801) perturbed only temperature as the case was a midlatitude event. Australian east coast lows occur in the subtropics, usually between 28°S and 38°S (Sydney is at 34°S) so it was decided to apply random perturbations to both the wind and the temperature. The perturbations were also applied directly to the grid-point model and not in spectral space as in MB89 and MB94. The size of the perturbation was determined from known analysis error characteristics as discussed below. The boundary conditions supplied by the coarse-resolution model forecasts at 12-h time intervals were not perturbed.

120   3. Atmospheric Modelling and Prediction

The procedure adopted here for the generation of the ensemble members is we believe a very effective one and needs some elaboration. The random perturbations were treated as if they were data and added to the other observational data available. This combination of observations and random noise was then used to generate up to 100 initial analyses for the forecast model using the operational regional data analysis scheme, which is basically a multivariate statistical analysis scheme. This method of generating the ensemble members has great appeal as it not only spatially correlates the perturbations in both the horizontal and vertical via the analysis scheme but it also renders, in the truest sense we are aware of, each member of the initial ensemble as "equally likely". Three initial states are shown in Fig. 3.21a-c, all of which would be regarded as defining the initial state to within analysis errors. Figure 3.21a is the unperturbed sea level pressure (SLP) analysis, Figure 3.21b is the initial ensemble mean SLP analysis, and Figure 3.21c is the perturbed analysis most remote from the ensemble mean in the sense of a simple RMS difference. The striking aspect of the initial analysis in this case is its structural simplicity, with basically one shallow major feature, namely, the trough over the central northern part of the chart. Finally, the perturbed initial fields were passed through a diabatic dynamic initialization procedure (Daley, 1991), which does little more than filter out spurious gravity waves. The procedure was very effective as the initialized fields were free from high-frequency gravity waves and the forecasts exhibited little or no initial shock.

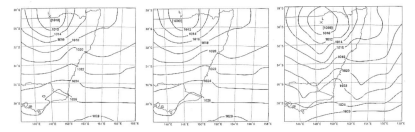

Fig. 3.21. (a) The unperturbed sea level pressure (SLP) analysis. (b) The initial ensemble mean SLP analysis. (c) The perturbed analysis most remote from the ensemble mean.

Size of initial perturbations. The data density imposes a constraint on the envelope of the amplitude of the perturbations, the envelope being larger over the data-sparse oceans. For this study, the errors for analyzed winds over the Australian region typically are 7 ms^{-1}, 5 ms^{-1}, and 3 ms^{-1} for the upper, middle, and lower levels (300 hPa, 500 hPa, and 850 hPa), respectively. Analysis temperature errors were 1.5°C at each of these levels. No attempt was made in this study to test the sensitivity to perturbations of the moisture field. The 100 initial states were generated by adding Gaussian noise at each grid point, constrained within the above-mentioned error limits for the wind and temperature fields, A full multivariate analysis is performed using the perturbed data and then each perturbed analysis was initialized using HIRES.

In summary, the procedure consisted of generating the 100 model forecasts using HIRES at 15-km horizontal resolution centered on NSW. Initial and boundary conditions were provided by the coarse mesh LAPS model interpolated from 75 km to 15 km. The 15-km horizontal resolution was deliberately chosen for two main reasons: it is considered to be close to the current practically feasible horizontal resolution at which the model will be run locally in real time on the recently acquired BoM computer upgrade; and it does not violate the nonhydrostatic and cumulus parameterization assumptions (Golding and Leslie, 1993).

The impact of lateral boundary conditions. It is well known that there are potential problems associated with lateral boundary conditions in limited-area forecast models. A recent article by Warner et al. (1997) provides a comprehensive overview of these problems and suggests practical ways in which they can be minimized. The domain used here was chosen very carefully after experimentation and many years of prior knowledge of the region. However, it still is instructive to look at the error patterns at 36 hr for the domain used and for a domain with the lengths of the sides doubled in each of the horizontal directions. Given that the interest was on the short-range period out to 36 hr only, our expectations were that the domain was large enough. The integration domains are shown in Fig. 3.22a and are labeled C for the Australian region coarse mesh model domain (50-km horizontal resolution), L for the large 15-km resolution domain, and S for the small 15-km resolution domain used for the experiments described in this study. That the smaller domain is large enough, at least for the present event, is confirmed in Figs. 3.22b and 3.22c, which show the error fields near the surface and at 500 hPa. Here, the error fields are defined, conventionally, as the difference between the forecasts over domains L and S, both nested in the coarse mesh model and both run at the same resolution of 15 km. The smaller domain shows some evidence of larger errors penetrating the edges of the area of interest but they are quite small, being everywhere less than 2.0 hPa at SLP and less than 20 m at 500 hPa. Not surprisingly they are largest in the southwest corner of the domain. The larger domain was not selected because the additional computational cost was not justified by the relatively minor influence of the boundary conditions. In other regions, for example in the mid-latitudes, a larger region might be required but it would be assessed before a final configuration was chosen.

3.8.3 Variational Data Assimilation

It has long been recognized that NWP is essentially an initial value problem, especially on the globe where there are no lateral boundaries. The problem of meteorological data analysis is the determination of the "optimal" initial

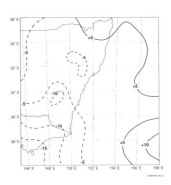
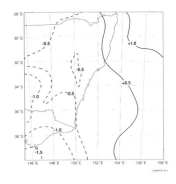

Fig. 3.22. (a) The Course mesh model domain and the two domains (L) large and (S) small for testing the sensitivity of the forecasts to the lateral boundary conditions. (b) The difference between the 36 hour forecasts of Sea Level Pressure for the large (L) and small (S) domains. (c) As in (b) except for 500 hPa.

state. There have been many attempts to do so and perhaps the most promising present procedure is four-dimensional variational data assimilation. The variational data assimilation technique may be expressed very simply and has its origins in work carried out long ago by Sasaki (1970). The basic idea is to use a weighted least squares procedure to minimize a cost function that measures the "distance" between a variable and the observations of that variable and the distance between the variable and the first guess for the variable (background field). The weights are the inverses of the observation error and the background error covariances. The minimum of the cost function is obtained for a value of the state variable that is taken to be the best analysis. The variational data assimilation approach has a number of different variants. It is commonly used either in 3-dimensional or 4-dimensional form, the latter involving forward and backward integrations in time. It therefore uses the model forecast.

Another distinction is between strong and weak variational data assimilation procedures. The strong constraint approach assumes that the model is perfect so that there is no residual term in the cost function involving the model formulation error. The weak constraint approach assumes that the model is not perfect and the cost function includes a term that represents the residual in the model formulation. Considerable success has been obtained by the introduction of the variational data assimilation approach (Leslie and LeMarshall, 1998; Kalnay et al., 2000). Leslie and LeMarshall applied a weak constraint four-dimensional data assimilation scheme to a number of tropical cyclones in various basins around the world. One example is presented here in Fig. 3.23 that shows forecast tracks for Hurricane Opal which developed in the Gulf of Mexico and crossed the coast near New Orleans.

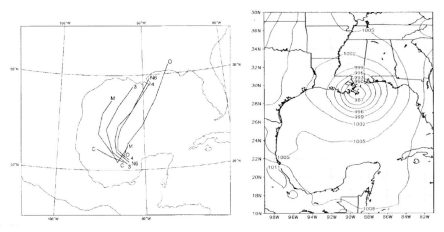

Fig. 3.23. (a) The 72 hour forecasts of Hurricane Opal track (O) with the control (M), 3D variational assimilation (3), 6-hour nudging (N6), 4D variational assimilation and CLIPER forecasts (C) commencing at 00 UTC October 2, 1995. (b) The 72-hour verifying SLP analysis, valid at 00 UTC October 5, 1995.

The impact of the 4-dimensional variational scheme was very large, as seen in the figure. Both the CLIPER (climatology-persistence) and the operational forecasts were very poor, pushing the hurricane track to the west instead of to the NNE. The forecast using the initial state derived from the four-dimensional variational data assimilation scheme produced a forecast that was very accurate both in the location and timing of landfall. The technique has also been applied by the authors to the western North Pacific and the Australian region, with similar success.

4. Ocean Modelling and Prediction

Matthew H. England and Peter R. Oke

4.1 Introduction

The ocean plays a vital role in our environment. As such, an ability to model and predict its circulation can be of enormous value. Modelling and prediction of ocean currents in coastal regions is important for many reasons, including influences on recreation, navigation, algal bloom formation, effluent dispersion, search and rescue operations, and oil spills. Ocean currents near the coast also affect beach conditions that impact upon the near-shore zone. Severe wave climates and storm surges can cause enormous destruction of the built environment. At larger scales, vast ocean currents carry heat around the globe, affecting climate and weather patterns such as those associated with the El-Niño event and the North Atlantic Oscillation. The oceans also have a vast capacity to absorb and redistribute gases such as carbon dioxide. They will therefore play a crucial role in determining our future climate.

To understand and predict the way the ocean affects our environment, a number of ocean models have been developed during the past half century. Some are computationally simple and predict a limited number of oceanic variables, such as tidal models or a wave climate model. Others, such as primitive equation ocean general circulation models (OGCMs), are computationally expensive and solve several equations in order to predict three-dimensional ocean currents and temperature-salinity $(T-S)$. In this chapter we describe the state-of-the-art in ocean modelling.

4.1.1 What Is an Ocean Model?

The World's oceans can be viewed as a turbulent stratified fluid on a rotating sphere with a multiply-connected domain and an uneven bottom bathymetry. More simply, the rotating earth has an ocean system divided by land masses and with varying water density and ocean depth. The external forcing of the ocean occurs through the mechanical forcing of the winds, the so-called "thermohaline" forcing via heat and freshwater fluxes across the air-sea interface, and through planetary forces manifest in tides. An ocean model is simply a computational solution to this problem: using physical conservation laws for mass, momentum, heat and so on, and some estimation of the forcing fields, the computer model predicts ocean currents and other properties such as temperature, salinity, and optionally chemical tracers or biological parameters.

4.1.2 Mean Large-scale Ocean Circulation

The global scale ocean circulation can be viewed in a number of ways. In the horizontal plane, mean circulation is dominated in the upper ocean by wind-driven flow. Figure 4.1 shows surface mean wind stress over the oceans, and the upper ocean circulation pattern. Large-scale gyres dominate at mid-latitudes, with intensified western boundary currents (WBCs) due to the Earth's rotation, carrying tropical heat poleward. At the tropics, easterly trade winds and the doldrums drive a tropical current/counter current system (for more details see Tomczak and Godfrey, 1994). At higher latitudes thermohaline circulation is manifest in the surface flow; for example, the North Atlantic Current extends into the Greenland/Norwegian Sea to feed North Atlantic Deep Water formation. In the Southern Ocean, a latitude band free of continental land masses permits the eastward flowing Antarctic Circumpolar Current (ACC) to circle the globe.

In the meridional plane a completely different view of the ocean circulation is obtained (Fig. 4.2). Water masses of different density classes ventilate the interior of the ocean in a complex manner. Around Antarctica dense bottom water is formed over the continental shelf by salt rejection during sea-ice formation and wintertime cooling. Further north, Circumpolar Deep Water (CDW) is upwelled under the subpolar westerlies, flowing either northward to form intermediate and mode waters (after the addition of freshwater via precipitation or sea-ice melt), or southward towards the Antarctic continent. In the Northern Hemisphere, a saltier North Atlantic accommodates deep water production whereas the North Pacific remains too fresh to see deep water convection. The water masses formed in the World Ocean subsequently recirculate and are either "consumed" by diapycnal mixing or when they resurface in the upper mixed layer. In both cases, $T-S$ properties are altered or reset and the water-mass is converted.

4.1.3 Oceanic Variability

The dominant picture of ocean circulation at the large-scale was one of steady flow until drifting buoy technologies in the 1960s revealed variability of flow patterns at rather small spatial scales. Near-shore variability was long known to exist and was thought to be controlled by fluctuations in tidal flows and local winds. At the large-scale, oceanic variability is evident in phenomena such as El-Niño and the Antarctic Circumpolar Wave; as well as in western

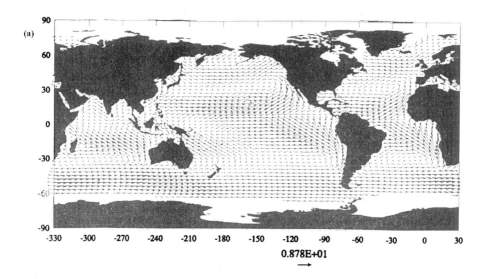

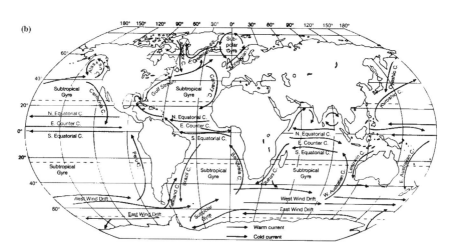

Fig. 4.1. (a) Surface mean wind stress over the oceans (from Barnier, 1998); and (b) Schematic of upper ocean circulation patterns (from Thurman, 1991).

boundary current flow pathways. In addition, oceanic eddies of size 30–100 km are seen near intense surface currents (e.g. the ACC and WBCs) and near the Equator, and play a key role in transporting climate properties around the globe. Figure 4.3 shows the kinetic energy spectrum estimated for the ocean and atmosphere as a function of horizontal wavenumber, revealing substantial oceanic energy at the length scale of these mesoscale eddies. Unfortunately, most present day coupled climate models do not resolve these scales of mo-

128   4. Ocean Modelling and Prediction

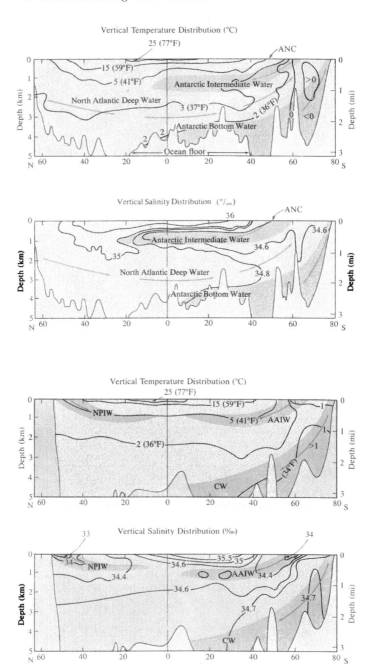

Fig. 4.2. Meridional latitude-depth ocean circulation schematics for the Pacific and Atlantic Oceans (from Thurman, 1991). AAIW refers to Antarctic Intermediate Water, NPIW to North Pacific Intermediate Water, and ANC to the Antarctic Convergence.

tion as computational requirements are too high at sub-eddy scale resolutions. These models adopt large-scale eddy parameterisations to permit reasonable model integration times. Coastal or regional models, as well as global simulations over shorter integration periods, can resolve mesoscale eddy variability (see, e.g., Semtner and Chervin, 1992; Webb et al., 1998).

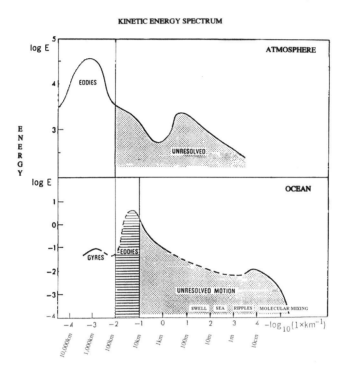

Fig. 4.3. Kinetic energy spectrum estimated for the ocean and atmosphere as a function of horizontal wavenumber (after Woods, 1985).

4.1.4 The Oceans, Climate, and Forcing

The oceans play a vital role in the global climate system via their capacity to absorb heat in certain locations, transport this heat vast distances, then release some of it back to the atmosphere at a later time. This is depicted in Fig. 4.4 which shows the global mean transport of heat by the oceans. This pattern of heat transport reflects the effects of western boundary currents carrying warm water poleward, as well as the net flux of heat towards deep water formation sites, particularly NADW. The ocean exhibits variability on a range of space-time scales, and some of this variability is at a large enough spatial scale to affect global weather systems. For example, the El-Niño/Southern Oscillation (ENSO) involves a massive redistribution of heat in the tropical Pacific Ocean via anomalous surface circulation patterns (see Philander, 1990 for details). In the Southern Ocean, the Antarctic Circumpolar Wave (ACW) advects heat anomalies around the globe, affecting wind patterns, pressure systems and sea-ice extent. Ocean modelling for climate studies has arisen from the need to understand and predict the way ocean circulation can vary and affect weather/climate. Given that the ocean circulation is determined by both wind and thermohaline factors, ocean climate models generally include both these forces whilst neglecting high frequency waves such as tides and swell.

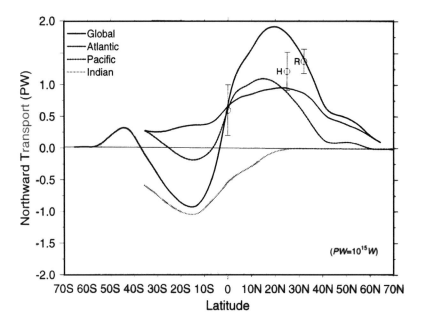

Fig. 4.4. Global mean transport of heat by the oceans (from Mathieu, 1999; using observations of da Silva et al., 1994).

Coastal ocean currents are affected by a number of different factors including winds, surface heating, buoyancy effects, tides, deep ocean forcing (e.g. WBCs) and coastal trapped waves (CTWs). The time scales on which many of these factors vary are from weather-band (3–7 days) to seasonal scales. Winds tend to be the most dominant force, locally affecting ocean currents, temperature and salinity (through upwelling/downwelling) and sea-level. Buoyancy effects often become important in response to wind-driven events (through geostrophic adjustment), in the presence of river outflows, or in regions where a significant amount of heat is gained or lost at the ocean surface. In many coastal regions tides are important [e.g., the North West Australian shelf (Holloway, 1984)] and in others deep ocean forcing is predominant [e.g., off eastern Australia (Oke and Middleton, 2000)]. Modelling the coastal ocean is therefore a challenging task, and consideration of what factors are most important for any particular region is necessary in order to adequately represent the true variability of the coastal ocean. At smaller scales, such as flows in harbours or bays, circulation patterns are often dominated by tidal flows, so their modelling can be simplified somewhat.

4.2 A Brief History of Ocean Modelling

A number of simple analytic and linear vorticity models of the basin-scale ocean circulation were developed prior to the proliferation of computing machines, including the so-called Sverdrup model of wind-driven flow (Welander, 1959), the Stommel-Arons model of abyssal circulation (Stommel and Arons, 1960), and Wyrtki's (1961) simple model of thermal overturning circulation. For a review of these early analytic modelling efforts the reader is referred to Weaver and Hughes (1992). Similarly, analytic models of wind-driven coastal jets (Allen, 1973), continental shelf waves (Allen, 1980) and stratified flows over sloping topography (Chapman and Lentz, 1997) have given modellers great insight into the dynamics of coastal ocean flows.

The first real progress towards a primitive euqation ocean circulation model came with the work of Kirk Bryan and Michael Cox in the 1960s (Bryan and Cox, 1967, 1968), in pioneering work towards a coupled climate model. They developed a model of ocean circulation carrying variable $T - S$ (and therefore density) based upon the conservation equations for mass, momentum, heat and moisture, and the equation of state. It is not surprising this work was completed at an institution where atmospheric modelling was already well-established (the GFDL), as the ocean and atmosphere have a number of similarities, and their modelling requires many analogous techniques. Bryan-Cox assumed the ocean had negligible variations in sea-level (i.e. the "rigid-lid" approximation), so that high-frequency gravity waves are ignored in the model formulation, and the depth-averaged component of velocity (the "barotropic" mode) is solved using an iterative technique. Their model was configured in a variety of ways: a 2-D model, a 3-D basin model, and a full World Ocean model (Cox, 1975).

The computational requirements of this early model were relatively high, enabling only short integrations from initial conditions, and therefore solutions that were not in thermodynamic equilibrium with the model forcing. Nevertheless, Bryan-Cox achieved global simulations with realistic continental outlines, rough bottom bathymetry, prognostic equations for $T - S$, an elimation of high frequency modes, and approximate closure schemes for the effects of mixing, friction, and eddies. Their work can be seen as the genesis of modern-day ocean modelling.

Since the early efforts of Bryan and Cox, a great number of ocean model developments have occurred (for an outstanding review, see Griffies et al., 2001). These include the exploration of different grid systems. The GFDL model operates on a Cartesian grid with geopotential (i.e., horizontal) layers in the vertical. Models have now been developed with terrain-following (Haidvogel et al., 1991) and density-layer coordinates (Bleck and Boudra, 1986). In the horizontal plane, models have been developed with curvilinear coordinates to follow a local coastline (Blumberg and Mellor, 1987).

In addition to different grid systems, a great variety of model options have been developed. In models that do not explicitly resolve mesoscale eddies, their effects can be parameterised in a number of ways (e.g., Cox, 1987; Redi, 1982; Gent and McWilliams, 1990; Gent et al., 1995; Griffies et al., 1997). Free surface formulations were also developed to enable direct prediction of the height and pressure of the ocean surface (e.g., Killworth et al., 1991; Dukowicz and Smith, 1994). This enabled direct comparison with satellite-derived data products as well as eliminating the need to solve the barotropic streamfunction iteratively (which becomes costly in higher resolution model domains and when multiple islands are involved).

Ocean model development has now proliferated due to improved numerical techniques, better global ocean data sets, diversity of model applications, and perhaps most dramatically, faster computers with ever increasing processing capacity.

4.3 Anatomy of Ocean Models

4.3.1 Governing Physics and Equations

Ocean models are capable of predicting a number of variables, normally the three components of velocity (u, v, w), temperature (T)–salinity (S) and therefore density (ρ). They also usually predict either the depth-integrated transport streamfunction or the sealevel pressure. These "predicted" variables are known as the model "prognostic" variables. To build an ocean model requires a number of governing equations in order to solve for the prognostic variables. To have a well-determined system requires the number of equations to be the same as the number of prognostic variables. For large-scale or regional coastal ocean models the governing equations are derived from the conservation laws of mass, heat and salt as well as the Navier-Stokes equations for flow of fluid on a rotating earth. Typically modellers reduce the latter to the so-called "primitive equations" by adopting the *Boussinesq* and *hydrostatic* approximations, meaning respectively that density variations do not affect the momentum balance except via the vertical buoyancy force (that is, density variations ρ are much less than the total density ρ_0), and that the buoyancy force is balanced solely by the vertical pressure gradient (therefore it is assumed that vertical velocities are small compared to horizontal velocities). These assumptions are valid in almost all oceanic circulation regimes; a notable exception being oceanic convection of unstably stratified waters (discussed later) wherein nonhydrostatic flow occurs.

The primitive equations operating in ocean models are depicted in the schematic diagram of Fig. 4.5, which shows how the equations are interrelated as well as what surface forcing is required (see also Sect. 4.3.4).

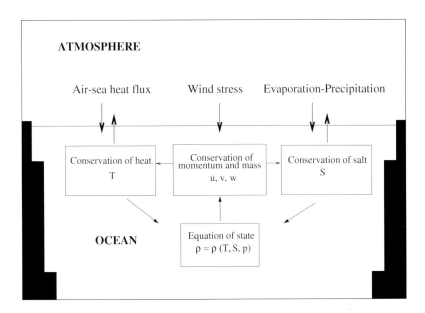

Fig. 4.5. Schematic diagram showing the conservation laws, prognostic variables, and air-sea property fluxes used in ocean models.

The model equations can be written in Cartesian coordinates as follows.

Horizontal momentum equations:

$$\frac{du}{dt} - fv = \frac{\partial u}{\partial t} + u\frac{\partial u}{\partial x} + v\frac{\partial u}{\partial y} + w\frac{\partial u}{\partial z} - fv = -\frac{1}{\rho_0}\frac{\partial p}{\partial x} + F_u + D_u \quad (4.1)$$

$$\frac{dv}{dt} + fu = \frac{\partial v}{\partial t} + u\frac{\partial v}{\partial x} + v\frac{\partial v}{\partial y} + w\frac{\partial v}{\partial z} + fu = -\frac{1}{\rho_0}\frac{\partial p}{\partial y} + F_v + D_v \quad (4.2)$$

Hydrostatic approximation:

$$\rho g = -\frac{\partial p}{\partial z} \quad (4.3)$$

Continuity equation:
$$\frac{\partial u}{\partial x} + \frac{\partial v}{\partial y} + \frac{\partial w}{\partial z} = 0 \tag{4.4}$$

Conservation of heat:
$$\frac{dT}{dt} = \frac{\partial T}{\partial t} + u\frac{\partial T}{\partial x} + v\frac{\partial T}{\partial y} + w\frac{\partial T}{\partial z} = F_T + D_T \tag{4.5}$$

Conservation of salt:
$$\frac{dS}{dt} = \frac{\partial S}{\partial t} + u\frac{\partial S}{\partial x} + v\frac{\partial S}{dy} + w\frac{\partial S}{\partial z} = F_S + D_S \tag{4.6}$$

where (x, y, z) is Cartesian space, t is time, (u, v, w) the three components of velocity, f is the Coriolis parameter, ρ_0 the mean ocean density, p is pressure, ρ is density, g is gravity, T is potential temperature and S salinity. The terms denoted by F and D represent, respectively, forcing and dissipation terms, discussed below. The Coriolis parameter $f = 2\Omega \sin \phi$ where Ω is the angular velocity of the Earth's rotation (7.3×10^{-5} sec^{-1}) and ϕ is latitude. T, the potential temperature (often also denoted as θ), is the temperature a given fluid element would have if it were moved to a fixed reference pressure (normally the sea surface). This quantity is approximately a conserved property, unlike *in situ* temperature. Density ρ is a function of potential temperature θ, salinity S and pressure p through the non-linear Equation of State [see Gill (1982) appendix for details].

The configuration of an ocean model involves solving (4.1)–(4.6) for u, v, w, p, T and S over a given grid. The spatial-scale of the grid chosen determines to what extent various processes are resolved. Motion in the ocean occurs at a variety of scales, from molecular diffusion processes (scales of 10^{-6} metres) right through to oceanic gyres (scales of 10^7 metres).

Figure 4.6 shows these processes as a function of spatial extent. It turns out that a significant component of oceanic energy resides at the scale of the external Rossby radius R, which is the length-scale at which rotation effects are as important a restoring force on motion as gravitational (or buoyancy) effects.

$$R = \sqrt{gh}/|f| \tag{4.7}$$

where h is the depth of the ocean. Oceanic eddies are typically of the scale of the external Rossby radius. R varies from around 100 km at tropical latitudes down to 10 km at high latitude. Coarse resolution ocean models adopt spatial grids of increment $\approx 100 - 400$ km, well above the Rossby radius of deformation. In such models, mesoscale eddy effects must be parameterized in some way (discussed below).

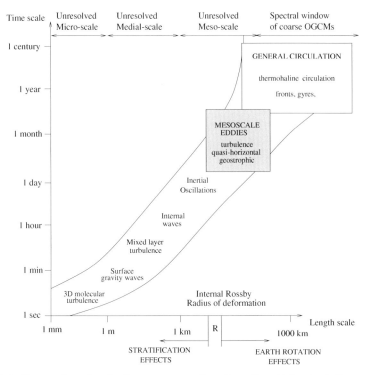

Fig. 4.6. Types of motion in the ocean as a function of spatial extent (from Mathieu, 1999).

4.3.2 Model Choice of Vertical Coordinate

Ocean modellers adopt a variety of numerical techniques for the treatment of the vertical coordinate. The three most common vertical schemes used in large-scale ocean models are depicted in Fig. 4.7. The first one uses geopotential or horizontal z-levels, reducing the observed bathymetry to a series of steps (e.g., MOM, DieCast). The second uses isopycnal layers as the vertical coordinate system, where the layer-averaged velocities and layer-thicknesses are the dependent variables (e.g., MICOM). The third uses a terrain-following sigma-coordinate system through the transformation of the water column depth from $z = 0$ to the bottom into a uniform depth ranging from 0 to 1 (e.g., POM, SPEM).

Each vertical coordinate system has its own advantages and disadvantages. The "z-level" model is flexible in a number of applications and lowest in computational requirements, although its grid orientation can result in excessive diapycnal mixing. In addition, downslope plume flows normally require some form of parameterisation to preserve water-mass signatures. Similarly, upwelling through thin bottom boundary layers in coastal regions is

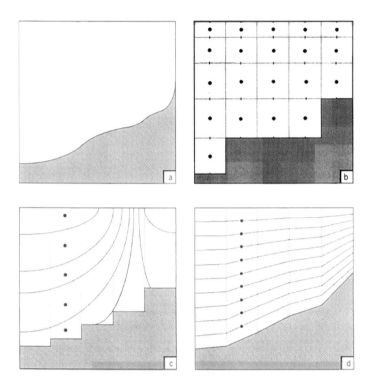

Fig. 4.7. The three most common vertical schemes used in large-scale ocean models (**a**) Observed bathymetry; (**b**) z-level model topography; (**c**) isopycnal layer; and (**d**) sigma-coordinate topography (from DYNAMO, 1997).

not well resolved by z-level models. Isopycnal layer models are ideal for simulating water-mass spreading and eliminating unphysical diapycnic mixing, although the use of single potential density values is dynamically inconsistent and can lead to errors in high latitude water mass distributions. The sigma-coordinate, with terrain-following levels, is most likely to realistically capture bottom boundary flows, such as bottom water plumes. Its coordinate is ideal for flow dominated by topographic effects, although it can result in excessive diapycnic mixing near strong topographic or isopycnal slopes. Sigma-coordinate models also require strongly smoothed bathymetry or high horizontal resolution to avoid numerical errors associated with the calculation of the pressure gradient terms.

4.3.3 Subgrid-scale Processes and Dissipation

Given the enormous range of spatial scales apparent in ocean circulation processes (order 10^{12}), all ocean models need to treat subgrid processes to some degree. For coarse resolution models, this includes mesoscale eddies and their effect on the large-scale flow.

The dissipation or mixing of momentum in the ocean is required to balance the continual input of mechanical wind energy at the air-sea interface. Kinetic energy in the ocean is transferred from large-scales to smaller scales (eventually molecular). The standard approach to parameterising the mixing of momentum is to relate the subgrid-scale dissipation to large-scale properties of the flow via a Fickian equation of the form

$$D_u = \frac{\partial}{\partial x}\left(A_H \frac{\partial u}{\partial x}\right) + \frac{\partial}{\partial z}\left(A_V \frac{\partial u}{\partial z}\right) \tag{4.8}$$

$$D_v = \frac{\partial}{\partial y}\left(A_H \frac{\partial v}{\partial y}\right) + \frac{\partial}{\partial z}\left(A_V \frac{\partial v}{\partial z}\right) \tag{4.9}$$

$$D_w = \frac{\partial}{\partial z}\left(A_V \frac{\partial w}{\partial z}\right) \tag{4.10}$$

where A_H, A_V are the horizontal/vertical eddy viscosity coefficients, respectively. Typically ocean modellers adopt values for A_H, A_V that greatly exceed their estimated magnitude to avoid numerical instabilities. Under a hydrostatic assumption, as typically used in coastal and larger scale models, D_w is considered negligible compared to ρg [as per (4.3)]. The horizontal viscosity terms represent a very ad-hoc parameterisation for the exchange of horizontal momentum from sub-grid scales up to the model grid-scale. The term is required to maintain numerical stability as the viscosity approximation acts to dissipate energy, without causing spurious sources of momentum. The most common approach is a Laplacian operator.

An example of a flow and resolution-dependent parameterization for A_H is

$$A_H = C\Delta x \Delta y \frac{1}{2}\left(\left[\frac{\partial u}{\partial x}\right]^2 + \left[\frac{\partial v}{\partial x} + \frac{\partial u}{\partial y}\right]^2 + \left[\frac{\partial v}{\partial y}\right]^2\right)^{\frac{1}{2}}$$

where C is the Smagorinsky constant (typically < 0.2) and $\Delta x, \Delta y$ are the horizontal grid spacings. This formulation determines A_H as a function of grid resolution and horizontal velocity gradients (e.g., Smagorinsky, 1963). For coastal applications with high horizontal resolution, A_H is often set to small constant values [e.g., for $\Delta x = 0.5\,\text{km}$, $A_H = 2\,\text{m}^2\text{s}^{-1}$ (Allen et al., 1995)]; whereas for coarser horizontal resolution, A_H can have magnitudes of the order of 10 to 100 m^{2}s^{-1}.

For different applications the vertical eddy viscosity A_V must be suitably defined in order to adequately model factors such as the frictional effects of the wind on the ocean surface, and the frictional drag associated with flow over the ocean floor. Many applications assume that A_V is constant in space and time [e.g., in most coarse global models, with $A_V \simeq 2 \times 10^{-3}$ m^{2}s^{-1} (Toggweiler et al., 1989; England, 1993)]. Others assume that A_V does not change with time but varies over the water column according to some predefined shape function (e.g., Lentz, 1995), while still others parameterise A_V as a function of the stability of the water column (e.g., Pacanowski and Philander, 1991; Mellor and Yamada, 1982). This aspect of ocean modelling remains uncertain since modellers are trying to capture the effects of processes that occur on scales of the order of centimetres to metres using vertical grids that have scales of the order of 10–1000 m.

Viscosity processes in the horizontal occur on scales of 10's to 100's of kilometres, such as mixing by eddies, whereas in the vertical, they are dominated by vertical shear instabilities, convective overturning and breaking internal waves over rough bathymetry. This accounts for viscosity coefficient values with $A_H \approx 10^6 A_V$. This is to ensure numerical stability, and given that model simulations are relatively insensitive to this parameter, few other approaches have been tested. An exception is the argument by Holloway (1992) that the ocean, in the absence of momentum input from the wind, would spin down not to a state of rest but to a state of higher system entropy. This is because eddies interacting with bottom topography can exert a large-scale systematic force on the mean ocean circulation. Under this so-called "topographic stress" parameterisation, the horizontal viscosity terms of (4.8)–(4.9) are rewritten with (u, v) replaced by $(u - u_*, v - v_*)$, where (u_*, v_*) represent the maximum entropy solution velocities (see Holloway, 1992; Eby and Holloway, 1994 for further details).

The mixing of scalars (such as T, S, and chemical tracers) has received much attention in recent years. The traditional formulation for subgrid-scale tracer mixing adopted by Bryan (1969) and Cox (1984) was of the form

$$D_t = \frac{\partial}{\partial x}\left(K_H \frac{\partial T}{\partial x}\right) + \frac{\partial}{\partial y}\left(K_H \frac{\partial T}{\partial y}\right) + \frac{\partial}{\partial z}\left(K_V \frac{\partial T}{\partial z}\right) \quad (4.11)$$

with a similar equation for salinity. Here, K_H, K_V are the eddy diffusivities in the horizontal/vertical directions. Normally K_H, K_V are taken to be either constant or some simple depth-dependent profile (e.g., Bryan and Lewis, 1979). Like the viscosity coefficients, K_H and K_V are typically chosen to ensure numerical stability with $K_H \approx 1 \times 10^7$ m^{2}s^{-1} and $K_V \approx 0.2 - 1.0 \times 10^{-4}$ m^{2}s^{-1}. In the real ocean, however, mesoscale eddies are known to diffuse scalars more efficiently *along* surfaces of constant potential density. The work involved in mixing tracers *across* density surfaces is order $10^6 - 10^7$ more than that required to stir *along* density surfaces. Thus, high horizontal eddy diffusivities will be unrealistic in regions of steeply sloping density surfaces. England (1993), Hirst and Cai (1994) and others have demonstrated the problems of using strictly Cartesian mixing coefficients in regions such as the Southern Ocean. Model artifacts include unrealistic water-mass blending, spurious vertical velocities and excessive poleward heat transport across the ACC.

A solution to the problem of tracer mixing in ocean models was first proposed by Redi (1982) and implemented by Cox (1987), wherein the eddy diffusivity tensor $K_{H,V}$ was oriented along density surfaces rather than in a Cartesian system. However, to ensure numerical stability, this so-called "isopycnal mixing" scheme originally required a background horizontal diffusivity, meaning that whilst it simulated increased isopycnal mixing, it also maintained some spurious horizontal diffusion. Theoretical developments in more recent studies have identified solutions to this problem. One solution is to construct a model grid that orients its surfaces along isopycnals rather than along a Cartesian coordinate system (e.g., Bleck et al., 1992). Such models have now been configured over global domains and have quite successfully captured the ocean's thermohaline circulation and climate processes (e.g., Bleck et al., 1997; Sun, 1997).

Mesoscale eddies affect tracers not only by increasing mixing rates along density surfaces, but also by inducing a larger-scale transport rather like an adiabatic advection term (e.g., Rhines, 1982; Gent and McWilliams, 1990; McDougall, 1991). Gent and McWilliams (1990) and later Gent et al. (1995) proposed a parameterization for this process wherein the large-scale density field is used to estimate the magnitude of the eddy-induced advection, (u_*, v_*), namely:

$$u_* = \frac{\partial}{\partial z}\left(K_e \frac{\partial \rho/\partial x}{\partial \rho/\partial z}\right) \qquad (4.12)$$

$$v_* = \frac{\partial}{\partial z}\left(K_e \frac{\partial \rho/\partial y}{\partial \rho/\partial z}\right) \qquad (4.13)$$

and since the eddy-induced advection field is non-divergent (Gent et al., 1995), w_* can be derived from the equation

$$\frac{\partial u_*}{\partial x} + \frac{\partial v_*}{\partial y} + \frac{\partial w_*}{\partial z} = 0$$

yielding

$$w_* = \frac{\partial}{\partial x}\left(K_e \frac{\partial \rho/\partial x}{\partial \rho/\partial z}\right) + \frac{\partial}{\partial y}\left(K_e \frac{\partial \rho/\partial y}{\partial \rho/\partial z}\right) \qquad (4.14)$$

It turns out that the Gent et al. (1995) mixing scheme results in a positive definite sink, on the global mean, of available potential energy. This means that model runs of coarse resolution can be integrated with zero background horizontal diffusion and yet remain stable, as the Gent et al. (1995) advection terms act as a viscosity or dissipative term on the model scalar properties. Successful simulations with zero K_H have been achieved with minimal numerical problems (see Hirst and McDougall, 1996; England and Hirst, 1997).

Recent estimates of vertical diffusion rates show very low values ($\approx$ 0.1 cm^{2}s^{-1}) in the upper ocean, elevated values near regions of rough bottom bathymetry (up to 10 cm^{2}s^{-1}) and much weaker values in the ocean interior over regions of smooth bottom bathymetry such as abyssal plains (Ledwell et al., 1998; Polzin et al., 1997). Models have traditionally adopted constant or simple depth-dependent profiles of K_V. Unfortunately, key ocean model parameters such as the meridional overturn and poleward heat transport are controlled to a large extent by the magnitude of K_V (e.g., Bryan, 1987). Recent efforts have been made to estimate K_V as a function of bottom bathymetry roughness and ocean depth in global ocean models (e.g., Hasumi and Suginohara, 1999). It turns out the meridional overturn and poleward heat transport in an ocean model can be vigorous with zero K_V over the ocean interior and only enhanced K_V over rough terrain (e.g., Marotzke, 1997). This gives increased confidence in the capacity of ocean models to realistically capture the large-scale ocean circulation without fully resolving smaller-scale physical processes.

In most applications of coastal and tropical ocean models, flow-dependent vertical mixing schemes are used to represent enhanced mixing in the frictional surface and bottom boundary layers. These schemes are typically dependent on the local Richardson number:

$$Ri = \frac{N^2}{(\partial \bar{u}/\partial z)^2}$$

where $N^2 = \dfrac{g}{\rho_0}\dfrac{\partial \rho}{\partial z}$ is the buoyancy frequency, and $\bar{u}$ the mean horizontal flow speed. As such, Ri quantifies the vertical stability of the water column in relation to the velocity shear. Examples of Richardson number dependent schemes are described in detail by Mellor and Yamada (1982), Pacanowski and Philander (1991) and Kantha and Clayson (1994) to name a few.

It is now known that significant vertical mixing occurs over rough bottom bathymetry as barotropic tides agitate internal wave breaking (e.g., Toole et al., 1997). Future parameterizations of K_V should take account of global tidal flow fields and bathymetry roughness in order to incorporate these effects in some way.

Another subgrid-scale oceanic process is vertical convection, wherein surface buoyancy loss (via cooling, evaporation, or sea-ice formation) leads to vertically unstable waters and overturn to depths up to 1000 m or so. Examples include Mode Waters (McCartney, 1977), 18°C water in the North Atlantic (Worthington, 1976), and Weddell Sea Bottom Water. Since convection is intimately tied to water-mass formation, representing it in ocean models is critical. Because the horizontal scale of convection is order kilometres, no greater than its vertical scale, non-hydrostatic processes are involved. Present parameterisations of vertical convection simply mix $T - S$ and other scalars completely over the unstable portion of the water column, removing the vertically unstable layer outside any calculation of vertical motion. Model simulations of nonhydrostatic convection by Send and Marshall (1997) indicate that to first order this vertical mixing approach approximates the integral effects of vertical convection on the simulated $T-S$ fields. The problem remains, however, that the horizontal extent of convection in coarse resolution models is necessarily at least the dimension of a model grid box, which can be about 100–400 km.

4.3.4 Boundary Conditions and Surface Forcing

Ocean models have to be given explicit boundary conditions for motion and temperature-salinity (see also Fig. 4.5). These include boundary conditions at the air-sea interface, bottom boundary conditions (for momentum), as well as lateral boundary conditions for regional models. Side boundary conditions at land masses are the most simple, including no-slip non-normal flow, and zero fluxes of heat and salt. In coarse models, bottom boundary layers normally adopt some simple relationship to approximate the effects of frictional drag on the deep ocean flow (see, for example, Toggweiler et al., 1989).

Surface forcing is a crucial aspect of boundary conditions in ocean models. Firstly for motion, modellers may choose between a "rigid-lid" approximation and a free surface condition. Under the rigid-lid approximation, the vertical velocity w is zero at the sea surface, thereby excluding surface gravity waves and allowing a longer model time step (Bryan, 1969). This means further that the total volume of the ocean remains constant and that freshwater fluxes across the air-sea interface must be represented as effective salt fluxes. Other surface pressure gradients, such as those due to large scale geostrophic flow, are allowed under a "rigid-lid" approximation, but not predicted directly. The free surface condition, on the other hand, carries sealevel height or pressure as a prognostic variable, thereby eliminating the need to predict the barotropic velocity field [which becomes costly in a domain of high resulution (Dukowicz and Smith, 1994)]. Techniques have emerged to handle a free surface condition without significantly shortening the model time step; either by using many small steps in time for solving the free-surface during each single time-step of the full 3D model (Killworth et al., 1991), or by solving the free-surface equations using an implicit method (Dukowicz and Smith, 1994). Benefits of adopting these approaches include improved computational efficiency in high resolution domains, an ability to model ocean flow over unsmoothed topography, and a natural prognostic variable for assimilation of satellite height data into ocean models (see, e.g., Stammer et al., 1996).

Surface forcing fields are required in ocean models for momentum, temperature and salinity [for an excellent review on these topics, see Barnier (1998)]. Surface pressure and barotropic motion will adjust freely to the model simulated $T-S$ and 3D motion, so direct surface forcing fields are not required.

Tidal forcing is also required when sub-diurnal scales of motion are important, such as for coastal ocean models or for flow in harbours and bays. Momentum input into the oceans is via mechanical wind forcing, normally expressed as a wind stress vector τ, where

$$\tau = \rho_a c_D |U_{10} - U_W| \mathbf{u}_{10} \tag{4.15}$$

with ρ_a the density of air, c_D a turbulent exchange drag coefficient, U_{10} the wind speed at anenometer height, 10 m above the ocean surface, U_W the ocean current speed at the sea surface, and $\mathbf{u}_{10}$ the wind velocity at anenometer height. This wind-stress is then converted into a forcing term in the momentum equations (4.1)–(4.2) via the expression

$$F_{u,v} = \frac{\partial}{\partial z}(\tau/\rho_0) \quad \text{at} \quad z = 0 \tag{4.16}$$

Direct observations of wind stress over the ocean are relatively sparse, though a number of long-term global climatologies exist (e.g., Hellerman and Rosenstein, 1983), as well as others derived from more recent remote sensing technologies (e.g., Bentamy et al., 1997). Another technique for model wind forcing is to adopt output from numerical weather prediction models (NWPs); normally these products are derived from a combination of observations and forecasts via data assimilation (e.g., Kalnay et al., 1996; Gibson et al., 1997). The advantage of NWP products is that they have global high density coverage, use available observations, and are dynamically consistent. They are, in addition, provided in real time which facilitates ocean hindcasting.

Surface forcing conditions for temperature (T)–salinity (S) in ocean models can be formulated in a number of ways. The equations are [refer to (4.5) and (4.6)]:

$$F_T = \frac{Q_{\text{net}}}{\Delta z_1 \rho_0 c_p} \tag{4.17}$$

$$F_S = \frac{S_0}{\Delta z_1}(E - P - R) \tag{4.18}$$

where Q_{net} is the net heat flux into the surface layer (W m^{-2}), Δz_1 is the upper model level thickness, ρ_0 density of seawater, c_p the specific heat of seawater, S_0 mean ocean salinity, E evaporation rate, P precipitation rate, and R river run-off rate (E, P and R are all in m s^{-1}). The value S_0 is required to convert the net freshwater flux into an equivalent salt flux in rigid-lid models [see Barnier (1998) for further details]. So, in formulating heat and salt fluxes, ocean modellers require some knowledge of Q_{net}, E, P and R. Q_{net} is comprised of several components;

$$Q_{\text{net}} = Q_{\text{SW}} - Q_{\text{LW}} - Q_{\text{LA}} - Q_{\text{SENS}} - Q_{\text{PEN}} \tag{4.19}$$

where Q_{SW} is the net shortwave radiation entering the ocean, Q_{LW} is the net longwave radiation emitted at the air-sea interface, Q_{LA} is the latent heat flux, Q_{SENS} is the sensible heat flux, and Q_{PEN} the penetrative heat flux from the base of model level 1 into model level 2.

The Q_{PEN} term can generally be neglected except when shallow surface layers are adopted (e.g., Godfrey and Schiller, 1997). Bulk formulae can be used to estimate the heat flux components of (4.19) (e.g., Barnier, 1998), in particular, Q_{SW} depends on latitude, time of year, cloud cover and surface albedo; Q_{LW} depends primarily on ocean surface temperature, Q_{LA} and Q_{SENS} depend on surface wind speed, humidity (for Q_{LA}), and air-sea temperature difference. Satellite and *in situ* observations of these variables can be used to construct bulk estimates of heat and freshwater fluxes across the air-sea interface.

Unfortunately many of these quantities are only sparsely observed over the ocean, and in addition they are characterised by high-frequency variability, making it extremely difficult to construct long-term climatologies. Also, variables such as precipitation and evaporation are not easily quantified using satellite technologies. Nevertheless, a number of global heat and freshwater flux climatologies exist (e.g., Esbenson and Kushnir, 1981; Josey et al., 1999; Baumgartner and Reichel, 1975). Unfortunately, direct forcing with these fluxes can lead to significant model errors (e.g., Moore and Reason, 1993), as small errors in fluxes can accumulate into large errors in model $T - S$ over a sufficient period of integration time. Because of this, large-scale ocean modellers have commonly adopted grossly simplified formulations of thermohaline forcing, for example, restoration to observed surface T and S:

$$F_T = \gamma_T \left(T_{\mathrm{obs}} - T_{\mathrm{model}}\right) \tag{4.20}$$
$$F_S = \gamma_S \left(S_{\mathrm{obs}} - S_{\mathrm{model}}\right) \tag{4.21}$$

where γ_T, γ_S are time-constants determining how long heat/salinity anomalies persist before they are damped by air-sea forcing, $(T_{\mathrm{obs}}, S_{\mathrm{obs}})$ are the observed climatological $T - S$, and $(T_{\mathrm{model}}, S_{\mathrm{model}})$ the model surface $T - S$. Haney (1971) justified this style of formulation for temperature so long as T_{obs} is replaced by T_{EFF}, where T_{EFF} is the temperature the ocean would obtain if there were no heat transported by it. Various techniques exist to estimate the distribution of T_{EFF} (e.g., Rahmstorf and Willebrand, 1995; Cai and Godfrey, 1995). There is, however, no such justification for salinity, as salinity at the sea-surface has no significant role in controlling air-sea freshwater fluxes. This lead many modellers to use so-called "mixed boundary conditions", (e.g., Bryan, 1987; Weaver et al., 1991), wherein a model run is integrated with restoring salinity conditions, diagnosed for the effective salt flux distribution, and re-run with this flux condition on S; thereby allowing the model to exhibit variability in S independent of the surface forcing fields. More recently, the Haney (1971) heat flux formulation has been extended to include simple parameterizations for the atmospheric dispersion of ocean heat anomalies (Rahmstorf and Willebrand, 1995; Power and Kleeman, 1994), to allow, for example, the simulation of oceanic variability otherwise suppressed by a boundary condition of the form in (4.20).

Data sets for surface thermohaline forcing include flux climatologies (refer to citations above and Woodruff et al., 1987, DaSilva et al., 1994), re-analyses of NWP simulations (e.g., Barnier et al., 1995; Béranger et al., 1999; Garnier et al., 2000) and satellite derived data products (e.g., Darnell et al., 1996). However, in practical terms, any direct flux forcing technique can lead to substantial errors in simulated $T - S$ because of possible model bias and errors in the flux fields themselves. For example, an error in heat flux as small as 1 W m^{-2} (which is at least an order of magnitude less than the typical error associated with heat flux climatologies) would result in an error in upper level T of 7.5°C after 50 years of run time (assuming a surface level of thickness 50 m). To address this problem, it is becoming common practice for modellers to adopt surface thermohaline forcing of the form

$$F_T = \frac{Q_\text{net}}{\Delta z_1 \rho_0 c_p} + Q_\text{CORR}$$

and

$$F_S = \frac{S_0}{\Delta z_1}(E - P - R) + S_\text{CORR} \qquad (4.22)$$

with

$$Q_\text{CORR} = \gamma_T (T_\text{obs} - T_\text{model})$$

$$S_\text{CORR} = \gamma_S (S_\text{obs} - S_\text{model})$$

corresponding to the Newtonian restoring terms described earlier in (4.20) – (4.21).

Typically, these heat flux and salt flux correction terms adopt time-scale values for γ_T, γ_S that give weak restoring towards observed $T - S$, thereby enabling thermohaline variability (see also Wood et al., 1999).

Lateral boundary conditions in regional ocean models represent a major area of uncertainty in ocean modelling. In order to resolve oceanic features with scales of the order of kilometres to 10's of kilometres, high horizontal resolution is required. However, limited computational resources mean that global coverage at such resolution is not feasible. Therefore either regional models have to be nested inside global models, where the global model provides boundary conditions for the regional model, or open boundary conditions must be incorporated into the regional model. Open boundary conditions generally assume limited physics at the boundary. There are several detailed studies on open boundary conditions (e.g., Orlanski, 1976; Chapman, 1985; Palma and Matano, 1997), where most attempt to represent the advection or propogation of modelled disturbances into or out of the domain. These conditions are referred to as passive conditions since they are designed to have minimal impact on the model solution. Often sponge layers (Chapman, 1985), which are designed to slow model disturbances down near open boundaries, are employed in order to further reduce any unwanted reflection. For cases where a typical state at the boundary is known or assumed, e.g. tidal flows, non-passive boundary conditions are often used, where a relaxation term is added to the passive conditions mentioned above (e.g., Flather, 1976; Blumberg and Kantha, 1985). Many applications relax temperature and salinity to their climatological values (e.g., Stevens, 1991; Gibbs et al., 1997; Oke and Middleton, 2001), with velocities being relaxed to their geostrophic boundary values that match the $T-S$ climatology at the boundary. Barotropic boundary flows may be estimated using a Sverdrup relationship. The uncertainty associated with the choice of open boundary condition means that a substantial amount of testing and model validation should be performed before confidence is shown in a regional simulation.

The above discussions of ocean model forcing are in the context of "ocean-only" model integrations. However, even when coupling to an atmospheric GCM, some integration of an ocean-only model is required prior to coupling. In that case, a number of spin-up strategies are possible (as discussed by Moore and Reason, 1993). This is detailed further in Chap. 2.

4.4 Some Commonly Used Ocean Models

There are a vast number of ocean models used around the world today. Some of these models are designed for a very specific use, limited to a certain geographic region and only resolving the dominant components of the equations of motion. Other more generalised models have been configured to be operational over a range of space and time-scales with a variable geographic domain. Such models are becoming widely used with literally hundreds of applications across many institutions. A limited set of such ocean models are described here.

4.4.1 The GFDL Modular Ocean Model

The GFDL Modular Ocean Model (MOM) is the most widely used ocean model in large-scale coupled climate simulations. It is a finite difference realisation of the primitive equations governing ocean circulation. These equations are formulated in spherical coordinates. An identifying feature of the GFDL model is that it is configured with its vertical coordinate as level geopotential surfaces (i.e., so-called z-level). The MOM grid system is a rectangular Arakawa staggered B grid. Further model details can be found in Pacanowski (1995) and Bryan (1969). Applications range from global at coarse and eddy-resolving resolution, down to regional and idealised process-oriented models. A full bibliography of MOM-related papers appears in an appendix in Pacanowski (1995).

As an example, Figure 4.8 shows the upper ocean circulation in the South Pacific Ocean from such a model (from England and Garçon, 1994). This model has resolution and geometry typical of state-of-the-art coupled ocean-atmosphere models used to study climate variability and anthropogenic climate change. The resolution is coarse (1.8 degrees in longitude and 1.6 degrees in latitude) so the eddy field is not resolved explicitly. There are 33 unequally spaced vertical levels with bottom topography and global continental outlines as realistic as possible for the given grid-box resolution. The model is driven by Hellerman and Rosenstein (1983) winds. The model temperature and salinity are relaxed to the climatological annual mean fields of Levitus (1982) at the surface. The figure shows vectors of ocean currents simulated at 70 m depth, with topography shallower than 3000 m depth shaded. Apparent is the bathymetric steering of the Antarctic Circumpolar Current (ACC), even in the upper ocean well above the main topographic features in the region.

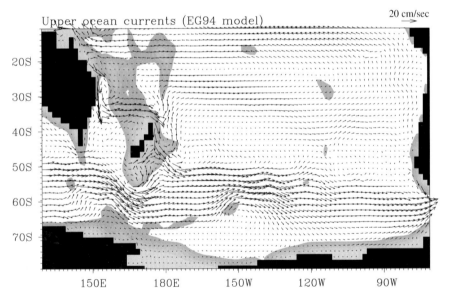

Fig. 4.8. Upper ocean circulation in the South Pacific Ocean, from England and Garçon (1994). Topography features shallower than 3000 m are shaded.

The tendency for flow in the ACC to be along constant depth contours is not difficult to understand. The reason for this is that the flow is mostly barotropic so that the potential vorticity

$$Q = \frac{\zeta + f}{h + \eta}$$

will be conserved, where ζ is the relative vorticity, f the Coriolis parameter or planetary vorticity, h the ocean depth and η is the relative sealevel elevation. At the latitude of the ACC, apart from mesoscale eddies, the large-scale flow is such that its relative vorticity $\zeta \ll f$. In addition, sealevel variations are of the order of $\simeq 1-2$ metres compared with an ocean depth of $h \simeq 2000-4000$ metres. So the conservation of potential vorticity implies that $f/h \simeq$ constant. With the ACC flow being primarily west to east we can take f as approximately constant, so a fluid column tends to have the same value of depth h and is thus steered along isobaths.

4.4.2 The Princeton Ocean Model

The Princeton Ocean Model (POM) adopts curvilinear orthogonal coordinates and a vertical sigma-coordinate to facilitate simulations of coastal zone flows (Mellor, 1998). The horizontal time-differencing is explicit whereas the vertical differencing is implicit, allowing a fine vertical resolution in the surface and bottom layers. This is important in coastal ocean models as the near-shore ocean can be a region of substantial surface and bottom boundary gradients. POM can be integrated with a free-surface and using split time-stepping (as in MOM). Another positive feature of the POM is the embedded turbulence sub-model (Mellor and Yamada, 1982), which is designed to provide realistic, flow-dependent vertical boundary layer mixing. This sub-model, along with the terrain-following sigma-coordinates, allows the modeller to resolve bottom boundary layer flows that are often associated with coastal upwelling, a feature that z-level models cannot adequately represent. Applications of the POM range from coastal (Mellor, 1986) through to regional (Middleton and Cirano, 1999) and basin-scale studies (Ezer and Mellor, 1997).

An example of the POM configured with idealised continental shelf and slope topography, and forced with constant (0.1 Pa) upwelling favourable winds is shown in Fig. 4.9 (taken from Oke and Middleton, 1998). This figure shows the evolution of a cross-shelf slice of temperature, alongshelf velocity and cross-shelf streamfunction over a 15-day period. This sequence shows the initial response to the wind in the streamfunction field, where an upwelling circulation is generated. By Day 5 a coastal jet has formed in the direction of the wind and isotherms have been upwelled to the surface. By Day 10 the upwelling is concentrated in the bottom boundary layer, as indicated by the concentration of the cross-shelf streamlines over the topography, and the upwelling front has been advected off-shore. This type of idealised configuration modelling has given oceanographers great insight into the overall dynamics of coastal upwelling (e.g., Allen et al., 1995), as it may enable the upwelling dynamics to be isolated from more complicated continental shelf processes, such as the effects of topographic variations.

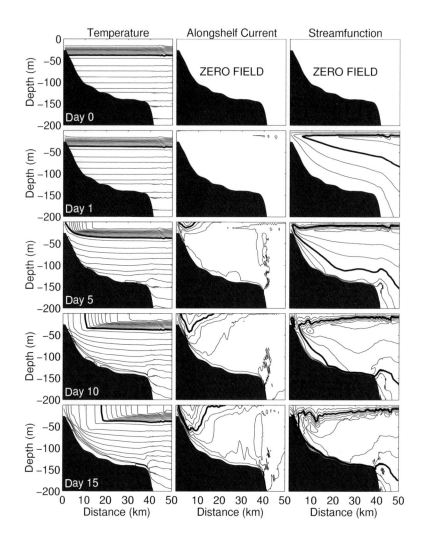

Fig. 4.9. A sequence of cross-shelf slices of temperature, along-shelf velocity and cross-shelf streamfunction (left-right; contour intervals = $0.5°C$, $0.1\,\mathrm{m\,s^{-1}}$, $0.1\,\mathrm{m^2 s^{-1}}$ respectively) showing their evolution in response to constant ($0.1\,\mathrm{Pa}$) upwelling favourable winds (Oke and Middleton, 1998).

4.4.3 The Miami Isopycnic Coordinate Model (MICOM)

MICOM is a primitive equation "isopycnic" ocean model that uses equations that have a coordinate of potential density in the vertical direction (Fig. 4.7c) instead of the traditional vertical coordinate of depth. That is, whereas z-level and sigma coordinate models predict the density at a fixed depth, MICOM predicts the depths at which certain density values are encountered. Thus, the traditional roles of water density and height as dependent and independent variables are reversed. The surface mixed layer (with different isopycnal values) sits over the subsurface isopycnic domain. The horizontal coordinate system is the Arakawa C grid. The model accommodates a user specified, horizontal geographic zone. Further model details can be found in Bleck et al. (1992). Recently, MICOM has been configured for global ocean simulations (Bleck et al., 1997). A map of the simulated surface ocean currents from a 2-degree by $2\cos(\phi)$-degree global MICOM simulation (Sun, 1997) is shown in Fig. 4.10.

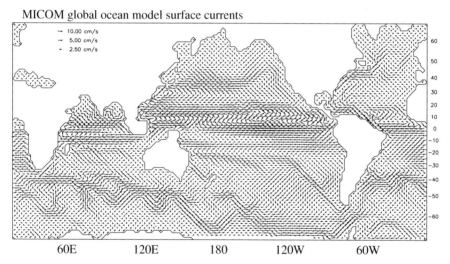

Fig. 4.10. Simulated surface ocean currents from a 2-degree by $2\cos(\phi)$-degree global MICOM simulation (after Sun, 1997).

4.4.4 The DieCast Model

The "DieCast" model is a z-level model like the GFDL MOM, only with different numerical and horizontal grid schemes. The DieCast model combines aspects of the Arakawa A and C grids. DieCast uses fourth order interpolations to transfer data between the A and C grid locations in order to combine the best features of the two grids. This procedure eliminates the A-grid "null space" problems and reduces or eliminates the numerical dispersion caused by Coriolis term integration on the C grid. The modified A grid model includes a fourth order approximation for the baroclinic pressure gradient associated with the important quasi-geostrophic thermal wind. Further discussion of the DieCast grid system is given by Dietrich (1997). The model geometry is in most ways identical to the GFDL MOM. The key difference between DieCast and the GFDL MOM is that DieCast defaults to higher order numerical schemes and incorporates a merged Arakawa A and C grid system, as described above.

4.5 Ocean Model Applications

4.5.1 A Global Coarse Resolution Model

Perhaps the most widely applied global coarse resolution model over the past two decades has been that configured initially by Bryan and Lewis (1979) based on the GFDL Bryan-Cox primitive equation numerical model (Bryan, 1969; Cox, 1984). The Bryan and Lewis (1979) simulation has been used in a wide-ranging series of coupled climate models (e.g., Manabe and Stouffer, 1988, 1993, 1996), ocean-only models used to understand water-mass formation processes (e.g., England, 1992; England et al., 1993; Toggweiler and Samuels, 1992, 1995) and in models of the oceanic carbon cycle (Sarmiento et al., 1998). It has also undergone extensive assessment using geochemical tracers such as radiocarbon (Toggweiler et al., 1989) and chlorofluorocarbons (England et al., 1994; England and Hirst, 1997).

A particular example of the utility of the coarse resolution model of Bryan and Lewis (1979) is exemplified in the England et al. (1993) assessment of the formation mechanism for mode and intermediate waters in the Southern Ocean. Realistic representation of the low-salinity tongue of Antarctic Intermediate Water (AAIW) was achieved by England (1993). He found that the AAIW tongue can quite successfully be simulated provided appropriate attention is taken to observed wintertime salinities near Antarctica, and so long as an isopycnal mixing scheme is incorporated into the model. A diagram showing the observed and modelled salinity is included in Fig. 4.11.

England et al. (1993) found that AAIW is not generated by direct subduction of surface water near the polar front as had been the traditional belief (e.g., Sverdrup et al., 1942). Instead, the renewal process is concentrated in certain locations, particularly in the southeast Pacific Ocean off southern Chile (see Fig. 4.12). The outflow of the East Australian Current progressively cools (by heat loss to the atmosphere and assimilation of polar water, carried north by the surface Ekman drift) and freshens (due to the northward Ekman transport of low salinity Subantarctic Surface Waters) during its slow movement across the South Pacific towards the coast of Chile.

4.5 Ocean Model Applications   151

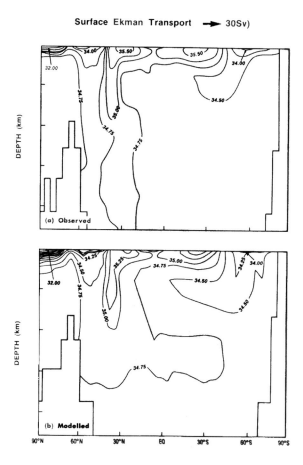

Fig. 4.11. Observed and modelled salinity in the depth-latitude zonal mean. Observations from Levitus (1982), model simulation is that of England et al. (1993).

This results in progressively cooler, denser, and fresher surface water, leading to deeper convective mixed layers towards the east. Off Chile, advection of warmer subsurface water from the north (at 100–900 m depth) enables more convective overturn, resulting in very deep mixed layers from which AAIW is fed into the South Pacific (via the subtropical gyre) and also into the Malvinas Current (via the Drake Passage). This formation mechanism for AAIW was first proposed by McCartney (1977) based on observations; although a detailed dynamical framework was not clear until the England et al. (1993) study.

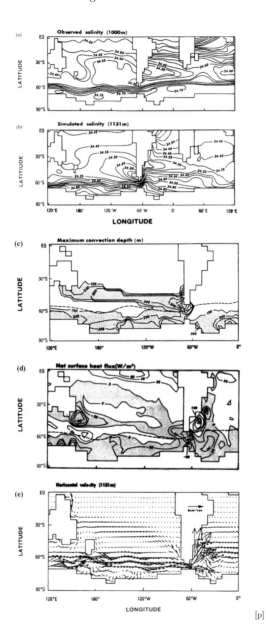

Fig. 4.12. (a) Observed annual-mean salinity at 1000 m depth redrafted from Levitus (1982); (b)–(e) Simulated properties in the England et al. (1993) ocean model: (b) salinity near 1000 m depth; (c) net surface heat flux (W m^{-2}) into the ocean; (d) maximum depth (m) of convective overturn; and (e) horizontal velocity near 1000 m depth.

4.5.2 Global Eddy-permitting Simulations

The first global domain eddy-permitting ocean model was integrated by Semtner and Chervin (1992). Presently, several groups are moving in this direction, although the huge computational cost of an eddy-permitting global model limits applications to multi-decadal runs. Two prominent examples of global eddy-permitting models include the Ocean Circulation and Climate Advanced Modelling Project (OCCAM) (de Cuevas et al., 1998) and the Parallel Ocean Climate Model (Stammer et al., 1996; Tokmakian, 1996).

The OCCAM project has developed two high resolution (1/4 and 1/8 degree) models of the World Ocean – including the Arctic Ocean and marginal seas such as the Mediterranean. Vertical resolution has 36 depth levels, ranging from 20 m near the surface, down to 255 m at 5500 m depth. OCCAM is based on the GFDL MOM version of the Bryan-Cox-Semtner ocean model but includes a free surface and improved advection schemes. A regular longitude-latitude grid is used for the Pacific, Indian and South Atlantic Oceans. A rotated grid is used for the Arctic and North Atlantic Oceans to avoid the convergence of meridians near the poles. The model was started from the Levitus annual mean $T - S$ fields. The surface forcing uses ECMWF monthly mean winds and a relaxation to the Levitus seasonal surface $T - S$ climatology. This initial model run was integrated for 12 model years. The OCCAM model has been run with high resolution forcing using the six-hourly ECMWF re-analysis data from 1992 onwards. The OCCAM simulation in the region of the Agulhas Retroflection is shown in Fig. 4.13. Agulhas eddies are spawned south of Africa, transporting heat and salt from the Indian Ocean into the Atlantic, contributing to the global transport of properties between ocean basins. This has been linked with the global thermohaline transport of North Atlantic Deep Water (NADW) (Gordon, 1986). Coarse resolution ocean models cannot explicitly resolve oceanic eddies, and so they do not include the heat and salt fluxes associated with Agulhas rings.

The Parallel Ocean Climate Model (POCM) has nominal lateral resolution of $1/4°$ (Stammer et al., 1996; Tokmakian, 1996). The POCM domain is nearly global running from 75°S to 65°N. The actual grid is a Mercator grid of size 0.4° in longitude yielding a square grid everywhere between the Equator and 75° latitude (Stammer et al., 1996). The resulting average grid size is $1/4°$ in latitude. Model bathymetry was derived by a grid cell average of actual ocean depths over a resolution of $1/12°$. Unlike most coarse models, the fine resolution model includes a free surface (after Killworth et al., 1991) that treats the sea level pressure as a prognostic variable.

The POCM is integrated for the period 1987 through to June 1998 using ECMWF derived daily wind stress fields, climatological monthly mean ECMWF sea surface heat fluxes produced by Barnier et al. (1995), and some additional T, S surface restoring terms. The surface restoring of T and S adopts the Levitus et al. (1994) monthly climatology with a 30-day relaxation time scale. Subsurface to 2000 m depth $T - S$ are also restored towards

154   4. Ocean Modelling and Prediction

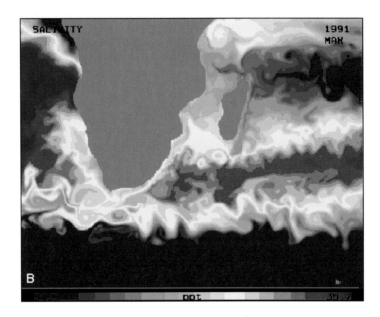

Fig. 4.13. OCCAM simulation of sea surface salinity in the region of the Agulhas Current and leakage into the Indian Ocean (from Semtner, 1995).

Levitus (1982) along artificial model boundaries north of 58°N and south of 68°S to approximate the exchange of water properties with those regions not included in the model domain.

An example of the POCM simulation is shown for the South Atlantic upper ocean in Fig. 4.14. Only velocity vectors at every third grid box are drawn, with no spatial averaging, otherwise the current vectors are difficult to visualise. The Brazil-Falkland confluence is close to the location described from hydrographic and satellite observations. Also, the South Atlantic Current is simulated to the north of the ACC (and separate from it), unlike coarser models which tend to simulate a broad ACC and no distinct SAC. At the tropics, the POCM resolves some of the meridional structure in zonal currents observed (as discussed in Stramma and England, 1999). Eddy-resolving simulations capture much more spatial structure than their coarse resolution counterparts, including more realistic western boundary currents, frontal dynamics, and internal oceanic variability.

4.5.3 Regional Simulations in the North Atlantic Ocean

A large variety of simulations of the North Atlantic Ocean have been carried out within the World Ocean Circulation Experiment (WOCE) Community

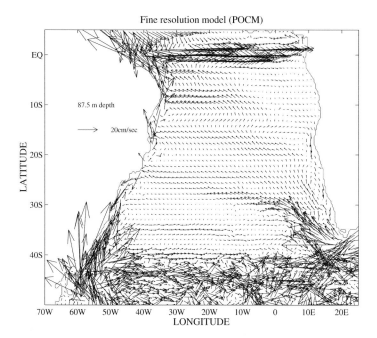

Fig. 4.14. POCM simulation for the South Atlantic upper ocean near 100 m depth (from Stramma and England, 1999). Only velocity vectors at every third grid box are drawn.

Modeling Effort (CME) and more recently by other modelling groups. An overview of some of the CME North Atlantic models is given by Böning et al. (1996), with particular reference to the sensitivity of deep-water formation and meridional overturning to a number of model parameters. It turns out that surface thermohaline forcing (England, 1993), model resolution (Böning et al., 1996), mixing parameterisation (Böning et al., 1995), as well as the resolution of certain subsurface topographic features (Roberts and Wood, 1997) all control model NADW formation rates.

More recently, the Dynamics of North Atlantic Models (DYNAMO) study intercompared a number of simulations in the region using models with different vertical coordinate scheme; namely a model with horizontal z-levels, another with isopycnal layers and thirdly one that used a dimensionless sigma-coordinate (as per Fig. 4.7). The goal of the DYNAMO project was to develop an improved simulation of the circulation in the North Atlantic Ocean, including its variability on synoptic and seasonal time-scales. The study included a systematic assessment of the ability of eddy-resolving models with different vertical coordinates to reproduce the essential features of the hydrographic structure and velocity field between 20°S and 70°N.

Figure 4.15 shows the poleward heat transport in all three ocean models compared with observations. The northward heat transport in the z-level simulation is markedly lower than the observations and the sigma and isopycnal cases south of 40°N. This is due to mixing in the outflow region and spurious upwelling of NADW at midlatitudes (DYNAMO, 1997). On the other hand, the isopycnal run simulates a more realistic heat transport pattern, although its low eddy kinetic energy is compensated by an enhanced deep NADW outflow. Thus, examination of the integral measure of poleward heat transport aliases more subtle dynamical discrepencies in that model. The sigma model also captures a realistic heat transport pattern, although the formation site for NADW is not concentrated in the subpolar region (figure not shown). Overall, the intercomparison of models in the DYNAMO project underscores the relative merits and shortcomings of different vertical coordinates in the context of basin-scale modelling.

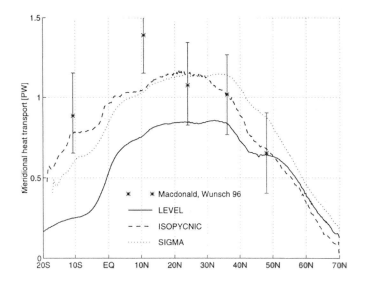

Fig. 4.15. Poleward heat transport in the DYNAMO experiments with z-level, isopycnal and sigma coordinate vertical schemes (from DYNAMO, 1997). Observations are included from McDonald and Wunsch (1996).

4.5.4 ENSO Modelling

Modelling the upper ocean dynamics in tropical waters is crucial for an improved understanding of the El-Niño/ Southern Oscillation (ENSO). ENSO ocean models are ultimately coupled to atmospheric GCMs in order to predict the climate impact of changes in tropical SST. Ocean models used in the context of ENSO simulations have ranged in complexity from shallow water models representing tropical upper ocean dynamics (e.g., Cane and Sarachik, 1977; McCreary, 1976) and modified shallow water models (e.g., Schopf and Cane, 1983), through to three-dimensional general circulation models (e.g., Philander and Pacanowski, 1980). A similar hierarchy of atmospheric models also exists, from those employing simple damped shallow-water dynamics (e.g., Gill, 1980) through to full atmospheric GCMs. In turn, the coupled ocean-atmosphere models used can range from simple models and intermediate coupled models through to 3D coupled GCMs (for a review of the former see Neelin et al., 1998).

Ocean GCMs used to study ENSO dynamics are normally constructed with enhanced horizontal resolution in the tropics, and enhanced vertical resolution in the upper 300–400 metres. This is done to optimise model performance in the equatorial zone without unduly increasing computational costs. Upper ocean vertical mixing schemes are also generally more sophisticated than those used in standard global GCMs. For a review of ocean and coupled GCMs used to study ENSO dynamics, the reader is referred to Delecluse et al. (1998).

4.5.5 A Regional Model of the Southern Ocean

The Fine Resolution Antarctic Model (FRAM) is a primitive equation numerical model of the Southern Ocean between latitudes 24°S and 79°S. The model was initialised with $T = -2°C$ and $S = 36.69$ psu and relaxed to Levitus annual mean $T - S$ over 6 years. Surface wind forcing is that of Hellerman and Rosenstein (1983). Various strategies for gradual imposition of these fluxes is adopted to minimise numerical instability in the model spin-up phase (for details see de Cuevas, 1992, 1993). Model mixing schemes include a mixture of harmonic and biharmonic terms and a quadratic bottom friction stress. The total model run time is 16 years. The open boundary condition used in FRAM is a combination of a Sverdrup balance in the barotropic mode and a simple quasi-geostrophic balance and Orlanski radiation in the baroclinic terms (see Stevens, 1990, 1991 for further details).

In the final seasonal cycle phase of the model run, the Antarctic Circumpolar Current transport through Drake Passage oscillates between 195 and 200 Sv ($1\,\text{Sv} = 1 \times 10^6 \text{ m}^3\text{s}^{-1}$). This overly strong transport is a problem chronic to global-scale high resolution models. The main regions of eddy formation in the FRAM are in the Agulhas Current and along the path of the Circumpolar Current. The FRAM streamfunction simulated after the initial spin-up phase of the model run is illustrated in Fig. 4.16. Clearly apparent is the resolution of the ACC and its associated frontal dynamics and meanders. In addition, a substantial amount of eddy kinetic energy (figure not shown) is simulated in regions where the ACC encounters topographic features, such as the Campbell and Kerguelen Plateaus.

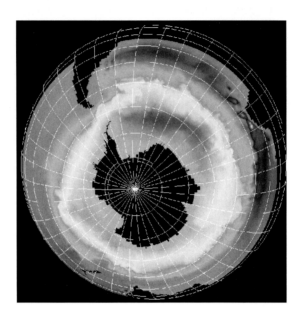

Fig. 4.16. Simulated barotropic streamfunction in the Fine Resolution Antarctic Model (FRAM) after the initial spin-up phase (from Webb et al., 1991).

4.5.6 A Coastal Ocean Model off Eastern Australia

An example of the POM configured for the EAC region is the NSW shelf model, which was developed at the UNSW Oceanography Laboratory (e.g., Gibbs et al., 1997; Oke and Middleton, 2001). The model utilises a curvilinear orthogonal grid with horizontal resolution of 5–20 km and extends along the entire coast of NSW. Observed surface and 250 m temperature fields are combined with Levitus climatology to produce the initial density field which is then used to determine the initial velocity field, via dynamic height calculations. With a constant inflow at the northern boundary and open boundary conditions which are relaxed to climatology at the east and south, the model has proved to be very useful for investigating the role that the EAC plays in nutrient enrichment of NSW coastal waters (e.g., Gibbs et al., 1997; Oke and Middleton, 2001). An example of the surface temperature and velocity fields modelled by the NSW shelf model for a period during January 1997 (from Oke and Middleton, 2001) is shown in Fig. 4.17. The modelled fields, which were qualitatively similar to observed temperature fields at the sea-surface and at a depth of 250 m, indicate that over a period of about a week, the EAC intensified over the continental shelf and extended southwards along the coast to the south of Sydney. A localised upwelling occurred immediately to the south of Port Stephens (indicated by the cold water mass near the coast). Through an analysis of the model fields it was hypothesised that the acceleration of the EAC over the narrow continental shelf near Smoky Cape resulted in uplifting of colder water which ultimately reached the surface in the vicinity of Port Stephens. The January 1997 period corresponded to a time when an algal bloom formed off Port Stephens. As a result of the modelling study it was suggested that topographically induced EAC-driven upwelling plays an important role in the nutrient enrichment of New South Wales coastal waters.

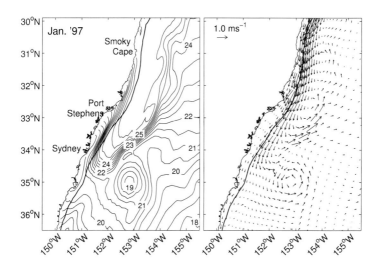

Fig. 4.17. Simulated surface temperature (left) and velocity (right) from the NSW shelf model, a configuration of the POM, showing fields for January 1997 (Oke and Middleton, 2001).

4.5.7 A Coastal Model of a River Plume

An example of a version of the POM utilised for a process-oriented study of the response of a river plume during upwelling favourable winds (Fong and Geyer, 2001) is outlined below. The model is configured for the northern hemisphere ($f=10^{-4}$ s^{-1}) with a rectangular basin and idealized moderately steep nearshore bathymetry, that is typical of many narrow continental shelves. Freshwater is discharged via a short river/estuary system at the coast in the upper left hand corner of the 95 km × 450 km domain (Fig. 4.18). In the region of interest the resolution is less than 1 metre in the vertical, 1.5–3 km in the cross-shore direction and 3–6 km in the alongshore direction. A modified Mellor-Yamada turbulence sub-model is utilized (Mellor and Yamada, 1982; NunuzVaz and Simpson, 1994) and a recursive Smolarkiewicz advection scheme (Smolarkiewicz and Grabowski, 1990) is used to advect salt and temperature. A steady inflow of $0.1\,\mathrm{m\,s^{-1}}$ is imposed at the northern boundary to model the ambient continental shelf currents and a combination of clamped and radiative boundary conditions are employed at the offshore and southern boundaries. The salinity was initially 32 psu throughout the domain and the river plume is simulated by discharging freshwater (0 psu) from a point source at the coast at a constant rate of $1500\,\mathrm{m^3 s^{-1}}$. The plume is established in the absence of wind over a 1 month period (Fig. 4.18a), after which time constant 0.1 Pa upwelling favourable winds are applied over a 3-day period. The simulations demonstrate that, in response to upwelling favourable winds, the surface-trapped river plume widens and thins, and is advected offshore by the cross-shore Ekman transport (Fig. 4.18b–d). The thinned plume is susceptible to significant mixing due to the vertically sheared horizontal currents. Fong and Geyer (2001) utilise this configuration to investigate how the advective processes change the shape of the plume and how these advective motions alter the mixing of the plume with the ambient coastal waters.

4.6 Exploiting Ocean Observations

4.6.1 Model Assessment

The assessment of ocean models involves the comparison of a set of model variables or diagnostics with observations. Model assessment techniques range from simple qualitative comparisons through to statistical significance tests. It is convenient to look at large-scale and coastal models separately, as different quantities and time-scales are involved in such assessments.

Large-scale models. For large-scale models – those of an ocean basin-scale or greater – much model assessment is focused on long-term climatological hydrographic properties, particularly $T - S$, as well as integrated transport quantities like the net flow in the ACC. This is particularly the case for coarse resolution models, such as those incorporated into climate simulations. In such non-eddy-resolving models, ocean current speeds are slow and have none of the high-frequency variability associated with eddies or tides in the real ocean. Thus, direct comparison with observed current meter records is inappropriate. Instead, integrated transport measures, both in the horizontal and meridional plane, provide a more meaningful assessment of the model.

Water-mass formation is also most often assessed indirectly in coarse models; that is, by analysing model and observed $T - S$ rather than the model

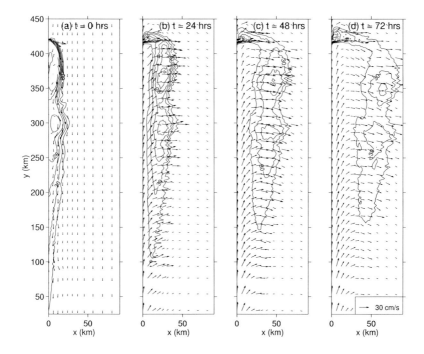

Fig. 4.18. Surface velocity and salinity for the evolution of a river plume during upwelling favourable wind conditions. (**a**) after 1 month of buoyancy forcing with no wind; (**b**)–(**d**) after 24, 48 and 72 hours of 0.1 Pa upwelling favourable winds. The scale for velocity is shown in panel (d) and the contour intervals are 1 psu for salinity (adapted from Fong and Geyer, 2001).

subduction/convection processes. This is partly because the processes that are linked with water-mass formation, such as convection, mixing, and deep currents, are extremely difficult to measure directly. In addition, such processes are subgrid-scale to a coarse resolution model, so assessment of them is best achieved by analysing their parameterised effects on model $T - S$ (e.g., England, 1993; Hirst and Cai, 1994; Hirst and McDougall, 1996).

Because temperature and salinity are prognostic variables in global ocean models and intrinsic in any definition of a water-mass, it is tempting to rely solely on them in the assessment of model water-mass formation. However, they provide only limited information on model water-mass formation rates, such as indicating the depth of rapid ventilation associated with surface mixing. Geochemical tracers, on the other hand, provide detailed information on the pathways and rates of water-mass renewal beneath the surface mixed layer, and therefore provide a stringent test of model behaviour. The main

162   4. Ocean Modelling and Prediction

tracers that have been used in this context include tritium (Sarmiento, 1983), chlorofluorocarbons (England et al., 1994; England, 1995), and natural and bomb-produced radiocarbon (Toggweiler et al., 1989).

Assessment of large-scale eddy-resolving ocean models relies on quite different data sets to those discussed above. These models capture some degree of high-frequency variability, such as meanders in boundary currents as well as mesoscale eddy activity. They are also only run over interannual to decadal time-scales, so they do not simulate the long-term climatological water-mass properties of the ocean interior. A more stringent test of model behaviour in this situation is to compare properties such as the global eddy kinetic energy density or meridional fluxes of heat/freshwater. Figure 4.19 shows a comparison of the surface-height variability observed by satellite with that simulated in a 1/6° model (Maltrud et al., 1998). The model and observed eddy kinetic energy are in overall agreement, with high eddy activity where the ACC interacts with topography and in western boundary currents.

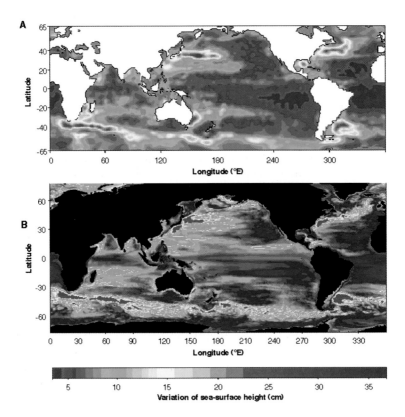

Fig. 4.19. Comparison between the surface-height variability (**a**) observed (Wunsch, 1996) with that (**b**) simulated in a 1/6° model (that of Maltrud et al., 1998) (from IPCC, 1995).

Regional and coastal ocean models. Many observational studies focus on continental shelves and coastal regions since this is where most recreational and commercial marine activities are focussed. Such studies typically involve spatially and temporally intense measurements of specific oceanic regions. As a result the validation of regional coastal ocean models is feasible for periods when observations are available. Typical observations of the coastal ocean for any given experiment include *in situ* measurements of currents and $T - S$ from an array of moored instruments at fixed locations or from shipboard surveys. Additionally, remotely sensed sea surface temperatures are often available. Together these data provide an incomplete picture of the coastal circulation which can be compared and contrasted with output from ocean models in order to either validate the model, or gain insight into the dominant dynamical processes that determine the circulation of the particular region. An example of a cross-section of temperature, salinity and potential density off Newport, Oregon observed during the OSU NOPP summer field season of 1999 is shown in Fig. 4.20 (Austin and Barth, 2001).

These sections show the isohals and isopycnals outcropping near the coast and a subsurface temperature maximum (e.g., the 9° isotherm) being subducted under the shallow surface mixed layer. The mechanism by which this subsurface temperature maximum is formed is unclear. Model analysis needs to be undertaken to determine the dynamical balances operating in the region.

Model validations usually involve a comparison between point source measurements and model output either qualitatively, by identifying similarities and differences in modelled and observed features, or quantitatively through statistical comparisons in the time domain (e.g., means, variances, correlations, empirical orthogonal functions and so on) or in the frequency domain (e.g., coherence, phase, gain). For a given application the validation requirements may be different. For example, if the timing and magnitude of temperature fluctuations are of interest in order to detect upwelling, then prediction of the mean temperature may be less important than the prediction of the variance, and it would be important for the model to be well correlated with observations. However, if coupling with the atmosphere is a concern, then the magnitude of the surface heat flux will depend on the mean surface temperature compared to the atmospheric temperature through the formulation of the sensible heat flux defined in (4.19).

A relatively recently developed method for observing the coastal ocean environment is the use of land-based high frequency (HF) Coastal Ocean Dynamics Application Radar (CODAR) arrays (e.g., Paduan and Rosenfield, 1996). CODAR measurements provide maps of near-surface ocean currents with high spatial ($\approx 1\,\text{km}$) and temporal ($\approx 10\,\text{minutes}$) resolution. The availability of these measurements provides modellers with an ideal opportunity to test the validity of their models. An example of a model-data comparison between an idealised configuration of the POM with CODAR data on the

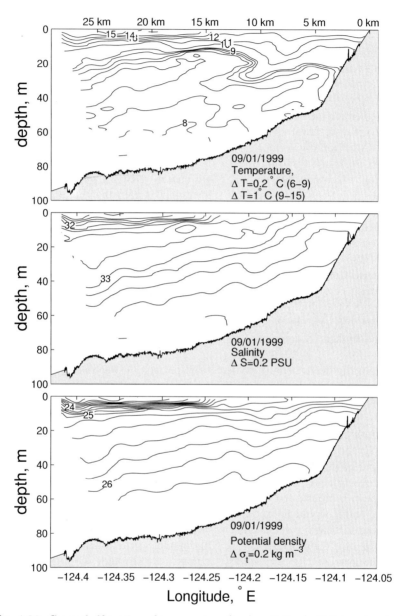

Fig. 4.20. Cross-shelf section of temperature (top), salinity (middle) and potential density (bottom) off Newport, Oregon observed during the OSU NOPP summer field season of 1999 (Austin and Barth, 2001).

Oregon continental shelf for the summer of 1998 is shown in Fig. 4.21 (from Oke et al., 2001). This figure shows a comparison between the modelled and observed mean surface currents (A and B), which indicates that the mean model fields are similar in structure and magnitude to the observed means. A comparison is also shown between the dominant spatial modes, obtained from an empirical orthogonal function (EOF) analysis of the modelled and observed fields (C–F). The spatial modes of an EOF analysis represent the structures of the variance fields. The percentage of the variance represented by each mode is shown in each panel indicating that 73% of the modelled variance and 48% of the observed variance is represented by these two EOFs. The first mode represents the acceleration of the coastal jet in response to upwelling favourable winds. The second mode represents the flow associated with upwelling relaxation (Gan and Allen, 2001), where a northward countercurrent is generated in response to the alongshore pressure gradients induced by the wind. Although the details of the modelled and observed means and the EOF modes differ, it is clear that the model is capturing a substantial amount of the true variability of the ocean in this region.

4.6.2 Inverse Methods and Data Assimilation

Data assimilation refers to the methods by which the inverse problem of the ocean circulation can be addressed. The inverse problem for the ocean circulation is the problem of inferring the state of the ocean circulation through a quantitative combination of theory and observations (Wunsch, 1996). Consider the system of equations that approximately and incompletely describes the ocean:

$$D\phi/dt = F + f \qquad (4.23)$$

with

Initial Conditions: $\phi_i = I + i$,

Boundary Conditions: $\phi_b = B + b$,

Observations: $\phi_m^o = D_m + d_m$

where ϕ represents the model state space, which consists of every model variable (e.g., u, v, w, T, S) at every model grid location; F, I, B and D represent the model forcing, initial conditions, boundary conditions and observations respectively; and f, i, b and d represent the errors in the model forecast, initial conditions, boundary conditions and observations respectively (Bennett, 1992). The inverse problem is to combine this information to obtain the most accurate and complete depiction of the ocean circulation that we can, given the resources available.

The forecast error f may be due to the approximations made by discretising the governing equations, or due to missing physics, if for example

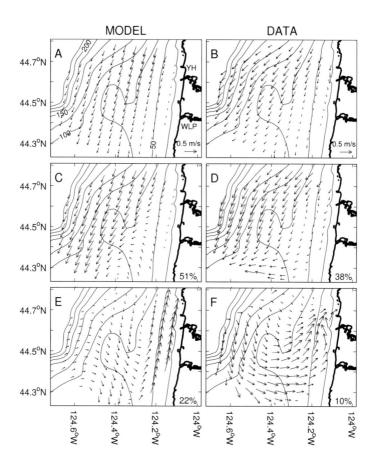

Fig. 4.21. Modelled (left) and observed (right) fields of surface currents off Oregon during the summer of 1998 showing the means (A–B) and the dominant spatial modes (C–F) calculated from an EOF analysis (Oke et al., 2001).

the non-linear terms in the equations are neglected. The error in the initial conditions i is a consequence of the fact that we can not know the precise state of the ocean at any moment in time. A model's initial condition will typically be derived from climatological data, or from an interpolation from sparse observations. The error in boundary conditions b may be due to the uncertainties in the applied wind or heat flux forcing; or in the case of regional models, due to the uncertainty in the lateral boundary conditions as well.

Finally, the observation error d_m represents the system noise which is the result of instrument error, measurement error and uncertainties in the observability of a particular oceanic variable. In order to produce the best

4.6 Exploiting Ocean Observations

possible depiction of the ocean circulation the method of least-squares is employed, where one attempts to find the solution to the governing equations, given the initial conditions, boundary conditions and observations, that is within the given error bounds of each component.

The simplest method for solving this inverse problem involve the application of sequential methods, namely optimal interpolation (e.g., Cohn et al., 1998) or the Kalman Filter (e.g., Miller, 1986). These methods attempt to optimally combine the observations and the dynamics to produce an analysis of the ocean ϕ^a. The analysis is produced by combining the model's forecast ϕ^f and the observations d_m using the so-called analysis equations:

$$\phi^a = \phi^f + \mathbf{K}(d_m - \mathbf{H}\phi^f) \tag{4.24}$$

with

$$\mathbf{K} = \mathbf{P}^f \mathbf{H}^T (\mathbf{H}\mathbf{P}^f \mathbf{H}^T + \mathbf{R})$$

where $\mathbf{K}$ is the Gain matrix and $\mathbf{H}$ is an interpolation matrix that interpolates the model state onto the space of observations. The Gain matrix depends on the forecast error covariance matrix $\mathbf{P}^f$ and the observation error covariance matrix $\mathbf{R}$. For sequential methods f, i and b are considered together and labelled the forecast error ϵ^f. The forecast and observation error covariances are given by

$$\mathbf{P}^f = \langle \epsilon^f \epsilon^{fT} \rangle$$

and

$$\mathbf{R} = \langle \epsilon^o \epsilon^{oT} \rangle$$

respectively, where $\langle \cdot \rangle$ denotes a time average and ϵ^o is the observation error. For sequential methods an analysis of the model state is produced at each assimilation cycle which will depend on the time scales of the problem that is under investigation. In addition to the analysis equations presented above, the Kalman Filter also solves an equation for the time evolution of $\mathbf{P}^f$. The implementation of the analysis equations involves the estimation of the forecast and observation error covariance matrices. In practise both ϵ^f and ϵ^o, and hence their covariances are unknown. The estimation of these covariance matrices involves assumptions about decorrelation length and time scales, and often about the homogeneity and isotropy of the model error fields. These matrices must be estimated prior to assimilation and their validity tested through a series of objective statistical tests after each assimilation (Bennett, 1992).

As an alternative to sequential methods, the generalized inverse method (e.g., Bennett et al., 1993; Errico, 1997) may be used. This approach involves the formulation of a quadratic penalty functional or cost function $\mathcal{J}$:

$$\mathcal{J} = W_f \int \int f^2 dt dx + W_i \int i^2 dx + W_b \int b^2 dt + W_d \sum d_m^2 \tag{4.25}$$

where W_f, W_i, W_b and W_d are positive weight functions for each component of the system described above. These weights are related to the above-mentioned forecast and observation error covariances and must also be chosen prior to assimilation. The cost function $\mathcal{J}$ is a single number for each depiction of ϕ. The cost function must be minimized by identifying the smallest values for f, i, b and d_m in the weighted least-squares sense. Once the global minimum of $\mathcal{J}$ is obtained, the optimal solution to ϕ is obtained. For a detailed discussion of the generalized inverse method and other methods for data assimilation the interested reader is referred to Bennett (1992) or Wunsch (1996).

The development and implementation of practical data assimilation techniques is vital if operational forecasting of the ocean circulation is going to become a reality. The development and maturity of remote sensing and *in situ* observing systems, the advances in scientific knowledge of the global and regional ocean circulation, and the development of sophisticated ocean models has made real-time observing and forecasting systems feasible. As outlined above data assimilation enables available observations derived either remotely, from satellites or radar systems, or *in situ*, from moored instruments, drifters or shipboard surveys, to be combined with ocean models in order to produce a complete depiction of the ocean circulation at time scales of a few days and space scales of several tens of kilometres. The Global Ocean Data Assimilation Experiment (GODAE) is an experiment that is designed to demonstrate the practicality and feasibility of routine, real-time global ocean data assimilation and prediction (Smith and Lefebvre, 1998).

All weather forecasting systems that are presently in operation utilise data assimilation in some form, through either initialisation to an objective map of the atmospheric state or through more sophistocated assimilation techniques. Forecasting the ocean circulation presents all of the same difficulties and challenges as weather forecasting, except that ocean observations are much more sparse compared to observations of the atmosphere. Consequently the development of reliable and practical data assimilation systems for ocean forecasting is more crucial since we must endeavor to take full advantage of the limited observations that are available.

4.6.3 Applications of Data Assimilation to Coastal Ocean Models

One application of sub-optimal sequential data assimilation to a regional, primitive equation model of the Oregon continental shelf circulation is outlined below. This application involved assimilation of surface velocity data obtained from a land-based HF CODAR array during the summer of 1998 (Oke et al., 2000). The surface information was projected over the entire water column in order to correct velocities and density at depth. In order to demonstrate how well the surface information was projected over depth the depth-averaged velocity, in 80 m of water, obtained from a moored acoustic doppler profiler (ADP) is compared with the model hindcast with and without assimilation (Fig. 4.22). The magnitude of the complex correlation between the observations and the model without assimilation is 0.42, and with assimilation is 0.76, indicating that the assimilation improved the hindcast by approximately 50%, demonstrating its potential for coastal ocean modeling.

4.6.4 Application of Data Assimilation to Large-scale Models

An example of a project that is endeavouring to develop a forecast system for the global oceans is the GODAE. This development is a very challenging task, both from a scientific and technical perspective. Issues that must be overcome include development of advanced models; development of efficient and effective assimilation schemes; estimation of error statistics; data management and quality control; and access to large computer facilities and communication systems. The main benefits of a global nowcast and forecast system include the availability of reliable initial conditions for coupled ocean-atmosphere models which are used for climate and seasonal forecasting; reliable boundary conditions for high resolution regional ocean models; as well as applications to marine safety, fisheries, offshore industry and management of continental shelf and coastal areas.

4.6.5 Variational Data Assimilation, Example

Table 4.1. Comparison of various solutions for the M_2 and K_1 surface elevation in the Barents Sea: RMS errors (cm) (from Kurapov and Kivman, 1999).

Model	Resolution & Domain	M_2 RMS error (cm)	K_1 RMS error (cm)
Gjevik et al. (1994)	25×25 km	7.4	2.6
Non-linear, no assimilation	Barents Sea		
Kowalik and Proshutinsky (1995)	14×14 km	6.9	1.6
Non-linear, no assimilation	Arctic Ocean		
Kivman (1997)	1° × 1°	5.0	2.2
GIM, finite difference	Arctic Ocean		
Kurapov and Kivman (1999)	1–52 km	3.6	1.2
GIM, finite element.	Barents Sea		

The generalised inverse of a high-resolution finite element model of the Barents Sea, which is a part of the Arctic Ocean, based on the linearised shallow water equations was developed by Kurapov and Kivman (1999).

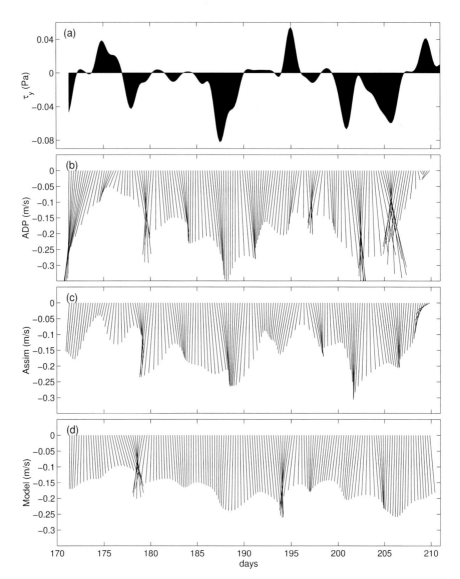

Fig. 4.22. (a) Alongshore wind stress from Newport, Oregon 1998; vector stick plots of depth-averaged velocities from (b) a moored ADP v^o (at 80 m depth), (c) model simulation with assimilation of surface velocity data v^a, and (d) model simulation with no assimilation v^m. Correlations: $C(v^o, v^m) = 0.42$; $C(v^o, v^a) = 0.76$ (from Oke et al., 2001).

This inverse model involved assimilation of tidal constituents from 47 coastal tide gauges. The resulting assimilation was validated by a comparison of the analysed tidal elevations with independent tide gauges in the interior of the domain and near the open boundaries. The results indicate that with a low weight given to the open boundary conditions in the cost function, the generalised inverse is capable of reproducing tidal elevations of the dominant tidal constituents with greater precision than other more complicated models with larger domains that did not utilise data assimilation. These comparisons are summarised in Table 4.1. While the linear model clearly has significant errors due to the neglected physics, particularly in the shallow waters, these comparisons demonstrate that its generalised inverse provides a very effective, dynamically based, interpolator for this region. This example demonstrates the power of inverse methods for ocean modeling, particularly for hindcasting and nowcasting.

4.7 Concluding Remarks

Ocean circulation models form an important component of oceanographic and climate research. Applications range from simulations of flow in bays and harbours through to coastal, regional, and global-scale models. In this chapter we reviewed the governing equations, model grid systems, boundary conditions, and the parameterisation of subgrid-scale processes. We also gave specific examples of a number of models, from large-scale climate related simulations, to coastal experiments, river plume models and tidal flows. The use of observational data was also reviewed, from model assessment to data assimilation and inverse model techniques. The future directions of ocean modelling research are farreaching; they include refinements of subgrid-scale parameterisations, use of higher resolution models, development of improved numerical schemes, applications of data assimilation towards predictive systems, and coupled modelling with climate and biological modules.

5. Land Surface Processes

Parviz Irannejad and Yaping Shao

5.1 The Importance of Land-surface Processes

At the interface between the atmosphere and land, known as the land surface, exchanges of energy, mass and momentum take place. These exchanges are governed not only by the turbulent and molecular motions in the atmosphere, but also by surface soil hydrological processes and the behaviour of the continental biosphere. From this point of view, the land surface can be defined as a layer comprising the lower part of the atmospheric boundary layer, the top few metres of the soil, the continental biosphere and surface water bodies.

The land surface profoundly influences our social and economic activities. Soil provides over 90% of human food, livestock feed, fiber and fuel. More than 2.6 billion people are involved in agriculture (Hurni, 1998), with productivity depending on land quality and water availability, as well as other factors, such as atmospheric conditions. The population increase has placed an excessive pressure on land resources. By the early 1990s, about 24% of the inhabited land exhibited human-induced soil degradation. Causative factors are deforestation and removal of natural vegetation, over-grazing, agriculture and industrial activities. Today, more than 900 million people in 100 countries are affected by desertification (Steer, 1998).

The land surface is a key component of the environmental system. While the sun is the ultimate energy source for the environment, two thirds of the immediate energy input for the atmosphere comes from the surface. Figure 5.1 shows the globally averaged energy balance of the Earth system. Water, which has a closed cycle in the environment, plays a fundamental role in capturing, utilizing and redistributing that energy. The surface is the sole source of atmospheric moisture. Evapotranspiration of water from the surface and its condensation in the atmosphere constitutes the link between the energy and water cycles in the environment. Land surface plays an important role in these energy and water cycles and the interactions between the atmosphere and the land surface is one of the main topics in environmental studies.

174   5. Land Surface Processes

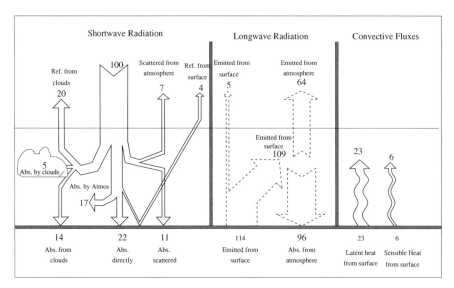

Fig. 5.1. Energy balance of the Earth system in percentage (data from Bryant, 1997).

On the one hand, land-surface processes significantly influence climate and weather. While the land surface occupies less than 1/3 of the Earth's surface, the thermodynamic contrast between ocean and land and the strong spatial and temporal variations of land-surface properties (e.g. topography, roughness, albedo and vegetation cover) make its contribution to the atmospheric dynamics very important. Large-scale changes in land surface properties have major consequences to weather and climate. For instance, large scale desertification can result in a decreased precipitation (Charney et al., 1977; Dirmeyer and Shukla, 1996), a weakening of the tropical easterly jet (Laval and Picon, 1986) and a weakening of monsoon circulation (Xue, 1996). Deforestation may lead to decreased precipitation and evapotranspiration, a change in moisture convergence and climate change (e.g. Lean and Rowntree, 1997). On local scales, in areas with different land-surface characteristics, net radiation is partitioned differently between sensible and latent-heat fluxes. As a consequence, horizontal temperature contrasts develop, resulting in mesoscale circulations as strong as sea breeze (Pielke and Segal, 1986). This leads to the formation and the maintenance of the low-level jets (Wu and Raman, 1997), the development of zones of convective clouds (e.g. Mahfouf et al., 1987) and enhancement of precipitation (Taylor et al., 1997) along land-cover-type and soil-type boundaries.

Soil water is the main moisture source for the atmosphere over the continents. The role of soil moisture in the atmospheric dynamics and continental climates has been the subject of many studies. Similar to the sea-surface temperature, but with shorter time scales, soil moisture acts as a low-frequency

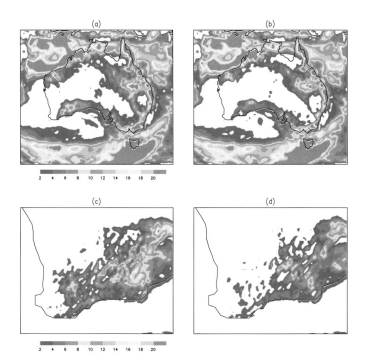

Fig. 5.2. (a) Simulated rainfall over the Australian region over a 10-day period in February 1996 with 50 km model resolution and high initial soil moisture; (b) as (a), but with normal soil moisture; (c) as (a), but for the Western Australian region with 10 km model resolution; (d) as (c), but with normal soil moisture.

forcing to the atmosphere (Delworth and Manabe, 1988). For example, the spring-time soil moisture might influence the summer precipitation in the interior of continents (Oglesby, 1991). Increased soil moisture enhances local rainfall when the lower atmosphere is thermally unstable and relatively dry, but may decrease rainfall when the atmosphere is humid and lacks sufficient thermal forcing to initiate deep convection (Pan et al., 1996). Figure 5.2 shows an example. Temperature, pressure and precipitation are sensitive to soil-moisture anomalies and a reduction in soil moisture prolongs the summertime drought (Shukla and Mintz, 1982; Oglesby and Erickson, 1989). Soil-moisture affects the mean and median near surface air temperature (Huang et al., 1996) and its spatial variations impacts the short-term atmospheric circulation patterns (Anthes and Kuo, 1986).

On the other hand, land-surface properties and conditions are strongly influenced by weather and climate. From the long-term perspective, climate patterns (mostly precipitation and temperature) determine the distribution of soil types, water resources and vegetation communities (e.g. subtropical deserts and tropical forests). Studies of these effects are beyond the scope of the environmental modelling and prediction considered in this book. On shorter time scales, annual and inter-annual climatic fluctuations lead to phenomena, such as drought and flood, which have major consequence for land productivity, surface soil characteristics and vegetation cover. Important land-surface processes, such as soil erosion and salinisation, are also exacerbated under these situations. On even shorter time scales, atmospheric quantities, such as radiation, air temperature and precipitation, act as external driving force for the variations in soil moisture, soil temperature and photosynthetic activities of plants. Under severe weather conditions, these land surface quantities may undergo sudden and dramatic changes.

In recent years, considerable efforts have been made to model land-surface processes. These efforts reflect the increasing awareness of the need to model the environment as an integrated system. Land-surface schemes, as will be described in this chapter, have been developed and implemented for improved climate and weather prediction, water resource management and ecological modelling. The following sections provide an overview of land-surface modelling.

5.2 Land-surface Parameterization Schemes

A land-surface parameterisation scheme, or simply a land-surface scheme, is an algorithm for determining the exchanges of energy, mass and momentum between the atmosphere and land surface. These exchanges are complex functions of many physical processes which have a range of temporal and spatial scales. It is impossible and unnecessary to incorporate all the details of these processes into a numerical scheme and hence, land-surface schemes have been developed based on various simplifications. Depending on these simplifications, land-surface schemes with different degrees of complexity have been constructed. Figure 5.3 is a schematic representation of the major processes to be modelled in a typical land-surface scheme.

The essence of a land-surface scheme is to provide a solution to the surface energy and water budgets over the land. Since both these budgets are strongly dependent upon the availability of moisture at the surface, modules for soil moisture and surface hydrology are the core of a land-surface scheme. In early atmospheric models, surface hydrological processes were not explicitly modelled, but a prescribed Bowen ratio was used to calculate sensible and latent-heat fluxes.

Manabe (1969) first introduced explicit treatment of surface hydrology into atmospheric modelling by including a single-layer (bucket) soil model in a general circulation model (GCM). That scheme was designed to represent seasonal cycles of moisture and energy fluxes on the continental scale, but did not account for the role of the biosphere in regulating these fluxes. The land surface was assumed to be homogeneous with identical properties

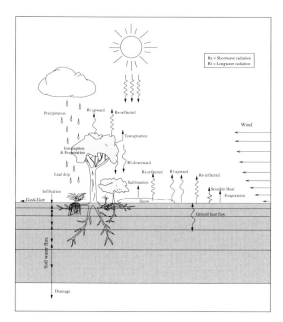

Fig. 5.3. Important processes of the land surface.

over all continents. Deardorff (1978) included a two-layer soil model and a canopy layer in a land-surface scheme. In his model, the canopy layer and the soil surface interact with each other and with the atmosphere. The two soil layers allow the prediction of both diurnal and seasonal cycles of surface soil temperature and moisture. More elaborate land-surface schemes have been developed later by Dickinson et al. (1986) and Sellers et al. (1986). The last decade has seen a rapid development in land-surface schemes and the expansion of their applications (see Table 5.1).

In most land-surface schemes used in atmospheric models, it is assumed that land-surface properties are homogeneous over a model grid box. In reality, land surface is highly heterogeneous, due to ecosystem diversity, complex morphology, soil variability and climatic forcing (Giorgi, 1997). As land-surface processes are mostly nonlinear, heterogeneity can profoundly affect the fluxes of momentum, water and energy at the surface. This understanding has triggered the development of a new generation of land-surface schemes which account for the subgrid-scale variations of surface properties and atmospheric forcing, especially precipitation (e.g. Avissar, 1992; Famiglietti and Wood, 1995). Section 5.7 reviews the strategies used for including the influences of land-surface heterogeneity.

178   5. Land Surface Processes

Table 5.1. List of land-surface schemes and some basic information.

Model	Number of layers for				Time Step	Spin-up Time[6]	Major Purpose	Philosophy for			Reference
	C[1]	T[2]	S[3]	R[4]				C[1]	T[2]	S[3]	
ALSIS	1	2	x	x	30 min	2 yr	flex	ad	f-r	Richards	Irannejad and Shao (1998)
BASE	1	3	3	3	20 min	5 yr	GCM	ad	h-d	Philip-de Vries	Desborough and Pitman (1998)
BATS	1	2	3	2	30 min	3 yr	GCM	P-M	f-r	Darcy	Dickinson et al. (1986, 1993)
BUCK	0	1	1	1	30 min	1 yr	GCM	implicit	e-b	bucket[5]	Manabe (1969), Robock et al. (1995)
CAPS	1	3	3	2	30 min	1 yr	GCM	P-M	h-d	Darcy	Mahrt & Pan (1984)
CAPSLLNL	1	2	3	1	5 min	10 yr	GCM	P-M	h-d	diffusion	Mahrt & Pan (1984)
CLASS	1	3	3	3	30 min	-	GCM	P-M	h-d	Darcy	Verseghy (1991) Verseghy et al. (1993)
CRIRO9	1	3	2	1	30 min	2 yr	GCM	ad	h-d	f-r	Kowalczyk et al. (1991
ECHAM	1	5	1	1	30 min	5 yr	GCM	ad	h-d	bucket[5]	Dümenil & Todini (19
GISS	1	6	6	6	30 min	13 yr	GCM	ad	h-d	Darcy	Abramopoulos et al. (1988)
IAP94	1	3	3	2	60 min	60 yr	GCM	P-M	h-d	Darcy	-
ISBA	1	2	2	1	5 min	-	GCM-meso	ad	f-r	f-r	Noilhan & Planton (19
MOSAIC	1	2	3	2	5 min	6 yr	GCM	P-M	f-r	Darcy	Koster & Suarez (1992
NMC	1	3	3	2	30 min	3 yr	GCM-	P-M	h-d	Darcy	Mahrt & Pan (1984)
PLACE	1	7	50	2	5 min	3 yr	flexible	ad	h-d	Darcy	Wetzel & Boone (1995
SECHIBA2	1	7	2	1	30 min	2 yr	GCM	ad	h-d	Choisnel	Ducoudre et al. (1993)
SEWAB	1	6	6	1	10-30 sec	3 yr	meso	ad	h-d	diffusion	-
SPONSOR	1	1	2	2	24 hr	3 yr	GCM	ad	e-b	bucket[5]	Shmakin et al. (1998)
SSiB	1	2	3	1	30 min	2 yr	GCM-meso	P-M	f-r	diffusion	Xue et al. (1991)
SWAP	1	2	1	1	24 hr	2 yr	meso	ad	h-d	bucket[5]	-
SWB	0	3	2	2	30 min	3 yr	meso	-	h-d	bucket[5]	Schaake et al. (1995)
UGAMP	1	3	3	2	30 min	14 yr	GCM	P-M	h-d	Darcy	-
UKMO	1	4	4	4	30 min	1 yr	GCM meso	P-M	h-d	Darcy	Warrilow et al. (1986)

[1] Conopy, [2] Soil Temperature, [3] Soil Moisture, [4] Roots, [5] bucket plus variation, [6] "-" represents no record of spin-up time

5.3 Modelling Soil Hydrology in Land-surface Schemes

The flow of water in an unsaturated soil involves the transfer of liquid water and water vapour under the influence of hydraulic, vapour pressure, temperature and osmotic gradients. In atmospheric models, soil hydrological processes are considerably simplified for the calculation of soil moisture.

5.3.1 Bucket Model

The simplest approach to the prediction of soil moisture is a single-layer soil-water budget model, known as the bucket model. Central to this approach are the concepts of field capacity, W_f, and wilting point, W_w. The water content, at which internal flow becomes negligible, is termed field capacity. The wilting point is the soil-water content below which water extraction by plant roots presumably ceases and plants wilt. Practically, field capacity and wilting point are taken as water content at matric potentials of -10 kPa to 34 kPa and 1500 kPa, respectively. W_f and W_w are known to be dependent on soil types, but are often simply assumed to be constant for all soils.

The bucket model (Fig. 5.4) treats the soil profile as a bucket into which rainfall flows until it is full (assuming that precipitation rate never exceeds the infiltration capacity of dry soils). Excess water is treated as either surface runoff or drainage at the bucket bottom and is not available for further evapotranspiration. This concept has been used extensively for water balance studies in agricultural farms and has been applied by Budyko (1956) for calculating the energy balance of the World. Manabe (1969) extended the application of bucket model to atmospheric models. In climate models, the depth of the soil layer is normally set to 1 m, mainly because most of the root system of plants (crops) are concentrated in this layer.

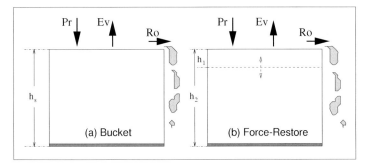

Fig. 5.4. The schematic representation for the bucket and force-restore model.

Another simplification often applied is to neglecte the interception of precipitation by vegetation. In reality, the importance of interception loss depends on the nature, type and density of vegetation cover, precipitation characteristics and season. Early estimates have shown that in annual time scales, interception loss may account for up to 25% the total evapotranspiration (Shuttleworth, 1988).

Suppose the available soil moisture, W_a, is defined by

$$W_a = W - W_w \tag{5.1}$$

where W is soil moisture. The evolution of W_a obeys

$$\frac{\partial W_a}{\partial t} = \begin{cases} P_r - R_o - E_v & S_{\text{now}} = 0 \\ P_r + S_m - R_o & S_{\text{now}} > 0 \end{cases} \tag{5.2}$$

with

$$R_o = \begin{cases} \max(0., P_r - E_v - \frac{(W_f - W_a)}{dt}) & S_{\text{now}} = 0 \\ \max(0., P_r + S_m - \frac{(W_f - W_a)}{dt}) & S_{\text{now}} > 0 \end{cases} \tag{5.3}$$

where P_r is precipitation, E_v is evapotranspiration, R_o is runoff, S_m is snow melt and dt is time interval. If the surface is snow covered or the available soil moisture is zero, no evaporation from the soil takes place. For the snow-free land surface, evaporation occurs at its potential rate, if $W_a > 0.75(W_f - W_w)$. Otherwise, the Budyko's moisture availability factor, β, is used to adjust the evaporation (See Sect. 5.5.2).

Although the bucket model does not explicitly model the vegetation impact on the land-surface hydrology, the assumped moisture storage available for evaporation is correct for a vegetation cover with a root system extending to a soil depth of about one meter. Bare soil evaporation takes place from a very thin surface soil layer. For this reason, the bucket model often highly overestimates evaporation, especially when applied to a bare land surface or to a land covered with a very shallow-rooted vegetation.

A modification to the bucket model is a multi-layer bucket, in which precipitation water cascades from the upper to the lower zones when the upper zones reach field capacity. The layered bucket is a considerable improvement over the single bucket model (Calder et al., 1983). Introducing vegetation to a bucket model and accounting for the impact of canopy resistance on regulating evapotranspiration may crucially improve the simulation of the annual cycles of soil moisture and surface fluxes (Zhang et al., 2000; Irannejad and Shao, 2000).

5.3.2 Force-restore Model

The force-restore model (Deardorff, 1977) is an approximation of the multi-layer convection-diffusion model for calculating soil moisture. Soil moisture in a thin top surface layer is forced by the upper boundary conditions (precipitation and evaporation) and restored by the moisture diffusion from the deep soil reservoir (Fig. 5.4). Deardorff's model also includes an explicit homogeneous, single-layer vegetation layer and separately computes energy budgets for the canopy layer and the ground surface. The canopy layer partly shields the ground surface and intercepts some of the solar radiation and precipitation. The intercepted precipitation wets the canopy and is available for evaporation. A canopy resistance is introduced to represent the combined effect of the canopy stomates on transpiration. For vegetated ground, the evolutions of soil moisture in the two soil layers can be expressed as

$$\frac{\partial \theta_1}{\partial t} = -C_1 \frac{E_s + d_1 E_t + R_o - P_e}{\rho_w d_1} - C_2 \frac{\theta_1 - \theta_2}{\tau_1} \tag{5.4}$$

$$\frac{\partial \theta_2}{\partial t} = \frac{E_s + E_t + R_o - P_e}{\rho_w d_2} \tag{5.5}$$

where ρ_w is the density of water, t is time, τ_1 is period of one day, θ_1 and θ_2 are soil moisture of the top soil layer of depth d_1 and the bulk soil layer of depth d_2. The depth d_1 corresponds to the depth to which diurnal soil moisture cycle extends, and d_2 corresponds to the depth below which the water flux is negligible. E_s is soil evaporation, P_e is effective precipitation at the ground surface, E_t is the transpiration rate per unit ground area, and C_1 and C_2 are the force and the restore coefficients, respectively. Runoff of precipitation reaching the ground occurs only when θ_1 exceeds the saturation soil moisture, θ_s. Drainage is limited to the cases when θ_2 exceeds field capacity. C_1 and C_2 may be calculated based on the water contents of the surface and deep soil layers (Noilhan and Mahfouf, 1996).

5.3.3 Multi-layer Soil Model

The prediction of soil moisture in a multi-layer soil model (Fig. 5.5) is frequently performed by numerically solving the one-dimensional Richards equation. This equation is a convective-diffusive equation which is derived by combining the modified Darcy's law for unsaturated soils, i.e.,

$$q = K(\frac{\partial \psi}{\partial z} - 1) \tag{5.6}$$

and the one-dimensional equation of conservation of water

$$\frac{\partial \theta}{\partial t} = -\frac{\partial q}{\partial z} - S \tag{5.7}$$

where K is hydraulic conductivity, ψ is matric potential of the soil water and S is the sink term for water due to evapotranspiration and horizontal discharge. This combination gives

$$\frac{\partial \theta}{\partial t} = \frac{\partial}{\partial z}(K - K\frac{\partial \psi}{\partial z}) - S \tag{5.8}$$

Assuming nonhysteretic flow condition for soils, the water content and matric potential are uniquely related ($\theta = \theta(\psi)$ and $\psi = \psi(\theta)$). Therefore, Equation (5.8) can be written as

$$C_w \frac{\partial \psi}{\partial t} = \frac{\partial}{\partial z}\left(K - K\frac{\partial \psi}{\partial z}\right) - S \tag{5.9}$$

or

$$\frac{\partial \theta}{\partial t} = \frac{\partial}{\partial z}\left(K - D\frac{\partial \theta}{\partial z}\right) - S \tag{5.10}$$

where $C_w = \dfrac{d\theta}{d\psi}$ is specific water capacity and $D = K\dfrac{d\psi}{d\theta}$ is soil hydraulic diffusivity. The Richards equation (5.8), (5.9) or (5.10) is highly nonlinear, as K, θ and D are normally highly nonlinear functions of ψ.

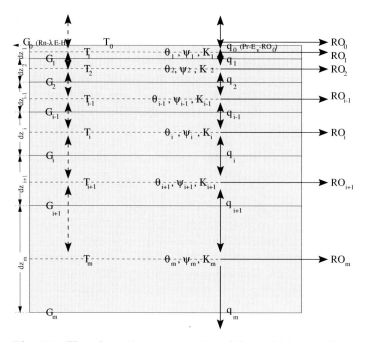

Fig. 5.5. The schematic representation of the multi-layer soil model.

Analytical solutions to the Richards equation exist only for certain initial and simple boundary conditions. Numerical solutions via the finite-element and finite-difference techniques are readily available. For finite-difference solutions the space domain (soil depth) is discretised to several parallel layers. The values of state variables, θ and ψ of each soil layer are updated at each time step and assigned to the centre of each soil layer. The water flux is calculated at the interface of the model soil layers. Due to the sharp gradients of

ψ, especially near the surface, and nonlinearity of the hydraulic functions, a very fine vertical discretisation is required. This requires very short time-steps to avoid numerical instability, especially under heterogeneous conditions, for infiltration into initially dry soils, or for coarse-textured soils that are characterised by sharp wetting fronts.

The most accurate numerical solutions are obtained using implicit finite-difference schemes with implicit or explicit evaluation of hydraulic properties (Haverkamp et al., 1977). The pressure-based form of the Richards equation (5.9) may create mass conservation problems (Milly, 1988) and is inaccurate for infiltration into dry soils (Kirkland et al., 1992). A mass-conserving solution can be obtained by modifying the capacity term to force the global mass balance. The θ-based equation (5.10), on the other hand, conserves mass. However, it is applicable only to homogeneous soils, where the hydraulic properties remain unchanged with depth. The mixed form (5.8) conserves mass accurately (Celia et al., 1990).

Transformations of the Richards equation have been introduced in order to deal with its nonlinearity. Redinger et al. (1984) applied the Kirchhoff transform

$$U = \int_{-\infty}^{\psi} K d\psi = \int_{0}^{\theta} D d\theta \tag{5.11}$$

to the diffusive term of the Richards' equation, so that it becomes

$$\frac{\partial \theta}{\partial t} = -\frac{\partial}{\partial z}(K - \frac{\partial U}{\partial z}) - S \tag{5.12}$$

The Kirchhoff transform reduces the high gradients of ψ to lower ones of U. This alleviates the problem of numerical instability, so that without losing accuracy a coarser node spacing can be used. Since the Kirchhoff transform depends on soil hydraulic properties, its application is limited to homogeneous soils (Campbell, 1985). However, it is possible to apply this technique to layered and gradational soils by making appropriate corrections due to the change in soil hydraulic properties with depth (Ross and Bristow, 1990). The advantage of transformation-based methods is that they reduce the truncation error using the same grid spacing and are numerically several orders of magnitude faster than the pressure head-based method (Pan and Wierenga, 1997).

The solution of the Richards equation requires closure relationships between hydraulic conductivity, soil water content, and matric potential. Many different functions have been proposed for the water retention and hydraulic conductivity curves (e.g. Brooks and Corey, 1964; Clapp and Hornberger, 1978 and van Genuchten, 1980). A comparison of these models is shown in Fig. 5.6. The Clapp and Hornberger model is most commonly used in land surface schemes. The hydraulic conductivity function of this model is

$$K(\theta) = K_s \Theta^{(2b+3)} \tag{5.13}$$

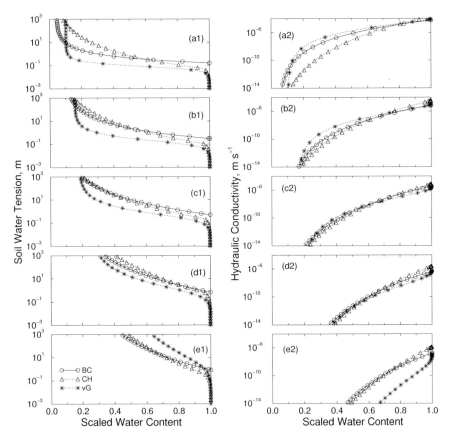

Fig. 5.6. Comparison of **(1)** retention curves and **(2)** hydraulic conductivity functions of the Brooks and Corey (1964), Clapp and Hornberger (1978) and van Genuchten (1980) hydraulic models for **(a)** sandy soil, **(b)** sandy loam, **(c)** loam, **(d)** clay loam and **(e)** clay.

or

$$K(\psi) = K_s \left(\frac{\psi_s}{\psi}\right)^{2+\frac{3}{b}} \quad (5.14)$$

and the water retention function is

$$\psi = \begin{cases} \psi_s \Theta^{-b} & \psi \leq \psi_i \\ -m(\Theta - n)(\Theta - 1) & \psi > \psi_i \end{cases} \quad (5.15)$$

where b is a dimensionless pore-size distribution index, ψ_s is air-entry matric potential and ψ_i is matric potential at the near saturation inflection point. The parameters m and n are calculated so that (5.15) passes through (Θ_i, ψ_i) and $(1, 0)$.

Land-surface simulations are sensitive to the choice of the soil hydraulic functions (Cuenca et al., 1996). However, this sensitivity may arise mainly from the inconsistencies of the soil hydraulic parameters in different functions (Shao and Irannejad, 1999). Figure 5.7 shows that different soil hydraulic functions may be equally appropriate, provided that accurate parameter values are available. The Clapp and Hornberger model is the most favorable because it requires less parameters and is numerically efficient.

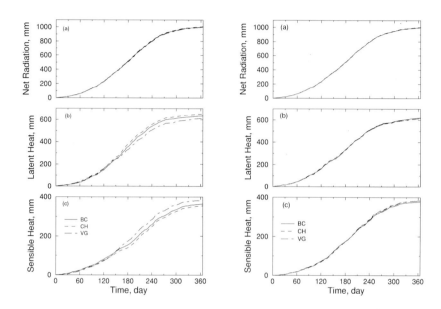

Fig. 5.7. Simulated (a) net radiation; (b) cumulative evapotranspiration; and (c) sensible heat flux for HAPEX-MOBILHY using four different soil hydraulic functions before (left-hand side panels) and after (righ-hand side panels) parameter adjustment. Net radiation and sensible heat flux are given in evaporation equivalent unit.

5.4 Soil Temperature

Heat can transfer in soils by conduction, by the flow of liquid water, by diffusion of vapour and by convection. Compared to the two former processes the two latter ones are usually negligible. For dry soils, where water flux is very small, the only significant process for heat flow is conduction. For these reasons, the theory based on heat conduction usually provides results comparable with observations (Kirkham and Powers, 1972). The heat flux caused by heat conduction within the soil profile can be expressed as

$$G = -K_h \frac{\partial T}{\partial z} \tag{5.16}$$

where K_h is thermal conductivity and T is temperature. Substituting (5.16) into the continuity equation for soil temperature, i.e.

$$\frac{\partial T}{\partial t} = -\frac{1}{C_s} \frac{\partial G}{\partial z} \tag{5.17}$$

gives

$$\frac{\partial T}{\partial t} = D_h \frac{\partial^2 T}{\partial z^2} \tag{5.18}$$

where C_s ($=\rho_s c_s$) is the volumetric heat capacity of the soil, ρ_s is soil density, c_s is specific heat capacity, and D_h ($=\frac{K_h}{C_s}$) is the soil thermal diffusivity. C_s, ρ_s, K_h, and subsequently D_h depend on the soil type and water content. Neglecting the density and heat capacity of the air in the soil, we obtain

$$C_s = (1 - \theta_s)\rho_q c_q + \theta \rho_w c_w \tag{5.19}$$

where subscripts q and w denote soil solid and water component, respectively.

K_h varies over several orders of magnitude in the range of field soil moisture. McCumber and Pielke (1981) have suggested the following empirical relationship between K_h and the matrix potential irrespective of the soil texture:

$$K_h = \begin{cases} e^{-P_f + 2.7} & P_f \leq 5.1 \\ 0.00041 & P_f > 5.1 \end{cases} \tag{5.20}$$

where K_h has units [cal s^{-1}cm^{-1}K^{-1}] and P_f is the base-10 logarithm of absolute value of the soil matric potential given in [cm]. Figure 5.8 illustrates K_h and D_h as a function of relative soil water content for three different soil types.

To numerically solve (5.17), heat fluxes are calculated at the model's soil layer interfaces and the state variable T_i is updated every time step and is assigned to the centre of each layer (Fig. 5.5). The lower soil boundary is usually assigned zero heat flux conditions. This assumption is acceptable for short time periods and for any time-scale when the soil depth is taken well below the damping depth of the annual temperature wave. The upper boundary conditions is defined by (5.16) with G_0 given as

$$G_0 = K_{h1} \frac{2(T_0 - T_1)}{dz_1} \tag{5.21}$$

where T_0 is the surface temperature, usually diagnosed by an iterative solution of the surface energy balance equation.

As for soil moisture, soil temperature can be modelled using a force-restore model (Bhumralkar, 1975; Blackadar, 1976). In such a model, the

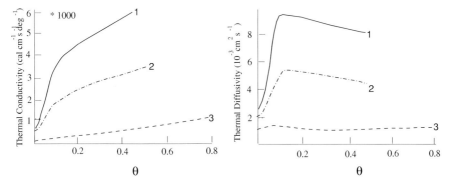

Fig. 5.8. Thermal conductivity and thermal diffusivity as functions of relative water content for (1) sand; (2) loam; and (3) peat (from Hillel, 1980).

temperature of a thin upper layer is forced by the ground heat flux. In analogy to Equations (5.4) and (5.5), we have

$$\frac{\partial T_1}{\partial t} = -C_1 \frac{G_0}{C_s d_1} - C_2 \frac{T_1 - T_2}{\tau_1} \tag{5.22}$$

$$\frac{\partial T_2}{\partial t} = -\frac{G_0}{C_s d_2} \tag{5.23}$$

Here, $d_1 \, (= \sqrt{D_h \tau_1})$ and $d_2 \, (= \sqrt{365 D_h \tau_1})$ are the damping depths of diurnal and annual temperature waves, and C_1 and C_2 are the force and restore coefficients, respectively.

Since the force-restore method is derived based on a single sinuidal periodic surface forcing, the model response to higher harmonics, resulting from atmospheric phenomena such as cloudiness, and warm and cold waves, may be distorted. This is especially profound if the upper soil layer has a thickness of less than one damping depth of diurnal forcing. However, because the diurnal forcing is usually much stronger than higher harmonics, the method is reasonably accurate for application purposes. The other shortcoming of the force-restore method is the use of constant force and restore coefficients, while the soil heat flow is highly dependent on heat capacity and moisture content of the soil.

5.5 Surface Energy Balance

Neglecting the small amount of energy involved in the plant photosynthesis, respiration and decomposition of organic materials, and assuming that the canopy has zero heat capacity, the land-surface energy balance can be written as

$$R_n - \lambda E - H_s - G_0 = 0 \qquad (5.24)$$

with

$$R_n = (1-\alpha)R_s\downarrow + \epsilon R_l\downarrow - \epsilon\sigma T_0^4 \qquad (5.25)$$

where R_n is the surface net radiation, σ is the Stefan-Boltzmann constant, $R_s\downarrow$ is downward shortwave radiation, $R_l\downarrow$ is downward longwave radiation, α is surface albedo, ϵ is the surface emissivity, λ is latent heat of vaporisation, E is the evaporation rate, H_s is sensible heat, and G_0 is the ground heat flux at the surface. In (5.24), the sign convention used is that non-radiative fluxes are positive away from the surface, and radiative fluxes are positive towards the surface.

The emissivity, ϵ, of a substance according to the Kirchhoff law is equal to its absorptivity for the same wavelength range. The emissivities of land surfaces range from 0.90 to 0.99. However, in land-surface schemes usually a universal emissivity is assumed for all surface types. A value of 0.98 can be taken as representative for both soil and green vegetation. Since $|R_l\downarrow - \sigma T_0^4|$ is usually small, this assumption does not lead to large errors in calculating the surface energy balance. The soil surface temperature, T_0, is commonly calculated by an iterative solution of (5.24) until the energy balance is achieved within a specified accuracy. All terms in (5.24) are either directly or indirectly dependent on surface soil moisture condition.

5.5.1 Surface Fluxes

The interactions between the atmosphere and land are realised through the exchanges of energy, mass and momentum. In the first instance, the energy, mass and momentum fluxes take place in the atmospheric surface layer where the Monin–Obukhov similarity theory applies. This theory is generally accepted for describing the surface layer in numerical models.

In land-surface schemes, the bulk aerodynamic method is widely used for the calculation of surface fluxes. For momentum flux, τ_m, sensible heat flux, H_s and latent heat flux λE (λ is the latent heat of vaporisation or sublimation), we have

$$\tau_m = \rho_a \frac{U}{r_{am}} \qquad (5.26)$$

$$H_s = \rho_a c_p \frac{T_0 - T_r}{r_{ah}} \qquad (5.27)$$

$$\lambda E = \lambda \rho_a \frac{q_0 - q_r}{r_{ae}} \qquad (5.28)$$

where r_{am}, r_{ah} and r_{ae} are aerodynamic resistances for the transfer of momentum, heat and moisture from the land surface to the reference height, with

$$r_{am} = (C_m U)^{-1} \tag{5.29a}$$
$$r_{ah} = (C_h U)^{-1} \tag{5.29b}$$
$$r_{ae} = (C_e U)^{-1} \tag{5.29c}$$

where ρ_a is air density, c_p is the isobaric specific heat of air, C_m, C_h and C_e are the bulk transfer coefficients for momentum, heat and moisture respectively; U, T_r and q_r are wind speed, temperature (potential temperature) and specific humidity at the reference height in the atmosphere; T_0 and q_0 are air temperature and specific humidity at the surface. The latter quantity can be further expressed as

$$q_0 = r_h q^*(T_0)$$

where $q^*(T_0)$ is saturated specific humidity at the given surface temperature and atmospheric pressure and r_h is relative humidity of air at the roughness height.

Evaporation is an important quantity, as it links the surface energy and surface water budgets. The total evaporation from the land surface is the result of evaporation from the soil surface, that from the wet fraction of canopy, and transpiration from the dry fraction of canopy. When the surface is wet (saturated soil or wet canopy), i.e., $r_h = 1$ and $q_0 = q^*(T_0)$, evaporation takes place at its potential rate. We obtain from (5.28) that

$$E = E_p = \rho_a \frac{q^*(T_0) - q_r}{r_{av}} \tag{5.30}$$

where E_p is potential evaporation.

An alternative to calculating E_p is the Penman combination method. The crucial step is

$$q^*(T_0) - q_r = s(T_0 - T_r) + \delta q \tag{5.31}$$

where $s = \dfrac{\partial q^*}{\partial T}$ is the slope of saturation specific humidity curve at $\dfrac{T_0 + T_r}{2}$. Combining (5.30) and (5.31) and eliminating temperature differences using (5.27) and (5.24) for saturated surface, and assuming that roughness lengths for humidity and heat are equal, we obtain

$$E_p = \Gamma \frac{R_n - G_0}{\lambda} + \frac{(1 - \Gamma)\rho_a \delta q}{r_{av}} \tag{5.32}$$

with

$$\Gamma = \frac{s}{s + \gamma}$$

and

$$\gamma = \frac{c_p}{\lambda}$$

A simplification is made to (5.32) by Priestley and Taylor (1972). They assumed that over vast water bodies or saturated land areas the effect of large-scale advection on evaporation is negligible, and evaporation occurs in an equilibrium rate that is dependent on available energy and local advection. Therefore

$$E_s = \alpha_{PT} \frac{\Gamma H_s}{r_{av}} \tag{5.33}$$

where α_{PT} is an empirical coefficient of about 1.26, accounting for the local advection effect. In general, α_{PT} is not a constant, but a variable depending on the surface roughness (Garratt, 1992).

5.5.2 Evaporation from Unsaturated Soil

As soil dries, a dry layer forms at the surface, with its thickness dependent on time and the evaporation rate. In this case, evaporation takes place at the interface between the dry and wet soil layers. This location is known as the "location of evaporation sites", "evaporation zone" or the "drying front". Therefore, in unsaturated soils, q_0 depends upon the distance of the drying front from the surface and the rate of vapour diffusion through the soil pore spaces. In this case, q_0 can be calculated as

$$q_0 = q^*|_{(T_s)} - \frac{r_s}{\rho_a} E \tag{5.34}$$

where $q^*|_{(T_s)}$ is the saturation specific humidity at the evaporation zone with temperature T_s, and r_s is the resistance to the vapour diffusion from the evaporation zone to the surface. Rearranging (5.28) for q_0 and denoting surface soil evaporation as E_s we obtain

$$q_0 = q_r + \frac{r_{av}}{\rho_a} E_s \tag{5.35}$$

By assuming steady state condition ($E_s = E$) and equating (5.34) and (5.35), q_0 can be eliminated from the calculation of soil evaporation. It follows that

$$E = \rho_a \frac{q^*|_{(T_s)} - q_r}{r_{av} + r_s} \tag{5.36}$$

The problem with (5.36) is the need to calculate T_s and r_s. A method for calculating r_s is given by Camillo and Gurney (1986). Usually, (5.36) is used to estimate the soil evaporation, with q_0 approximated by the so-called α and β methods (Mahfouf and Noilhan, 1991), namely,

$$q_0 = \begin{cases} \beta q^*|_{(T_0)} + q_r(1-\beta) & \beta \text{ method} \\ \alpha q^*|_{(T_0)} & \alpha \text{ method} \end{cases} \tag{5.37}$$

where $\alpha = r_h$. Therefore, (5.36) can be written as

$$E = \begin{cases} \rho_a \beta \dfrac{q^*|_{(T_0)} - q_r}{r_{av}} & \beta \text{ method} \\ \rho_a \dfrac{\alpha q^*|_{(T_0)} - q_r}{r_{av}} & \alpha \text{ method} \end{cases} \quad (5.38)$$

Several empirical functions have been proposed for β and α (see e.g. Mihailovic et. al., 1995).

5.5.3 Transpiration

Transpiration is the diffusive flux of water vapour from plant organs to the atmosphere. Similar to unsaturated soil, the evaporation zone is not at the surface, but is inside the "valve-like structure" on the leaves, called stomata. Therefore, in the calculation of transpiration it is important to include an additional resistance, called the bulk stomatal or canopy resistance. Figure 5.9 illustrates a simple resistance network against the water flux from the soil and canopy to an atmospheric reference height. Monteith (1965) modified (5.32) for the vegetation covered land by introducing the surface resistance. This modification accounts for the physiological resistance that vegetation imposes on the transfer of vapour from the internal organs to the air, therefore

$$E = \frac{s(R_n - H_g) + \dfrac{c_p \rho}{r_a} \delta q}{\lambda s + c_p(1 + \dfrac{r_s}{r_{av}})} \quad (5.39)$$

Equation (5.39) is applicable for homogeneous vegetation that extensively covers the surface, so that the vegetation can be treated as a big leaf. Under this condition the surface resistance is often dominated by the bulk stomatal resistance (Sellers et al., 1992). Evaporation from the interior of the stomata to the leaf surface can be written as

$$E = \rho_a \frac{q^*|_{(T_s)} - q_s}{r_{st}} \quad (5.40)$$

where r_{st} is the bulk stomatal resistance and T_s is the leaf temperature taken to be equal to the leaf surface temperature.

Correspondingly, evaporative flux from the dry leaf surface to the canopy source height level, $z_d + z_0$, and from $z_d + z_0$ to the reference height are

$$E = \rho_a \frac{q_s - q_0}{r_{0v}} \quad (5.41)$$

$$E = \rho_a \frac{q_0 - q_r}{r_a} \quad (5.42)$$

where r_{0v} and r_a are canopy air resistance and resistance to water vapour flux from the canopy source height to the reference height. Assuming continuity

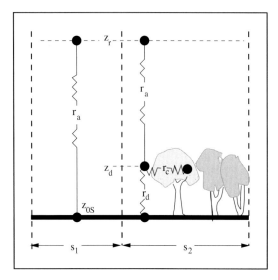

Fig. 5.9. Resistance network to vapour flux from the soil and dry canopy to the atmospheric reference height.

of the evaporative flux at the leaf surface and eliminating the unknown q_s by combining (5.40) and (5.41), we obtain

$$E = \rho_a \frac{q^*|_{(T_0)} - q_0}{r_{st} + r_{0v}} \tag{5.43}$$

Similarly, assuming continuity at canopy level and eliminating the need to q_0 between (5.42) and (5.43) yields

$$E = \rho_a \frac{q^*|_{(T_0)} - q_r}{r_{st} + r_{0v} + r_a} \tag{5.44}$$

Most land surface schemes, however, do not explicitly parameterise the canopy air resistance, and use the concept of bulk canopy resistance, r_c, the resistance from the leaf sub-stomatal cavities to the canopy source height, where $r_c = r_{st} + r_{0v}$.

5.5.4 Bulk Transfer Coefficients

The Monin-Obukhov similarity theory is widely used for estimating the bulk transfer coefficients (e.g. Stull, 1988). Based on this theory the bulk transfer coefficients, C_m and C_h can be expressed as

$$C_m = \frac{\kappa^2}{(\ln \frac{z_r - z_d}{z_{0m}} - \psi_1)^2} \tag{5.45}$$

$$C_h = \frac{\kappa^2}{(\ln \frac{z_r - z_d}{z_{0m}} - \psi_1)(\ln \frac{z_r - z_d}{z_{0h}} - \psi_2)} \tag{5.46}$$

where z_{0m} and z_{0h} are surface roughness lengths and ψ_1 and ψ_2 are stability corrections (e.g. Paulson, 1970) for momentum and heat transfer respectively, and κ (≈ 0.4) is the von Karman constant. The expressions for ψ_1 and ψ_2 given by Paulson are

$$\psi_1 = \begin{cases} 5\frac{z_r}{L} & \frac{z_r}{L} > 0 \\ \ln \frac{(1+x_1)^2(1+x_1^2)}{8} - 2\tan^{-1}(x_1) + \frac{\pi}{2} & \frac{z_r}{L} \le 0 \end{cases} \tag{5.47}$$

$$\psi_2 = \begin{cases} -5\frac{z_r}{L} & \frac{z_r}{L} > 0 \\ 2\ln \frac{1+x_1^2}{2} & \frac{z_r}{L} \le 0 \end{cases} \tag{5.48}$$

with

$$x_1 = (1 - 16\frac{z_r}{L})^{\frac{1}{4}} \tag{5.49}$$

The Monin-Obukhov length, L, is defined as

$$L = \frac{-u_*^3 c_p \rho_a T_a}{\kappa g H_s} \tag{5.50}$$

where g is acceleration due to gravity and the friction velocity, u_*, is given by

$$u_* = \frac{\kappa U_r}{(\ln \frac{z_r - z_d}{z_{0m}} - \psi_1)} \tag{5.51}$$

The bulk transfer coefficients are sensitive to the values of z_{0m} and z_d. These values are commonly assumed to be a constant fraction of the vegetation height, h_c, e.g. $z_{0m} = 0.13 h_c$, and $z_d = 0.64 h_c$ for crops and grass canopies and $z_{0m} = 0.06 h_c$, and $z_d = 0.8 h_c$ for forests. This assumption, however, is not realistic. Large differences exist in z_{0m}/h_c and z_d/h_c for different stands, suggesting that z_{0m} and z_d depend on both h_c and the density of the vegetation (Λ) (Raupach et al., 1991). While z_d/h_c increases with Λ, z_{0m}/h_c increases with vegetation roughness density, Λ, to a maximum and then decreases with Λ because of mutual sheltering of roughness elements. Raupach (1992, 1994) has presented analytical expressions for z_{0m} and z_d as functions of h_c and Λ for horizontally isotropic rough surfaces, namely,

$$z_d = h_c \left(1 - \frac{1 - e^{-x_2}}{x_2}\right) \tag{5.52}$$

$$z_{0m} = h_c \left[\frac{1 - e^{-x_2}}{x_2} \exp\left(1 - \ln C_w - \frac{1}{C_w} - \kappa \frac{U_h}{u_*}\right)\right] \tag{5.53}$$

with
$$x_2 = \sqrt{C_d \Lambda}$$
and
$$\frac{u_*}{U_h} = \min\left(\sqrt{(C_s + C_r \frac{\Lambda}{2}}, \left.\frac{u_*}{U_h}\right|_{\max}\right)$$

where C_r and C_s are drag coefficients of an isolated roughness element and of the substrate surface at canopy height in the absence of roughness elements, respectively. U_h is wind speed at canopy height, C_d is a free parameter, and $C_w = \frac{z_w - z_d}{h_c - z_d}$ with z_w the upper height limit.

The roughness lengths for heat and momentum are not equal (e.g. Garratt, 1992). Measurements show that the roughness length for heat, z_{0h}, is much smaller than that for momentum, z_{0m}. The difference between these two quantities arises from the fact that momentum transfer is augmented by local pressure gradients (from drag), while heat transfer near the surface is controlled by molecular diffusion. While z_{0m} has a marked dependence on the details of the upwind terrain, z_{0h} shows little variability and is mainly a function of the surface between the major roughness elements. Depending on the terrain and surface cover, the ratio of z_{0h}/z_{0m} changes considerably. For instance, it is approximately 0.5 for soybean crops (Hicks and Wesely, 1981), 0.2 for bean crops (Thom, 1975), 0.083 for a heterogeneous surface (Garratt, 1978), and between 1/7 and 1/12 for crop-lands (Brutsaert, 1982).

5.5.5 Canopy (Vegetation) Resistance

Bulk stomatal resistance, or canopy resistance, is a sub-grid scale variable which is difficult to measure. In most land-surface models, the stomatal resistance of a leaf, R_{st}, is calculated and the bulk stomatal resistance, r_{st}, is obtained by assuming all leaves of a canopy to operate in parallel using an analogy of Ohm's law. The common assumption is that

$$r_{st} = \frac{R_{st}}{LAI} \tag{5.54}$$

where LAI is leaf-area index.

It is known that R_{st} is a complicated function of many environmental factors. Jarvis (1976) proposed a model for R_{st} as function of radiation, ambient air CO_2 concentration, atmospheric vapour pressure deficit, leaf temperature and leaf water potential. The minimum values of R_{st} vary between 7–500 s m^{-1} with a mode of 167 s m^{-1} for herbaceous plants, and between 200–1000 s m^{-1} with a mode value of 333 s m^{-1} for woody plants.

Experiments show that R_{st} depends only weakly on leaf water potential, but much more strongly on soil moisture, especially in dry soils. This latter dependency is triggered by a metabolic signal from the roots (Schulze, 1986).

The Jarvis model has been modified by many others (e.g. Noilhan and Planton, 1989), mainly by replacing the the leaf-water potential function by one of soil moisture in the root zone. Most of the modified models also omit the impact of CO_2 concentration, assuming that it is almost constant with time. Following Noilhan and Planton (1989), we have

$$R_{st} = R_{s\,min} F_1 F_2^{-1} F_3^{-1} F_4^{-1} \tag{5.55}$$

In (5.55), F_1 represents the effect of the photosynthetically active radiation (Dickinson et al., 1986)

$$F_1 = \frac{1+f}{f + \frac{rs_{min}}{rs_{max}}} \tag{5.56}$$

with

$$f = \frac{1.1}{LAI} \frac{R_s}{R_{sl}} \tag{5.57}$$

where R_s is solar radiation and R_{sl} is a limiting value of $30\,\mathrm{W m^{-2}}$ for forest and of $100\,\mathrm{W m^{-2}}$ for crops and grassland. The factor F_2 describes the effect of water stress on the canopy resistance,

$$F_2 = \begin{cases} 1 & \theta > \theta_f \\ \frac{\theta - \theta_w}{\theta_f - \theta_w} & \theta_w < \theta < \theta_f \\ 0 & \theta < \theta_w \end{cases} \tag{5.58}$$

where θ is the mean soil moisture in the root zone, θ_w is the wilting point and θ_f is the field capacity. F_3 represents the effect of the water vapour deficit,

$$F_3 = 1 - 0.6\delta q \tag{5.59}$$

and F_4 is the factor describing the effect of the air temperature

$$F_4 = 1 - 0.0016(298 - T_a)^2 \tag{5.60}$$

5.5.6 Canopy Water Storage

Most land surface models treat the canopy water storage with the method proposed by Deardorff (1978) to calculate the wet fraction of a canopy, that is the 2/3 power law,

$$L_w = (W_{dew}/W_{d\,max})^{2/3} \tag{5.61}$$

where L_w is the fractional area of leaves covered by water, W_{dew} is the water stored on the canopy surface and $W_{d\,max}$ is the maximum water the canopy can hold, defined as

$$W_{d\,max} = \xi\, L_{SAI}$$

where ξ is a coefficient ranging from 0.05 to 0.5 and L_{SAI} is the leaf-stem area index.

5.5.7 Surface Albedo

Albedo depends on surface characteristics, solar geometry, and the spectral distribution of incident solar radiation. The typical spectral dependency of albedo of some of the common surfaces are shown in Fig. 5.10. The spectral properties of leaves make them highly absorbent in the visible, but reflective in the near-infrared wavelengths of the solar spectrum. As seen in Fig. 5.10, a sharp discontinuity exists in the green leaf albedo at the wavelength of about $0.72\,\mu$m.

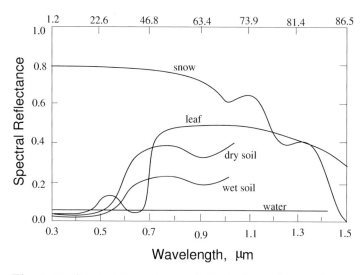

Fig. 5.10. Spectral dependency of albedo for various surfaces. The numbers at the top of the frame represent the cumulative percentage of solar radiation at wavelengths shorter than indicated (after Verstraete and Dickinson, 1986).

The parameterisation of surface albedo varies in different land-surface schemes. Splitting the solar spectrum into visible and near-infrared regions for calculating surface albedo is used, for instance, by Dickinson et al. (1993). Sellers et al. (1986) and Xue et al. (1991) not only consider the radiation wavelength, but also differentiate between the direct and diffused radiation in calculating surface albedo. However, since there is no evidence for a change in the mean spectral mix of incident solar radiation, most land-surface schemes (e.g. Irannejad and Shao, 1998) use a single all-spectrum albedo for each surface type.

The dependency of surface albedo on the solar angle for wet and dry sand is shown in Fig. 5.11. Although some land-surface schemes (e.g. Dickinson et al., 1993) take into account the variation of surface albedo with solar zenith angle, in most schemes (e.g. Noilhan and Planton, 1989) this variation is not considered.

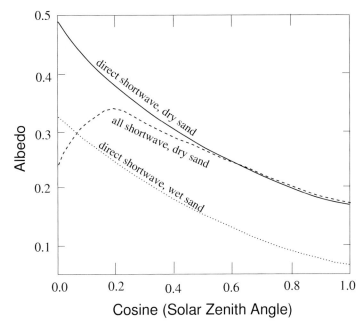

Fig. 5.11. Theoretical variation of albedo with solar zenith angle for a flat sand surface. The solid line is for dry and the dotted line is for wet sand. The dashed curve is the same as the soild line except that it includes the diffuse light contribution to the surface albedo (after Dickinson, 1983).

Surface albedo also depends on soil moisture. In land-surface schemes, this is usually considered by using the empirical relationship derived by Idso et al. (1975)

$$\alpha_s = \begin{cases} \alpha_{\mathrm{dry}} + \frac{\theta_1}{\theta_c}(\alpha_{\mathrm{sat}} - \alpha_{\mathrm{dry}}) & \theta_1 < \theta_c \\ \alpha_s = \alpha_{\mathrm{sat}} & \theta_1 \geq \theta_c \end{cases} \tag{5.62}$$

where α_{sat} and α_{dry} are the prescribed values of albedo at saturation and air-dry conditions respectively, θ_1 is soil moisture of the top soil layer, θ_c (~ 0.2) is a critical soil moisture above which α_s is independent of soil moisture.

The presence of snow significantly affects the surface albedo. There is once again a variety of different parameterisations of the snow effect on the surface albedo. Dickinson et al. (1993) proposed to parameterise α_s in the presence of snow by taking into account the spectral mix of incident radiation, solar zenith angle, soot loading of the snow, and snow depth, density and grain size. Robock et al. (1995) only considered it to be a function of snow depth and surface temperature.

5.6 Carbon-dioxide Budget

The terrestrial biosphere exchanges about 60 Gt CO_2-C with the atmosphere each year (Box, 1988; Potter et al., 1993). The concentration of atmospheric CO_2 has increased steadily since the beginning of the industrial era, and may continue to increase in the near future. As CO_2 is a principal greenhouse gas, changes to the global carbon cycle are important. The carbon cycle is linked closely to the water/energy cycles through the regulation of stomata during photosynthesis. Photosynthesis is the source of biomass production over land and the continental vegetation is one of the major CO_2 sinks for the atmosphere.

The net CO_2 flux from the surface is the difference between CO_2 uptake during plant production and CO_2 loss during respiration. The calculation of this net flux depends very much on the simulation of net primary productivity (NPP) and soil heterotrophic respiration (SHR). Several biospheric models have been developed in recent years to estimate NPP, SHR and net ecosystem productivity (NEP) (e.g. Foley, 1994). The seasonal variations of NEP are largely responsible for the variations of the atmospheric CO_2 concentration, which have been observed for several decades at a number of stations located at various latitudes (Nemry et al., 1996). This data can be used to test the performance of biospheric models if they are coupled with a 3-D atmospheric transport model to simulate the temporal and spatial distribution of atmospheric CO_2 simulated by the model NEP. Such a method has been applied by Fung et al. (1983) and Knorr and Heimann (1995) using remote sensing data to force the NEP calculation. Predictive models of terrestrial NPP have been developed by Warnant et al. (1994) and Bonan (1995). These models use submodels of leaf assimilation for C_3 species (Farquhar et al., 1980) and for C_4 species (Collatz et al., 1991). An example of coupling CO_2 uptake by photosynthesis and CO_2 loss during plant and soil microbial respiration with a land surface scheme for examining the global simulation of NPP, SHR and other CO_2 fluxes is presented by Bonan (1995).

5.7 Verification of Land-surface Schemes

In recent years, a number of field experiments have been carried out. These experiments have provided valuable data sets for the evaluation of land-surface schemes for given locations and regions. Some of these observational data sets that serve as the basis for evaluating the performance of land-surface schemes for different climatic regimes are listed here

1. ABRACOS (The Anglo-Brazilian Amazonian Climate Observation Study): Brazil (Gash and Nobre, 1997);
2. BOREAS (The Boreal Ecosystems Atmosphere Study): Canada (Sellers et al., 1995);

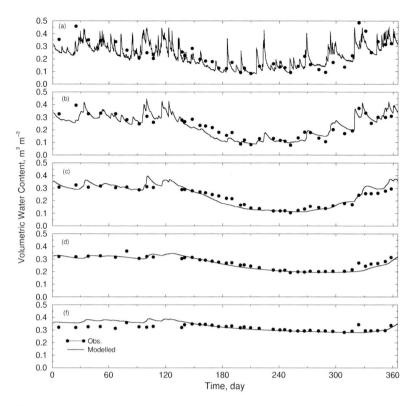

Fig. 5.12. Comparison of simulated annual cycle of soil moisture with observations from HAPEX–MOBILHY. The thicknesses of the 5 soil layers are 0.05, 0.15, 0.3, 0.5 and 0.6 m, respectively.

3. Cabauw: the Netherlands (Van Ulden and Wieringa, 1996);
4. FIFE: The First ISLSCP (International Satellite Land Surface Climatology Project) Field Experiment, United States (Betts and Ball, 1998);
5. HAPEX-EFEDA (Hydrological and Atmospheric Pilot Experiment-Echival Field Experiment in a Desertification-threatened Area): Spain (Bolle et al., 1993);
6. HAPEX-MOBILHY (Hydrological and Atmospheric Pilot Experiment – Modelisation du Bilan Hydrique): France (Andre et al., 1986);
7. HAPEX-Sahel (Hydrological and Atmospheric Pilot Experiment in the Sahel): Niger (Goutorbe et al., 1994);
8. Red-Arkansas River basin: United States (Wood et al., 1998);
9. SALSA (The Semi-Arid Land-Surface-Atmosphere Program): United States and Mexico (Chehbouni et al., 2000);
10. Valdai: Russia (Vinnikov et al., 1996).

Many existing land-surface schemes have been calibrated off-line with pre-specified atmospheric data and land-surface parameters. As recent studies have shown (e.g. Shao and Henderson-Sellers, 1996; Chen et al., 1997), most of these land-surface schemes perform well for the simulation of a wide range of land surface quantities, including surface-energy fluxes and soil moisture.

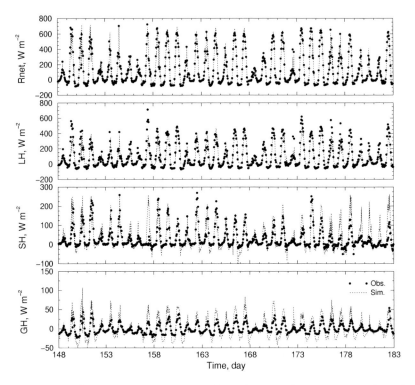

Fig. 5.13. Comparison of simulated diurnal cycle of net radiation (Rnet), latent heat (LH), sensible heat (SH), and ground heat flux (GH) with observations from HAPEX for the observational period of 28 May-30 June 1987.

As an example, Figure 5.12 compares the modelled soil moisture of the five soil layers with the HAPEX–MOBILHY observations (André et al., 1986). This data set has been used in the Phase 2b of the Project for Intercomparison of Land-surface Parameterization Schemes (Henderson-Sellers et al., 1995; Shao and Henderson-Sellers, 1996). It is first seen that the particular land-surface scheme correctly predicts the annual cycle of soil moisture of all layers. Frequent rainfall and low evaporation keep soil moisture close to the field capacity for the first four months of the year; as precipitation decreases and available energy for evaporation increases, soil water begins to decrease at the beginning of the growing season (early May) and reaches a minimum during August to October. Further, the land-surface scheme, correctly captures the diurnal variations of soil moisture in the top soil layers, as can be seen in Fig.5.12a, where both simulated and observed soil moisture fluctuate rapidly under the influence of precipitation and evapotranspiration. Figure 5.13 compares the simulated and observed diurnal cycles of surface net radiation and the non-radiative heat fluxes for the Intensive Observational Period (IOP) of HAPEX. The scheme correctly predicts the surface net radiation and the diurnal trend of sensible heat flux. However, the simulated sensible flux has slightly larger diurnal amplitude than that observed. This is in association with the model's underestimation of evaporation during the day and its overestimation at nights.

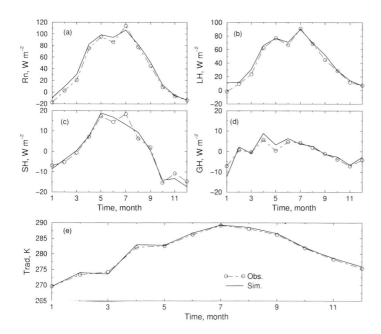

Fig. 5.14. Comparison of monthly mean surface energy fluxes and surface temperature simulated using a land-surface scheme with observations from Cabauw (from Irannejad and Shao, 1998).

The Cabauw data have been used for the evaluation of a number of land-surface schemes in PILPS Phase 2a (Chen et al., 1997). The data are collected at a 213 m tall meteorological mast in Cabauw, the Netherlands (51° 58′N, 4° 55′E) (Beljaars and Bosveld, 1997). This data set is most suitable for the evaluation of land-surface schemes when they are applied to grassland under a moderate maritime climate. Figure 5.14 compares the monthly means of simulated surface heat fluxes and surface temperature with observations from Cabauw. The scheme captures the annual trend of observed fields very well, apart from slight overestimations of net radiation and sensible heat flux. The simulated monthly latent heat is consistent with observations, with exceptions in January and May. The scheme slightly overestimates the annual range of ground heat flux. The simulated diurnal fluctuation of surface fluxes and temperature for the period of Julian days 253 to 262, are compared with observations in Fig. 5.15. Clearly, the scheme can also simulate observations once time spans. Figure 5.16 shows the good agreement between simulations and observations on different time scales. Except for the ground heat flux, the slope of regression lines are close to 1. The correlation coefficients vary between 0.990 to 0.996 for net radiation, 0.986 to 0.999 for surface temperature, 0.928 to 0.992 for latent heat flux and 0.890 to 0.929 for sensible heat flux over different time scales. The ground heat flux shows not only the steepest regression coefficient, but also has the largest deviations from the regression line. The correlation coefficients between simulated and observed ground heat vary between 0.770 to 0.929 from half-hourly to monthly time scales. The standard error of coefficients is generally small.

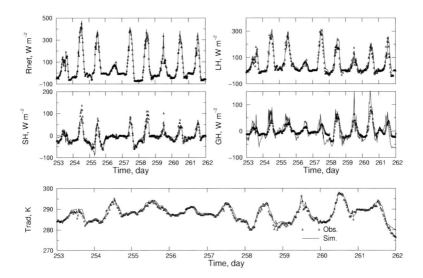

Fig. 5.15. Comparison of diurnal cycles of surface fluxes and surface temperature simulated using a land-surface scheme with observations from Cabauw for the Julian days 253 to 262 (from Irannejad and Shao, 1998).

5.8 Land Surface Modelling over Heterogeneous Surfaces

Most of the land surface schemes which are currently in use in atmospheric models are based on the 'big leaf' concept, which implies that a model grid cell is homogeneously covered by one (or sometimes more) big leaf. However, the land surface is rarely homogeneous over the resolvable scales of

5.8 Land Surface Modelling over Heterogeneous Surfaces

atmospheric models (~100 km × 100 km for GCMs and ~10 km × 10 km for the NWP model). This can be seen, for instance, in the maps of soils and vegetation (Fig. 5.17). For modelling purposes, effective parameters are defined for each grid cell either as those of the most frequently existing surface type or by averaging the properties of all types occurring within the grid cell based on the fractional cover of each surface type. Some averaging methods take local nonlinear processes into account (Mahrt et al., 1992). The use of effective grid-averaged parameters reduces the computational costs and can be useful for some practical applications, especially in conditions of moderate heterogeneity (e.g. Giorgi and Avissar, 1997).

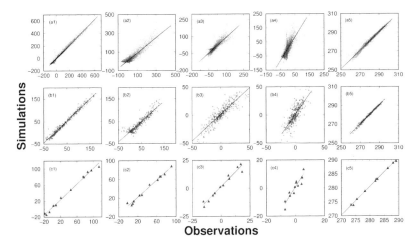

Fig. 5.16. Scatter diagrams of simulated vs observed net radiation (**1**), latent heat flux (**2**), sensible heat flux (**3**), ground heat flux (**4**) and surface temperature (**5**) at half–hourly (**a**), daily (**b**), and monthly (**c**) time scales for Cabauw.

Due to the nonlinear nature of many land-surface processes, heterogeneity may affect the exchanges of water and energy between the surface and the atmosphere. For instance, the high degree of soil moisture variability and the strongly nonlinear relationship between evapotranspiration and soil moisture may result in a relationship between regional evapotranspiration and area-averaged soil moisture that is fundamentally different from the relationship that applies at a point (Wetzel and Chang, 1987). In these situations the grid-averaged surface fluxes are substantially different from those calculated using the grid-averaged surface properties. Various approaches have been developed for examining the role of subgrid-scale heterogeneity in land surface characteristics. These can be classified into models of dynamic effects (e.g. Avissar and Schmidt, 1998) and aggregation effects (e.g. Avissar and Pielke, 1989).

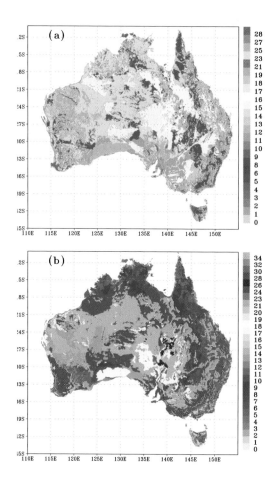

Fig. 5.17. (a) Soil and (b) vegetation types over Australia.

5.8.1 Models of Dynamic Effects

Models of dynamic effects aim at describing the effects of surface heterogeneity on microscale and mesoscale atmospheric circulations. Studies of the dynamic effects of surface heterogeneity can be found in, for instance, Mahfouf et al., (1987) and Seth and Giorgi (1996), among many others. The land surface discontinuity triggers the creation of meso-scale circulations (Fig. 5.18), which are affected by latitude and different background conditions such as wind velocity, thermal stratification and the humidity profile of the atmosphere (Chen and Avissar, 1994). These meso-scale fluxes may affect the cloud formation and the intensity and distribution of convective precipitation. Land surface heterogeneity in general enhances shallow convective precipitation due to two mechanisms which are controlled by spatial distribution of water availability for evapotranspiration, namely turbulence and meso-scale circulations. Lynn et al. (1998) used a two-dimensional cloud-resolving model to study the generation of deep moist convection over heterogeneous landscapes by prescribing different arrangements of dry and wet patches. They showed that rainfall occurred most intensely along sea-breeze-like fronts which formed at patches boundaries. Figure 5.19 shows the timing and distribution of rainfall over the heterogeneously wet/dry patches for a case study.

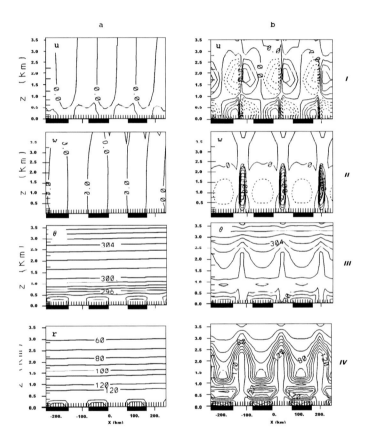

Fig. 5.18. Two-dimensional sections of (**i**) the horizontal wind component u parallel to the domain (ms^{-1}), positive eastward; (**ii**) the vertical wind component w (cms^{-1}), positive upward; (**iii**) the potential temperature θ (K); and (**iv**) the mixing ratio obtained at (**a**) 0900 and (**b**) 1800 LST. Irrigated areas are indicated by dark underbars. Solid lines indicate positive values, and dashed lines indicate negative values (after Chen and Avissar, 1994).

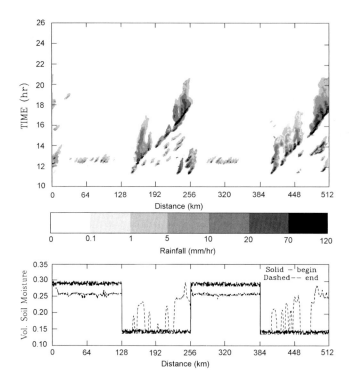

Fig. 5.19. Spatial distribution of soil moisture (bottom) and time and spatial distribution of rainfall (top). The soil moisture shown is for the top soil layer (0.01–0.1 m) at the beginning and end of the model simulation. The initial soil moisture distribution in the other soil layers was the same as in the top layer (after Lynn et al., 1998).

5.8.2 Models of Aggregation Effects

The objective of models of aggregation effects is to calculate the contribution of subgrid-scale heterogeneity of land surface characteristics in the grid-averaged energy, mass and momentum fluxes. Two methods are commonly used in land-surface schemes, namely, the continuous and the discrete methods.

Continuous methods. Continuous representation of surface heterogeneity involves using an analytical or empirical probability density function (PDF) to define statistically the spatial distribution of heterogeneous variables, and integrating relevant processes over the PDF. Avissar (1992) applied this method to describe the effect of heterogeneity in canopy resistance in a simple canopy model. To do this, Avissar integrated the surface energy balance equation as

$$\int f_{\text{PDF}}(r_c)(R_n - H_s - \lambda E - G_0)\mathrm{d}r_c = 0 \tag{5.63}$$

This equation was simplified to

$$a\,T_c^4 + b\,T_c + c\,r_c \mathrm{e}^{f(T_c)} + d = 0 \tag{5.64}$$

where T_c is the canopy temperature and a–d are coefficients.

To calculate the area-averaged surface fluxes, $f_{\text{PDF}}(r_c)$ can be divided into a number of intervals and Equation (5.64) can be solved iteratively using the stomatal resistance of each interval, $r_{c,i}$, for canopy temperature of the interval, $T_{c,i}$. $T_{c,i}$ is then used to calculate the integrals of each term in (5.63). Avissar (1992) used five different types of PDFs and found large differences between the results. The same approach, but using PDFs of stomatal resistance, leaf-area index, surface roughness, albedo and soil wetness (Johnson et al., 1993; Li and Avissar, 1994), has shown that latent and sensible heat fluxes were sensitive and radiative fluxes were not so sensitive to spatial variability.

Although the continuous PDF can be used to represent a wide range of heterogeneity scales which occur within a grid cell, its application to complicated land surface schemes encounters a number of difficulties. For numerical solutions, the PDF needs to be divided into different intervals, over each of which the land-surface simulation is performed. If the number of intervals is too small, the shape of the PDF will be distorted, while if the number of intervals is too large, the simulation is computationally too costly. A land-surface scheme usually incorporates a large number of processes with many input parameters. To account for the heterogeneity of all these processes requires a multi-dimensional joint PDF which can be extremely difficult to integrate numerically. Incorporating single variable functions for N variables, each divided into I intervals requires the solution of the land surface scheme's equations over I^N space nodes over a single grid box. This is computationally equivalent to solving the land-surface scheme over a very fine resolution space, but changing the geographical spacing to parameter spacing. Furthermore, the aggregated surface energy fluxes are sensitive to the choice of PDF as shown by Avissar (1992).

Discrete methods. In the discrete representation of subgrid-scale heterogeneity, it is assumed that the grid cell is covered by a number of homogeneous subgrids. The surface calculations are performed separately for each subgrid and averaged by weights of their relative surface areas to provide the grid-averaged state variables and fluxes. The simplest and computationally most efficient approach to represent surface heterogeneity is the so-called mixture method (Koster and Suarez, 1992). In this method, the surface is assumed to be covered by a homogeneous mixture of surface types (vegetation, bare soil, snow) with a tightly coupled energy balance. Different surface types simultaneously interact with the soil and with an interface layer in the atmosphere.

The usually nonlinear flux-state variable relationship suggests that the mixture method is not adequate when the state variables are very different in various fractions of the grid box. One way to overcome this shortcoming is the mosaic method (e.g. Avissar and Pielke, 1989), in which each subgrid area directly exchanges fluxes with the atmosphere independent of other subgrids. In the mosaic approach all the areas having the same properties are combined in a tile. This makes the inclusion of climatic distribution in the models difficult. Assigning an individual surface type and elevation to each subgrid (explicit method), based on high-resolution geographical information, may provide realistic results. However, this approach is computationally impractical for implementing into atmospheric models for global or regional simulations. Figure 5.20 illustrates the introduction of mosaic approach and explicit subgrid method to a atmospheric model grid box.

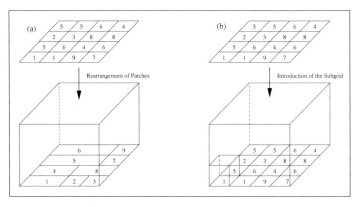

Fig. 5.20. Schematic plot of an atmospheric grid cell at the boundary Earth-atmosphere exemplary showing how the land-use types within a grid cell are rearranged in the mosaic approach (**a**) and how the explicit subgrid is introduced in the explicit subgrid strategy; (**b**) for land use-types within an atmospheric grid cell. The upper parts show the land-use types as given for the respective atmospheric grid cell by the land-use data set and the number indicate different land-use types. In the explicit subgrid strategy the small box represents a subgrid cell (adopted from Mölder et al., 1996).

Studies on the impacts of the inclusion of subgrid-scale heterogeneity of surface properties and climate conditions have provided different, and sometimes contradictory, results. Koster and Suarez (1992) found only small differences in the surface fluxes computed using the mixture and the mosaic methods. Mölders et al. (1996) compared the effects of the mosaic approach and the explicit subgrid-scale surface heterogeneity on the predicted hydrologically relevant variables. They found significantly different partitioning of atmospheric radiation and surface moisture forcing using two different strategies. Their results showed a reduction in the surface evaporation when the area-weighted near surface atmospheric forcing was used. Ghan et al. (1997) found that subgrid-scale variations in precipitation had the largest impact on the regional evaporation. While neglecting spatial variations in precipitation increased the summertime evaporation by 15%, the increase due to neglecting the spatial heterogeneity of vegetation and soil were on the order of 4% and 2% respectively. On the other hand, Mahrt and Sun (1995) used data of three field experiments and concluded that assuming spatially constant atmospheric forcing and spatially varying surface properties provided a close approximation of the area composite fluxes.

5.9 Land-surface Parameters

Depending on the complexity of land-surface schemes, the number of parameters required for specifying land surface properties varies from a few to the order of ten. Numerous studies have shown that the performance of land-surface schemes are sensitive to the choice of land-surface parameters (Ek and Cuenca, 1994; Shao and Irannejad, 1999). While different land-surface schemes may use different parameters, they can be divided roughly into categories for surface vegetation, aerodynamics, radiation, soil hydraulics and thermal properties, as listed in Table 5.2.

Table 5.2. Categories of parameters required by land-surface schemes.

Categories	Parameters (examples)
Vegetation	Leaf-area index
	Fractional vegetation cover
	Canopy height
	Root distribution
Aerodynamics	Surface and canopy roughness length
	Zero-displacement height
Radiation Properties	Surface and canopy albedo for visible light and NIR light
	Thermal emissivity
Hydraulic Properties	Field capacity, Wilting point
	Saturation matric potential, hydraulic conductivity
	Other model dependent parameters
Thermal Properties	Thermal conductivity, heat capacity

For improved parameterisation of land surfaces in GCMs the availability of global archives of soils and vegetation are an essential prerequisite. Most of the available global data sets have been compiled using the existing national atlases of soil and vegetation. Wilson and Henderson-Sellers (1985) described an archive that consists of current land cover types and a global soil data base that includes soil color, texture and drainage characteristics. A global ecotype data set has been developed by Olson et al. (1983) in which vegetation types are classified based on carbon density or biomass.

The most complete global land-surface dataset has been compiled by the International Satellite Land Surface Climatology Project (ISLSCP) Initiative I data collection (Sellers et al., 1992). This data set has a spatial resolution of $1° \times 1°$, a time span of 24 months (1987–1988) and a monthly temporal resolution for most of the parameters. The collection consists of datasets of vegetation, hydrology, soils, snow, ice, ocean, radiation, clouds, and near-surface meteorology.

Surface properties continuously vary with time. Therefore, it is also necessary to reactualise the surface parameters at time intervals from a few days (phenological events) to several years (anthropogenic changes). The heterogeneity of the surface makes it impossible to regularly re-establish new sets of large scale parameters by ground survey, or laboratory and field measurements. Remote sensing offers an attractive alternative. The existing satellite observations provide information about the general state of the surface at three different scales; the field scale (the Thematic Mapper of Landsat), and intermediate scale of 1 to 5 km (the AVHRR of the NOAA satellites) and regional or mesoscale of 20 to 100 km which is obtained by averaging the data taken at the two other scales.

Using the visible ($0.65\,\mu$m) and near infrared ($0.85\,\mu$m) radiance a distinction between bare and vegetated soils can be made at the field scale. It is then possible to determine the dominant nature of the surface at regional scale. This information can be used for estimating the canopy cover and, provided that there is a dominant vegetation type over the region, for setting parameters such as the maximum stomatal conductance. At regional scale, the normalised difference vegetation index (NDVI) calculated using the visible and near infrared images can be used to estimate the leaf area index.

5.10 The Impact of Land Surface on Climate and Weather: Some Case Studies

Sensitivity studies with numerical weather prediction models have shown that initially wet surfaces tend to produce more local rainfall and that evolution of meteorological conditions, e.g. cyclonic disturbances, differ for different land surfaces. Larger values of initial soil moisture may delay the onset of precipitation but increase its amount (Clark and Arritt, 1995). Observations have shown that high antecedent soil-moisture conditions increase the intensity of precipitation associated with storm activities and affect interstorm duration (Caporali et al., 1996). Analysis of a 14-year soil-moisture data set from the State of Illinois (Hollinger and Isard, 1994) has shown that extreme soil moisture availability/lack acts as either a feedback mechanism maintaining the wet/dry conditions established in the beginning of each summer or as a flag

5.10 The Impact of Land Surface on Climate and Weather: Some Case Studies

indicative of some large-scale process that affects both the soil moisture and the precipitation regime (Findell and Eltahir, 1997).

One of the great land-surface perturbations worldwide is the strong desertification in the Sahel and to a lesser degree in many semi-arid regions around the world. The impacts of desertification on climate have been studied mainly using numerical models. For instance, Charney (1975) studied the desertification impact on the African climate by increasing the surface albedo and found that this reduces precipitation. This finding has been confirmed by a number of studies using different global circulation models (e.g. Laval and Picon, 1986). Similar sensitivity studies on the impact of soil moisture on African climate revealed that initially drier soils would lead to less precipitation. In most of these studies, it has been assumed that desertification changes only one of the surface parameters. Apart from this unrealistic assumption, these studies also imposed changes on surface properties which were more or less arbitrary and usually far greater than what might be expected following desertification.

Xue and Shukla (1993) have conducted a set of numerical experiments for desertification in the Sahel by changing the natural broadleaf shrub to the bare soil. They altered all vegetation and soil parameters of braodleaf shrub to those of bare soil, according to their definitions given in the SiB land-surface scheme (Sellers et al., 1986). They also limited the magnitude of the changes to within a reasonable range, on the basis of remotely sensed data. For instance, they altered the surface albedo in the regions of severe desertification (as described by Dregne, 1977) by 0.09, much less than what was used by Charney (1975). The numerical experimets were conducted for three months (June–August) with five sets of initial and boundary conditions for the natural and deforested cases. The ensemble results of the two cases were compared to determine the possible impact of the Sahel desertification on climate. They found that the average rainfall decreases by more than 1.5 mm d^{-1} in the desertified area, especially in the western region, but increases to the south of that area (Fig. 5.21a). This variation in the rainfall pattern has been attributed to the changes in the wind field. One of the main moisture sources is the large scale, low level, moist, southwest airflow from the Atlantic Ocean, which becomes weaker due to the effect of desertification, while the hot dry flow from the northeast becomes stronger (Fig. 5.21b). These changes in the flow field lead to reduced moisture convergence and cloud cover over the desertified area. As a consequence, the mean latent heat flux (Fig. 5.21c) descreases accompanied with increased surface temperature (Fig. 5.21d).

Xue (1996) has also studied the impact of the desertification of the Mongolian grassland on the climate of the region, by considering the extent of the land-surface changes over a period of 40 years. Based on observations in the desertified area, the vegetation cover has decreased from 70–80% to 30–40% and the height of the predominant herbs from over 60 cm to about 40–50 cm. The simulation period was 90 days in summer (June–August). The rainfall in

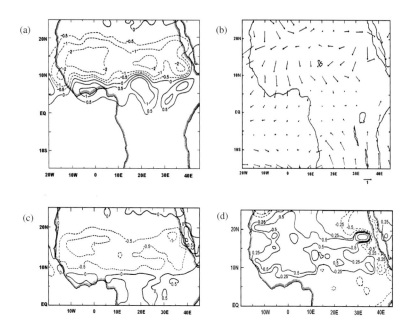

Fig. 5.21. Changes in the 3-month (JJA) mean of (**a**) precipitation rate, mm d^{-1}, (**b**) wind field at 850 mb, m s^{-1}, (**c**) latent heat flux, W m^{-2} and (**d**) surface temperature, K, in the Sahel due to desertification (adopted from Xue and Shukl, 1993).

the area and in China during this period is mainly due to the Indian and the East Asian summer-monsoon airflows. Any substantial changes in these flow patterns may affect all aspects of the regional climate. Figure 5.22a shows the wind vector differnce at 700 hPa between the desertification and natural grassland cases. The changes in the mean wind field have inceased the vertically-integrated moisture convergence, but not large enough to compensate for the reduction of evaporation from the region (Fig. 5.22b). As a result, rainfall (Fig. 5.22c) and soil moisture decreased, while surface temperature (Fig. 5.22d) increased in the desertified area.

5.10 The Impact of Land Surface on Climate and Weather: Some Case Studies 213

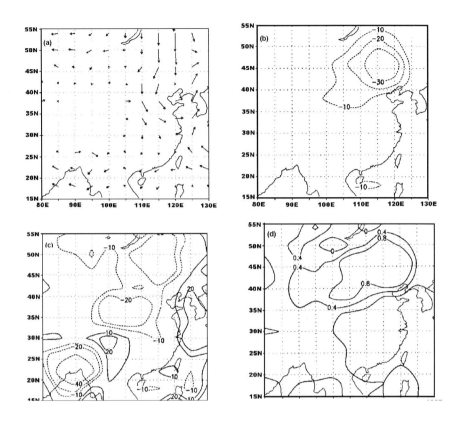

Fig. 5.22. Mean differences in the 3-month (JJA) (**a**) wind vector at 700 mb, m s^{-1} (**b**) latent heat flux, W m^{-2}, (**c**) precipitation, mm mo^{-1} and (**d**) surface temperature, K, between the natural grassland and desertification cases in Mongolia (adopted from Xue, 1996).

6. Hydrological Modelling and Forecasting

Suxia Liu

6.1 The Hydrosphere

The hydrosphere is one of the key components of the earth's environment. The unique properties of water make the hydrological system special. For instance, its high latent heat and high specific heat capacity mean it significantly influences the climate system. In the environment, water exists in four states (see Fig. 6.1), namely, as liquid and vapour in the atmosphere (A); surface water (SU) including river, ice, snow and the ocean; soil water (SO); and groundwater (G). The continuing transformation and movement of water between and within these four states, i.e., the hydrological cycle, takes place under the effects of gravity and solar radiation. It is increasingly understood that vegetation strongly modulates the hydrological cycle. Vegetation water (V) is now considered to be the fifth component in the hydrological cycle. Human activities (M) also substantially affect the movement and transformation of water, and hence the hydrological cycle in general should also include a sixth component, M.

There are small to medium scale hydrological cycles on continents and large scale hydrological cycles between land and ocean. For small to medium scale hydrological cycles, the hydrological processes occur either within or between the states of A, SU, SO, G and V as shown in Fig. 6.1. The examples of the former are such as water vapour transport in the atmosphere, flood routing, runoff, soil water movement, groundwater movement, water movement within vegetation. The examples of the latter include interception, infiltration, evaporation, transpiration, groundwater seepage and recharge, precipitation, root water uptake, hydraulic relationship between groundwater and surface water and capillary water upwarding. For large scale hydrological cycles as shown in Fig. 6.2, water is provided to the atmosphere by evaporation, mostly over the ocean, but also by evapotranspiration from plants and soil over the continents. Globally, this flux is balanced by the return of water to the surface through various types of precipitation. However, in the vapour phase, water can be transported several thousands of kilometres before condensing. As a consequence, precipitation over the continents exceeds

216   6. Hydrological Modelling and Forecasting

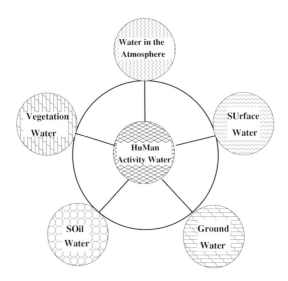

Fig. 6.1. Water transformation and movement between and within the six states, i.e., water in the atmosphere (A), vegetation water (V), soil water (SO), surface water (SU), groundwater (G) and human activity water (M).

evapotranspiration, but the budget is balanced by a surface flow of water (river, etc.) between land and sea. The lifetime of a water molecule in the atmosphere is estimated to be about 10 days.

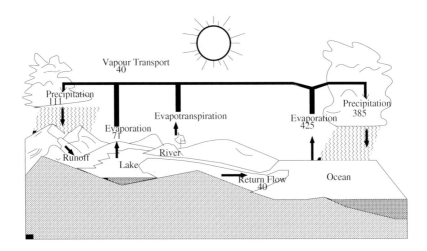

Fig. 6.2. The global water cycle. Water fluxes are expressed in thousands of cubic kilometres per year (after Thomas and Crutzen, 1995).

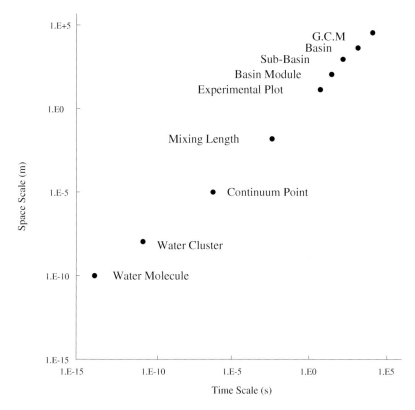

Fig. 6.3. Scales in hydrology (From Dooge, 1992).

Hydrological processes take place on a wide spectrum of temporal and spatial scales as shown in Fig. 6.3. These processes are complex and sometimes subject to random fluctuations. Therefore, the subject of hydrology differs from classic mechanics. In a sense, the latter deals with the processes in small complexity and randomness, which can be treated by analytical theory. It also differs from statistics, which deals with aggregates with large randomness. Hydrology, being too complex for analysis and too organised for statistics, falls into the region of system, the organised complexity, as shown in Fig. 6.4. In time the hydrological system shows both periodic and random behaviour. Mostly, a river has its flooding and dry season each year. However its hydrograph can never be repeated precisely with time. In space the hydrological system has both similarity and peculiarity. If the geographic and climatic conditions of two hydrological systems in two different areas are similar, then their hydrological processes may have similar behaviour. However, because the underlying conditions can never be the same in any two places, the two hydrological systems must also have their own characteristics.

6. Hydrological Modelling and Forecasting

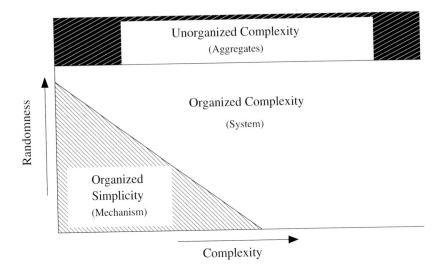

Fig. 6.4. Mechanisms, systems and aggregates (after Weinberg, 1975).

This complexity poses a considerable challenge to the development of hydrology. While small-scale hydrological processes can be precisely described by physical laws, it is often difficult to apply these laws to processes occurring on medium to large scales, e.g., the catchment scale. In dealing with these processes conceptualisation is important. The so-called conceptual hydrological modelling, which is popular in catchment hydrology, is a useful tool in hydrological forecasting. Hydrological models can contribute to the modelling of the global climate by providing estimates of evapotranspirative fluxes. They also provide information on regional variation of soil water availability, water supply, surface runoff, groundwater, and fresh-water inputs to the ocean. Hydrological forecasting is closely related to environmental problems such as flooding and drought. For environmental modelling and prediction, it is therefore important to understand the methods of hydrological modelling and hydrological forecasting, and how they can be used in the solutions of major problems.

For a hydrologist, forecasting generally means the extension of a time series in real time, whereas prediction means the generation of one or more realisations of time series representing a future of either a specified frequency of occurrence or a possible scenario under conditions of uncertainty (Dooge, 1992). As far as hydrologists can judge, forecasting and prediction frequently have the opposite meanings in the atmospheric sciences. In this chapter, in order to avoid confusion, we use one word of forecasting to cover the meaning of the two terms.

In this chapter, we first introduce a mathematical description for each single hydrological process (water movement in the same state and water transfer between different states) and its simplification for use in practice. We then explain the behaviour of world's popular hydrological models, and how they model all the hydrological processes in a catchment, which includes the classic Xinanjiang hydrological model, the widely-used semi-distributed TOPOMODEL and the forthcoming integrated hydrological modelling system. Finally, the basic methods of hydrological forecasting and its advances in recent years, to include the short-, mid- and long-term discharge forecasting, water quality forecasting as well as water supply and water demand forecasting, are generalised.

6.2 Mathematical Representation and Simplification

There seems no fundamental theorem in hydrology which could pervade the whole of the hydrological theorem. However, the lumped form of the continuity equation (inflow minus outflow equals rate of the change of storage) does qualify to be such a fundamental theorem (Dooge, 1992) and is invoked in all models of hydrological processes, as shown in the following sections. In order to simulate how water behaves in and between the states as shown in Fig. 6.1, a physical, simplified or empirical momentum equation is needed, which can be considered as another important basic theorem of hydrology.

6.2.1 Water Vapour

In the atmosphere, water exists as vapour, liquid and ice. One of the important forms of water transfer in the atmosphere is in the form of water vapour. Conceptual models have been developed to describe the water (vapour) balance in the atmosphere.

One kind of such models is the model of the precipitation recycling rate, a diagnostic measure of the potential for interactions between land surface hydrology and regional climate. It was used to describe the contribution of evaporation within a region to precipitation in that same region by Brubaker et al. (1993), Eltahir and Bras (1994) and Yi and Tao (1997). In their studies, two species of water vapour molecules are considered: those which evapotranspirate outside the region and those which evapotranspirate within the region. The definition of the word "region" includes all the area under study, say, the Amazon basin, or the Yangtze River basin. It is not restricted to the area of a single grid point. For a finite control volume of the atmosphere located at any point within the region, say, Manus in the Amazon basin, or Nanjing in the Yangtze River basin, the continuity equation of mass (water vapour content, W) reads:

220   6. Hydrological Modelling and Forecasting

$$\left.\begin{array}{l}\frac{\partial W_w}{\partial t} = I_w + ET - O_w - P_w \\ \frac{\partial W_o}{\partial t} = I_o - O_o - P_o\end{array}\right\} \quad (6.1)$$

where subscript w and o denote molecules which evapotranspirate within the region and outside the region respectively. I and O are inflow and outflow, P is precipitation, and ET is evapotranspiration (see Fig. 6.5). W, I, O, P and ET are variables in both space and time. Each of I and O is a summation of the components of flux in the two horizontal directions. For the whole of the region, I_w is equivalent to ET physically. However for a finite control volume of the atmosphere located at any point within the region, I_w and ET stand for different physical meanings. The molecules for I_W are evapotranspirated from the areas under other points within the region), while the molucules for ET are directly evapotranspirated form the local area right under that point instead.

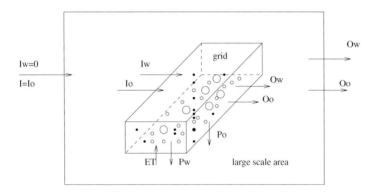

Fig. 6.5. The scheme of the atmospheric control volume in the precipitation recycling model (after Eltahir and Bras, 1994; Yi and Tao, 1997).

Assumed that water vapour molecules from the two species defined above are well mixed, the precipitation recycling ratio, ρ_T, is expressed as:

$$\rho_T = \frac{P_w}{P_w + P_o} = \frac{W_w}{W_w + W_o} = \frac{O_w}{O_w + O_o} \quad (6.2)$$

At any location within the region, ρ_T estimates the ratio of recycled precipitation to the total precipitation falling at that location. Recycled precipitation consists of molecules of water vapour which were in the atmosphere as a result of an evapotranspiration event at that location, or elsewhere in the region under study.

Assumed that the derivatives in (6.1) are zero compared with the flux and rearranging (6.1) give:

$$I_w + ET = O_w + P_w \brace I_o = O_o + P_o} \quad (6.3)$$

Substituting for O_w, P_w, O_o and P_o from (6.2) into (6.3)

$$I_w + ET = \rho_T(O_w + O_o) + \rho_T(P_w + P_o) \quad (6.4a)$$
$$I_o = (1 - \rho_T)(O_w + O_o) + (1 - \rho_T)(P_w + P_o) \quad (6.4b)$$

Combining (6.4a) and (6.4b) and rearranging lead to:

$$\rho_T = \frac{I_w + ET}{I_w + ET + I_o} \quad (6.5)$$

Equation (6.5) is the formula for estimating recycling in the model. Actually (6.5) can be directly derived from Fig. 6.5 with the analogy of O, or P, or W to I and ET by (6.2). In Eltahir and Bras (1994), the estimation procedure of ρ_T is as follows:

1. Interpolating the flux and evaporation data in a rectangular grid which covers the total area of the region. ρ_T is estimated at points which lie at half the distance between the grid points.
2. Estimating ρ_T in trial and error where ρ_T is first guessed at all points in the data grid. Equation (6.2) is then used in partitioning the flux at every point and in both directions into O_w and O_o. It is important to note that O_w and O_o for one grid point are I_w and I_o for adjacent point. These estimates of I_w and I_o together with ET are then used in (6.5) to arrive at improved estimates of the distribution of ρ_T. Following this procedure, ρ_T is estimated downstream for all the points. The new estimates of ρ_T are then compared with the previous estimates. The procedure is repeated until the new and previous estimates of ρ_T converge.

They estimated that about 25% of all the rain that falls in the Amazon basin is contributed by evaporation within the basin. The estimate of recycling of 25% is not sensitive to the resolution used in the study (2.5 °× 2.5 °) for describing the spatial distribution of precipitation recycling. This sensitivity was studied by considering the extreme case of treating the total area of the basin as one grid-cell. Using (6.5) with the data shown in Fig. 6.6 ($I_w = 0, I_o = 141, ET = 58$), ρ_T is computed as $\rho_T = (0+58)/(0+58+141) = 0.29$. This estimate of ρ_T is not significantly different from the conclusion of their paper, using 2.5 °× 2.5 ° as a resolution.

Yi and Tao (1997) calculated the precipitation recycling ratio over the Yangtze River basin (not including the head of the river). They found that 10% of the annual mean rain falling over the whole study region is contributed by evaporation within the same region, while 90% is contributed from other areas or the sea. The results also indicate that the fraction of precipitation that is locally derived varies substantially with location and season. The precipitation recycling ratio reaches as high as 0.41 or 0.46 in late summer in

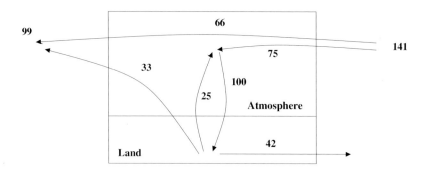

Fig. 6.6. Hydrological cycle of the Amazon basin, annual precipitation = 1.95 m = 100 units (after Eltahir and Bras, 1994).

two locations—the northwestern part of the Sichuan Basin, and the northern part of southern China.

Another model is used by Liu and Wang (1997) by using the internal hydrological cycle coefficient, K_I, which is defined as:

$$K_I = P_w/ET \qquad (6.6)$$

and the external hydrological cycle coefficient, K_E, which is defined as:

$$K_E = (W_o - W_{too})/P_o \qquad (6.7)$$

where W_{too} is water vapour content output to other areas. Other variables are defined the same as in (6.1). They compared the water balance for several regions, including China, Europe and Australia, all of which are about the same size. It has been shown that the external hydrological cycle coefficient for China is close to that for Europe, while the internal hydrological cycle coefficient is close to that of Australia. The topography of China is high in the west and low in the east, which allows the water vapour to escape easily from China via the Westerly Stream. This characteristic of water balance is similar to that in Australia, where the water in and out is very active. This is why, although the water vapour into China is similar to that into Europe, the internal hydrological cycle is not like that of Europe, being similar to Australia instead. Over China, just 17% of water evaporated from the land surface returns as precipitation, and 83% of that escapes the area.

6.2.2 Precipitation

Precipitation is usually assumed to be a known hydrological input variable. It is the water transfer from the atmosphere to land surface. Determination of the average amount of precipitation is a fundamental requirement for many environmental studies. A number of techniques for estimating mean areal precipitation have been developed, such as simple unweighted mean (UM), isohyetal (ISO), Thiessen polygon (TP) among others. Singh and Chouldury (Singh, 1989) compared 13 methods. Among them, ISO yielded higher estimates of mean daily and monthly areal rainfall than other methods in the research area. It was found that as the time period of rainfall grew from a day to a month to a year, the methods deviated less from one another. The simpler methods were as good as more complicate ones. Kriging (KG) method, with its ability to provide the assessment of the estimation error, should be superior to the traditional interpolation as long as the semivariance function can be accurately specified (Tabios and Salas, 1985). Usually precipitation needs to be measured at a number of sample points. Sometimes the records are not consistent; at other times, the record at one station may not complete. Double-Mass Analysis is frequently used for detecting heterogeneity of precipitation. As it is less than obvious to distinguish real changes from pure random fluctuations, statistical tests (Buishands, 1982) are necessary to evaluate the significance of departure from homogeneity.

The most common method for estimating a missing data point in a raingage network is the "normal ratio" method. Three index stations, next to the missing point location, are defined. The missing precipitation value is given by (Bras, 1990):

$$P_4 = \frac{1}{3}\left[\frac{N_4}{N_1}P_1 + \frac{N_4}{N_2}P_2 + \frac{N_4}{N_3}P_3\right] \tag{6.8}$$

where P_4 is the precipitation at the missing location, P_i the precipitation at index station i, and N_i the long-term normal precipitation at station i. To be useful, the index stations must be near and highly correlated to the location of interest. When the high spatial variability of precipitation occurs within a small area, skipping the missing value may show more accurate information of the precipitation distribution in space than the "normal ratio" method.

Tung (1983) compared five methods of estimating point rainfall among arithmetic average, normal ratio, modified normal ratio, inverse distance, modified inverse distance and linear programming methods. It was found that the arithmetic average and inverse distance methods did not yield desirable results, which might mean they are not suitable in mountainous regions. Ashraf et al. (1997) shows that of the three interpolation techniques among kriging, inverse distance and co-kriging, kriging interpolation provides the lowest root mean square interpolation error. Of the many techniques in use to determine integrated total precipitation, radar instruments are the most preferred which must be handled knowingly. While there certainly will be errors, with care a lot of valuable information can be obtained (De Troch et al., 1996).

6.2.3 Soil Water

Soil is one of the important reservoirs of water in the environmental system. Fluxes of water into the soil, from precipitation, snow melt and condensation, and out of soil, by evaporation, drainage and runoff, are important components of the hydrological cycle. Soil water is an important factor in accurate hydrological forecasting. Knowledge of water flow in the vadose zone is essential to derive proper management conditions for plant growth, and for environmental protection in agricultural and environmental systems. Hanks (1985) introduced a simple model to calculate soil water content, or soil moisture, in which the change in soil water storage (in depth) in the plant root zone, S, is calculated by the following continuity equation:

$$dS/dt = P + I_r - ET - Dr - Ro \tag{6.9}$$

where P is precipitation, I_r is irrigation, ET is evapotranspiration, Dr is drainage, and Ro is runoff. Assuming Dr and Ro to be zero, the estimation procedure of ET being related to plant yield, soil water can be calculated day by day with known daily precipitation. The flow chart of the calculation is shown in Fig. 6.7 (Hanks, 1985). It is more a patch model than a catchment model.

If considering the variation of soil water along the vertical depth, soil water calculation follows the soil water movement equation. Since water is incompressible, a combination of one-dimensional Darcy's law:

$$q_D = -K(\theta)\frac{\partial(\psi_m(\theta) - z)}{\partial z} \tag{6.10}$$

and one-dimensional soil moisture continuity equation:

$$\frac{\partial \theta}{\partial t} = -\frac{\partial q_D}{\partial z} - SS_{sw} \tag{6.11}$$

gives one dimensional equation of soil water movement, or the Richards equation:

$$\frac{\partial \theta}{\partial t} = -\frac{\partial K(\theta)}{\partial z} + \frac{\partial}{\partial z}\left[K(\theta)\frac{\partial \psi_m(\theta)}{\partial z}\right] - SS_{sw} \tag{6.12}$$

where q_D is Darcy specific discharge, θ is volumetric soil moisture, related with S in (6.9) by $S = \rho_s \theta d$ with ρ_s being soil density and d being depth of soil layer, ψ_m is the matric potential of soil water and z is vertical coordinate, taking positively downwards. The quantity K is hydraulic conductivity. Numerical experiments have shown that soil hydraulic functions, especially parameters for calculating soil hydraulic conductivity, play an important role in soil moisture calculation (Shao and Irannejad, 1999). The quantity SS_{sw} is the sink and source term of soil water movement, which includes runoff, evapotranspiration and root water uptake. Hanks (1985) gave a simple model to calculate root water uptake:

6.2 Mathematical Representation and Simplification

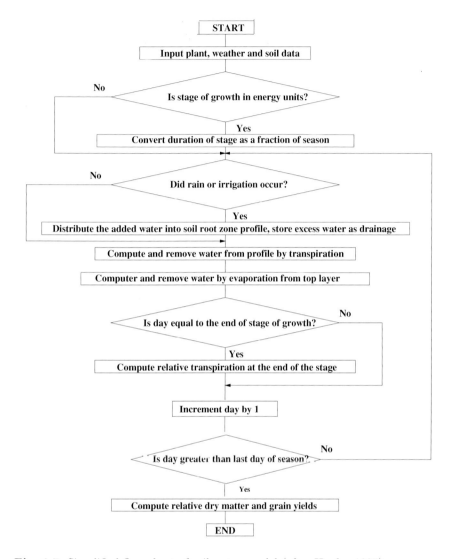

Fig. 6.7. Simplified flow chart of soil water model (after Hanks, 1985).

$$S_r(z) = \frac{RDF(z)K(\theta)}{\Delta x \Delta z} [\psi_{\text{root}} + 1.05 \cdot z - \psi_m(z,t) - \psi_o(z,t)] \qquad (6.13)$$

where ψ_{root} is the effective root water head at the soil surface. The term $1.05 \cdot z$ is a correction to the root water head at other soil depths, ψ_m is the matric head, ψ_o is the osmotic head which accounts for salinity, RDF is the root length density function, Δx is the distance from the root surface to the point in which the soil water potential $\psi_m(z,t)$ and $\psi_o(z,t)$ are measured (arbitrarily assumed to be 10 mm), and Δz is the depth increment. The model considers the most fundamental properties of soil and is thus able to simulate soil water flow under a wide variety of conditions.

When soil is assumed homogeneous, the initial soil water distribution is homogeneous, the analytic solution to the equation for soil water movement (6.12) can be obtained. The main difficulty is to treat the nonlinearity of the equation. Usually it is overcome by using a linearisation method, i.e., using the averaged value for those parameters, which are closely related with the variable of soil moisture or matric potential. The parameters then turn out to be a constant. When soil is non-homogeneous, or the distribution of initial soil moisture is non-homogeneous, it is much harder to get an analytical solution. Two-dimensional or three-dimensional saturated or unsaturated soil water movement equations are therefore necessary. The solution to these equations relies on numerical methods, such as finite difference or finite element techniques. The key problem again is still how to deal with the nonlinearity. The techniques of Explicit Linearisation (EL), Forecasting and Correction (FC) as well as Iteration (IT) method are sometimes effective for solving this problem (Lei and Yang, 1988).

In classic hydrology, soil water has been a weak part of modelling. In some previous rainfall-runoff models, soil water was not treated with care. The antecedent soil moisture condition is often pre-specified, or calculated using an empirical model. Improvements have been made recently. Jiang et al. (1999) improved the conceptual hydrological model by establishing a conceptual soil water model. The model was successfully used in the dynamic water distribution system of the Huoquan irrigation district, Shandong, China.

One of biggest driving forces behind the study of soil water in hydrology is due to the advances in research into land-surface parameterisation (Shao and Henderson-Sellers, 1996, see more in Chap. 5). Soil water models provide the hydrological basis for land-surface modelling (Camillio et al., 1983; Abramopoulous, 1988; Delworth and Manabe, 1989). One class of these models is the Soil-Water-Atmosphere-Plant (SWAP, Van Dam et al., 1997) model. Its main characteristics are accurate numerical solutions of the soil water flow equation and incorporation of solute transport, heat flow, soil heterogeneity, detailed crop growth, regional drainage at various levels and surface water management. Remote sensing techniques (Schmugge et al., 1980; Jackson et al., 1983; Biftu and Gan, 1999) are now used to estimate soil water content on a large scale, while in situ observations of soil water content are invaluable for use in developing land-surface parameterisation schemes, for studying the patterns of climate change (Holliger and Issard, 1994; Robock et al., 1995; Vinnikov et al., 1996; Liu et al., 2001), and validating hydrological models.

6.2.4 Infiltration

Infiltration is one of the most important and complex elements of the hydrological cycle, which occurs at the interface between atmosphere and land surface (see Fig. 6.1). It is closely related to soil type and is the major variable in determining soil erosion. Normally infiltration is quantified by infiltration rate, f, the rate at which water enters into the soil surface. It is expressed as volume per unit area per unit time and has the dimensions of length per unit of time. Cumulative infiltration, F, the volume of infiltration from the beginning of simulation or observation period, or the rainfall event, is also frequently used. It somtimetimes is called infiltration volume or accumulated infiltration and measured in centimetres. Infiltration capacity, f_c, is the maximum rate at which soil can absorb water through its surface and has the same dimension as that of infiltration rate.

There are two ways to research infiltration. One is based on the Richards Equation as described in the previous section and Chap. 5, another is to use empirical or conceptual models.

Based on the Richards Equation, infiltration is determined by three kinds of boundary conditions. The first is that the soil water content at soil surface is known. Soil surface is wetted, say, by irrigation, but without ponding water and the soil water is homogeneous at the surface. The boundary condition can be written as follows:

$$\theta = \theta_0, \qquad \text{when } t > 0 \text{ and } z = 0 \tag{6.14}$$

Infiltration here is controlled by soil infiltration capacity. The second boundary condition is that precipitation is less than soil infiltration capacity. There is no ponding water too, but the infiltration here is controlled by water supplied by precipitation, the so-called flux control. It can be expressed as:

$$K(\theta) - D(\theta)\frac{\partial \psi_m(\theta)}{\partial z} = P(t) \tag{6.15}$$

The third boundary condition is that the precipitation is greater than infiltration capacity and so ponding water is at soil surface. The soil water movement thus includes the movement in both the unsaturated zone and the saturated zone, the dependent parameter therefore should be changed from θ into water potential because only water potential is continuous across the two zones.

When looking at infiltration in a vegetated soil surface, the problem of infiltration becomes more complicated. With vegetation evapotranspiration, the sink term must include root water uptake. The solution is based on certain assumptions and catches the most important components, as shown in many soil-vegetation-atmosphere transfer models (Henderson-Sellers et al., 1993; Leuning, 1995; Pielke et al., 1998).

The empirical models, used in hydrological analyses, do not employ physics of flow through porous media and are derived from experimental data. Their principal characteristic is to express infiltration capacity as function of time. One of the best-known empirical infiltration models in hydrology is the Horton model. It was hypothesised that infiltration is similar to exhaustion process according to which the rate of performing work is proportional to the amount of work remaining to be performed. By integration with initial conditions, the model is expressed as:

$$F = f_c t + \frac{1}{k_f}(f_{ci} - f_c)[1 - \exp(-k_f t)] \qquad (6.16)$$

where f_c is infiltration capacity, f_{ci} is initial infiltration capacity, and k_f is a proportionality factor dependent on soil type and initial moisture content. The parameters f_{ci}, f_c and k_f have to be determined by data fitting, which is the principal weakness of this method.

The conceptual models employ physics of flow through porous media in a simplified form. Their principal characteristic is that they contain parameters that can be estimated from physically measurable properties of soil water. A simple model, based on Darcy's Law, is called Green and Ampt model or G-A model. In the model, the infiltration is proportional to the total gradient, i.e

$$f = K_s \frac{z_f + \psi_c + H_p}{z_f} \qquad (6.17)$$

where K_s is the saturated hydraulic conductivity, H_p is the ponding height above the soil surface, ψ_c is the capillary suction at the wetting front, and z_f is the increasing depth of the wet front. The total infiltration water is given by

$$F = (\theta_s - \theta_i) z_f \qquad (6.18)$$

where θ_i is the initial soil moisture content and θ_s is the saturated water content. The value of $\theta_s - \theta_i$ expresses the difference between volumetric water contents before and after wetting. The relationship between F and f leads to

$$f = \frac{dF}{dt} = (\theta_s - \theta_i) \frac{dz_f}{dt} \qquad (6.19)$$

Combining (6.17) and (6.19) gives

$$\frac{dz_f}{dt} = \frac{K_s}{\theta_s - \theta_i} \frac{z_f + \psi_c + H_p}{z_f} \qquad (6.20)$$

Integrating (6.20) with $z_f = 0$ when $t = 0$, the relationship between z_f and infiltration time t is obtained

$$t = \frac{\theta_s - \theta_i}{K_s}\left[z_f - (\psi_c + H_p)\ln\frac{z_f + \psi_c + H_p}{\psi_c + H_p}\right] \qquad (6.21)$$

The advantage of the G-A model is that its parameters K_s and ψ_c are physical properties of the soil-water system and can therefore be obtained from physical measurements. While the model could be used to treat infiltration independently, its assumption of an infinite sharp wetting front cannot approximate a realistic spatial distribution of soil water, except in coarse-texture soils. Further, it may also misrepresent significantly the infiltration flux at the surface. The other disadvantage of the model is that it relates the depth of penetration or infiltrating volume to time in implicit form.

Another conceptual model, which is based on the Fokker-Planck equation, the Philip Two-Term (PTT) model, can be expressed as:

$$F = st^{-5} + A_p \tag{6.22}$$

in which s is sorptivity and A_p is a parameter. Philip (1957–1958) has shown that $A_p = K_s/3$ where K_s is saturated hydraulic conductivity. Because the parameters A_p and s have physical meanings, the PTT model has frequently been incorporated in hydrological studies and satisfactory results have been obtained.

Although there are many other empirical and conceptual models of infiltration, there appears to be a revival of interest in G-A and PTT. It has been shown that the PTT model can be derived from the G-A model and the two are therefore related (Philip, 1966). When the parameter of diffusivity $D(\theta)$, which is defined as the ratio of hydraulic conductivity to the specific water content $d\theta/d\psi$, is replaced by a δ function, the G-A formula can be derived from the Richards Equation, that is:

$$D(\theta) = \alpha \delta(\theta - \theta_s) = \begin{cases} 0 & \theta \neq \theta_s \\ \alpha & \theta = \theta_s \end{cases} \tag{6.23}$$

where α is a parameter. In this way, it seems that the three methods, PTT, G-A and the Richards equation, are related with each other.

6.2.5 Depression Storage

When it rains, part of the precipitation is retained in depressions, which vary widely in area, depth, volume and number, with no possibility for escape as surface runoff. The retained water is called depression storage. Depression storage may play a significant role in modifying land surface responses. Due to difficulty in obtaining direct measurement, it has not been the subject of extensive investigation. Generally, its total volume is expressed as:

$$V = S_d[1 - \exp(-k_d P_e)] \tag{6.24}$$

where V is the volume of water stored, S_d is the maximum storage capacity, P_e is the effective rainfall volume [gross rainfall (P) – evapotranspiration (E) – interception (In) – infiltration (F)] and k_d is a constant or parameter.

Gross surface retention that includes interception by vegetated cover, depression storage and evaporation during precipitation, may be of sizable magnitude, ranging from 10 mm to 50 mm for cultivated fields, grasslands and forests. The rate at which depression storage is depleted by evaporation and infiltration during periods between storms determines the available capability at the beginning of any storm. While evaporation from the soil surface is relatively unimportant during rain periods, it can be of considerable magnitude immediately following the end of rain. In environmental modelling, for mountainous area, surface retention may be neglected. However, for other areas decreasing surface slope not only increases depression storage, but also reduces velocities of overland for given depth of surface detention. The maximum depression storage may reach 100 mm. Stock ponds, terracing, and contour farming all tend to moderate the runoff cycle in this manner.

Depressions has been a difficult task to deal with in raster digital elevation model (DEM) even with the availability of high speed computers. Methods, developed to process raster DEM automatically in order to delineate and measure the properties of drainage networks and drainage basins, are being recognised as potentially valuable tools for the topographic parameterisation in hydrological models (Martz and Garbrecht, 1998). One method is to stimulate breaching of the outlet of closed depressions to eliminate or reduce those expected to have been produced by elevation overestimates. Another method modifies flat surfaces to produce more realistic and topographically consistent drainage patterns than those provided under the assumption that flow will move along the shortest available path to the outlet. All of these methods ultimately rely on some form of overland flow simulation to define drainage courses and catchment areas and, therefore, have difficulties dealing with closed depressions and flat areas on digital land surface models. It seems more likely that depressions are caused by both under- and over-estimation errors and that the flow directions across flat areas are determined by the distribution of both higher and lower elevation surrounding flat areas. Much attention should be paid to the simulation when the resolution of DEM is rather course, where a lot of surface depressions may not be shown practically.

6.2.6 Vegetation Water

The hydrological cycle is strongly modulated by the earth's vegetation. In 1993, the International Geosphere-Biosphere Program (IGBP) established a core project entitled "Biospheric Aspects of the Hydrological Cycle" (BAHC), the theme of which is "How does vegetation interact with the physical processes of the hydrological cycle?" In the beginning of 1999 (BAHC News No. 6, Oct. 1998), BAHC entered an important period in its existence when it synthesised its results, as a part of the overall IGBP science synthesis. The theme is integrated as "How do changes in biospheric processes interact with global and regional climate, hydrological processes and water resources when driven by changes in atmospheric composition and land cover?". A more synthetic study on vegetation water, or the interaction between biosphere and hydrosphere is ongoing.

One of the most obvious characteristics of the transition from the former theme to the new theme of BAHC is that root water uptake has been stressed much more. Results from a numerical climate model (Milly, 1997) indicate that the summer dryness and related changes of the land-surface water balance are highly sensitive to possible changes of plant-available holding capacity of soil, which depends on plant rooting depth and rooting density. Root distribution may change when ecosystems respond to carbon dioxide from greenhouse warming, and this will result in changes to soil water availability. Responses may be constrained by nutrient availability and the depth of soil development. Reviews of root water uptake models have been given by Feddes (1981). In principle, two alternative approaches can be taken to the uptake of soil water by roots. The first is to consider the convergent radial flow of soil water toward and into a representative individual root, taken to be a line or narrow-tube sink, uniform along its length (i.e., of constant and definable thickness and absorptive properties). The root system as a whole can then be described by a set of individual roots, assumed to be regularly spaced in the soil at a definable distance that may vary within the soil profile. This approach, called the microscopic scale or single root model, usually involves casting the flow equation in cylindrical coordinates and solving it for the distribution of pressure heads, water contents and fluxes from the root outward. However, analytical and numerical solutions tend to ignore sectional differences in absorptive activity, which might result from differences in age or location of the roots. The alternative approach is to regard the root system in its entirety as a diffuse sink, which permeates each depth layer of soil uniformly, although not necessarily with a constant strength throughout the root zone. This approach, termed the macroscopic scale or root system model, disregards flow patterns toward individual roots, and thus avoids the geometric complications involved in analysing the distribution of fluxes and potential gradients on a microscopic scale. The major shortcoming of the macroscopic approach is that it is based on the gross spatial averaging of the pressure (and osmotic) heads, and takes no account of the decrease in pressure head (and increase in concentration of salts) in the soil at the immediate periphery of the absorbing roots. Feddes et al. (1981) have developed a simple macroscopic root water uptake model:

$$S_r = a(h) S_{\text{rmax}} \tag{6.25}$$

where S_r is the root water uptake (cm^3 H$_2$O cm^{-3}soil d^{-1}), and $a(h)$ is a dimensionless function of soil water pressure head h (cm). The quantity S_{rmax} is the root water uptake (cm^{3}cm^{-3}d^{-1}) under potential evapotranspiration conditions, which can be approached as

$$S_{\text{rmax}} = T_p / D_{\text{root}} \tag{6.26}$$

where D_{root} is the depth (cm) of the rooting zone and T_p is the potential transpiration rate (cm d^{-1}). The relationship between T_p and h is seen in Fig. 6.8. If root-length density distributions RDF (cm cm^{-3}) are known, (6.26) can be alternatively expressed as:

$$S_{\text{rmax}} = \left[RDF(z) / \int_{-D_{\text{root}}}^{0} RDF(z) \mathrm{d}z \right] \cdot T_p \tag{6.27}$$

where z(cm) is the vertical coordinate, taken positively upwards. It is believed that a simple macroscopic approach for root uptake as described by (6.26) and (6.27) is feasible for application in hydrological and climate models.

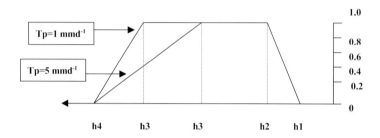

Fig. 6.8. Dimensionless sink term variable as a function of the absolute value of the soil water matric head (after Feddes, 1981).

6.2.7 Interception

Interception is the water transfer between vegetation and other states, as shown in Fig. 6.1. A portion of the precipitation falling in land surface is intercepted by vegetated cover and other above ground objects such as roofs, which is defined as interception. Part of the intercepted precipitation wets and adheres to these objects and then returns to the atmosphere through evaporation. This part is called interception loss. In studying major storms and floods interception loss is frequently neglected. It may, however, be significant in water balance studies, where its influence depends on the nature, type, and density of intercepted cover, precipitation characteristics and the season of the year. The interception is always high in humid forested regions. There are many empirical models to calculate interception loss (In), most of which are based on the continuity equation:

$$In = P_c - P_g - P_s - E_i \tag{6.28}$$

and

$$P_c = P_g + E_i + \Delta S_i/dt \tag{6.29}$$

where P_c is the amount of precipitation over the vegetated cover, P_s is the amount of precipitation reaching the ground from flow down the tree stems or stemflow, P_g is the amount of precipitation reaching the ground through the vegetated cover or throughflow, E_i is evaporation of intercepted water, and $\Delta S_i/dt$ is the change of storage which is due to interception.

There are two ways to develop monthly, seasonly and yearly models for computing interception loss. First, interception loss can be estimated for each storm in a given time period and then summed up. Second, relations can be derived directly for estimating interception loss for a given period (Rowe, 1983). In Land surface models (Dickinson, 1986), the law that interception is equal to 3/2 maximum interception, which is determined by the vegetation type, is often used to calculate interception.

6.2.8 Evapotranspiration

Evapotranspiration is the water transfer from soils and plants to the atmosphere. It is the sum of volume of water used by vegetation (transpiration) and that evaporated from soil and intercepted precipitation from vegetation. Nearly 75 percent of the total annual precipitation on land surface returns to the atmosphere by evaporation and transpiration. If soil water is not limiting, then evapotranspiration from a saturated soils is approximately equal to evaporation from a free-water surface and is called potential evapotranspiration. When soil water is limited, evaporation is known to occur in several stages and is controlled by climatic factors and then by soil characteristics, which is called actual evapotranspiration. Likewise if water is not a limitation and the area is completely shaded by vegetation, then evapotranspiration occurs at a potential rate.

As evapotranspiration is one of the main components of land surface processes–latent heat, it has been exhaustedly studied in land-surface process research by using physical, physiological and meteorological laws, as shown in Chap. 5.

In most hydrological models evapotranspiration is always calculated empirically or conceptually. In some models, such as RORB model (Laurenson and Mein 1993), evapotranspiration is lumped into a loss effect with other hydrological processes. In Xinanjiang (Zhao, 1992), it is calculated by three layer conceptual model. The total areal mean soil moisture capacity of a basin, WM, is divided into three portions: UM the upper part, LM the lower part and DM the deeper part. WU, WL and WD are the storages at any moment corresponding to the three layers. Evaporation occurs at the potential rate, equals to k_e multiplied by the pan evaporation rate E_p,

$$EU = k_e \times E_p \tag{6.30}$$

until the storage WU of the uppermost later is exhausted. On exhaustion of the upper layer, any remaining potential evaporation is applied to the lower layer. The efficiency is modified by multiplying by the ratio of the actual storage WL to the capacity storage LM of that layer:

$$EL = (k_e \times E_p - EU) \times WL/LM \tag{6.31}$$

When the lower layer storage WL is reduced to a specific portion, C_d, of LM the evapotranspiration is assumed to continue, but at a rate, ED, given by:

$$ED = C_d \times (k_e \times E_p - EU) \tag{6.32}$$

The parameter k_e is the product of three coefficients: k_1, k_2 and k_3. Coefficient k_1 is the ratio of the pan measurement to the measurement from larger water surface; k_2 is the ratio of evaporation capacity in land to water evaporation; k_3 considers the representation of the station to take measurements of water evaporation, mainly considering the problem of the altitude. The higher the altitude, the lower the atmospheric pressure, the smaller the potential evapotranspiration.

Another method, advection-aridity approach, or the complementary hypothesis is widely used for regional evapotranspiration calculation (Morton, 1983; Dooge, 1992) and mid- and long-term hydrological modelling and forecasting. In the approach a clear distinction is made between potential evapotranspiration under potential conditions, PE_o, determined by available energy, and the apparent potential evapotranspiration, PE', under non-potential conditions obtained by applying to measured variables a formula calibrated for potential conditions. Under potential conditions there is no shortage of soil moisture, and actual evapotranspiration, AE, is equal to:

$$AE = PE_o \tag{6.33}$$

If the water supply to the surface is limited, then the amount of the excess energy is given by:

$$Q_1 = \lambda(PE_o - AE) \tag{6.34}$$

where λ is the latent heat of water .

The atmospheric parameters used to estimate the potential evapotranspiration such as wind speed, relative humidity and cloudiness are different for these non-potential conditions. The net effect is that the application of a combination-type formula to the observed atmospheric parameters gives an apparent potential evapotranspiration (PE'), which is greater than PE_o for the same available radiation at the top of the boundary layer. The energy equivalent to this difference is given by:

$$Q_2 = \lambda(PE' - PE_o) \tag{6.35}$$

Assuming

$$Q_1 = Q_2 \tag{6.36}$$

then

$$AE + PE' = 2PE_o \tag{6.37}$$

This indicates that under water-limiting conditions the actual evapotranspiration (AE) and the apparent potential evapotranspiration (PE') vary in a complementary fashion.

Alternatively, Xiong and Guo (1999) developed a two-parameter model to calculate monthly actual evapotranspiration for mid- and long-term hydrological modelling:

$$AE = C_e \times E_p \times \tanh(P/E_p) \tag{6.38}$$

where E_p is the monthly pan evaporation, P is monthly rainfall, C_e is their proposed coefficient, relating the effect to the change of time scale, and $\tanh(\cdot)$ is the hyperbolic tangent function.

How to consider the heterogeneous land surface is a big challenge for regional evaporation estimation. Remote sensing is a useful tool to calculate the regional evapotranspirative flux. At present, meteorological instruments taking measurements at heights of several meters can only represent an area of 100 m square, which can not match with the resolution of NOAA/AVHRR being about 1~10 km square. One of the solution is to increase the height where meteorological measurements are taken. Another way is to get related information, say surface temperature, at the combine Convective Boundary Layer (CBL) with remote sensing techniques (Mo, 1997). The CBL, or mixed layer, is the layer which has a thickness of several meter to 1000 m, where the turbulent fluxes of land surface affect the atmosphere in daily time scale. The daily variation of moisture in this layer represents the evapotranspiration over a much larger scale, such as 10 km to 100 km square. So if we can measure or calculate this variation, then we could get the evapotranspiration over the larger area surface and get a more accurate estimate of regional evapotranspiration.

6.2.9 Channel Flow

Water movement over the land surface largely consists of two elements: channel flow and overland flow where water moves as a thin sheet. Channel flow modelling, or channel routing, is a mathematical method (model) to simulate the changing magnitude, speed and shape of a flood wave as it propagates through rivers. The Saint-Venant equations (1871) provide the basic theory for the one-dimensional analysis of flood wave propagation. It includes the conservation of mass equation, or simply, the continuity equation:

$$\frac{\partial Au}{\partial x} + \frac{\partial A}{\partial t} = 0 \tag{6.39}$$

and the conservation of momentum equation:

$$\frac{\partial u}{\partial t} + u\frac{\partial u}{\partial x} + g\left(\frac{\partial h_c}{\partial x} + S_f\right) = 0 \tag{6.40}$$

in which t is time, x is distance along the longitudinal axis of the waterway, A is cross-sectional area, u is velocity, g is the gravity acceleration constant, h_c is the water surface elevation above a datum, and s_f is the friction slope, which may be evaluated using a steady flow empirical formula such as Chezy or Manning equation. Saint-Venant equations are quasi-linear, hyperbolic partial differential equations with two dependent parameters (u and h), and two independent parameters (x and t). Due to the mathematical complexity of the Saint-Venant equations, simplifications are necessary to obtain feasible solutions. One of the simplifications is the hydraulic method. Its aim is to make an analytical solution to the Saint-Venant equations available, or to simplify the Saint-Venant equations so that they can be solved by explicit or implicit finite difference methods. The linearisation ignores the least important non-linear terms and/or linearisation of the remaining non-linear terms in the equations. It is only within the last three decades, with the advent of high-speed computers, that the complete Saint-Venant equations could be solved. The other simplification is a hydrological method known as the Muskingum routing model, first developed by McCarthy (1938) and now is still widely used all over the world.

6.2.10 Overland Flow

Generally speaking, there are two kinds of overland flow that occur in catchments. One is Horton overland flow, which is produced when the rainfall intensity exceeds the infiltration capacity. When Horton overland flow occurs, overland flow is generated more or less all over a catchment and its routing to stream channels dominates the form of the hydrograph. Another kind of overland flow is saturation overland flow. When saturation overland flow occurs, the contribution area is usually a narrow strip along valley bottoms, widening somewhat around stream heads. Travel distance and routing delays are therefore short. In this case, overland flow routing delays may, for most purposes, be ignored in forecasting slope-based discharge.

The continuity equation and momentum equation for overland flow are analogous to the equations for channel flow. One of the biggest differences between the two is that the equations for overland flow are usually two-dimensional, shown as follows:

$$\left.\begin{array}{l} \frac{\partial h}{\partial t} + \frac{\partial (uh)}{\partial x} + \frac{\partial (vh)}{\partial y} = SS_o(x,y,t) \\ \frac{\partial (uh)}{\partial t} + \frac{\partial (uuh)}{\partial x} + \frac{\partial (uvh)}{\partial y} + gh\frac{\partial h}{\partial x} = gh(S_{ax} - S_{fx}) + D_{lx} \\ \frac{\partial (vh)}{\partial t} + \frac{\partial (uvh)}{\partial x} + \frac{\partial (vvh)}{\partial y} + gh\frac{\partial h}{\partial y} = gh(S_{ay} - S_{fy}) + D_{ly} \end{array}\right\} \quad (6.41)$$

where h is the water depth of overland flow, t is time, u and v are the velocities of overland flow in the x and y directions, respectively, SS_o is a sink and source term for overland flow, g is gravity acceleration constant, S_{ax} and S_{ay} are the degrees of slopes in the x and y directions, respectively. The quantities S_{fx} and S_{fy} are the head losses in the x and y directions, which are closely related to slope roughness. The quantities D_{lx} and D_{ly} are the side inflow or outflow. The above equations are very useful in the distributed and dynamically hydrological numerical modelling (Huang, 1997).

Equation (6.41) has been found wide application in the research of soil erosion in slope scale. By considering the input of precipitation, P, (6.41) can be written (Thomas et al., 1994; Qi and Huang, 1997):

$$\frac{\partial h}{\partial t} + u\frac{\partial h}{\partial x} + h\frac{\partial u}{\partial x} = P \tag{6.42}$$

$$\frac{\partial u}{\partial t} + u\frac{\partial u}{\partial x} + g\frac{\partial h}{\partial x} = g(S_{ax} - S_{fx}) - \frac{uP}{h} \tag{6.43}$$

The head loss S_{fx} is calculated by:

$$S_{fx} = n^2 u/(h^{4/3}) \tag{6.44}$$

where n is Manning roughness. The sediment transport equation is needed when applying the above equations to soil erosion research. The general sediment equation is as follows (Thomas et al., 1994):

$$\frac{\partial h C_S}{\partial t} + \frac{\partial (hu C_S)}{\partial x} = D_{rd} + D_i \tag{6.45}$$

where C_S is sediment concentration, D_i is the soil loss from slope induced by overland flow, and D_{rd} is the soil loss from slope induced by rainfall dripping. The above method has advantages over the Universal Soil Loss Equation (USLE, Wischmeier et al. 1958) in its capacity to forecast erosion based on rainfall events.

As overland flow is irregular in space, it is difficult to obtain an analytic solution to (6.41). There are many simplified methods for overland flow routing, one of which is the model suggested by Beven and Kirby (1979). In their view, the average depth of water in depressions, h_0, is assumed to be stationary, and depths in excess of h_0 are assumed to travel at a constant velocity, u_0. Thus, the discharge of overland flow per unit width, q, is:

$$q = (h - h_0)u_0 \tag{6.46}$$
$$u/u_0 = 1 - h_0/h \tag{6.47}$$

This approach has been found to be valid for rivers with discharges of up to $10\,\text{m}^3\,\text{s}^{-1}$ for a limiting velocity of u_0 being $1\,\text{m}\,\text{s}^{-1}$.

6.2.11 Groundwater

Streamflow may be comprised of surface flow, interflow, bank flow, or groundwater flow. Groundwater flow is sometimes called groundwater runoff, base flow, sustainable flow, low flow, or percolation flow. Streamflow recession represents withdrawal of water from storage with no inflow. Baseflow recession is the lowest part of the falling limb of the hydrograph. Baseflow recession has long been a subject of considerable interest and inquiry. The governing equations for groundwater flow are the continuity equation or law of conservation of mass and Darcy's law or a storage-discharge relation. The governing equations can be expressed in both lumped and distributed forms.

The distributed continuity equation is written, for example, as follows:

$$\left[\frac{\partial}{\partial x}(\rho_w q_x) + \frac{\partial}{\partial y}(\rho_w q_y) + \frac{\partial}{\partial z}(\rho_w q_z)\right] dxdydz = -\frac{\partial M}{\partial t} \qquad (6.48)$$

where ρ_w is water density, q_x, q_y and q_z are Darcy's velocities in flow directions x, y and z, respectively, and M is mass inside the control element $dxdydz$ below the water table. The three-dimensional Darcy's law is:

$$q_x = -K_s \frac{\partial H}{\partial x} \qquad (6.49)$$

$$q_y = -K_s \frac{\partial H}{\partial y} \qquad (6.50)$$

$$q_z = -K_s \frac{\partial H}{\partial z} \qquad (6.51)$$

where H is piezometric head, and K_s is saturated hydraulic conductivity. By combining the continuity equation and Darcy's law, and then specialising in the resulting equation, several flow equations can be derived, such as the Boussinesq equation under the Dupuit-Forchheimer approximation for flow in an unconfined aquifer. In unconfined aquifers, the upper surface of the zone of saturation is represented by the water table and is under atmospheric pressure. In confined aquifers, groundwater is confined at a pressure greater than atmospheric pressure by overlying impervious strata.

As the lumped equations are always simpler than distributed equations, by neglecting the spatial variation of groundwater flow parameters, they find wider application in hydrology than the distributed equations described above. The lumped continuity (water balance) equation is:

$$Q_{gI} - Q_{gO} = s_c \frac{dV}{dt} \qquad (6.52)$$

where s_c is the storage coefficient, and V is the total volume of the aquifer. During a time interval Δt, the inflow to the groundwater system, Q_{gI}, equals the outflow from the system, Q_{gO}, plus the rate of the change of water storage in the aquifer. The inflow to the groundwater system might consist of natural recharge, drainage, artificial recharge, irrigation return flow, stream flow, and deep percolation. Outflow from a groundwater system may include pumping, artificial drainage, springs, substance outflow, and phreatophyte transpiration. The familiar lumped storage-discharge relation is the one that applies to a linear storage element expressed as:

$$Q_{gO} = aV = aA_a h_w \qquad (6.53)$$

where a is a storage constant, A_a is the area of the aquifer, and h_w is the mean water level in the aquifer. If we consider both the spatial and temporal variation, the storage-discharge relation is expressed as:

$$Q_{gO} = q_D A_a = -K_s A_a \frac{dH}{dt} \qquad (6.54)$$

where q_D is a specific discharge or Darcy velocity. The quantity H, defined in (6.51), is equal to $Z + p/(\rho g)$, where p is water pressure, ρ is water density and Z is the static head above a given datum.

Bardsley and Campbell (1994) presented their preliminary results from a groundwater experiment carried out at a field site near Matamata (New Zealand). It reveals that confined aquifers can act like giant weighing lysimeters, with pore water pressure providing real-time measures of changes in amounts of surface and near-surface water. This allows for the possibility of water balance studies on the scale of hectares using measurements from a single site. Given suitable confined aquifers, the technique can be applied in various environments to provide hydrological measurements as diverse as quantification of evapotranspiration loss, areal precipitation measurement, monitoring the water content of an accumulating snowpack, and net lateral groundwater transfers in unconfined aquifers.

6.2.12 Water Quality

As part of a general concern for the environment, water quality is an important issue in hydrological processes. The raising of cattle and hogs on dispersed farmsteads is being replaced by large, confined feeding operations. Many of these are sited on lots adjacent to streams so that the rainfall washes away the manure. Mining and petroleum operations are also major pollutes. The quantities of wastes, infiltrated from fertilised farmland, increase in the nitrate content in the soil and wells of rural towns, and exceed the self-purification capacity of many rivers, groundwater and soil water. Water quality has thus been receiving renewed attention. A number of mathematical models have been applied to simulate water-quality conditions in streams, lakes or reservoirs, urban areas, aquifers, soils, estuaries and fluvial networks (James, 1993).

Water Quality in Streamflow. The one-dimensional water quality model is given by:

$$\frac{\partial C}{\partial t} + u\frac{\partial C}{\partial x} = D\frac{\partial^2 C}{\partial x^2} + SS_{wq} \tag{6.55}$$

where u is velocity, C is the concentration of the constituent, x is the distance along the river, t is time, D is the diffusion coefficient, and SS_{wq} is a sink and source term for stream water quality. For the concentration of different constituents, such as dissolved oxygen (DO) (the concentration is O) or biochemical oxygen demand (BOD) (the concentration is L), the source and sink terms are different. Street-Phelps (1925) built a steady stream water quality model as follows:

$$u\frac{\partial L}{\partial x} = D\frac{\partial^2 L}{\partial x^2} - K_1 L \tag{6.56}$$

$$u\frac{\partial O}{\partial x} = D\frac{\partial^2 O}{\partial x^2} - K_1 L + K_2(O_s - O) \tag{6.57}$$

where K_1 and K_2 are the coefficients of aeration and re-aeration, and O_s is the saturated concentration of dissolved oxygen. The model is the earliest stream water quality model. Many of the early streamflow quality models involved DO and the development of algorithms are based on the Street-Phelps model. Potential sources include re-aeration and photosynthesis, while potential sinks include both carbonaceous and nitrogenous BOD, benthal bottom deposits demand and respiration.

Another famous water quality model is QUAL-II. It is the extension of the stream water QUALity model developed by Texas Water Development Board in 1971, which considers as many as 15 constituents (Roesner, 1977). A relatively new stream water quality model is WASP4 (the updated version of the Water Quality Analysis Program (WASP), developed by Hydroscience, Inc. (Ambrose et al., 1987; USEPA, 1988). The estimation of parameters in multi-constituent stream water-quality models is critical and has been the subject of detailed investigation. With the aid of computers, there are many optimising mathematical methods for parameter estimation. Examples of these are the Marquardt method and the Gauss-Newton method (Liu, 1994).

In order to facilitate the site-specific, problem-specific development and application of water quality models in Great Lakes watersheds, a modelling support system that links water quality models with a geographic information system (GIS) (ARC/INFO) has been developed (DePinto et al., 1994). This system, which is called Geo-WAMS (Geographically-based Watershed Analysis and Modelling System), automates such modelling tasks as: spatial and temporal exploratory analysis of watershed data; model scenario management; model input configuration; model input data editing and conversion to appropriate model input structure; model processing; model output interpretation, reporting and display; transfer of model output data between models; and model calibration, confirmation, and application. The design of Geo-WAMS and its feasibility and utility are demonstrated by a prototype application to the Buffalo River (Buffalo, NY) watershed. The prototype provides a modelling framework for addressing watershed management questions that require quantification of the relationship between sources of oxygen demanding materials in the watershed and distribution of dissolved oxygen in the Buffalo River. It includes a watershed loading model, the output from which is automatically converted to input for a modified version of EPA's WASP4 and a dynamic eutrophication and/or dissolved oxygen model for simulation of dissolved oxygen in the river. In this way the impact of regulatory or remedial actions in the watershed on the dissolved oxygen resources in the lower river can be examined. Ma and Zhang (1998) integrated WASP4 and GIS, and obtained interesting results in China.

Another group of water-quality models for streams is the surface water solute transport models, which incorporate the effects of transient storage that several investigators (e.g., Kim, 1990) have proposed. These transient storage (or dead zone) models consider a physical mechanism wherein solute mass is exchanged between the main channel and an immobile storage zone. This occurs in small streams when portions of transported solute become isolated in zones of water that are immobile relative in the main channel (e.g., pools, gravel beds). Runkel (1993) paid much attention to the numeric aspect of this approach. Of particular interest is the coupled nature of the equations describing mass conservation for the stream channel and the storage zone by using implicit finite difference techniques.

Water Quality in Below-ground. The continuity equation for any solute in water is:

$$\frac{\partial(\theta C)}{\partial t} = -\frac{\partial J}{\partial z} + SS_{\text{solute}} \tag{6.58}$$

where, for below-ground solute movement, the total dispersion flux, J, is:

$$J = J_d + J_a = -D_{sh}(u,\theta)\frac{\partial C}{\partial z} + q_D C \tag{6.59}$$

Here, diffusion flux, the term J_d, obeys Fick's Law:

$$J_d = -D_{sh}(u,\theta)\frac{\partial C}{\partial z} \tag{6.60}$$

and advection flux, the term J_a, obeys the advection equation:

$$J_a = q_D C \tag{6.61}$$

It follows that the basic equation of solute movement in below-ground is:

$$\frac{\partial(\theta C)}{\partial t} = \frac{\partial}{\partial z}\left[D_{sh}(u,\theta)\frac{\partial C}{\partial z}\right] - \frac{\partial(q_D C)}{\partial z} + SS_{solute} \tag{6.62}$$

In the above equations, θ is soil moisture, u is the velocity of solute movement, C is the concentration of solute, q_D is soil water flux, or Darcy velocity, $D_{sh}(u,\theta)$ is the dispersion coefficient (including molecular and mechanical), and SS_{solute} is a sink and source term of solute in below-ground.

When considering the problem of dead zones, or transient zones, as in stream water quality modelling, (6.62) becomes:

$$\frac{\partial(\theta_m C_m + \theta_{im} C_{im})}{\partial t} = \frac{\partial}{\partial z}\left[D_{sh}(u_m,\theta_m)\frac{\partial C_m}{\partial z}\right] -$$
$$\frac{\partial(q_D C_m)}{\partial z} + SS_{msolute} \tag{6.63}$$

where m denotes mobile water, and im represents immobile water. Modelling below-ground water quality often requires large amounts of data. Such data is limited by current observation techniques.

Solute transport in heterogeneous space has been studied with increasing intensity over the past few decades as a result of growing concerns about groundwater quality and pollution (Yang et al., 1997).

Based on the equation of heat conduction, Fourier's law, the one-dimensional heat flux in soil, q_h, can be written:

$$q_h = -K_h \frac{\partial T}{\partial z} \tag{6.64}$$

where K_h is heat conductivity, and T is soil temperature. The continuity equation of heat is:

$$c_v \frac{\partial T}{\partial t} = -\frac{\partial q_h}{\partial z} \tag{6.65}$$

Substituting (6.64) into (6.65), we arrive at the basic equation of soil heat transfer:

$$c_v \frac{\partial T}{\partial t} = \frac{\partial}{\partial z}\left(K_h \frac{\partial T}{\partial z}\right) \tag{6.66}$$

where c_v is specific heat capacity. When it is necessary to consider the coupling between soil heat and soil water movement, the soil water movement equation (6.12) should be written as:

$$\frac{\partial \theta}{\partial t} = -\frac{\partial K(\theta)}{\partial z} + \frac{\partial}{\partial z}\left[K(\theta)\frac{\partial \psi_m}{\partial z}\right] + \frac{\partial}{\partial z}\left(D_T \frac{\partial T}{\partial z}\right) - SS_{swh} \tag{6.67}$$

where ψ_m is matric head, D_T is the diffusion coefficient of temperature, and SS_{swh} is a sink and source term for soil water and heat movement. The third term in right hand of the equation is the water flux induced by temperature gradient. Under frozen conditions part of the soil water becomes ice. The ice content θ_{ice} change is equal to the change of the dynamic storage of liquid water. Thus, the soil water movement equation (6.12) becomes:

$$\frac{\partial(\theta + \frac{\rho_{ice}}{\rho_w}\theta_{ice})}{\partial t} = -\frac{\partial K(\theta)}{\partial z} + \frac{\partial}{\partial z}\left[K(\theta)\frac{\partial \psi_m}{\partial z}\right] - SS_{swice} \tag{6.68}$$

and the soil heat equation (6.66) becomes:

$$c_v \frac{\partial T}{\partial t} = \frac{\partial}{\partial z}(K_h \frac{\partial T}{\partial z}) + \lambda_{ice}\rho_{ice}\frac{\partial \theta_{ice}}{\partial t} \tag{6.69}$$

where ρ_{ice} is the density of ice, ρ_w is the density of water, λ_{ice} is the latent heat of the melting of ice (335 J g^{-1}, or 79.7 cal g^{-1}), and SS_{swice} is the sink and source term for soil water movement in frozen condition.

Water Quality in Reservoirs. Research on the modelling of water quality in reservoirs is not as advanced as the research on water quality of stream water. The main reasons are: 1) reservoir ecosystems are more complex than stream ecosystems; 2) models of reservoirs are usually at least two-dimensional, while a one-dimensional water quality model is frequently used for most problems in stream management; and 3) changes in meteorology and climate have a larger effect on reservoirs than on streams.

One of the earliest reservoir water quality models was established by Roesner et al. (1974), which is still used widely (Xie, 1996). In their model, the reservoir is divided into vertical segments. For segment j, the time rate of temperature change, $\dot{T}_j$, is calculated by:

$$\frac{a_j + a_{j+1}}{2}\dot{T}_j = \frac{h_{swo}}{c_p \rho_w \Delta z}\{a_{j+1}\exp[-\alpha(z_0 - z_j - \Delta z)] -$$
$$- a_j \exp[-\alpha(z_0 - z_j)]\} + \frac{A_{j+1} a_{j+1}}{\Delta z^2} T_{j+1} -$$
$$- T_j \left(\frac{A_j a_j + A_{j+1} a_{j+1}}{\Delta z^2} + \frac{Q_{vj+1} - Q_{ij} + Q_{oj}}{\Delta z}\right) +$$
$$+ \left(\frac{A_j a_j}{\Delta z^2} + \frac{Q_v}{\Delta z}\right) T_{j-1} \qquad (6.70)$$

where a_j is the surface area of plane (segment) j, h_{swo} is the short wave solar radiation flux, C_p is heat capacity, ρ_w is the density of water, T_j is the temperature of the segment j, and α is the extinction coefficient for short wave solar radiation in water. For clear water, α can be taken as 0.3 m^{-1}. The quantity z_j is the distance from the segment j to the water surface of the reservoir, Δz is the thickness of the segment j, Q_{ij} is the flow into the segment j (horizontal), Q_{oj} is the flow out of the segment j (horizontal), Q_v is the vertical heat flux, and A_j is the effective diffusion coefficient at the segment j. The model can be solved numerically. Much work has been done on the solution to the equation (Foster and Charlesworth, 1996).

The model of CE-QUAL-W2, standing for Corps of Engineers water QUALity model, Waterbody 2D, is a newly developed two-dimensional (longitudinal/vertical) hydrodynamic and water quality model for rivers, estuaries, lakes, and reservoirs (Cole and Buchak, 1995). Because the model assumes lateral homogeneity, it is best suited for relatively long and narrow waterbodies exhibiting longitudinal and vertical water quality gradients. The model has been applied to rivers, lakes, reservoirs, and estuaries in the United States (Bales and Giorgino, 1998).

Two- and three-dimensional mathematical modelling of the hydrodynamics and pollution distribution in lakes and reservoirs is generally based on a grid with a relatively large cell size due to hardware limitations. Geographic information system (GIS) has advanced capabilities for interpolation and visualisation. Chen et al. (1998) loosely coupled a grid-based GIS with a three-dimensional reservoir quality model. It is shown that even smaller grids (2-m^2) of Fe and Mn concentrations in the area of water intake can be obtained by GIS coupling method as such. These are very useful for a detailed study of the quality of a water supply.

Water Quality in Estuaries and Fluvial Networks. The estuary is the interactive water body between the stream and ocean, which is largely affected by tides. The delicate "channel-node-channel" fluvial network model is widely used in China (Ye, 1986; Chu and Xu, 1992; Zheng et al., 1997; Jin and Han, 1998). It is relatively simple and can satisfy the required accuracy. The basic equation of the model is written based on the concept of "channel-node-channel" as follows:

$$V_i \frac{\partial C_i}{\partial t} = -\sum_j Q_{i,j} C_{i,j} + \sum_j \left(D_{i,j} A_{i,j} \frac{\partial C_i}{\partial x} \right) - k v_i C_i + S S_i \quad (6.71)$$

where V_i is the volume of element i, C_i is the concentration of water quality at element i, $Q_{i,j}$ is the flow discharge, $A_{i,j}$ is the area of the cross-section between elements i and j, $D_{i,j}$ is the diffusion coefficient, k is the coefficient of decay, v_i is the velocity of water in element i, and SS_i is the source and sink term. The equation describes water quality transport in esturies and fluvial networks. It shows that the change in the concentration of element i depends on the net flux of the element, the diffusion flux and the source and sink term, which includes the decay term and the other source and sink terms.

6.3 Hydrological Modelling

The previous section gives detailed mathematical description for various hydrological processes. Theories of many hydrological processes such as infiltration, evaporation, overland flow, sediment transport and subsurface water movement have been developed at small space-time scales. However, hydrological modelling and forecasting are also required at much larger space and time scales. It is questionable if the basic theories can be applied in a larger scale where the effect of spatial and temporal heterogeneity come into play. The powerful tool in hydrology to solve these problems is conceptualisation. The previous section has shown the simplification to the mathematical description of water movement and transfer for each single hydrological processes. In this section, we will first display one of the popular hydrological models – the Xinanjiang model, showing how catchment hydrological models express the whole of the hydrological cycle in a conceptualised way. The TOPMODEL, one of the widely used semi-distributed hydrological model, is then discussed. Finally an integrated hydrological modelling system is given, which is suited to assess environmental impacts of land use changes.

6.3.1 Xinanjiang Model

A variety of popular hydrological models are discussed in Singh (1995). Most of hydrological models are in the realm of civil engineering applications of catchment hydrological models, such as HEC (Hydrological Engineering Centre) flood hydrograph package (HEC, 1990) and the Precipitation-Runoff Modelling System (PRMS) (Leavesley, 1990) in the U. S., RORB runoff and stream routing program in Australia (Laurenson and Mein, 1993), Tank model in Japan (Sugawara, 1995), Xinanjiang model (Zhao, 1992) in China, UBC watershed model in Canada (Quick and Singh, 1992), and HBV model in Sweden (Bergström, 1992). Large scale hydrological models are appearing since 1990 by embodying Geographical information system and remote sensing technology into hydrological modelling. The most basic characteristics of hydrological modelling is conceptualisation. This characteristic is beautifully shown in Xinanjiang model.

Xinanjiang model was developed in 1973 and published in 1980 (Zhao et al., 1980). It is a rainfall-runoff basin model for use in humid and semi-humid regions. Its main feature is the concept of runoff formation on repletion of storage, which means that runoff is not produced until the soil water content of the aeration zone reaches field capacity, and thereafter runoff equals the rainfall excess, or effective rainfall without further loss. A model of three soil layers represents the evapotranspiration component. The variation of the capacity value is assumed to obey a distributed function throughout the basin as shown in Fig. 6.9, because of which, Xinanjiang model sometimes is classfied into semi-distributed model. This idea is used in many land surface parameterisation schemes to account for heterogeneity (Wood, 1991; Liang et al., 1994; Abdulla and Lettenmaier, 1997; Wood et al., 1997). In Fig. 6.9, f/F represents the proportion of the pervious area of the basin whose tension water capacity is less than or equal to the value of the ordinate $W'M$. For the tension water capacity at a point, $W'M$ varies from zero to a maximum MM (a parameter) according to the relationship:

$$(1 - f/F) = (1 - W'M/MM)^B \tag{6.72}$$

where B is a parameter.

Prior to 1980, runoff in the Xinanjiang model was separated into surface and groundwater components using Horton's concept of infiltration. Subsequently, the concept of hill-slope hydrology was introduced with an additional component, interflow, being identified (Zhao, 1992). Runoff concentration to the outflow of each sub-basin is represented by a unit hydrograph or by a lag-and-route technique. The damping or routing effects of the channel system connecting the sub-basins are presented by Muskingum routing model. The model structure is shown in Fig. 6.10. The basin is divided into a set of sub-basins. The outflow hydrograph from each of the sub-basins is first simulated and then routed down the channels to the main basin outlet. The inputs to the model are P, the measured areal mean rainfall depth on the sub-basin,

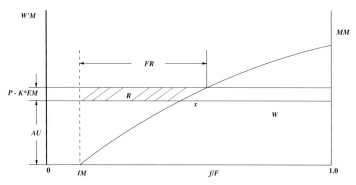

Fig. 6.9. The distribution of the tension water capacity in the Xinanjiang model (after Zhao, 1992).

and EM, the measured pan evaporation in the same units. The outputs are TQ, the outlet discharge from the whole basin, and E, the actual evapotranspiration from the whole basin, which is the sum of the evapotranspiration from the upper soil layer EU, the lower soil layer EL and the deepest layer ED. The state variables are: W, areal mean tension water storage having components WU, WL and WD in the upper layer, lower layer and deepest layer; S, areal mean free water storage; R, runoff from pervious areas having components RS, RI and RG as the surface, interflow and groundwater runoff, respectively; RB, runoff from the impervious area; Q, the discharge from each sub-basin having components QS, QI and QG as surface runoff, interflow and groundwater flow, respectively; FR, the variable runoff producing area; and T, the total sub-basin inflow to the channel network, having components TS, TI and TG.

There are 15 parameters in all, the physical meanings of which are explained in Table 6.1.

Generally, the model is particularly sensitive to the six parameters K, SM, KG, KI, CG, CS and L. Optimisation of parameters is achieved with different objective functions according to the nature of each parameter. The model has been widely used in China since 1980 in almost all areas except the loess area. For the loess area of China, another model, based on the concept of runoff formation in excess of infiltration would be necessary. Such a model would generate surface runoff only, without interflow or groundwater. In semi-arid and semi-humid areas, the Xinanjiang model cannot always find successful application. Li et al. (1998a) improved the Xinanjiang model by adding the mechanism of the runoff formation in excess infiltration and obtained satisfactory results in semi-arid areas as shown in Fig. 6.11. The Xinanjiang model is being expanded and developed to meet with miscellaneous surface conditions such as snow cover, karst, large plains, swamps. It was mainly used in flood forecasting. More recently, it has also been used

6.3 Hydrological Modelling   247

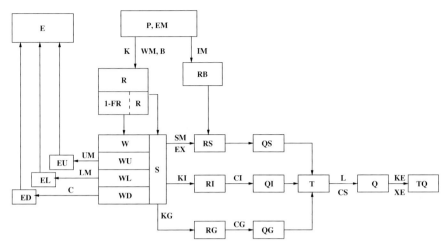

Fig. 6.10. Xinanjiang model structure (after Zhao, 1992).

for other purposes such as water resources estimation, flood design and field drainage, and water quality accounting.

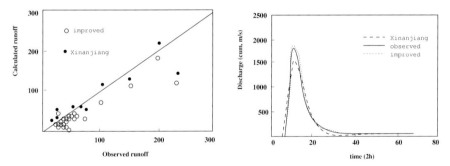

Fig. 6.11. Simulated results of flood forecasting using the original Xinanjiang model, and the improved model by Li (1998a), and a comparison with the observed.

A major problem in hydrology modelling is to simulate groundwater runoff on the basis of specified climatological inputs (precipitation, temperature, pressure, and the like). The linear storage method first proposed by Dooge (1960) has found wide application to this problem (Biddiscombe, 1985). In the approach, a subsurface basin may consist of one or more groundwater reservoirs. Each reservoir can be considered as a linear system composed of one or more storage elements. Thus, the basin is composed of linear reservoirs (or storage elements) combined in a particular manner that best represents the natural system. Almost all rainfall-runoff models, including Xinanjiang

Table 6.1. The parameters used in the Xinanjiang model.

symbol	The physical meaning
K	the ratio of potential evapotranspiration to pan evaporation
$W'M$	water capacity at a point in the catchment
MM	the maximum of the water capacity at point scale
WM	the areal mean tension water storage capacity
UM	the areal mean tension water storage capacity in the upper soil layer
LM	the areal mean tension water storage capacity in the lower soil layer
DM	the areal mean tension water storage capacity in the deeper soil layer
C	a factor, less than unity, by which any remaining potential evaporation is multiplied in application to the deepest soil layer
IM	the impervious area of the sub-basin
SM	areal mean free water storage capacity
EX	a parameter in the distribution of free water storage capacity for three component water sources (surface water, interflow and groundwater)
B	the parameter in the distribution of water storage capacity for two component water sources (surface water and groundwater)
MS	maximum free water storage capacity (related through EX to SM)
KI	a coefficient relating RI, a contribution to interflow storage
KG	a coefficient relating RG, a contribution to groundwater storage
CI	the interflow reservoir constant of the sub-basin
CG	the groundwater reservoir constant of the sub-basin
L	the lag parameter of the flow concentration within the sub-basin
CS	the route parameter of the flow concentration within the sub-basin

model, consider groundwater as such and use the relative empirical formula to describe the recession of groundwater.

6.3.2 TOPOMODEL

Distributed hydrological models have been largely established in the last two decades. A distributed model usually involves descriptive equations, expressible in formal mathematical terms, the solution of which must be found using approximate numerical methods in some form of discretisation. For some hydrological processes, the equations of flow through the system are not well understood, and resort must then be made to empirical generalisations that are not explicitly distributed. A further characteristic of distributed models is that they are expensive to run. As the models forecast what happens at a large number of points within a catchment, and do not merely deal with the conceptual average of the lumped models, the study of distributed catchment models leads hydrological forecasting, from the traditional to the modern way, to meet changing needs in complicated environmental research. A number of such models include: the System Hydrologique European (SHE) model developed in Denmark, France and Britain (Jonch-Clausen, 1979); the Institute of Hydrology Distributed Model (IHDM, Morris, 1980); and Areal Nonpoint Source Watershed Environment Response Simulation (ANSWERS) model that can consider the influence of agricultural practices and soil characteristics (Bouraui, 1997).

There are semi-distributed models that are more conceptual in nature, in postulating functional forms relating storage to catchment response. Some are physically-based, or process based model in the sense that they have parameters that are measurable in the field. TOPMODEL (Beven and Kirkby, 1979) is such an example. Originally developed at the University of Leeds, UK, in the middle of 1970's, it has found wide application all over the world. Although it has never been a definitive model structure, a set of specific concepts it used, make it be applied to forecast different types of hydrological response, particularly including the distributed variable contributing area response, both surface and subsurface. The model made use of the idea that the distributed nature of catchment response could be indexed on the basis of an analysis of topography. It assumes that the saturated throughflow system is in a quasi steady-state, so that the downslope flux per unit contour width at any point i, q_i, can be expressed as:

$$q_i = a_i r \tag{6.73}$$

where a_i is the drainage area per unit contour width at point i (the dimension is m), r is a nominal recharge rate (m s^{-1}), q_i has the unit of square metres per second. Another assumption is that the hydraulic conductivity at each point i decreases exponentially with depth:

$$K_{si(z)} = K_{soi} \exp(-c_f z) \tag{6.74}$$

where $K_{si(z)}$ is the saturated hydraulic conductivity at depth z below the surface (m s^{-1}), K_{soi} is the saturated hydraulic conductivity at the soil surface (m s^{-1}), z is the vertical depth below the surface (m), and c_f is a parameter related to soil structure (m^{-1}). Combining Darcy's Law with the above equations and assuming that the water table is nearly parallel to the soil surface, the throughflow flux at a point i can be expressed as

$$q_i = T_i \exp(-c_f z_i) \tan \beta_i \tag{6.75}$$

where T_i is the transmissivity of the soil, and $\tan \beta_i$ is the surface slope at the point i. Recently, the original form of the exponential transmissivity function was re-evaluated using linear and parabolic transmissivity functions (Ambroise et al., 1996). Fuzzy estimates of saturated areas were incorporated into the calibration processes on constraining the forecasting of TOPMODEL (Franks et al., 1998).

Dimensional methods have been extrapolated from classical field scale problems in soil heterogeneity to larger domains, compatible with the gridsize of large-scale models, to solve the scale problem that is easily met by many distributed models (Kabat et al., 1997). Taking the Swedish HBV hydrological model as an example of the research, the parameters in the distributed approach are found to be relatively stable over a wide range of scales from the macro to the continental scale (Bergström and Graham, 1998). The distributed model shows a promising ability to reflect the spatial variability of the rainfall runoff processes (Merz and Bardossy, 1998). Remote sensing and Geographical Information System have provided a wider perspective in the application of distributed hydrological models (De Troth, 1996). For example, based on the Digital Elevation Model (DEM), a spatially distributed unit hydrograph is developed using GIS (Maidment, 1993). Remote sensing data is used to model the spatial distribution of evaporation on different scales (Mauser and Schadlich, 1998). A GIS-based variable source area hydrological model is established to identify the saturated area required in the effective control of non-point source pollution from contaminants transported by runoff (Frankenberger et al., 1999).

In a study in a semi-arid watershed, Michaud and Sorooshian (1994) compared a lumped conceptual model (SCS-Soil Conservation Service model), a distributed conceptual model (SCS with eight sub-catchments, one per rain gauge), and a distributed physically-based model (KINEROS-a KINEmatic runoff and EROSion model) for simulations of storm events. They found that with calibration, the accuracy of the two distributed models was similar. Without calibration, the distributed physically-based model performed better than the distributed conceptual model, and in both cases the lumped conceptual model performed poorly. Refsgaard and Knudsen (1996), based on the above test experiences, compared a lumped conceptual rainfall-runoff model (NAM), a semi-distributed hydrological model (WATBAL-WATer BALance model), and a distributed physically-based hydrological model (MIKE-SHE) on a relatively larger catchment. It is concluded that all models performed equally well when at least one year's data were available for calibration, while the distributed models performed marginally better for cases where no calibration was allowed.

6.3.3 Integrated Hydrological Modelling System

Human-induced activities influence the basic hydrological processes, and therefore one of the key aspects in modern hydrological modelling is to simulate hydrological processes under the catchment changes by human activities. Catchment changes can be distinguished as point changes and non-point changes. Non-point changes include agricultural change, such as increasing litter and agricultural drainage; urbanisation, such as the changing size of the impervious area; forest activities, such as deforestation and afforestation; road developments, such as landscape restoration; or mining activities, such as the disposal of mined materials. Point changes include reservoir storage, channel improvement, and sewage treatment, to name a few. During the past two decades, field experimental studies, the synthetic unit hydrograph method and the Soil Conservation Service method (Soil Conservation Service, 1971) have been used to investigate the effects of catchment changes on hydrological processes (Singh, 1989). The Soil Conservation Service method computes the volume of effective rainfall as a function of curve number (CN), which in turn, depends on the soil-vegetation land-use complex. By routing this volume through the SCS (Soil Conservation Service) unit hydrograph method, the direct runoff hydrograph can be produced, thus reflecting the effect of land use/cover change, one of the main catchment changes, on hydrological processes. Later on catchment models are used to simulate the effect of land use change on hydrological processes. The earliest models to simulate the effects of land use on hydrological effects are the model developed by Onstad and Jamieson (1970) and the model by Alley et al. (1980), which contain parameters related to the soil,vegetation and land-use which had attempted to focus attention on the use of mathematical models as a forecasting tool.

In order to explore the mechanism of the response of hydrological processes to the catchment changes, an integrated hydrological modelling system is needed. The model firstly must be physically based. A physically-based model is necessarily distributed because the equations on which they are defined generally involve one or more space coordinates. Secondly, the model must consider not only the water transfer in the horizontal direction (river routing), but also the water transfer in the vertical direction. The theory of water transfer in the Soil-Vegetation-Atmosphere Transfer (SVAT) system is especially needed at this juncture. Last, but not least, the model should be applied to medium to large scale basins, instead of only in small scale basins, as most distributed models are at present (Zhang et al., 1996).

Almost all classical hydrological models use empirical or conceptual formulae to calculate evaporation and soil moisture. The parameters involved are not directly related to physical parameters, such as roughness lengths and leaf area index, which can explicitly describe catchment changes. On the other hand, research into the land-atmosphere interaction scheme, from the simple Bucket model (Manabe, 1969) to the complicated SiB model (Sellers et al., 1986) and the BATS model (Dickinson, 1986), shows the description of runoff has been the weakest link. To forecast the effects of catchment changes on hydrological processes, the combination of a land-atmosphere interaction scheme and a hydrological catchment model is necessary, which itself may also improve the research in land-atmosphere interactions. Figure 6.12 shows such a scheme (Wigmosta et al., 1994). In the model, the model representation is a drainage basin. Digital elevation data are used to model topographic controls on incoming solar radiation, precipitation, air temperature and downslope water movement. Linked one-dimensional soil water movement equation and energy balance equations are solved independently for each modelling grid cell. Grid cells are allowed to exchange saturated subsurface flow with their eight adjacent neighbours. A recent research in this aspect is shown in Wood et al. (1998). Liu et al. (1998) applied the BATS model to simulate runoff and soil temperature for mountain and plain areas in Haihe River during the flood season in 1991. In order to consider the lumped role of geography (plain and mountain) in runoff generation, the sensitivity analyses of the power value, n, of soil water content in the runoff formula in the BATS model was conducted. The interesting results are shown in Table 6.2. It shows that for mountain areas, when $n = 1.5$, the simulated total runoff amount, is closer to the observed than the case for $n = 4$. For the plain catchment, when $n = 4$, the simulated soil moisture is more in agreement with the observed than the case for $n = 1.5$. The larger the value of n, the flatter the topography. It is seen that the results of simulated soil temperature, for both the mountain and plain areas, correspond to the results of runoff. The simulated precision of runoff is closely related to that of soil temperature in the upper layer. All illustrates that the improvement of runoff simulation in the BATS model is important for appropriate estimation of latent heat and sensible heat fluxes between land surface and the atmosphere.

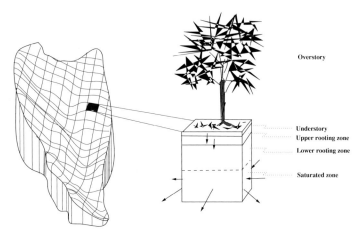

Fig. 6.12. A distributed hydrological model (after Wigmosta, 1994).

Research on low flow (Smakhtin, 2001), a very hot topic in the international program of Flow Regimes from International Experimental and Network Data (FRIEND), has greatly pushed forward the research of integrated hydrological modelling system. Tallaksen and Erichsen (1994) established a new model, HBVMOR, composed of two independent models, a rainfall-runoff model (HBV), and a specialised evapotranspiration model (AMOR).

Table 6.2. The simulated total runoff amount and the standard deviation of the error between the simulated soil temperature and the observed soil temperature (T) in the upper layer (after Liu et al., 1998).

	simulate total runoff amount (mm)		standard deviation of err of T (°C)	
	$n=1.5$	$n=4.0$	$n=1.5$	$n=4.0$
mountain	551	439	1.3	1.7
plain	53	13	2.1	1.9

Rainfall-runoff models traditionally have a very simple representation of evapotranspiration, whereas the specialised evapotranspiration models often have a simplified treatment of soil and groundwater interaction and lack verification on natural, heterogeneous catchments. As the surface is given an explicit model representation, the linked model AMOR and HBV, allows for an evaluation of possible consequences due to land use and climatic changes.

One of the key problem in the coupling between hydrological models and land-surface interaction schemes is scale (Panagoulia and Dimou, 1997; Yates, 1997). Catchment hydrological models usually have an hourly or daily time scale. For the study of the large-scale hydrological cycle in GCMs, the time scale is usually monthly or yearly. Liu (1997b) established a multi-scale hydrological model including monthly, daily and hourly data based models. The model is able to use a wide range of data, from longer series of monthly data, to more detailed hourly data, and thus may abstract more information from the data available. The consistency of calibrated parameters among the hierarchy of models gives sound reliability and reduce uncertainty. Moreover, the results from the multi-scale models provide an insight into the scale effect on hydrological processes. In her model, the Xinanjiang model is selected for the construction of the hierarchy of the multi-scale model. The multi-scale modelling has been carried out for the Huaihe River Basin with an area of 120 000 square kilometres. The parameters from three levels (day, month and year) are matched with each other and the relationship between the scales is replaced with geographical conditions to a certain degree.

An integrated hydrological modelling system should be, as mentioned above, distributed, dynamic, and with the human dimension considered. However, at present, this kind of model has not been fully established. According to the importance of over-riding problems, for practical use the strategy may be to consider more on certain aspects, and less on others. For simple flood forecasting, we may only need conceptual models, statistical models, or even just simple empirical models.

6.4 Hydrological Forecasting

Hydrological forecasting concentrates the forecast of hydrological elements in the near future. The objectives of hydrological forecasting mostly refer to river discharge, which include the discharge at the outlet of a watershed based on known precipitation processes, or the river flow discharge and/or water stage in the lower reach based on corresponding information in the upper reach. Discharge forecasting can again be divided into short-term, mid-term and long-term forecasting. When leading time (the gap between the time to forecast and the time when forecasting object appears) is shorter than the maximum routing time in a catchment, the forecasting belongs to short-term forecasting. When the leading time is longer than the maximum routing time, the forecasting is called mid- and long-term forecasting. Based on the principles of World Meteorological Organisation (WMO, 1992), the leading time for short-term forecasting is two days, for mid-term forecasting 2–10 days, and for long-term forecasting 10 days after. In China, the leading time for mid-term forecasting is usually defined as 3–5 days, for long-term forecasting usually from 15 days to one year, and for super long-term forecasting usually over one year.

The methods available for discharge forecasting range from empirical methods to more advanced methods requiring the usage of advanced computing, Geographical Information System (GIS), remote sensing, and hydrological models. Short-term hydrological forecasting relies greatly on physical models. For mid- and long-term forecasting, it is hard to obtain the forecasting results through runoff formation and routing by making use of observed rainfall data or observed discharge data upstream. The main aim of the forecasting thus becomes establishing the relationship with antecedent circulation, sea surface temperature, solar activities or other geophysical factors and historic data by using various statistical methods.

In the early stages of human civilisation, water was considered an infinite resource all over the world, except in arid regions. With the development of economies and technological progress, water demand by human activities has increased substantially. A crisis between water demand and supply has occurred even in areas where water is naturally abundant. In addition, increasing water pollution has exacerbated water shortages. The objectives of hydrological forecasting thus also include water quality, the effects of environmental change on hydrological processes and water demand and water supply. At present, water quality forecasting is in its infancy. State-of-the-art forecasting of the effects of environmental change on hydrological processes and the water demand and water supply is in the form of scenario forecasting.

The last but not the least objective of hydrological forecasting is the hydrological elements related with atmospheric circulation, i.e. larger scale forecasting, which deals with the effects of climate change and atmosphere-ocean interactions on hydrological processes. High-resolution numerical weather forecasting modelling (Gauntlett and Leslie, 1975; see more in Chap. 3) is promising in this field to provide valuable forecasts for catchment over periods ranging from a few hours to two or three days. The large scale hydrological forecast is shown in detail in Yu et al. (1999) on the simulating the river-basin response to an atmospheric forcing by linking a mesoscale meteorological model [Penn. State-National Center for Atmospheric Research (NCAR) Mesoscale Meteorological Model (MM5)] and hydrological system, and Droegemeier et al. (2000) and Sugimoto et al. (2001) on quantitative precipitation forecasting (QPF), ensemble forecasting, spatial downscaling (applying more sophistical statistical techniques to numerical modelling) and data assimilation.

6.4.1 Short-term Hydrological Forecasting

Generally speaking, there are three kinds of methods for short-term forecasting: namely, the empirical method, the mathematical method and conceptual catchment modelling.

Empirical Method. The empirical method has been used widely since the very beginning of the history of hydrology. The emphasis of this method is on establishing an empirical relationship between rainfall and runoff. For the forecasting of large rivers, especially for forecasting discharge in the lower reaches based on the data of the upper reaches, this method is effective. Because the empirical method contains few or no physical principles, it has almost no up- and down-scale ability in time or space, which very much limits its application in many situations.

Mathematical Method. The mathematical method, or the so-called systematic approach, aims at establishing the relationship between rainfall and runoff, or discharge (stage) in the upper stream and down stream. A simple example is the Simple Linear model which is expressed as (Liang, 1995):

$$y_t = \sum_{k=1}^{p} h_k x_{t-k+1} \tag{6.76}$$

or in continuous expression:

$$y_t = \int_0^t h(t-\tau) x_t \mathrm{d}\tau \tag{6.77}$$

where y_t is the time series of discharge of river flow, x_t is the time series of rainfall, τ is the time lag, and h_k is kernel function, or unit hydrograph, which can be calculated by the ordinary least square method, constrained least squares method, or Rosenbrock Search method. The quantity p is the length of unit hydrograph.

To define x_d as rainfall seasonal mean, and y_d as discharge seasonal mean, by replacing x_t in (6.76) with $x_t - x_d$, and replacing y_t with $y_t - y_d$, the model then becomes the Linear Perturbation model.

By defining gain factor (G) as:

$$\sum_{k=1}^{p} h_k = G \tag{6.78}$$

The model becomes the Simple Linear Variable Gain Factor model. The details of the mathematical models can be found in Special Issue of J. of Hydrology, Vol. 133, 1980. Mathematical models are very handy and useful for operational river flow forecasting.

Conceptual Catchment Modelling. With an increase in the knowledge of the hydrological system, hydrologists have used a large number of conceptual models for short-term forecasting. Using concepts to explain the physical mechanisms within the hydrological system, conceptual models make forecasting more scientific. There is a big gap between the models for pure scientific research and the models for operational forecasting. This implies that models should require only input data fields which exist, that they should be flexible, such that they also work if parts of the required input data are not transmitted; the consideration of risk and uncertainty should be realistic and not spectacular. In the normal situation, simple model is especially welcome for real-time forecasting.

Application of Water Transfer Equation. Based on the fluid dynamics of porous media, Liu (1997a) established a water transfer equation for a groundwater/surface water interface for a short-term river flow forecasting. The equation is as follows:

$$\frac{\partial}{\partial t}(b\epsilon\theta_c) + \frac{\partial}{\partial x}(b\epsilon\theta_c u) = P'_e - FC' \tag{6.79}$$

in which u is velocity in the x direction, b is the depth of the interface, ϵ is the porosity, and θ_c is the saturation coefficient. The quantity P'_e is the effective rainfall at a point and FC' is the infiltration to groundwater at that point, both of which are the normal components of mass transfer at the two sides of the interface. Effective rainfall equals precipitation minus evapotranspiration. A groundwater/surface water interface was conceptualised as a wedge ABC, as shown in Fig. 6.13.

The water balance at the interface was written as:

$$W_2 - W_1 + Ro = P_e - FC \tag{6.80}$$

where W_1 and W_2 represent soil water content of the interface at time t_1 and time t_2, respectively, Ro refers to runoff, P_e means effective rainfall over the interface, and FC refers to the interaction of the interface with groundwater. The quantities P_e and FC equal the averages of PE' and FC', respectively. By comparing the water transfer equation with the water balance equation, the first term of the left-hand side of the water transfer equation refers to the temporal variation of water capacity at the interface. It is called term I, which closely relates to the moisture θ, the depth b, and the porosity ϵ. The second term refers to the spatial variation of water transfer, and is called term II. Water transfer increases with increasing ϵ, θ_c, b, and velocity u. The first term and the second term of the right-hand side of the equation are called term III and term IV, respectively. If the angle $\angle ABC$ of the interface as shown in Fig. 6.13 is reduced to zero, then b corresponds to zero. From the water transfer equation it is seen that P'_e then equals FC', and from the water balance equation that Ro equals zero. This means that there is no water transfer at the interface. This is an unusual concept to classic hydrology, in which, if the groundwater table rises to the land surface, surface runoff (Ro) equals precipitation. However, for this case, using the concept of the above mentioned transfer theory, there is no water transfer. The groundwater is viewed as a reflector mirror, and so-called runoff is simply the behaviour of precipitation at the land surface only. The interface in this case has no water transfer function. The depth of the interface b is defined as the index of water transfer function. The water transfer function increases with increasing value of b.

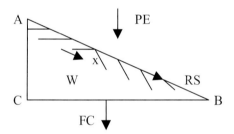

Fig. 6.13. The conceptual groundwater/surface water interface (after Liu, 1997a).

By calculating terms I, III and IV of the water transfer equation, the term II, water transfer is obtained. To solve the water transfer equation (6.79), other equations are often needed, such as momentum transfer and intrinsic equations as described in fluid dynamics. In many cases, in order to make the solution practical, conceptualisation is still needed. However, the difference between the use of the water transfer equation with conceptualisation, and the use of catchment modelling is obvious, as shown in the case study described as follows. The case study involves forecasting floods of the Yanghe Reservoir Basin located at Qinhuangdao, Hebei Province of China, as shown in Fig. 6.14. Flood forecasting in this basin is challenging because of the irregular shape of the basin and the lack of data.

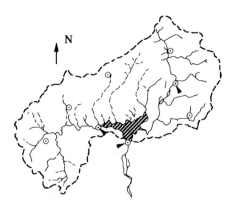

Fig. 6.14. Map of the Yanghe catchment (after Liu, 1997a).

On the one hand, using comparative analysis, the values are obtained to substitute into the water transfer equation for the water balance equation of the interface. On the other hand, it is the value of each term of the water transfer equation that explains the meaning of the relevant water balance term. Therefore, the question of how to calculate the relevant water balance terms can be solved by calculating appropriate terms of the water transfer equation.

In the study, although data for point precipitation is available, the average precipitation from the concept of term III must be calculated. It is needed to consider how to obtain the average precipitation. The simplest way is to divide the area into sub-areas according to the weights of the hyetal station. In this way, the more sub-areas divided, the less the difference between them. With an increasing number of sub-areas, they eventually reach the point where the differences within the sub-areas become meaningless. On the other hand, if there are too few sub-reaches then the representation of the average value of precipitation over the few sub-areas is of questionable accuracy. A scale exists, called the Representative Elementary Length (REL) (Wood et al., 1990; Liu and Liu, 1994) that is useful in this context. When the sub-area scale is smaller than REL, the effect of the spatial variability within the sub-areas on the hydrograph is evident, and it is impractical to average the weight of the point data as the average input. When the scale is larger than REL, the representation of the data as an average is questionable. Therefore, only when REL has been calculated is it possible to correctly determine the input term. Using semi-variance analysis, the maximum correlation length, equivalent to REL, is equal to $17.81\,\text{km}$ (the area of the basin is $755\,\text{km}^2$). Based on this REL, and using stepwise regression analysis, principal component analysis, grey system theory and the hyetal map, the input term is calculated by weighting the average of the precipitation. This is repeated at the eight hyetal stations. If water movement is assumed to be unidirectional, then term II represents the hydrograph at the cross-section of the basin. After comparing the generic algorithm (Wang, 1991) and the Hooke-Jeeves algorithm, the two methods were combined to create a powerful and efficient method to estimate parameters for calculating term II. Eight floods are used to create optimal object functions. The qualified peak of floods is 62.5%, which to a certain extent identifies the capability of using this method to solve the problem.

The Xinanjiang model was chosen as an example for comparison. It is found that, when the unsaturated soil is considered as one layer and the river reach is not divided into sub-reaches, the simulation result from Xinanjiang model is in agreement with that of the equation solution. This result is reflected in the Frugality Principle. Simply stated, when using the equation to solve flood forecasts, the basics are captured. The forecast resulting from either the water transfer equation or a model cannot reflect more information than the data provides. The flow chart of the water transfer equation solution is shown in Fig. 6.15, where P is precipitation, $W(\theta)$ is soil water content, IMP is impervious area, R is runoff, RS is surface runoff, RG is groundwater flow, t represents the hydrograph and Q is discharge at outlet. The parameters WM, B and KG are as given in the Xinanjiang model in Table 6.1. The quantity N is the number of linear reservoirs for routine, Fc is the infiltration capacity, NK is the scale coefficient, and KG is the coefficient. From Fig. 6.15 it is seen that the water transfer equation solution avoids confusing concepts often incurred by using the common watershed simulation. These concepts include hydrograph separation and runoff formation identification. Moreover, all the assumptions to deduce the water transfer equation are the same as those conceptualised in the model.

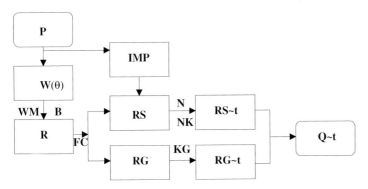

Fig. 6.15. The flow chart of the solution to the water transfer equation (after Liu, 1997a).

New Methods for Channel Flood Forecasting. Based on the time series analysis and optimisation control theory, as show in Fig. 6.16, Rui et al. (1998)

proposed a water level forecasting model with the consideration of backwater effects. It basically overcomes the shortage of not directly dealing with backwater effects in a hydrological way, and complicated calculation in a hydraulic way. Rui et al. (1999) proposed a flood routing method based on an analytical solution to the linear diffusion wave equation with free downstream conditions. In the model, using the series of the unit rectangle impulse function with constant coefficients represents the discretisation of inflow hydrograph. The model is successfully applied in flood forecasting at the six reaches of the lower Yellow River, where the looped depth-discharge relationship exists. Another tendency in channel flood forecasting is to apply new methods, such as artificial neural networks (Yuan et al., 1999; Campolo et al., 1999).

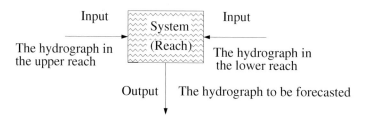

Fig. 6.16. The system of water level forecasting with backwater effect (after Rui et al., 1998).

The Techniques of Short-term Forecasting. How to use one or more of the aforementioned models to solve a real problem in short-term hydrological forecasting usually depends on the following factors:

- the accuracy of input data and its resolution in space and time;
- the availability of real-time data, which greatly depends on the data retrieval system;
- the group's capability to use a complex model and the training environment;
- the function of the equipment to process data;
- the aim of forecasting; and
- the results, if forecasting is over or below that expected.

For short-term forecasting, there are two stages to follow. First stage: the aim is to find a suitable model or forecasting method based on input data. This stage is usually off-line and conducted by theoretical researchers. They can, and want to, try any method either deterministic or statistical. They can use any new method from decision-making, systematic analysis, control theory and from other sciences. They do not take part in such jobs as data retrieval, transformation or processing. They always use historical data for model calibration and verification. They are allowed to make mistakes. If necessary, they can ask a computer to repeat calculations until they are satisfied

6.4 Hydrological Forecasting

that the parameters are sorted out. Second stage: the aim is to announce a real forecast, or warning, to make the losses caused by either flood or drought as small as possible. This stage is usually on-line and conducted by experienced teams. They are always very careful to change a scheme, which has existed already, into a scheme which is supposed to be more advanced, but is still new to them. They are directly responsible for data retrieval, data transformation and data processing. They are not allowed to make any mistakes. This makes them reluctant to use new forecasting methods, which usually involve bigger risks. In order to extend the leading time, using precipitation data directly observed by radar or satellite for input is helpful to raise the quality of forecasting (Kraijenhoff and Moll, 1986; Schultz, 1988). Because of the errors in both model and data, sometimes forecasting results are far away from the observed. These errors include random errors and systematic errors.

How to make errors as small as possible is a big problem. Real-time forecasting based on the new observed information was developed in 1980 (Kitanidis and Bras, 1980). Kalman filtering, proposed prior to the 1960s by Venna, is a very useful tool to process real-time correction of forecast results (Hino, 1970; O'Connel and Todini, 1977; Zhu, 1993). Li et al. (1998b) developed an updating semi-auto-adaptive Kalman filter model of channel routing. In the model, the measurement error covariance matrix was used as a real-time estimate based on the innovation information sequence. The model is useful to consider the inflow in the channel routing as shown in Fig. 6.17. Real-time forecasting has been welcomed by forecasters, as this effectively avoids the details of the model and source of errors, and mainly concentrates on the bias between the simulated and observed. However, of all the hydrological models, only linear input-output models, such as the ARMA model, are easily used in real-time forecasting, which is especially efficient when applied to reach forecasting for large catchments. Real-time forecasting is difficult to modify the results of a conceptual model by the observed output data because of the complicated relationship between the input and the output. Real-time forecasters must be very careful when the hydrograph varies very much near a flood peak, where the new forecast value is influenced a lot by the real-time information, which may cause bigger errors. All in all, real-time forecasting is just a technology to modify the forecasting results. The key to raising the accuracy of forecasting still further relies on the establishment of a reasonable hydrological model.

There is another important factor controlling the accuracy of forecasting—people's experience, especially in some special circumstance. A good example is in the forecasting of discharge at Luoshan Station (Hydrology Bureau of Yangtze Water Resources Commission, 1999), where a river and lake combine to make the water balance difficult to achieve and operational experiences helped a lot in the forecasting.

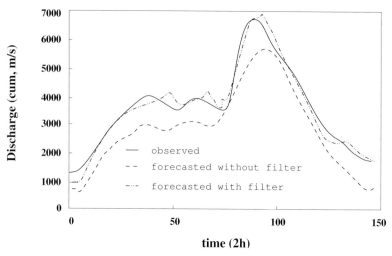

Fig. 6.17. The comparison of the discharge between the observed, forecast with filter, and forecast without filter (after Li, 1998b).

6.4.2 Mid- and Long-term Hydrological Forecasting

Generally speaking, there are three kinds of methods for mid- and long-term hydrological forecasting, which are described as follows:

Synoptic Method. The variation in runoff mainly relies on precipitation, which is greatly controlled by the atmospheric circulation. The anomaly in runoff, which causes either drought or flooding, is always closely related to the anomaly in the circulation. Based on these facts, synoptic method aims to establish the relationship between runoff and the pressure at 500 mb, 200 mb, 100 mb or sea level, from the North Hemisphere to the South Hemisphere. This method is the mainstream for mid-and long-term hydrological forecasting. With the development of more and more mature numerical weather models, synoptic methos is advanced from statistical analysisi to the numerical forecasting, see more in Chap. 7.

Underlying Surface Heat Status Method. By mainly absorbing the longwave radiation from the underlying surface, the atmosphere exchanges water and heat with the underlying surface. Underlying surface heat status is therefore one of the most important factors for long-term hydrological forecasting. In addition, the ocean, with its enormous scale and heat inertia, cannot be neglected in the forecasting. Earlier in 1970s, scientists in China found that the anomaly of sea surface temperature at the equator in the east Pacific Ocean was closely related to the anomaly in precipitation in the Yangtze

River. Different phases of the ENSO (El-Niño/Southern Oscillation) have different influences on the status of flood and drought in China. When ENSO is in a warm phase, that is, when the sea surface temperature at the equator in the east Pacific Ocean increases, the precipitation over the Huai River is high, while other areas, such as the Yangtze River Basin, southern China, and Yellow River Basin, are dry. Correlation analysis shows (Qian, 1990) that there exists connections between the Antarctic sea-ice extent from February to March and runoff for the upper Yangtze River from August to September. These results are useful to reveal the physical background of the variation in runoff for the river. Among the features of the underlying surface, the effects of the Qinghai Plateau on long-term hydrological forecasting are also obvious in China. Results of numerical studies show that the heat variability of the Qinghai Plateau in summer is the main cause of the variability in runoff on the plateau and in the upper reaches of the Yellow River, as shown in Fig. 6.18 (Li, 1998c).

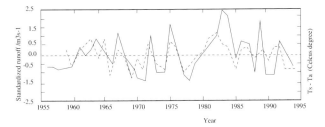

Fig. 6.18. Curves of standardised runoff of Tangnaihai in July (solid line) and departure of $T_s - T_a$ of Hezo in early February (dashed line). Here, T_a is surface temperature, and T_o is average temperature (after Li, 1998c).

Geophysical Method. This method is also called the method of factor analysis. The aim of the method is to establish a relationship between long-term hydrological processes and geophysical factors, including solar activity, space dust, planet movement, tide, moon activity, self-rotation of the earth, geomagnetism and volcanic activity, among others. Many results show that such relationships exist. Figure 6.19 shows that there exists a good relationship between the annual maximum discharge and strong solar spot activities. In double week of solar activity, the frequency to appear low discharge (8000 m³ s^{-1}) is 73%, which is twice of the frequency of single week (40%). In single week of solar activity, the frequency to appear large flood (10000 m³ s^{-1}) is 32%, which is also twice of that in double week (15%). Huayuankou station easily incurs serious floods when it is close to the peak or trough of a single week of solar activity (Huo et al., 1999).

No matter which above-mentioned methods are used, the probability and statistical techniques are main tools for long-term hydrological forecasting.

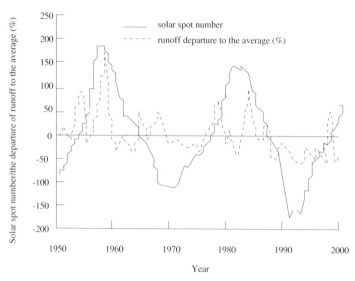

Fig. 6.19. The relation of solar spot number and the departure of runoff compared to the average (after Huo, 1999).

Among them, multiple regression such as stepwise regression, principal component analysis or cluster analysis, and time series analysis, such as the autoregression model, are used mostly. Recently, many new techniques have been introduced for mid- and long-term forecasting. For example, Zhao (1998) applied the algorithm of the wavelet network model of chaotic phase space to mid- and long-term forecasting. The results are satisfactory in both the calibrated period and the verified period, while the results of the traditional multiple regression method are only good in the calibrated period and poor in the verified period, as shown in Fig. 6.20.

6.4.3 Water Quality Forecasting

The purpose of modelling water quality based on the equations in Sect. 6.2.12 is to forecast whether, or to what extent, a change in land or water use will impair the utility of the water. The methods for water quality forecasting also range from the simple to the complex. The simple method focuses on those aspects that have overriding importance in the entire water quality system. For example, because the most severe pollution occurs in the dry season of a river, most water quality models are simplified as time-invariant. Forecasting with these kinds of models is actually just reach forecasting. That is, the concentration at the lower reach of the river is forecasted based on the concentration at the upper reach of the river by the model. The solution to these kinds of models is usually analytical (Liu, 1994).

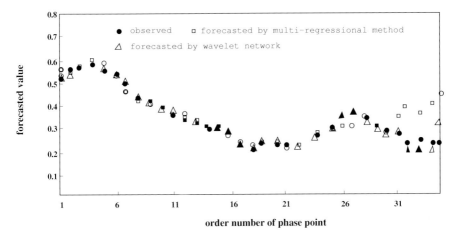

Fig. 6.20. Comparison between the observed, the forecast by multiple regressive method, and forecast by wavelet network (after Zhao, 1998).

Another simplified example is the forecasting of water quality in reservoirs. Usually, the model for water quality in reservoirs is three-dimensional. However, the vertical variation in the variables, for example, water temperature, is prominent. Therefore in practice, a one-dimensional water temperature model is often used to forecast the vertical distribution in the reservoir, as the famous WRE (Water TempeRaturE, developed by Water Resources Engineers, Inc., Walnut Creek, California) model did. The model is time variant. The solution is usually numerical.

Remote Sensing and Geographic Information System (GIS) were used to forecast daily suspended sediment levels in Rivers, combined with WESTEX Water Turbidity model (developed by the U.S. Army Corps of Engineers Waterways Experiment Station and the University of TEXas), where observed data was unavailable. Such quantitative methods were used to forecast potential cumulative impacts on environmental resources by U.S. Army Corps of Engineers (Canter, 1996).

Besides the above deterministic water quality model, non-deterministic water quality models have been used over the last two decades, such as the stochastic water quality model, along with the application of the Grey System theory and fuzzy theory.

The response time, t_c, is another alternative often used for water quality forecating, especially in the forecasting of eutrophication of a lake or reservoir. It is the time to reach the new balance between the concentration of the water body and its actual load, which can be expressed as:

$$\frac{dC}{dt} = \frac{C_{eqV} - C}{t_C} \qquad (6.81)$$

266   6. Hydrological Modelling and Forecasting

where C is concentration at time t, and C_{eqV} is the concentration at balance status which can be estimated by a conceptual model.

In practice, most of the methods to forecast water quality is equivalent to scenario simulation, as is most of the forecasting of water supply and water demand, which will be described in next section. Usually, three typical years with high, average and low discharge are chosen. The output of the model using the data from these three years is the result for forecasting.

A well-established non-point pollution model usually consists of a rainfall-runoff sub-model, an erosion and sediment transport sub-model, a pollutant transport sub-model, and a water quality sub-model in the water body that receives pollutants. Non-point source pollution is hard to forecast. Figure 6.21 shows that the most precise model for forecasting is the hydrological model over a small impervious basin. The error in the pollutant transport sub-model is the largest. Even so, there are still successful case studies in non-point pollution forecasting (Young et al., 1989; Rinaldo et al., 1989; Wen and Liu, 1991).

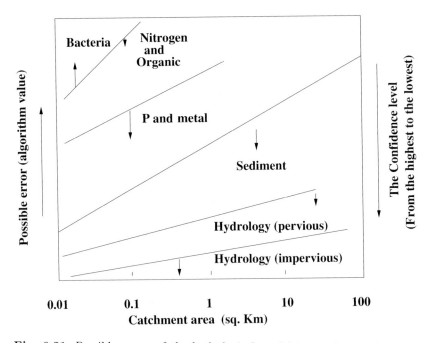

Fig. 6.21. Possible errors of the hydrological model in pervious and impervious surface, the model of sediment transport and the pollutant transport model with catchment data.

6.4.4 Water Supply and Water Demand Forecasting

General Methods of the Forecasting. Generally, the methods for water demand and water supply forecasting (Gardiner and Herrington, 1986) can be classified into three categories as follows:

- *Judgmental forecasting* Judgmental approaches rely upon the experience of an individual, and may be either entirely subjective in nature, or a modification of more objective results derived from other approaches. Perhaps one of the greatest challenges in forecasting is in successfully combining the elegance of sophistical statistical forecasting techniques with judgmental adjustments made necessary because future trends may be different from past patterns.
- *Causal forecasting* Causal, explanatory or analytical forecasts are those in which an attempt is made to forecast the variable of concern, by reference to other variables which, it is assumed, control or influence it. This kind of approach has been increasingly favoured by utility forecasters in recent years. It may, of course, be based on relationships estimated through simple or multivariate regression. Severe problems may arise, however, as regression models used for forecasting often make several assumptions concerning data. For example, it is often assumed that the errors are normally distributed and that the variables for forecasting are independent. However, this assumption rarely holds absolutely for water demand data. In particular, variables used as forecasters, such as population growth and industry activity, may well have theoretical links.
- *Extrapolative forecasting* The method derived from time series involves consideration of only the variable of concern, and the forecasting of future values from the projection of past trends. It is noted that more complex or sophistical methods are not necessarily more accurate than simple ones. Accuracy is not, of course, the only criterion of interest. The choice of method must also depend upon the type of data, type of series, the time horizon of the forecast, as well as upon budgetary constraints.

No matter which method is used, the process of water supply and water demand forecasting cannot be separated from decision-making (Jenkins, 1982; Whipple, 1996; Wood, 1998). A typical frame of water resources decision support system is shown in Fig. 6.22 (Jiang, 2000).

Hydrological Replacement Time. Water is normally regarded as a renewable resource. Quantitative renewal through the hydrological cycle depends essentially on rainfall, with wide quantitative and periodic variations as well as considerable regional differences. The concept of a hydrological replacement time, which is defined as the average time for water to complete a trip through a particular phase of the hydrological cycle, is introduced to determine the water availability from groundwater and surface water sources. The hydrological desirable amount of water utilisation from groundwater and surface water

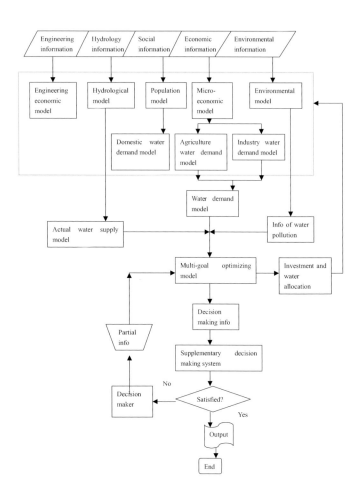

Fig. 6.22. A typical water resources decision support system (after Jiang, 2000).

is proportional to the amount available from these sources. The available water is calculated on the basis of hydrological replacement, volume and any allowable water source depletion. Slutsky and Yen (1997) conducted a simple and idealised long-term analysis of water availability based on the concept of hydrological replacement time in nature to forecast water supply, i.e. hydrologically desirable water usage from surface and groundwater sources. They concluded that on the basis of the hydrological characteristics of groundwater and surface water sources, it is desirable to utilise more surface water than groundwater at both the global and continental scale. Long-term water usage can be no more than that naturally available through the hydrological cycle.

Ecological Water Demand. The concept of ecological water demand (Tubman, 1996; May, 1996; Petts and Calow, 1996) needs to be emphasised for water demand forecasting. When making a plan for both the short-term and the long-term, water demand always includes more than water demand by agriculture, industry and daily life. Water demand should also encompasses the requirements demanded by fish and wildlife, by transporting sediments to reach a natural balance between water and sediments, and by the riverine system itself.

New Characteristics of Water Resources Management. Water resources management is a vital tool of infrastructure policy and is particularly important for protecting and preserving natural habitats by satisfying ecological water demand and measures to protect areas of water (Biswas, 1987). Recently water resources management have changed greatly. Its new characteristics include a non-point control; entity function for the whole of the basin's ecosystem; preparedness to combat pollution other than just treatment; application of non-engineering acts to water resources management; utilisation of wetlands; partnership among government, the public and enterprises; sharing of information and data about water resources; water resources education; and comprehensive basin-wide management of water, air and eco-resources, to name a few. Controlling measures, i.e. targeted water resources management (Munasinghe, 1990), is highly needed. It signifies the regulation of all human uses of, and effects on, surface water and groundwater, thereby necessitating the development of objectives and general conditions covering all sectors for the utilisation of water resources by competing users, guaranteeing the environmental compatibility of water resources management activities. Overuse, changes in land-use, climatic shifts etc. may in the long-term impair the capacity for renewal, thereby reducing water supplies and their use. A distinction should be made between those uses which do not involve actual water consumption (e.g. water used for cooling purposes) and those in which water is wholly or partially consumed or contaminated. Measures to protect against the damaging effects of water (e.g. flood protection) should be regarded as uses. Peripheral conditions of water resources management are thus determined not only by technical and economic considerations but also by sociological, socio-cultural, legal, medical/hygienic and political aspects.

Forecasting Models.

1. *IWR-MAIN model*
   Perhaps the simplest approach for forecasting future water use is the use of projected gross per capita water use rates and projected population. Gross per capita use can be observed from known water consumption and known population. Projections can be made for different population growth scenarios combined with varying assumptions about future

changes in the capita use. IWR-MAIN (Version 6.1), a PC-based software tool, developed in 1960's, updated by the U.S. Army Corps of Engineers' Institute for Water Resources recently, used a special approach for water use forecasting. It is based on observed relationships between water use and causal factors, or determinants, of urban demand for water. Causal relationships have been developed separately for residential and non-residential use. Forecasting relationships used in IWR-MAIN for the residential sector are based on the integration of approximately 60 studies of residential water demand, which contained about 200 empirically estimated water use equations. For the residential sector, the generalised form of the equation for projected average use is (IWR-MAIN User's Manual and System Description, April 1995, p. D-2):

$$Q = aI^{d1} MP^{d2} e^{(FC)(d3)} H^{d4} HD^{d5} T^{d6} R^{d7} \tag{6.82}$$

where
Q = forecasted water use in gallons per day;
I = median household income (\$ 1,000 s);
MP = effective marginal price (\$ /1,000 gal);
e = base of the natural logarithm;
FC = fixed charge (\$);
H = mean household size (persons/household);
HD = housing density;
T = maximum-day temperature (degrees F);
R = total seasonal rainfall (inches);
a = intercept in gallons/day; and
$d1, ..., d7$ = elasticity values for each independent or explanatory variable.

The above relationship can be approximated for any residential subsector, season, purpose (e.g., indoor, outdoor), depending on data availability. Elasticity is a measure of the relationship between water use and a given explanatory or independent variable. For example, an income elasticity of +0.5 would indicate that an 1% increase in income would result in a 0.5 % increase in water use.

Forecasting relationships used in IWR-MAIN for the nonresidential sector are based on the relationship between employment and water use in over 7,000 establishments representing the eight major industry groups throughout the United States. For the non-residential sector, the generalised form of the equation for projecting water use is:

$$Q = aPR^{d1} MP^{d2} CDD^{d3} OTH^{d4} \tag{6.83}$$

where
Q = water use in gallons/employee/day;
PR = labour productivity;
MP = marginal price (\$ /1,000 gal);

CDD = cooling degree days (number of days);
OTH = others (user added);
a = model intercept (gallons/employee/day); and
$d1, ..., d4$ = elasticities for independent/explanatory variables.

The above relationship is designed to be approximated by Standard Industrial Classification (SIC) group. While this model is operational within the IWR-MAIN model, the elasticities for all of the explanatory and independent variables are currently set to zero, since there are currently no available elasticities for them. As a result, this version of IWR-MAIN uses a single coefficient equation to calculate water use by industry group:

$$Q = GED \times E \tag{6.84}$$

where
Q = water use in gallons per day;
GED = gallons per employee per day; and
E = number of employees.

The updated IWR-MAIN model is widely used in the forecasting of water use in many places in the U.S.

2. *WATFORE model*

   WATFORE model is one of the several new models to forecast water use and water demand. It is currently used by the City of Austin and the Edwards Underground Water District of the U.S. to assist officials in making water policy decisions. The model provides a preview of expected water consumption that can range from as little as two days to as much as two months in advance. The program requires daily data on water pumpage, daily rainfall, maximum air temperature, and expected weather conditions. WATFORE is most frequently used to anticipate water use in the near future and to estimate the chance that extreme use may occur. For example, the City of Austin has a water conservation ordinance that requires mandatory water conservation if water use goes beyond a specified level for several consecutive days. The city uses WATFORE to estimate the chance this will occur within a two-week period. The model is also helpful in scheduling water deliveries and studying the impact of demand management programs. Another model for water demand forecasting (Baumann et al., 1998) is used to guide water conservation plan in Environmental Protection Agency of the U.S.

3. *DCSE model*

   Diba Consulting Software Engineers (DCSE) recently has developed an automated geographic information system (GIS)-based water demand forecasting tool to rapidly estimate the water demands in response to changes in the land use and the related use factors. The changes in land

272   6. Hydrological Modelling and Forecasting

use include varying the build-out and the phasing plans to evaluate different growth scenarios. The automated tool is used to generate and analyse the water demand component of the Water Resources Master Plans (WRMPs). The water demand data generated by the forecasting tool is used to evaluate the system capacity and to plan future system expansions. ArcView is used as the front-end graphical user interface (GUI), as well as a tool for displaying post-processed data. This facilitates changing the water use factors and updating water demand forecasts. Data are stored in a relational database. The algorithms for computing water demand are bundled in a Forecast Generator Engine (FGE), using Java.

4. *Using Hybrid Neural Network method to forecast water demand*
Neural networks are simulation models that consist of processing elements or nodes that are interconnected in a massive, parallel fashion. The basic structure consists of an input layer of nodes, that receive external inputs, hidden layers, and an output layer. Each node in the hidden layer processes a weighted sum of incoming input signals. Each node in the output layer generates an estimate of its contribution to the output signal. The neural network method has been used in short-term forecasting as mentioned in previous sections. In the study of Heller and Wang (1996), a HyBrid Neural Networks is used to forecast water demands to improve the accuracy of portable water demand forecasts. The hybrid method of forecasting is proved to be superior to conventional linear forecasting tools and to pure neural networks. The hybrid neural network shows marked improvement in interpreting and training complex data sets because it is able to identify seasonal lags. The time required to train and test the neural network using the hybrid method is reduced significantly because it allows the use of smaller and more appropriate network input structures. The method can be implemented on-line and specialised circuits already exist. The neural network forecasting could operate with Supervisory Control And Data Acquisition systems (SCADA) systems, permitting water utilities with opportunities for real time integrated water resources management.

5. *Water supply model*
Water supply forecasting is divided into two cases. One is in natural runoff forecasting, another is snowfall forecasting. The forecasting of natural runoff is shown in the previous sections of this chapter. They are based principally on measurements of precipitation, snow water equivalent, and antecedent runoff. Forecasts become more accurate as more of the data affecting runoff are measured.

The forecasting of snowfall needs special attention. The snowfall accumulates during winter and spring, several months before the snow melts and appears as streamflow. Since the runoff from precipitation as snow is delayed, estimates of snowmelt runoff can be made well in advance of its occurrence. Fall precipitation influences the soil moisture conditions prior

6.4 Hydrological Forecasting

to the formation of the snowpack and explains, in part, the effectiveness of the snowpack in producing runoff. Sometimes the forecast needs to assume that climatic factors during the remainder of the snow accumulation and melt season will interact with a resultant average affect on runoff. Early season forecasts are therefore subject to a greater change than those made on later dates. Precipitation and snowfall accumulation of known probability as determined by analysis of past records are utilised in the preparation of probability runoff forecasts. The forecasts include an evaluation of the standard error of the forecasting model. The forecasts are presented at three levels of probability as follows:

a) Most Probable Forecast. Given the current hydrometeorological conditions to date, this is the best estimate of what the actual runoff volume will be this season.
b) Reasonable Maximum Forecast. Given current hydrometeorological conditions, the seasonal runoff that has a ten percent chance of being exceeded.
c) Reasonable Minimum Forecast. Given current hydrometeorological conditions, the seasonal runoff that has a ninety percent chance of being exceeded.

Runoff forecasts at all points are for full natural or unimpaired runoff corrected for evaporation, upstream diversions, and adjusted for other hydrologic changes as they are developed.

An important program called "Programme on Forecasting and Applications in Hydrology" is implemented from 2000 to 2003 by the World Meteorological Organisation. It relates hydrological modelling and forecasting to the application of hydrology in the studies of global change. The programme mounts activities in support of water resources development and management, hazard mitigation, studies of climate change and environmental protection, and is linked with the World Climate and Tropical Cyclone Programmes.

7. Simulation and Prediction of Ice-snow Cover

Gongbing Peng

The cryosphere is one of the basic components of the Earth's environment, existing together with the atmosphere, the hydrosphere, the lithosphere and the biosphere. All these components are closely related and interact with each other. The study of the cryosphere and its interactions with the other environmental components is one of the important tasks of modern environmental science.

The area covered by ice and snow can extend to high and middle latitudes in some regions, which is a key indicator of the cryosphere. The total mass of ice and snow is great and it has significant temporal and spatial variations. Furthermore, ice-snow cover bears the characteristics of an intensive cold source and high albedo. Its influence on other environmental components and even on the whole environment system cannot be ignored. In environmental modelling and prediction, the ice-snow cover must be carefully considered. The ice-snow cover can be considered to be a prediction indicator of the environment and its components. Due to the relatively conservative character of the temporal change for small time intervals and the significant inter-monthly and inter-annual variations of the cover, it is particularly valuable for long-range environmental predictions. The global warming effects due to anthropogenic emissions of gases can influence the direction and intensity of the ground disturbance. The effects of warming due to atmospheric climate change may only become apparent over many decades. In the short-term, they will be masked by other surface temperature changes due to microclimate effects, and due to the inter-annual variability of climate and weather. A tendency from anomalies of ice extent persisting for several months is apparent in the lagged autocorrelations of the amplitudes of the dominant ice eigenvectors. The monthly persistence of the ice anomalies is considerably greater than the persistence of anomaly fields of sea-level pressure, surface temperature and low-level geopotential height. These features make sea-ice a meaningful forecasting indicator in the long-term weather prediction.

7.1 Basic Characteristics of the Global Ice-snow Cover

The quantitative distribution of ice and snow on the Earth can be seen from Table 7.1. The total mass of the cryosphere is about 2.456×10^{22} g. The continental glacier and ice cover, including the ice cover of the Antarctica and Greenland, is the greatest contributor to the total mass of the cryosphere. It represents 97.72% of the total value. The global snow cover reaches a maximum area of 72×10^6 km^2, which is 14% of the Earth's surface. The data presented here should have included more recent observations, but the basic points made by the author are still valid.

Table 7.1. Quantitative characteristics of the basic components of the Earth's cryosphere (after Showmsky and Krenke, 1964).

Kinds of Cryosphere	Mass		Extent Area	
	Gram	%	10^6 km^2	Surface Area (%)
Continental Glacier-ice	2.4×10^{22}	97.72	16	11(continent al)
Underground Ice	5×10^{20}	2.04	32	25 (continental)
Sea-ice	4×10^{19}	0.16	26	7 (ocean)
Snow	1×10^{19}	0.04	72	14 (earth)
Iceberg	8×10^{18}	0.03	64	19 (ocean)
Ice in Atmosphere	2×10^{18}	0.01	–	–
Total	2.456×10^{22}	100		

7.1.1 Sea-ice Cover

Ice cover includes both sea-ice cover and continental ice cover, but the former undergoes more temporal and spatial variations and, therefore, its influence on the environment is more significant than that of the latter.

Antarctic Sea-ice Cover. The total area of the Antarctica is 14.16×10^6 km^2, most of which, about 97%, is covered by snow and ice. The average area of the snow and ice cover over the Antarctica is 25.7×10^6 km^2, which is about 10% of the total area of the Southern Hemisphere. The maximum area of the snow and ice cover in the polar region is observed in September, where it can reach about 13% of the total area of the Southern Hemisphere. Based on the analysis of monthly sea-ice data from 1973 to 1982, Peng and Domroes (1987a, b) suggested that the average sea-ice area of the Antarctica is about 12.1×10^6 km^2, but the seasonal variation is significant. For example, in September, the maximum sea-ice area can reach 18.95×10^6 km^2, occupying 7.5% of the total area of the Southern Hemisphere. In February, however, the area of sea-ice may be around 3.554×10^6 km^2, so that the difference between September and February can be 80% or more of the maximum value.

The location of the northern border of the Antarctic sea-ice depends on its location within the polar area. Normally, the northern border of the sea-ice reaches its most northern position, about 55°–65°S, during August and September. In February, the northern border withdraws to a region between 65°–71°S, but the distributions at different locations vary considerably.

Observations also indicated significant inter-annual variation. For instance, during the mid 1970s the area of sea-ice cover was larger than that during the late 1970s and early 1980s, as can be seen from Fig. 7.1b. The inter-annual variability of the Antarctic sea-ice area is more significant in winter than in the other seasons. For example, the positive departure of the Antarctic sea-ice in August 1974 is 6.1×10^6 km^2, while it is -7.2×10^6 km^2 in August 1981. The difference between the two departures is 1.33×10^6 km^2.

The Arctic Sea-ice Cover. The characteristics of the Arctic sea-ice are quite different from those of the Antarctic sea-ice. The Antarctic sea-ice has a close relationship with the wind field, since its movement is mainly driven by wind stress. The Arctic sea-ice, however, is mainly controlled by thermal processes. Budyko (1974), Walsh et al. (1979) and Wang et al. (1990) have studied the variation regulation of the Arctic sea-ice and have found that the total Arctic ice cover is 13×10^6 km^2, including 11×10^6 km^2 of sea-ice and 2×10^6 km^2 of continental ice. The maximum sea-ice area is in February (14.3×10^6 km^2), and the minimum is in August (7.0×10^6 km^2). On average, the Arctic sea-ice area occupies 72% of the total area of all the Arctic seas. The average thickness of the sea-ice is 3–4 metres, decreasing from the centre to the border. In summer, the ice thickness is less because of melting at the ice surface. In winter, it is greater because of the freezing of seawater under the ice. The seasonal variation in sea-ice thickness follows the variation in the total area of the Arctic seas; maximum in February or March and minimum in August.

Figure 7.1a also shows obvious inter-annual variations of the Arctic sea-ice extent. During the period from the late 1970s to the late 1980s, this extent was above average, except for the mid 1980s. In the mid 1970s and the early 1990s, lower values were observed.

It was also found that around the Arctic region, variations in the volume of ice in the Pacific and Atlantic are in retro-correlation phase. For example, in the 1960s the volume of ice in the Atlantic increased, while in the Pacific the volume of ice decreased. This situation is caused by variability in the air pressure gradient and wind direction.

7.1.2 Snow Cover

Snow Cover over the Northern Hemisphere. Continuous and widespread snow cover data was not available until autumn 1966, when satellite-observed

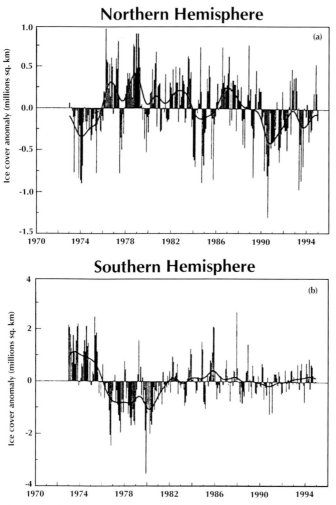

Fig. 7.1. Sea-ice extent anomalies relative to 1973–1994 for: **(a)** the Northern Hemisphere; and **(b)** the Southern Hemisphere (after Nicholls et al., 1996).

data began to be applied in this area. Robock (1980) produced snow cover distribution diagrams from January to December, with monthly-averaged snow cover data over the Northern Hemisphere for 10-12 years from 1966. The annual average snow cover area for the hemisphere was $25.3 \times 10^6 \text{ km}^2$ and the snow cover area in February reached $46.7 \times 10^6 \text{ km}^2$. The analysis by Robock shows a significant distribution difference of snow cover. The snow cover around 32.5°N was 10 times smaller than that around 82.5°N. With respect to seasonal variations, the snow cover in high latitude regions experiences little change. The same is true at lower latitudes. In contrast however,

Li (1990) and Zhou (1992) found that the snow cover in middle latitude regions varies considerably.

Barry (1987) produced annual variation curves of snow area over the Northern Hemisphere and continents for the period from 1967 to 1984. His results suggest the basic characteristics of the identical variation of average snow cover over the Northern Hemisphere and Eurasian continents, with a decrease during the 1970s and after that an increase towards the peak, before decreasing again during the 1980s. The difference between the maximum (1979) and the minimum (1970) Northern Hemisphere snow cover could be over 5×10^6 km^2. The annual average maximum was 28.2×10^6 km^2 and the minimum was 23×10^6 km^2. According to satellite observations over the Northern Hemisphere for the period 1973–1994 (Fig. 7.2), there has been a decreasing trend in seasonal and annual variation of land-surface snow cover extent. This was particularly significant in September and October. The annual mean extent of snow cover over the hemisphere has decreased by about 10% during the 21 year period from 1973 to 1994. The decrease in the extent of snow cover is closely connected to increases in surface temperature.

There are many researchers studying snow cover over each region of the Northern Hemisphere, such as the research on snow cover in America by Walsh et al. (1982), research on snow cover in China by Li (1990), and research on snow cover over the Tibet plateau by Zhou (1992).

Snow Cover over the Southern Hemisphere. The snow cover over South America and the Antarctic continent was analysed by Kotliakof et al. (1989). The snow cover over South America, mainly concentrated over high mountains with seasonal snow storage around 390 km^3 (water equivalent), is divided into three regions: 1) northern region over the continent, lying around 5°–10°N, with snow cover around 500 km^2 area, 0.06 km^3 capacity and 12 cm thickness; 2) Atlantic-middle America region, around 10°N-30°S, with snow area about 2.6×10^5 km^2, 30 km^3 capacity and 12 cm average thickness; and 3) Pacific-South America region, with stable seasonal area over 1 million km^2 and 30 cm average thickness.

Seasonal snow cover spread across the Antarctic continent, with 2.3×10^3 km^3 general capacity and 16.5 cm average thickness. Snow accumulation increases outwards from polar areas, with the thinnest around polar areas (less than 5 cm) and the thickest around the fringe (over 80 cm). Snow accumulation provided by evaporation over the Pacific is 5.25×10^6 km^2 in area, 1020 km^3 capacity and 19 cm in thickness, while that provided by evaporation over the Atlantic is 3.79×10^6 km^2 in area, 700 km^3 capacity and 18 cm in thickness. In addition, that provided by evaporation over the Indian Ocean is 4.96×10^6 km^2, 580 km^3 capacity and 12 cm in thickness. The annual snow cover area of the Southern Hemisphere is about 14.0×10^6 km^2. The maximum snow cover is in September and the minimum in February.

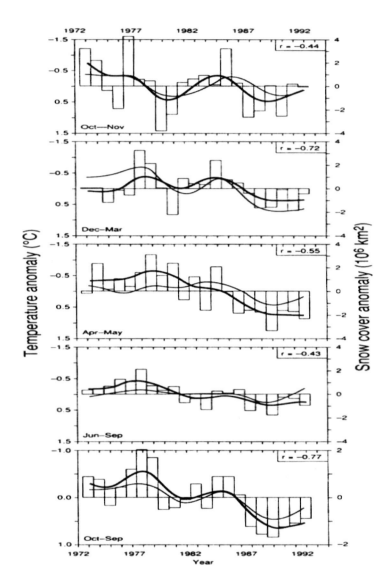

Fig. 7.2. Seasonal and annual variations of Northern Hemisphere land-surface snow cover extent (Greenland excluded) and the surface temperature over regions of transient snow cover. Yearly anomalies (shown as bars) are given for snow cover extent. Smooth curves were created using nine-point binomalies (thin) with scale reversed. "r" indicates correlation between annual values. Note bottom panel is for "snow year" (October-September) (from Nicholls et al., 1996).

7.2 Ice-snow Interactions with Other Environmental Components

7.2.1 Influences of Ice-snow Cover on the Atmosphere

The influences of ice-snow cover on the atmosphere can be discussed from the following three aspects.

The three processes affecting the atmosphere by ice-snow cover. Ice-snow cover on land or sea is closely related to global environmental changes. For example, the climate system is affected by ice-snow in several ways, as described below.

- Because of its high albedo (0.6 to 0.7 for bare ice and up to 0.9 for snow-covered ice), sea-ice reduces the proportion of solar energy absorbed by the surface. This effect is crucial in the ice-albedo positive feedback process. It also replaces the relatively warm ocean surface (above -2°C) by a much colder one, which prevents the heat flux from the ocean heat reservoir to the atmosphere. The temperature of the surface covered by sea-ice can drop to $-30°C$ or lower in winter;
- Obviously, ice-snow cover represents an obstacle to air-sea exchange of sensible heat, momentum, water vapour, carbon dioxide and other gases. Maykut (1978) calculated that the total heat loss through 3 m of the Arctic ice in winter is only 1.5% of that from the open ocean;
- Melting and freezing of ice and snow, accompanied by heat absorption and release, affect the atmosphere temperature field and result in the adjustment of its general circulation.

Influence of ice-snow surface temperatures on atmospheric temperature. After simplifying the energy balance equation of the earth-atmosphere system and the thermal balance equation over the underlying surface, and using some experience relationship, Li et al. (1996) have deduced the relation equation between surface temperature T_s and atmosphere temperature T_a:

$$\Delta T_a = A_a^{-1}(\varepsilon_a \Delta LE - \varepsilon_a \Delta R_S - S_0 \Delta \alpha_p) + \\ + \Delta T_s[1 - 2\sigma T_s(2T_s^2 + 3T_s \Delta T_s)] \qquad (7.1)$$

where $A_a = -\rho_a C'_p C_d |V| \varepsilon_a$, ρ_a is air density, C'_p planetary specific heat, C_d drag coefficient, V wind speed, ε_a infra emission rate in the atmosphere, LE evaporation latent heat, R_s short-wave radiation over ground, S_0 solar constant, α_p planetary albedo, and σ radiation constant of black body.

The interaction between ground temperatures and atmospheric temperatures at different underlying surfaces can be calculated by (7.1), which can also be used for sensitivity experiments through changing a kind of factor. Then the effect of various factors on the interaction between ice-snow cover and the atmosphere can be analysed.

282   7. Simulation and Prediction of Ice-snow Cover

Evidence for the influence of ice-snow cover on general circulation and synoptic-climatic events. Peng et al. (1996a) and Wang et al. (1992) pointed out the change of the general circulation of the atmosphere over the Northern Hemisphere associated with sea-ice area variation over the Arctic. The available data is monthly-averaged for the period 1973–1990, with statistical and synoptic-climatological analysis. The results are presented in (Fig. 7.3). It can be seen that there are significant influences of the Arctic sea-ice area on the Arctic polar vortex, including the vortex over Asia, the sub-tropical highs of the Northern Hemisphere, and SST in the eastern equatorial Pacific. The research (see Sect. 7.4.3) also showed that the Arctic sea-ice not only affects the atmospheric circulation, but also atmospheric precipitation and river flow rate. For example, the sea-ice area has a significant effect on the precipitation of the Yangtze River Basin and the flow rate of the river.

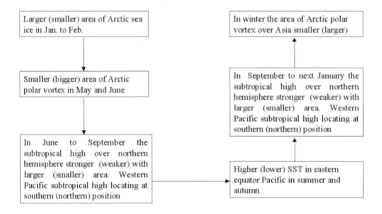

Fig. 7.3. Framework of the influence of the Arctic sea-ice changes on the general atmosphere circulation over the Northern Hemisphere (Peng et al., 1996a).

In terms of sea-ice monthly data over the Antarctic for the period 1973–1990, Peng and Wang (1989) and Peng et al. (1996a) conducted research into the effect of the sea-ice area over the Antarctic on the general atmosphere circulation of the Northern Hemisphere. The results are shown in Fig. 7.4. They have been obtained on the basis of synoptic-statistical analyses. Obviously, the Antarctic sea-ice area is an important contributing factor to the development of the sub-tropical highs of the Northern and Southern Hemispheres, SST in the eastern equatorial Pacific, Walker Cell, Hadley Cell, and the Arctic polar vortexes over Asia and northern America. According to the further research of Peng et al. (1996), if the total area of the Antarctic sea-ice is divided into areas of eight sectors, we could also find the influence of the sea-ice areas for different sectors on the 500 hPa heights of the Northern Hemisphere.

7.2 Ice-snow Interactions with Other Environmental Components

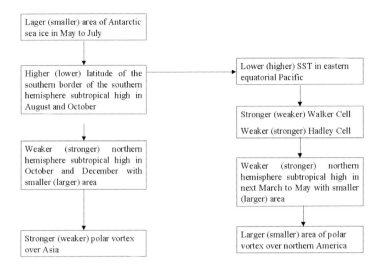

Fig. 7.4. Framework of the influence of the Antarctic sea-ice changes on the general atmosphere circulation over the Northern Hemisphere (Peng et al., 1996a).

Using monthly-averaged speed and temperature data, Qian et al. (1996) calculated longitudinal heat exchange at different latitudes over the Southern Hemisphere. Heat exchange in per unit atmosphere is $-C'_p V \partial T / \partial y$, among which C'_p is specific heat at constant pressure of the atmosphere, V longitude wind speed composition, and $\partial T / \partial y$ longitudinal temperature gradient. With the expression, monthly-averaged exchange through per unit latitude distance on every grid in the area of 0°–48°S and 0°E–0°W has been calculated. Considering observation data, the calculation was conducted only from sea surface to 850 hPa; V and T are obtained from 850 hPa fields.

Figure 7.5 shows the differences of longitudinal heat exchange of the same period in July 1974 (when sea-ice area was large) and 1977 (when sea-ice area was small) and that of three months later (October) in 1974 and 1977. From Fig. 7.5, we can see great differences between the two years, whether in the same period or three months later. This result provides a physical explanation of the sea-ice effect on the general atmospheric circulation.

7.2.2 Effects of the General Circulation of the Atmosphere and Climate on Ice-snow Cover

The development and variation of ice-snow cover are determined by the general circulation of the atmosphere and climate, as well as other associated factors. Although ice-snow cover can produce feedback effects on the general circulation and climate, the action of the atmosphere on ice-snow is the dominate process in ice-atmosphere interactions. Many researchers have suggested

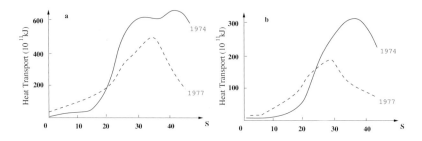

Fig. 7.5. The longitudinal heat transport at different latitudes in 1974 and 1977 (Qian et al., 1996).

the variation of ice-snow cover is an indicator for climatic change. More snow years and less snow years must be related to the general circulation pattern, and the factor directly affecting ice-snow cover is the atmospheric temperature field.

The influences of the atmosphere on ice-snow cover can be both thermal and dynamic. They are closely connected to each other. The formation and development of sea-ice is associated with the heat balance. Sea-ice forms when the temperature over the upper sea drops to freezing point. The freezing rate is mainly determined by the atmosphere temperature and ice thickness. The melting rate is also determined by the heat balance over the ice surface. According to Budyko (1974), the heat balance equation is:

$$R = LE + H_Q + A + lh\rho_i \tag{7.2}$$

Here, R is the radiation balance, L evaporation latent heat, E ice surface evaporation, H_Q turbulent heat fluxes between the ice surface and the atmosphere, A vertical heat flux between ice surface and the underneath of the ice, l ice melting latent heat, h ice thickness change caused by melting, and ρ_i is the ice density. Further analyses suggest that the atmospheric temperature and total radiation greatly affect melting speed. Overall, melting over the mid Arctic is of the order of tens centimetres per year, whereas ice consumption caused by evaporation is several centimetres per year. Ground wind speed is the main factor driving the flow of sea-ice. All of the factors mentioned above show that the general circulation of the atmosphere and climate, directly or indirectly, affect sea-ice cover.

7.2.3 Main Physical Processes of Interactions among Ice-Snow Cover, Atmospheric Circulation and Water Cycle

The interactions among ice-snow cover, atmospheric circulation and the water cycle are mainly analysed by the exchanges of their mass, energy and momentum. The physical processes concerning the influence of ice-snow cover on the atmosphere have been described in Sect. 7.2.1. It is important to understand

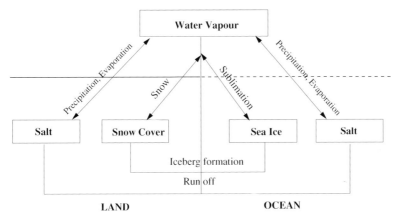

Fig. 7.6. Water cycle involved in the climate system (from Kosjin et al., 1957).

the role of high albedo on atmospheric processes. Ice-snow cover also plays a significant role in the water cycle. The main processes of the cycle in the climate system are evaporation and runoff. Both depend on the condition of ice-snow cover. The cover restrains evaporation from surfaces of oceans, lakes and rivers. In many cases the runoff is also related to the variation of the cover because of its melting and freezing. In Sect. 7.2.2, the influence of wind on the motion of sea-ice has been mentioned. Surface air temperature of the atmosphere and the surface temperature of the oceans control the development of sea-ice cover. The snow cover condition is also connected to the air temperature over the cover. Furthermore, ocean currents bring about the motion and change of sea-ice cover. Figure 7.6 illustrates the basic physical processes, and their interactions, of the water cycle that are involved in the climate system (Kosjin et al., 1957). Figure 7.7 expresses different physical processes that indicate the interaction between the atmosphere and the hydrosphere, including the effect of the cryosphere. Different equations describing the interactions are also presented.

7.3 Numerical Simulation

In numerical models of the climate system, the radiation process, dynamic processes and ground layer processes should be addressed. The surface processes play a more important role in climate models, because much of the incoming solar energy is absorbed by the ground layer, not directly by the atmosphere. The high albedo of the ice-snow cover leads to a large decrease in the solar radiation absorbed by the ground, so ice-snow factors must be fully considered in climate models.

There are many kinds of ice-snow models, or ice-ocean-atmosphere models. We try to sum them up in a few types and introduce the selected works.

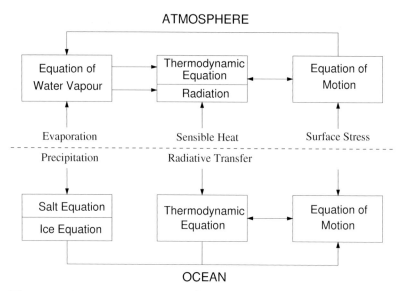

Fig. 7.7. The coupled ocean-atmosphere system (from Manabe et al., 1990).

7.3.1 Energy Balance Models

The One-Dimension Model of Sellers. The first energy balance model was established by Budyko (1969) and Sellers (1969). The model was applied to study the relation between ice and atmosphere. It is a one-dimension model in which the temperature varied with latitude. The model was used to study the influence of the solar constant, carbon dioxide and the ice-snow interaction on temperature as well as ice age formation. It was also used to study the distributions of ocean, continent and ice-covered regions.

Sellers (1973) established an energy balance model considering land-atmosphere-ocean interactions, which included seasonal coupling and the interaction between land and ocean. In this model, an idealized water and land distribution similar to a realistic one is adopted. The temporal-spatial average heat energy equation was used to apply average surface temperatures on ocean and land. The time integration step was one month. There were some parameterizations on the vertical variation of temperature/humidity and the wind components in the latitude and longitude direction. The surface speed components were obtained by two kinematic equations. This model comprises the variation of snow and ice cover areas, meridional transportation of moisture and heat caused by the general circulation, heat storage and transportation by ocean, influences of aerosol, moisture, carbon dioxide, O_3, cloud and so on.

The energy in the model was calculated by the following formula:

$$R' = \frac{1}{g}\int_0^{P_0} \frac{\mathrm{d}}{\mathrm{d}t}\overline{(Lq + C_pT + gh')}\mathrm{d}p + C\int_0^d \frac{\mathrm{d}\overline{T}_E}{\mathrm{d}t}\mathrm{d}z \qquad (7.3)$$

Here, R' is the net effective radiation from soil or water depth d to the top of atmosphere, g is gravitational acceleration, P_0 ground surface atmosphere pressure, L latent heat of condensation, and q specific heat. C_p is specific heat AT constant pressure, T_E atmosphere temperature, h' height over sea surface level, C heat capacity of soil or water, and T the temperature of soil or water.

On the basis of the model, Sellers (1973) changed some parameterization schemes and conducted three experiments concerning CO_2, variation of atmospheric optical thickness and solar constant. The results suggest that high latitude areas of the Northern Hemisphere are very sensitive to the factors resulting in climate change. Although this model is more complete and reasonable than that by Sellers (1969), the parameterization scheme requires further improvement.

The Two-Dimensional Model of Chao and Chen. The model of Budyko and Sellers not only averaged the zonal circle, but also integrated vertically. This is an approximate climate description of the global atmosphere. Many physical processes in the atmosphere were smoothed. Chao and Chen (1980) re-studied the questions proposed by Budyko and Sellers with a two-dimensional energy balance model, which is a zonal-averaged one and includes vertical structure. The basic equation is the equation of energy balance between radiation and turbulence transportation.

$$\frac{K}{a}\frac{\partial(1-x^2)}{\partial x}\frac{\partial T}{\partial x}+\frac{\partial}{\partial z}(k_t\frac{\partial T}{\partial z})+\sum_j a'_j\rho_c(A_j+B_j-2E_j)+a''\rho_c Q_{sr} = 0 \quad (7.4)$$

In the above equation, T is atmosphere temperature, a'_j the long-wave radiation absorption coefficient for wave length at λ_j, a'' average absorption coefficient for solar radiation, A_j upward long-wave radiation flux for wave length within $\delta\lambda_j$, B_j downward long-wave radiation flux for wave length $\delta\lambda_j$, Q_{sr} solar radiation flux, E_j radiation energy of black body with $\delta\lambda_j$, ρ_c absorbing medium density, K, k_t horizontal and vertical turbulence exchange coefficients, and a the Earth radius, $\sum$ sum symbol for the whole absorbing spectrum and z vertical coordinate, $x = \sin\theta$, θ is latitude. It is easy to derive the following equations:

$$\frac{\partial A_j}{\partial z} = a'_j\rho_c(A_j - E_j) \quad (7.5)$$

$$\frac{\partial B_j}{\partial z} = a'_j\rho_c(E_j - B_j) \quad (7.6)$$

$$\frac{\partial Q_{sr}}{\partial z} = a''\rho_c Q_{sr} \quad (7.7)$$

The two-dimensional model has the following features. Suppose there are no downward long-wave radiation flux and turbulence processes above the atmospheric upper limit. There is a global radiation balance. The limit of the upper level is an isotherm level. The heat balance requirement above the global surface is:

$$A_n + (1-\alpha)Q_{sr} = B - k_t \frac{\partial T}{\partial z} \tag{7.8}$$

where A_n is net solar radiation flux coming to the atmosphere and B–long-wave radiation flux going to outer space. The calculated result indicates that the dependence of the ice boundary on the solar constant for the two-dimensional energy balance model is much more stable than that for the one-dimensional energy balance model.

When the solar constant is decreased by 15% the ice boundary can move to 50°N from 72°N. However, for the one-dimensional model, the ice boundary can move to 50°N when the solar constant is decreased by only 2%. It is obvious that the first result is more reasonable.

One-Dimensional Multi-Layer Model of a Snow Cover Based on the Balance of Mass and Energy. Loth and Oberhuber (1993) have established a one-dimensional multi-layer model of snow cover. The feature of the model is in combining relatively accurate model physics with minimal computer time. It is based on the balance of mass and energy including the important internal processes (the diffusion of temperature and water vapour, melting and freezing, the extinction of short-wave radiation and the retention of liquid water). It is intended to avoid the use of tuned parameters.

Four snow types (cold snow, wet snow, firm snow, and ice) are distinguished. The forcing terms only include standard meteorological parameters. An efficient numerical scheme makes the model suitable for long-term climate studies. For the implementation of the model in a general circulation model (GCM), parameterizations are added to relate the snow albedo, the long-wave emission, the turbulent heat fluxes and the mass fluxes to the different vegetation and surface types given in modern GCMs for every grid square.

Ignoring the lateral fluxes, the energy balance equation used in this model has the following form:

$$Q^* = Q_s^* + Q_L^* + Q_H + Q_{ES} + Q_B + Q_{HPR} \tag{7.9}$$

where Q^* is the energy balance, and Q_s^* short-wave radiation balance which could be calculated from the global radiation, Q_s, and the albedo α; Q_L^* long-wave radiation balance; Q_H sensible turbulent heat fluxes; Q_{ES} latent turbulent heat fluxes; Q_B ground heat flux; Q_{HPR} the energy input due to rain.

The energy fluxes Q depend on the surface snow temperature, T_s, and are formulated because T_s reacts instantaneously to anomalies in Q.

$$Q(T_s^{n+1}) = Q(T_s^n) + \left.\frac{\partial Q}{\partial T}\right|_n (T_s^{n+1} - T_s^n) \tag{7.10}$$

where Q is one of the energy fluxes Q_L^*, Q_H, or Q_{ES}. A similar approach could be used for the ground heat flux and the temperature of the bottom layer.

The mass balance M^* depends on the snowfall rate M_{PS}, rainfall rate M_{PR}, turbulent water vapour flux M_V, and runoff M_A.

$$M^* = M_{PS} + M_{PR} + M_V - M_A \qquad (7.11)$$

where M^* is in $\mathrm{kg\,m^{-2}s^{-1}}$.

The model is designed for long-term climate simulations and the snow cover is considered as a multi-layer system consisting of ice, liquid water, water vapour and air, with a minimum depth of 1 mm. The internal processes of heat diffusion, water vapour diffusion and absorption of short-wave radiation, and the processes of freezing, melting and ageing are simulated using an integration time step of two hours. Energy and mass fluxes, which depend on the snow temperature, are implicitly introduced. In order to guarantee numerical stability, even with extremely thin layers, the equation system is written Euler backward and is solved by applying the direct technique of Richtmyer and Morton (1967), which provides the new temperature values in one step. In the melting and freezing scheme, a transformation of the coordinates to the mass fluxes of the single components is incorporated to make mass and energy conserved.

A series of sensitivity tests has been conducted during different years for Potsdam, which is located at 52°N, 13°E. Long-term integration for six winters carried out with data sets from the station demonstrate a good correspondence between the observations and the simulated values of snow depth and water equivalent. All phases of snow cover change could generally be reproduced well. The model is able to cope with deep snow cover ($> 0.50\,\mathrm{m}$) as well as with rapidly occurring shallow covers ($< 0.05\,\mathrm{m}$). The times of the accumulation and the end of ablation were exactly modelled. The simulation results for the winters of 1975/1976 and 1978/1979 were especially good because there was no tuning performed, local parameterizations were avoided, and no corrections were applied to the standard synoptic measurements used. The simulation results were less encouraging for November-December of 1980, but still generally correspond to the observations, as can be seen from Fig. 7.8. The contrast between the simulated and observed values still reflect the high sensitivity of the model. The model is not only potentially applicable to global studies but also well suited to local simulations.

This can be viewed as a good model because it has several special features. Firstly, the model can be used in long-term climate simulations. Most of the existing numerical snow models are used only for runoff studies. Some other models have higher vertical resolution, but only reveal the internal processes of snow, and these models require extensive computing time. This model pays attention to the study of a few important physical processes, so the computation is more convenient, using an effective numerical scheme.

290   7. Simulation and Prediction of Ice-snow Cover

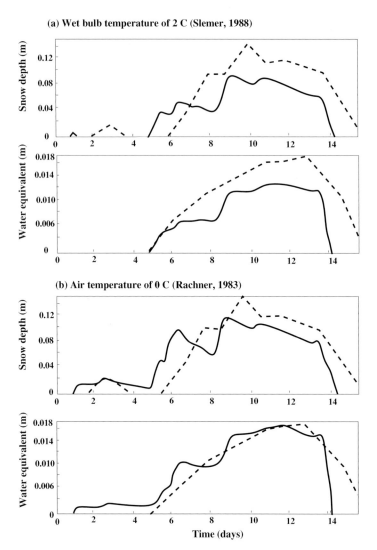

Fig. 7.8. Simulated snow depth and water equivalent for Potsdam during the time interval 27 November 1980 to 12 December 1980, applying different criteria for the distinction of snowfall and rain. Dashed curve: observations; solid curve: simulations (after Loth and Oberhuber, 1993).

Secondly, the model considers a multi-layer system of snow cover, allowing an accurate simulation of the heat conduction within the snow. Thirdly, water vapour diffusion, liquid water storage, snow density and thermal conductivity have also been considered. On the other hand, the parameterization process has been given with many details. The simulation results of this model are better than those of Verseghy (1991) and Dickinson et al. (1986). However, considering only the snow layer, the snow cover model of Verseghy can also be ranked among the best models.

The disadvantage of Loth's snow cover model is that the criterion for determining the aggregational state of precipitation is not very accurate, which can have a negative effect on the accuracy of the simulated snow depth and water equivalent.

7.3.2 Dynamic-Thermodynamic Models

The Hibler model. Atmospheric fields of wind and temperature are the main factors determining sea-ice dynamic-thermodynamic change. Hibler (1979) designed a dynamic-thermodynamic model to simulate sea-ice circulation and thickness, and to study the ice dynamic effect of the Arctic sea-ice thickness and sea-atmosphere flux. Later, Hibler and Flato (1992) reinterpreted the model. The main idea of the model is to couple dynamic process and ice thickness. Dynamic process will lead to high heat energy loss over regions of ice divergence and low heat energy loss over areas of ice convergence. Ice thickness and compactness are determined by the continuity equation, which includes variation of ice mass and percent of open water caused by advection, ice melting and thermal effects. Hibler's model contains the following four components:

Momentum balance. In the Cartesian coordinate system, the equation of sea-ice motion for the two-dimensional case is:

$$m D\mathbf{u}/Dt = -mf\mathbf{k} \times \mathbf{u} + \tau_a + \tau_w - mg \cdot \nabla H + \mathbf{F} \qquad (7.12)$$

where $D/Dt = \partial/\partial t + \mathbf{u} \cdot \nabla$ is the total time derivative, $\mathbf{k}$ is a unit vector normal to the surface, $\mathbf{u}$ ice velocity, f the Coriolis parameter, m the ice mass of unit area, τ_a and τ_w the stresses from the atmosphere and ocean, respectively, H sea-surface dynamic height, g the gravity acceleration, and $\mathbf{F}$ the stress inside ice variation.

Constitutive law. Taking ice as a non-linear viscous compressible fluid obeying the constitutive law:

$$\sigma_{ij} = 2\eta(\dot{\varepsilon}_{ij}, P)\dot{\varepsilon}_{ij} + [\zeta(\dot{\varepsilon}_{ij}, P) - \eta(\dot{\varepsilon}_{ij}, P)]\dot{\varepsilon}_{kk}\delta_{ij} - P\delta_{ij}/2 \qquad (7.13)$$

Here σ_{ij} is the two-dimensional strain tensor, $\dot{\varepsilon}_{ij}$ the strain rate tensor, $P/2$ a pressure term connected with the ice thickness, and ζ and η non-linear bulk and shear viscosities.

7. Simulation and Prediction of Ice-snow Cover

Ice thickness feature. For overall mean thickness h, and compactness A, the following continuity equations are used:

$$\frac{\partial h}{\partial t} = -\frac{\partial (uh)}{\partial x} - \frac{\partial (vh)}{\partial y} + S_h \tag{7.14}$$

$$\frac{\partial A}{\partial t} = -\frac{\partial (uA)}{\partial x} - \frac{\partial (vA)}{\partial y} + S_A \tag{7.15}$$

where $A < 1$; u and v are the components of the ice velocity vector, and S_h and S_A are thermodynamic terms given by:

$$S_h = f_g\left(\frac{h}{A}\right) A + (1-A) f_g(0) \tag{7.16}$$

$$S_A = \frac{f_g(0)}{h_0}(1-A)\delta_1 + \frac{A}{2h} S_h \delta_2 \tag{7.17}$$

where
$$\begin{array}{llll} \delta_1 = 1 & \text{if } f_g(0) \geq 0; & \delta_1 = 0 & \text{if } f_g(0) < 0 \\ \delta_2 = 0 & \text{if } S_h > 0; & \delta_2 = 1 & \text{if } S_h < 0 \end{array}$$

with $f_g(h)$ being the growth rate of ice, and h_0 a fixed demarcation thickness between thin and thick ice.

Thickness-dynamic couple. The ice pressure $P/2$ is taken to be a function of compactness and thickness according to:

$$P = P^* h \exp[-C(1-A)] \tag{7.18}$$

Here, P^* and C are fixed experimental constants, and h is calculated in metres.

Hibler (1979) and Walsh et al. (1985) applied this dynamic-thermodynamic model to undertake extensive research of sea-ice variation and global climate change. Hibler (1979) integrated the model over a long time period and obtained ice oscillations varying with season, by means of wind field data varying with season. Standard experiments have been made with an eight-year integration, stepped using one day. Comparisons were made between the calculated results and observations, including ice displacement, geological and temporal variation of ice thickness, ice ridge statistics, ice density charts, and mass balance statistics.

The simulation results show that many circulation features and thickness features of the Arctic ice cover can be found. His model was one of the best dynamic-thermodynamic models at that time, and his ideas and framework are still being used at present. This model was established 20 years ago and, unavoidably, has some shortcomings. For example, some qualitative analyses are not very accurate.

Walsh et al. (1985) produced simulations of sea-ice variation over the Northern Hemisphere using Hibler's model. The purpose was to analyse mass absorption-emission over high latitude ocean regions. This analysis depends on model atmosphere fields of ice thickness, density and speed. Walsh's simulation model considers two procedures: with dynamics and without dynamics. The results show that long-term oscillations of sea-ice are very sensitive to the model. In the case of the lack of correct disposition of an ice-dynamic process in the model, incorrect conclusions can result from the simulations. Normal seasonal circulation is controlled by thermodynamic processes, but the thickness abnormality for most polar sea-ice in winter, summer and autumn is restricted by dynamic process. The most serious deviation of simulation results is the excess ice over the northeast Atlantic in winter. This is produced by the model not considering ocean flow effects. Therefore, ocean-coupled procedures must be considered. The coupling of a dynamic thermodynamic sea-ice model to an ocean general circulation model has been carried out by Hibler and Bryan (1987), and Semtner (1987). The main relevance of the variable thinness distribution to climate modelling is its more precise treatment of the growth of ice. Flato (1991) made a comparison of the two-level and multi-level approaches to modelling the dynamic-thermodynamic evolution of an ice cover.

7.3.3 The Holland Model

Holland et al. (1991) have developed a dynamic thermodynamic sea-ice model of the Arctic sea-ice. It is an uncoupled one, extracted from Oberhuber's (1990) coupled sea-ice-mixed layer-isopycnal general circulation model and is written in spherical coordinates. In this model the thermodynamics are a modification of those of Parkinson and Washington (1979), while the dynamics use the full Hibler (1979) viscous-plastic rheology. The model is fully described in a report by Holland et al. (1991). The thermodynamic and dynamic aspects of the model are interrelated.

The main feature of the model is that for the momentum balance, the ice is considered to move in a two-dimensional spherical plane. The model adopted the momentum balance numerical diffusion, Coriolis force, sea surface heights, and wind and ocean current stresses. The internal ice stress and ice strength are combined by a constitutive law. The ice motion can cause local changes in the thickness. On the other hand, the ice thickness can also influence features of the ice's strength, so that the thermodynamic and dynamic characters are connected in this way.

In view of thermodynamic effects, the model has included ocean flux from the deep ocean under the mixing layer and the ocean current stress. The ocean density is taken as a constant, and the ocean currents in the model are specified as well as the convective overturning, so that the horizontal and vertical variations in density can be ignored.

Some sensitivity experiments have been conducted. In the control run, the initial ice field consisted of 0.0 m thickness of stationary sea-ice. Atmospheric forcing was applied to the model regularly, in all totalling 10 years' model time. The model quickly reached the equilibrium state within five years' model time. The simulated results show that there was a significant seasonal cycle in the variation in ice thickness, with the greatest ice thickness found in the north of the Canadian Archipelago. With regard to the spatial variation of the simulated ice thickness, it was found that the compactness also concentrated along the coastline of the Canadian Archipelago, which is identical to the observations, but in summer there was excessive melting in the model along the coast. The interactions between sea-ice and atmosphere could be found. The simulated results indicate that low atmospheric temperatures can result in low ice temperatures and an increasing ice thickness. Conversely, increasing atmospheric temperatures produce the opposite effect. The ice thickness can also be 40 cm less than the control run when the atmospheric temperature is 5°C lower than the control one.

Generally speaking, the model results can give a reasonable distribution of the sea surface temperature and ice temperature. Particularly in the central Arctic, the simulated temperature was quite close to the real situation. In some regions the model gave higher temperatures than those observed, which means that the thermodynamic calculations need to be improved.

7.3.4 Treatment of Ice-snow Cover in GCMs

General Introduction. At present, the most advanced models displaying the interaction between atmosphere, ocean, land and ice-snow are so-called general circulation models (GCMs). Of these, the most representative are the models by NCAR and ECMWF, though they must be improved in order to make them perfect. The implementation of a GCM requires a numerical solution technique, algorithms for the different physical parameterizations, and boundary data sets for predetermined vertical and horizontal resolution. Improvements to the simulation of atmospheric general circulation could be realized through physical parameterization, numerical methods, computational software and hardware. The most important problem to address in improving the model is to clarify the interactions between the components of the climate system. This includes the study of land surface processes, hydrology, biology, snow cover, sea-ice and ocean-atmosphere interactions.

Ice-snow cover plays an important role in the general circulation model. It is a cool source for the atmosphere above, and in turn this cooling effect can make the ice thicker. In this regard, sea-ice has a positive feedback mechanism on the whole system, since the existence of sea-ice can make the climate cooler, resulting in more sea-ice. Ice-snow cover is one of the components of a GCM, and it is natural to apply GCMs to simulate ice-snow evolution and the interaction between the atmosphere and ice.

Peng et al. (1992) introduced the numerical simulation of GCMs on ice-atmosphere processes and ice-period climate in the book "Climate and ice-snow cover" (1992). The following are some of the representative works.

Herman and Johnson (1978) conducted a series of numerical experiments using the Goddard general circulation model. The task is to study the effect of local variations in the Arctic sea-ice boundary on monthly-averaged output from the model.

Gates (1976) studied ice-period climate (18,000 years ago), with a global atmosphere circulation model. He simulated ice-period climate and modern climate in July with a two-level GCM. The results show that, globally, the ground temperature during the ice period was 4.9°C lower than that during the modern period, together with a slight decrease in relative humidity and 20% less precipitation over the Northern Hemisphere. Anti-cyclone circulation over the main ice level was less intensive, with a weaker summer monsoon and more intensive Westerlies over mid-latitudes, and a decrease in average meridional circulation in comparison with the modern period.

Manabe and Ahn (1977) used a nine-level global GCM to simulate tropical climate during the ice period. The results suggest that, during the ice period, the tropical continental climate was much drier and that continental temperatures decreased with more intensity than ocean temperatures. In addition, an increase in albedo is found to be the main reason leading to the weakening of the Asia monsoon.

Recent Advances. Simmonds and Budd (1991) applied a GCM in some numerical experiments to investigate the sensitivity of the Southern Hemisphere circulation to advances in the Antarctic pack ice. It has been found that significant warming of the troposphere and weakening of the zonal Westerlies occurred, but their magnitudes were not linearly related to ice concentration. In the experiments, much attention was paid to the effects of the sea-ice concentration. Mean sea-level pressure changes were neither monotonic nor zonally symmetric, but in the zonal mean, the circumpolar trough was displaced several degrees to the south, without any change of maximum depth. Simmonds and Wu (1993) have examined the behaviour of the synoptic systems that contribute to the climate state. It has been found that as the amount of open water in the Southern Hemisphere was increased, the area of maximum cyclone density shifted by up to 5° of latitude to the south. There was a decrease of baroclinicity and lows, and although now more numerous in the sea-ice area, they were less deep in a relative sense. The location of the circumpolar trough is determined by the number of cyclones present, but not by their intensity.

Murray and Simmonds (1995) have examined the sensitivity of the winter climate and cyclone behaviour to changes in sea-ice concentration in the Northern Hemisphere, using the GCM simulation for a perpetual January. It was related to the concentration of the Arctic sea-ice. In this study, Murray and Simmonds used a GCM, which is described by Simmonds (1985). The model is rhomboidally truncated at 21 waves and includes nine vertical levels. Surface fluxes and atmospheric physics are computed on a $5.75° \times 3.33°$ Gaussian grid at each time step. Climatological January-forcing included isolation, clouds, CO_2, land albedo, sea surface temperature and ice-snow cover. The control was run for 60 days after spin-up. Separate flux calculations were performed for the open water and ice-covered parts of each grid box in the sea-ice area, and each surface flux was taken to be the area-weighted average of the two components. The surface temperature was determined by the heat balance over ice and set at 1.8°C over water. Five different open water fractions were employed: 5, 20, 50, 80 and 100%. Each experiment was integrated for 300 days after a 90-day period of stabilization.

The results of the simulations show that the polar vortex was stronger and more zonal than it is found in reality. The main qualitative features of the cyclone density and flux distributions have a good agreement with observations. The primary effect of sea-ice removal was a monotonic, but non-linear increase in surface fluxes and lower level temperatures in the sea-ice zone. The reason for this is the raising of the geopotential surface and a winding down of the polar vortex at 500 hPa. Earlier modelling studies have shown the same results. At high latitude, pressures tended to fall in regions of the greatest lower-troposphere warming and to rise elsewhere. These pressure reductions and temperatures were biased toward the Canadian Basin. Pressure anomalies could also be found at lower latitude regions. There was a significant decrease in the speed and intensity of cyclone systems north of 45°N. Some density maxima were displaced toward regions of opened-up sea-ice, or in response to circulation changes. A number of aspects of the response were connected with changes in thermal steering and baroclinicity, or with non-linear effects. There was no significant expansion of cyclonic activity in the area of the central Arctic. A relative constancy of cyclone numbers and storm tracks occurred in the Arctic Ocean. This is different from that found in a similar study of the Southern Hemisphere. Now two cases are selected for the response to reductions in sea-ice concentration. The progressive removal of sea-ice through the increase of the leads fraction was accompanied by a monotonic increase in lower tropospheric temperatures in the Arctic region. Anomalies in the ice-free case were significant to a height of 500 hPa in the zonal average and throughout the sea-ice and adjacent areas of lower levels. Temperature anomalies at 850 hPa (Fig. 7.9) were distributed in a manner which is similar to those shown by Royer et al. (1990), but the amplitudes were greater and the cold anomaly over Russia was not found. The region of maximum warming at 850 hPa was centred over the Canadian Basin. At greater elevations the warming was less intensive, but the shift toward the Western Hemisphere was more marked. At these levels, anomalies of comparable magnitude and opposite sign were present over Siberia and the eastern Pacific.

Most of the changes noted in the sequence of experiments were not linearly related to f_w (the fraction of open water exposed over the ice region). This can be seen in the case of the mean surface temperature anomaly over the sea-ice area (poleward of 50°N), which rose to 40% of its ultimate value after only 20% of the sea surface had been exposed (Fig. 7.10). This value is comparable to the South Hemisphere winter value obtained by Simmonds

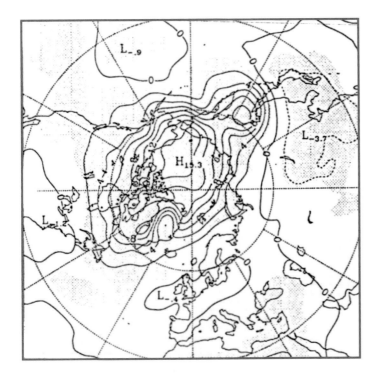

Fig. 7.9. The 850 hPa temperature anomaly (100% open water minus control) (contour interval, 2°C, with stippling for significance at the 95% confidence level) (from Murray and Simmonds, 1995).

and Budd (1991). The non-linearity arises because the temperature increase, as calculated by the model, is proportional to both f_w and the ice surface temperature rise, which is itself a function of f_w. The penetration of the warming to higher levels proceeded with diminishing amplitude and, owing to the retarding effect of the polar inversion, more linearly with f_w. Due to greater static stability in the central Arctic, the warming at 500 hPa (not shown) occurred there at a more advanced stage of ice removal ($f_w = 80\%$) than in the Sea of Okhotsk (20% and 50%).

Yang and Huang (1992) studied the effect of the Arctic sea-ice area over the Northern Hemisphere in winter and summer on global atmospheric circulation and climate, with a global circulation spectrum model of nine-level rhomboidal truncation of 15-wave. The experimental results show that when the sea-ice area over eastern Greenland increases, a wave train over the North-

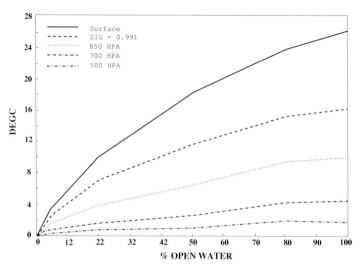

Fig. 7.10. Area-weighted average temperature anomalies for the sea-ice region north of 50°N at the surface, at $\sigma = 0.991$, and at the 850 hPa, 700 hPa, and 500 hPa levels (from Murray and Simmonds, 1995).

ern Hemisphere will spread towards the lower latitudes of Asia, and another wave train will develop from the lower latitudes of Asia to North America. These are very clear in the 200 hPa geopotential height field. If the three sea-ice regions, including eastern Greenland, Bering Sea and west of Greenland, increase together with the equatorial east Pacific SST, the amplitudes of the wave trains will be intensive, i.e., they have a consistent effect on the atmosphere. An experiment of the Arctic sea-ice abnormality in winter over the Northern Hemisphere focused on the northern Atlantic polar area and the northern Pacific polar area. The results indicate that the two sea-ice areas have significant influences on atmosphere wave trains over North America and Eurasia, resulting in a relative change in atmosphere circulation types or intensity.

Chen et al. (1996) studied the effect of an ice-cover abnormality over the Antarctic on the atmospheric circulation of the Northern and Southern Hemispheres. The models are CCM1 (1987) developed by NCAR with 12 levels in the vertical direction, rhomboidal truncation of 15-wave and 4.4° × 7.5° grid space in the horizontal direction. The model was integrated for 15 years at first, and another 10 years have been integrated by the author. The results of various sensitivity experiments clearly indicate the importance of the Antarctic ice-snow cover to global atmospheric circulation.

Using a three-level GCM model, Chen and Miao (1996) made some numerical experiments on the effects of the Antarctic sea-ice cover and ocean conditions on climate changes, and experiments of monsoon climate changes under the influence of the Antarctic sea-ice cover and sea surface temperature. Both experiments suggest that changes in the Antarctic sea-ice cover can effect the general atmospheric circulation over the Southern and Northern Hemispheres, including seasonal variation of high-level zonal wind over Asia.

7.3.5 Other Models

Besides the above three models, there are other kinds such as the ice-ocean coupled model, primitive equation model and random model. In recent years, ice-ocean coupled models have been developing at a rapid pace. The coupled sea-ice ocean general circulation model of Oberhuber (1993) is a good example. This model consists of several coupled models. The sea-ice is represented by a dynamic, thermodynamic model with viscous-plastic rheology and includes a snow model, where the mixed layer is represented by a turbulent kinetic energy model, and the deep ocean by an isopycnal layer model. The models interact via the exchange of momentum, mass, heat and salt. Forcing occurs via the specification of monthly mean atmospheric fields.

Holland et al. (1995) applied this model to conduct a numerical simulation of the sea-ice cover in the northern Greenland Sea. The coupled model is forced using monthly climatological atmospheric forcing. The model domain includes both the Arctic Ocean and the North Atlantic. The simulation results show that the Oberhuber (1993) coupled sea-ice-mixed layer-isopycnal ocean model is capable of producing a reasonable simulation of the ice characteristics in the northern Greenland Sea. Of particular interest in the simulation was the occurrence of a polynya in the same geographical area as the observed new polynya. An ice trough could also be well simulated. Cheng and Preller (1994) have finished a work concerning an ice-ocean coupled model for the Northern Hemisphere. Fichefet et al. (1994) also suggested their own ice-ocean coupled model for both the Northern and Southern Hemispheres.

7.4 Prediction of Sea-ice and Related Components

Ice-snow cover forecasting is a part of environmental forecasting. Forecasting snow cover is a requirement for industry, agriculture and land traffic, while sea-ice cover forecasting serves sailing, fishing and oil exploration in the oceans. Ice and snow cover can also be used to forecast the weather and climate development, oceans, and hydrology. Snow cover forecasting can be viewed as part of precipitation forecasting, so in this section attention will be paid to a discussion of sea-ice forecasting and to the application of ice-snow cover factors to weather and climate predictions. The forecasting of sea-ice employs three methods: factor analysis; numerical methods; and statistical methods. In fact, the three methods are closely connected. The basic methods of numerical forecasting are based on the equations of fluid dynamics, solving the equations and obtaining the forecast results.

The purpose of factor analysis is to reveal the controlling factors of sea-ice development through calculation and analysis, and then to find the dominant factors. Generally, the following factors are considered: 1) the air temperature around the sea-ice area; 2) general atmospheric circulation, especially the changes of atmospheric activity centres; 3) solar activities, geophysics factors and soil temperature at different layers; and 4) oceanic factors, such as El Nino and ocean currents which can influence sea-ice distribution. Zhang and Zhou (1994) have established the following relationship between the sea-ice thickness and air temperature near Bayuquan on the Bohai Sea.

$$h = \alpha(FDD - 3TDD)^{1/2} \tag{7.19}$$

where h is sea-ice thickness in cm, α is ice growth coefficient [$(cm/°C·d)^{1/2}$], FDD is accumulated freezing degree-days below $-2°C$, (°C·d), and TDD is accumulated thawing degree-days above $0°C$, (°C·d). In the sea zone near Bayuquan, the ice growth coefficient is 2.35. This formula is mainly suitable for calculating the annual maximum ice thickness.

Zhang and Zhou (1994) pointed out that during the years of the maximum sunspot relative number, the winters around the Bohai Sea were cold with accompanying heavy sea-ice. In addition, the surface temperature of the Pacific Ocean apparently impacts on the sea-ice development. On the other hand, the temperature field over the Bohai Sea can be affected by the general circulation pattern in the previous year.

At the present stage, statistical forecasting is the most popular method used for the operational forecasting of sea-ice. In statistical forecasting, the first task is to set up a statistical model in order to identify those factors that have a close connection to the subject of the forecast, while at the same time these factors must have some physical meaning. Then, it is possible to set up the equations to conduct the actual forecast. Take the work of Zou et al. (1996) as an example. They divided the Antarctic region to five areas, and used sea-ice data during the period 1983–1993 to complete the statistical analysis, utilizing five statistical forecasting equations for the five areas, respectively. The ice area forecast formula for the entire the Antarctic is

$$Ice = -275317.6 - 213354.5 \times N_3 - 14 + 210696.3 \times N_3 - 132 + \\ + 306576.3 \times N_4 - 179 \times S_3 \tag{7.20}$$

where N_3 indicates the sea temperature departure for the third sea area, and N_4 that of the fourth sea area; S is the sunspot number; and the number behind the negative sign "$-$" is the month number of time lag of sea-ice change behind SST change.

The advantage of factor analysis and statistical methods is that the forecast time can be short-range or long-range, and the factor analysis method highlights the physical background. The calculation processes are relatively simple and can save much computational time. The disadvantage is that they cannot be used to investigate the physical processes in detail, and the forecast results, in most cases, are the development tendency. At present, middle- to long-term sea-ice forecasting is mainly achieved through factor analysis and statistical methods.

7.4.1 Numerical Forecasting of Sea-ice

The numerical forecasting of sea-ice is based on numerical simulations. Forecast models can be divided into thermodynamic models, dynamic-thermodynamic models, and ice-ocean coupled models. Of these, the dynamic-thermodynamic model is the most popular.

Thermodynamic Ice Model. The basic features of thermodynamic models have been discussed by Peng et al. (1992) and Wu et al. (1992). Toyota and Sato (1994) have improved aspects of this kind of models. Such a sea-ice model has been implemented for sea-ice prediction in the Sea of Okhotsk for the winter season. The model forecasts the sea-ice for a range of seven days, and has been provided operational forecasts twice a week since December 1990.

Thermodynamic processes include solar, atmospheric and ice radiations, heat fluxes within sea-ice, and sensible and latent heat fluxes between the ice surface and the ocean. In the model mentioned above, the heat flux was assumed to be transferred from the ocean to the sea-ice constantly during all seasons, to represent the fact that sea-ice is apt to melt in some special portions of the Sea of Okhotsk. This assumption seems to be unrealistic, and the model did not include the seasonal change in the heat flux. A new thermodynamic process has been introduced into the model to make the process more realistic. Toyota and Sato (1994) introduced a new process, in which the ocean is assumed to be composed of a surface layer and a mixed layer. The heat exchange between the ocean and sea-ice is parameterized using the surface layer flow characteristics and the temperature difference between the surface and mixed layers. Since there was no heat flux given to the sea-ice prior to the new model, it is expected to be physically more consistent. The depth of the surface layer is set to 1 m because the role of the surface layer is the heat exchange between the ocean and sea-ice and its depth is throught to be the same order as ice thickness. The ocean current distribution used in this model is fixed throughout the seasons. Meteorological data are given by the output from the Japan Meteorological Agency atmospheric numerical model. Actual sea-ice distributions are decided on the basis of satellite, aeroplane, ship and coastal radar data of Hokkaido. In order to confirm the justification of the introduction of this new process, the authors have carried out long-term calculations. The results showed that this new process can be introduced efficiently to the numerical sea-ice model if the appropriate distribution of the depth of the mixed layer is given. There are still differences between analysis and forecast in some areas, such as around Sakhalin and the area around 45°–46°N, but the general sea-ice distribution is well predicted.

Dynamic-Thermodynamic Model. Since Hibler presented the first dynamic-thermodynamic sea-ice model in 1979, this kind of model has been applied and developed in numerical simulations, with many different models appearing. The following is an introduction to this model family.

- *U.S. Navy sea-ice forecast* According to Preller (1994), the three existing operational forecasting systems are: the Polar Ice Prediction System (PIPS) covering the Arctic basin, Barents and Greenland Seas; the Regional Polar Ice Prediction System – Barents Sea (RPIPS-B); and the Regional Polar Ice Prediction System – Greenland Sea (RPIPS-G). RPIPS-B covers the Barents and eastern Kara Seas. RPIPS-G covers the entire ice-covered region adjacent to the eastern coast of Greenland. PIPS has a grid resolution of 127 km, while the two regional models use higher resolution, 25 km for RPIPS-B and 20 km for RPIPS-G. Each of the forecasting systems are based on the dynamic-thermodynamic equations defined by the Hibler ice model (1979, 1980). The version of the model used in these systems is composed of five major components: 1) a momentum balance consisting of wind and water stresses, the Coriolis force and inertial forces; 2) a viscous-plastic ice rheology which relates ice stress to ice deformation; 3) a seven-level ice thickness distribution; 4) an ice strength equation; and 5) a heat budget, which uses the balance of atmosphere-ice-ocean heat fluxes, so the growth and decay of sea-ice could be determined.

The models are driven by wind stress, air temperatures and atmospheric heat fluxes, which are provided by the Navy's global atmospheric forecast model, the Navy Operational Global Atmospheric Prediction System. Ocean forcing is in the form of monthly mean geostrophic ocean currents and heat fluxes derived from a coupled ice-ocean model.

All three systems are used daily to make a 120-hour forecast of ice drift, ice thickness, ice concentration and the convergence/divergence of ice. The models are initialized from the previous day's 24-hour forecast. If the 24-hour forecast is not available, each model is initialized from a model-derived monthly mean climatology. Each of the regional models require ice-thickness information from PIPS, which is applied as a boundary condition at the models' open boundaries.

Once a week, the models are initialized from data in the form of a digitized ice concentration field provided by the JIC. This digitized field is derived from a number of different sources of satellite data, including the following: infrared-imagery from the Advanced Very High Resolution Radiometer on the NOAA polar orbiting satellite; visible-band imagery from the Defense Meteorological Satellite Program's Optical Line Scanner; and passive microwave data from the Special Sensor Microwave Image on board the Defense Meteorological Satellite Program's " morning" satellite. The

digitized ice concentration field also uses observations from both ships and aircraft.
- *Numerical sea-ice prediction in China*
  Wu et al. (1998) described the numerical sea-ice prediction in China. The model used is mainly for the Bohai Sea ice forecast. The model combines dynamic and thermal processes (Wu 1991; Wang and Wu 1994). The thermodynamic part of the model is based on the principle of the conservation of heat and determines the freezing and melting of ice (Wu and Wang 1992). The dynamic part, based on the principles of the conservation of mass and momentum, determines ice drift and deformation (Wu et al., 1997). Both parts contribute to determining the variation in ice thickness. The dynamic part of the model is based on the steady-state momentum equation and mass continuity equation, which read as follows:

$$\tau_a + \tau_w - mf\mathbf{k} \times \mathbf{V}_i + \mathbf{F} = 0 \tag{7.21}$$

$$\frac{\partial m}{\partial t} + \nabla \cdot (\mathbf{V}_i m) = \Phi \tag{7.22}$$

where τ_a and τ_w represent the air and water stresses, respectively. They can be obtained by using the ordinary quadratic stress law from the atmospheric and tidal current models. Here, m is the ice mass per unit area, $\mathbf{V}_i$ is ice velocity, k is the unit vector upward normal to the surface, f is the Coriolis parameter and Φ is a thermodynamic source or sink term. The force components due to internal ice stress have been written as

$$F_i = \frac{\partial \sigma_{ij}}{\partial x_i} \tag{7.23}$$

where σ_{ij} indicates the two-dimensional stress tensor within the ice. The stress depends mainly on the strain rate and the thickness distribution, in particular, the open water fraction. Sea-ice is considered as a compressible viscous-plastic material. The ice stress is expressed as

$$\sigma_{ij} = 2\eta \varepsilon_{ij} + [(\zeta - \eta)\dot{\varepsilon}_{kk} - p/2]\delta_{ij} \tag{7.24}$$

where ε_{ij} or ε is the strain rate tensor, $\dot{\varepsilon}_{kk} = tr\dot{\varepsilon}_{ij}$ denotes the trace, δ_{ij} is the Kronecker operator, and ζ and η are non-linear bulk and shear viscosity coefficients, respectively. The pressure term $p/2$ is taken to be a function of ice thickness and compactness as follows:

$$p = p_0 \bar{h} \exp[-c(1-A)] \tag{7.25}$$

where p_0 and c are taken as empirical constants. The ice thickness distribution is described with three idealized levels: open water $1 - A$ (A is ice compactness), level ice thickness h_1, and rubble thickness h_r (Wu, 1991, 1997). It follows that

$$m = \rho_i(h_1 + h_r)A \tag{7.26}$$

where h_1 and h_r are related to the average over a grid cell, and ρ_i is ice density. The continuity equation can be written for the three levels, as follows:

$$\frac{\partial A}{\partial t} = -\mathbf{V}_i \cdot \nabla A + \Psi_A + \Phi_A \tag{7.27}$$

$$\frac{\partial h_1}{\partial t} = -\mathbf{V}_i \cdot \nabla h_1 + \Psi_1 + \Phi_1 \tag{7.28}$$

$$\frac{\partial h_r}{\partial t} = -\mathbf{V}_i \cdot \nabla h_r + \Psi_r + \Phi_r \tag{7.29}$$

where the second terms on the right side describe mechanical deformation, and the third terms describe thermodynamic effects. When the ice density is constant, these are the forecasting equations for sea-ice. From (7.22), the mechanical redistribution functions, Ψ_A, Ψ_1 and Ψ_r, have to satisfy

$$h\Psi_A + A(\Psi_1 + \Psi_r) = -hA\nabla \cdot \mathbf{V}_i \tag{7.30}$$

where $h = h_1 + h_r$, and the functions (Ψ_A, Ψ_1 and Ψ_r) depend on the sign of ice velocity divergence and compactness (Wu 1991). The thermodynamic growth rates Φ_A, Φ_1 and Φ_r satisfy

$$\rho_i(h\Phi_A + \Phi_1 + \Phi_r) = \Phi \tag{7.31}$$

The thermodynamic calculations are based on heat budgets at the air-ice, air-water and ice-water interfaces. The fluxes include solar radiation, incoming and outgoing long-wave radiation, sensible and latent heat, conduction through the ice layer, and absorption and emission of energy due to phase changes (Parkinson and Washington, 1979; Wu, 1991, 1992). The ice model has been discretized using the Arakawa-type B-grid. The grid size is 8.64 km × 11.11 km and the time step is three hours.

Wu et al. (1998) pointed out that the routing forecasts during the winters from 1990 to 1996 showed that the present Bohai Sea ice model, proposed on the basis of properties of climate and ice conditions, is suitable for the short-range prediction of ice drift, growth and decay. The forecasting results were good at showing the evolution of ice conditions and were very useful to the oil industry. The statistical verification showed the reliability of the ice model, which could be applied to sea-ice prediction.

In order to improve operational numerical ice prediction, the accuracy of the initial ice fields has to be greatly improved, as do the atmospheric and oceanic models and the sea-ice model. Improvements in sea-ice prediction can be expected with the use of high-resolution satellite data. However, the analysed charts from the satellite imagery must be further verified on the basis of field observations in the Bohai Sea and the northern Huanghai Sea. In addition, more routine ice observations must be carried out at stations and from ships. Application of assimilative techniques to numerical sea-ice prediction can make significant improvements in the initial values for the ice-model. The development and application of a coupled sea-ice-ocean model will improve the model results and forecasts (Hibler and Bryan, 1987; Lemke, 1990; Cheng and Preller, 1994; Fichefet et al., 1994).

Generally speaking, the model of Wu et al. is a good one. A sea-ice model with three levels for simulating ice growth, decay and drift in the Bohai Sea is developed on the basis of the ice conditions in the Bohai Sea and existing sea-ice models. The model is described with level ice, rubble and open water. Both dynamic and thermodynamic processes of sea-ice are incorporated. The viscous-plastic constitutive law is applied in estimating the internal ice stress. The deformation equation is used to make the forecast equations consistent with the continuity equation and to express the effect of non-uniform dynamic forcing on sea-ice. The thermodynamic growth rates are used to express the effect of thermodynamic forcing from the atmosphere and ocean, and are parameterized according to the heat budgets at the air-ice and air-water interfaces. These mathematical treatments are able to improve our understanding of the complex dynamic and thermodynamic processes. The model is coupled with oceanic models and is linked to a numerical weather prediction model to forecast ice conditions in the Bohai Sea and the northern Yellow Sea. Statistical verification has also been used to objectively assess the model and the forecasting system. The forecast results were basically satisfying. Figure 7.11 shows the ice-covered areas (ICA) of the three-day forecasts from the model were near the analysed values from NOAA satellite imagery.

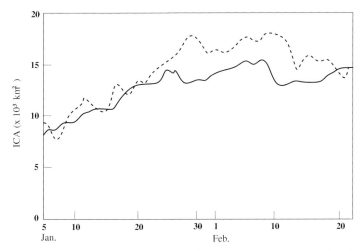

Fig. 7.11. The ice-covered area ICA in the Bohai Sea from the three-day forecasts (dashed line) and from the analyses of NOAA satellite imagery (solid line) during the period from 5 January to 22 February 1996 (after Wu et al., 1998).

- *Sea-ice forecasting in Canada and Denmark*

Besides the above sea-ice forecasting work in Japan, USA and China, many other countries, such as Canada, Finland, Sweden, Denmark and Russia, also have their own sea-ice forecasting methods. The thermodynamic and dynamic model is the most popular. The following are some examples. Neralla et al. (1994) discussed the influence of ice rheology on mesoscale ice forecasting, giving some features of the regional ice model (RIM).

This model is used by the Ice Centre of Environment Canada for operational ice forecasting. It is based on the rheological model of Hibler (1979). For the present study, a two-category ice thickness characterization is used, i.e., the ice cover consists of ice of uniform thickness, or areas of open water. A multi-category ice thickness model is also available in RIM (Intera 1985). Details of RIM are given by Neralla et al. (1988, 1993) and by Intera (1985). For the development of the rheology model, see Hibler (1979). This model is based on a viscous-plastic idealization. Stresses are related to strain rates by non-linear viscosity coefficients. Ice cover properties in this model are characterized by a strength parameter P^*, a transition strain rate $\dot{\varepsilon}_0$ which separates the regions of viscous and plastic deformation, and a ratio between the lengths of the axes of the elliptic yield envelope, e.

It has been shown that changing ice properties over a relatively wide range of values has very little effect on the forecasts. This conclusion brings into question the accuracy of determining material parameters from comparisons with particular data sets. Predicted ice concentrations in the vicinity of the boundaries appeared to depend on the conditions imposed at those boundaries. Further away from the boundaries, ice concentrations and velocities were not affected by the boundary conditions.

Christensen and Lu (1994) introduced a numerical sea-ice model for industrial application. The Danish Hydraulic Institute has implemented a nested grid facility within the two-dimensional, dynamic and thermodynamic sea-ice model, and investigated its applicability at high resolution (1 km) in the Greenland Sea. Initial sea-ice conditions were obtained at a 1 km resolution from Landsat MSS and NOAA-AVHRR and at a 15 km resolution from Nimbus 7-SMMR images. A conventional sea-ice model consists of a dynamic part and a thermodynamic part, which are coupled through a group of continuity equations. The main governing equation is the momentum balance of sea-ice, which reads as follows

$$m\frac{D\mathbf{U}}{Dt} = F_w + F_a + F_c + F_t + F_i \tag{7.32}$$

where m is the ice mass, and $\mathbf{U}$ is the ice velocity vector. On the right-hand side of equation 7.32, F_w and F_a are the water and air drag force, respectively, F_c the Coriolis force, F_t the ocean tilt force, and F_i is the internal sea-ice force. The ice velocity field can be obtained by solving the momentum equation. In the thermodynamic part, the governing equation is a heat flux balance on the top surface of the sea-ice or open water. The surface temperature field can be obtained by solving the heat flux balance equation. The surface temperature field determines sea-ice growth rates. The third part of a conventional three-level sea-ice model is a group of continuity equations. The continuity equations describe the evolution of the ice thickness and ice compactness characteristics caused by dynamic and thermodynamic effects. The equations are expressed as

$$\frac{\partial A_{tk}}{\partial t} = -\nabla \cdot (U\, A_{tk}) + S_{Atk} + \text{diffusion} \qquad (7.33)$$

$$\frac{\partial A_{tn}}{\partial t} = -\nabla \cdot (U\, A_{tn}) + S_{Atn} + \text{diffusion} \qquad (7.34)$$

$$\frac{\partial h_{tk}}{\partial t} = -\nabla \cdot (U\, h_{tk}) + S_{htk} + \text{diffusion} \qquad (7.35)$$

$$\frac{\partial h_{tn}}{\partial t} = -\nabla \cdot (U\, A_{tn}) + S_{htn} + \text{diffusion} \qquad (7.36)$$

where A is the ice compactness, h is the mean ice thickness, S is the thermodynamic sink or source term, the subscripts tk and tn represent thick ice and thin ice, respectively, and the subscripts A and h represent ice compactness and ice thickness, respectively. The diffusion terms are used for numerical stability. The terms S_{htk} and S_{htn} are the net growth or melting of thick ice and thin ice, respectively, and S_{Atk} and S_{Atn} are the effects of thermodynamics for thick and thin ice compactness, respectively. A simulation was undertaken of sea-ice conditions using a nested grid model on the North East Greenland shelf in early 1987. The results demonstrated the feasibility of a nested grid approach to sea-ice simulation, but a high resolution description of local processes is necessary, particularly for the description of total ice concentrations. A higher resolution of the raw input data, as well as modified numerical descriptions of the physical deformation, freezing and decay processes will be needed in further work in order to improve the simulation results.

Ice-Ocean Coupled Model. Cheng and Preller (1994) discuss an ice-ocean coupled model for the Northern Hemisphere. They describe the features of the Cox ocean model and an ice-ocean coupled model. The oceanic forcing for the ice model, such as ocean currents, temperature and salinity, and the oceanic heat fluxes, are provided in the ocean model of Cox. The oceanic forcing fields in the mixed-layer are represented by values from the first top level of the ocean model. The "robust" ocean model described by Sarmiento and Bryan (1982) was used to constrain the temperature and salinity fields toward climatology. The Cox ocean model has been spun up alone for five years to reach a quasi-steady state with a 30-day constraint on the mixed-layer and with a three-year constraint in the other levels. A distorted-physics method was used to allow a larger time step for temperature and salinity than for ocean current and stream function (Bryan, 1984). The Cox ocean model was spun up with a 30-minute time step for temperature and salinity. For ocean current and stream function a three-minute time step was applied.

At the time of coupling the ice and ocean models, it is necessary to relax the robust constraints, especially for the top mixed-layer, so that atmospheric forcing can interact with the layer to either heat or cool it. During calculation of the coupled model, the robust constraints were reset for all levels to be 250 days, which was weak enough to allow the mixed-layer to interact with atmospheric forcing, but strong enough to keep major distributions of ocean temperature and salinity.

It was the intention to maintain the independent integrity of both the PIPS2.0 ice model and the Cox ocean model, and simply couple the models by exchanging necessary information. In the ice model, the mixed-layer temperatures, the variable freezing temperatures, the oceanic heat fluxes and the ocean currents are provided from the ocean model. Two equations for computing water temperature and salinity from Cheng and Preller (1994) were used, except for two terms relating to atmospheric heating or cooling: $f_A \delta(z) R_0 \theta(T - T_f)/Z_{\mathrm{mix}}$ and $0.035 S_f \delta(z) Z_{\mathrm{mix}}$. The variables are defined as follows: $\delta(z)$ equals unity in the mixed-layer and zero otherwise; $\theta(T - T_f)$ equals unity when the mixed-layer temperature T is greater than the freezing point T_f, and zero otherwise; f_A is the ice growth/melt rate in open water due to atmospheric forcing only; R_0 is a ratio of the latent heat of fusion of sea-ice to the heat capacity of water; Z_{mix} is the mixed-layer thickness; and S_f the total ice growth/melt rate of open water and sea-ice due to atmospheric forcing only. Cheng and Preller (1994) used the growth/melt rate of sea-ice in open water, which includes the influence not only from atmospheric forcing, but also from the oceanic heat flux. Similarly, the total ice growth/melt rate of open water and sea-ice affected by the heat flux of the ocean was used here.

Cheng and Preller modified Hibler's ice model and adapted it to a domain that includes most of the sea-ice covered regions in the Northern Hemisphere. This model has been developed as an upgrade to the U.S. Navy's sea-ice forecasting system, PIPS, and is called PIPS2.0. PIPS2.0 uses a newly oriented spherical coordinate system, with the 170°W–10°E meridian as the "Equator", and the intersection of the 100°E meridian and the Equator as the "North Pole". The new orientation was introduced to avoid the numerical singularity at the Pole and the possible numerical instability in high latitudes. PIPS2.0 has been coupled with the Cox ocean model and spun up for five years to estimate the steady-state ice motion and ice growth/decay by using the 1986 monthly GAPS atmospheric forcing, the Levitus Climatological data, and Naval DBDB5 bathymetry data. The modelled ice edge is mainly consistent with the JIC weekly analysis, and the modelled ice thickness distribution agrees with submarine sonar data in the central Arctic basin. Inconsistencies in the Yellow Sea and the Baltic Sea could be caused by river runoff, which is not presently included in the coupled model, and may dominate the ice growth.

Fichefet et al. (1994) proposed a coupled sea-ice-upper-ocean model intended for coupling with AGCMs and ocean general circulation models, which is suitable for climate studies. It is composed of a global sea-ice model coupled to a one-dimensional mixed-layer-pycnocline model extending to a maximum depth of 300 metres in the ocean. The sea-ice model includes thermodynamic and dynamic components linked by advection processes. At the surface, the coupled model is forced by climatology derived from the output of a 10-year run (1979–1988) of the atmospheric general circulation model.

The atmospheric surface-layer parameterization used in the AGCMs provides the surface heat and momentum fluxes, given the atmospheric, cryospheric and oceanic fields. The temperature and salinity of the upper ocean, and annual mean geostrophic currents are derived from Levitus' (1982) atlas. A 20-year seasonal cycle simulation with the model of Fichefet et al. has been performed for both the Arctic and the Antarctic ice covers. The results show that the model reasonably well reproduces the observed sea-ice and upper-ocean characteristics of both hemispheres. However, a number of important shortcomings have been detected. Figure 7.12 shows the spatial distribution of the ice concentration predicted by the model in the Northern Hemisphere for the month of maximum ice extent (March) and the month of minimum ice extent (September). The thick line indicates the mean position of the ice edge as derived from satellite observations (Parkinson and Cavalieri, 1989). The thin line represents the predicted ice edge. It can be seen that the simulation of the ice edge in March was reasonably satisfactory. However, the ice extent was somewhat overestimated in the Bering Sea. The simulation of the ice edge in September was relatively satisfactory in the Greenland Sea, but the ice edge was located too far to the north in the Kara, Lapter and East Siberian Seas. In the Southern Hemisphere, the predicted ice concentrations were generally lower than in the Northern Hemisphere in comparison with the satellite observations (Gloersen et al., 1992).

Flato and Hibler (1991) have set up a two-level model, combined with simple cavitating fluid rheology, an oceanic boundary layer, and a thermodynamic model, which allows most of the relevant sea-ice processes to be included in global climate models. Figure 7.13 shows the spatial patterns of ice thickness build-up produced by such a model for both the Arctic and the Antarctic. The main features of the modelled thickness patterns basically agree with observations (Hibler and Flato, 1992).

On the other hand, it is meaningful to know the future developments in sea-ice numerical forecasting. The U.S. Navy sea-ice forecasting system is one of the representatives. Preller (1994) pointed out that future plans for the Navy's forecasting systems include the development of models for the partially ice-covered seas of the western Pacific, the Sea of Okhotsk, the Yellow Sea and the Sea of Japan. Work has already begun on an ice-ocean

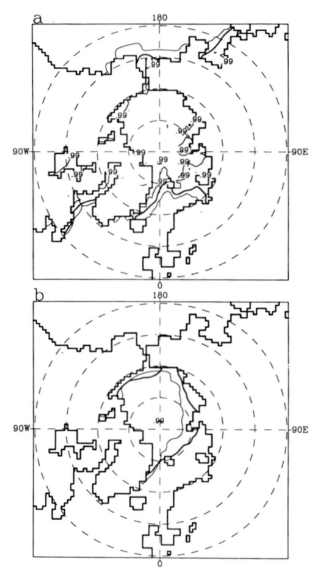

Fig. 7.12. Contours of ice concentration in the Northern Hemisphere for: **(a)** March; and **(b)** September. The thick line corresponds to the observed 15% concentration, as determined from SMMR Nimbus-7 data (Parkinson and Cavalieri, 1989). The thin line is the simulated 15% curve (from Fichefet et al., 1994).

model for the Sea of Okhotsk using the Hibler ice model coupled to the Bryan-Cox ocean model. An ocean model is also in preparation and that will cover the Yellow Sea and East China Sea. A version of the sigma coordinate, three-dimensional, primitive equation model (Blumberg and Mellor, 1987) used by the Naval Oceanography Center (Horton et al., 1992) is applied to this region. A multi-layer model has been tested in the north Pacific including the Sea of Japan (Huriburt et al., 1992). In recent years, there have been many advances in this field because of rapid developments in computer science and technology.

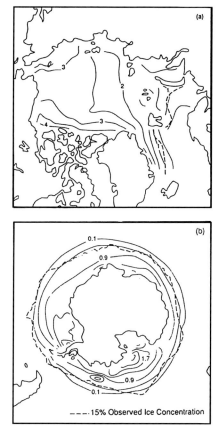

Fig. 7.13. Mean thickness contours in metres calculated by the cavitating fluid model for: **(a)** the Arctic at the end of March 1983; **(b)** the Antarctic at the end of July based on climatological forcing. The dashed lines indicate observed 15% ice concentration contours (from Hibler and Flato, 1992).

7.4.2 Meteorological and Hydrological Predictions using Ice-snow Cover as an Indicator

On the basis of the feedback effects of ice-snow cover on meteorological and hydrological conditions, numerical and synoptic-statistical methods are used to predict these conditions. We can cite many examples.

Numerical Method. Using the CCMOB version of the community forecast model (CCFM) of NCAR, Walsh and Ross (1988) have conducted a series of 30-day simulations to evaluate the sensitivities of climate to continental snow cover over North America and Eurasia. The analyses have been undertaken for specific dates during the winters of 1976/77 through to 1983/84. The model is a global spectral model in a σ coordinate system with nine vertical layers. The model step is 30 minutes and the length of each simulation is 30 days. The results of the experiments showed that the large-scale atmospheric circulation is more sensitive to Eurasian snow cover than to North American snow cover. The sensitivities to Eurasian snow cover are quite strong and systematic. It is very valuable for long-range meteorological predictions over the continent, whereas the observational correspondence of the snow boundary over eastern North America had no impact on the accuracy of the 30-day forecast. In 10 pairs of midwinter forecasts, the major effect of extensive snow in eastern North America is a reduction of near-surface air temperature in the vicinity of the snow anomaly. There are also changes in sea-level pressure and precipitation following changes in the snow cover. In a set of six cases for March, positive anomalies of Eurasian snow cover reduced air temperature by at least several degrees Celsius throughout the lower half of the troposphere in the region over, and downstream of, the snow anomaly. The positive Eurasian snow anomalies also systematically change pressures over different regions of the continent. In the Eurasian experiments, the 30-day forecast of pressure for the Eurasian continents varied with snow cover in a manner consistent with the observed pressure fields in the same months.

Ross and Walsh (1986) have assessed the influence of snow cover and sea-ice on the development of synoptic-scale cyclones in three regions: eastern North America; the North Atlantic Ocean; and the North Pacific Ocean. Daily observational data for 30 winters (1951–1980) were collected over the ice-covered areas in the North Atlantic and the North Pacific. Daily observations of sea-level pressure and 500-hPa geopotential height for the Northern Hemisphere were included. The sea-level pressure forecasts were obtained mainly by the persistence method. The 500-hPa forecasts were obtained by three different methods: a one-level barotropic vorticity model; persistence; and analogy evolution. The results indicate that the enhanced baroclinicity near the snow or ice margin contributes to stronger intensification, and/or to motion parallel to the snow or ice margin, in eastern North America and in the North Atlantic. A weaker signal was found in the North Pacific. A possible reason is the larger sea-ice area over the North Atlantic. The situation was similar for sea-level pressure and at the 500 hPa height, but the differences were statistically significant only at the sea-level pressure. In cases of extreme snow or ice cover, weekly or monthly forecasts might be improved by a consideration of snow or ice anomalies, if the anomalies persist through the forecast period. The weakness in the above experiments is that they are simple, short-term forecasting methods, so the authors undertook additional sensitivity experiments with CFM of NCAR. The results suggest that CFM is no better than the above three simple methods, although CFM considers much more complex physical processes. The problem is related to a combination of uncertainties involving parametric sensitivities, the formulation of the planetary boundary layer and vertical mixing processes, and the choice of the cases for the forecast experiments.

To investigate the influence of sudden large-scale snow cover removal in middle and high latitudes of the Southern and Northern Hemispheres on short-term climatic and hydrologic change, Yeh and Manabe (1983) used a simplified GCM with a limited computational domain and idealized geography. They found that a temporary but complete removal of snow cover can bring about a significant reduction in zonal mean soil moisture for the following spring and summer seasons. It has been found that there is a large negative difference in soil moisture between the removal of snow and standard experiments during the month of May. It is caused by two factors: 1) the removal of snow cover and, therefore, elimination of moisture available for snow melt, which would otherwise have filled the soil to saturation, or else produced runoff; and 2) increased evaporation from the soil surface due to the removal of snow cover. They also found that the negative difference in soil moisture is present for the entire spring and summer seasons in high latitudes. An initially induced anomaly in soil moisture persists in high latitudes for a long time period. An instantaneous, complete removal of snow cover in early spring can produce a hydrological effect that can last as long as four to five months. Section 3 of this chapter indicated that changes in sea-ice, in both the Antarctic and the Arctic, can result in changes in the general circulation and climate over the corresponding regions, so that sea-ice can be used to forecast weather and climate changes. Chao and Chen (1980), Herman and Johnson (1978), and Chen et al. (1996) have carried out numerical experiments and obtained interesting results.

Synoptic-statistic Method. Ice-snow cover should be taken as a major factor when undertaking predictions with hydro-meteorological synoptic-statistic methods. Some examples are given below. The study of Dey and Kumar (1982) illustrated that when the Eurasia snow cover area in winter and spring increased and snow in the spring melted slower, India monsoon was weaker and its motion was slower (and vice versa). Dey and Kumar (1983) also illustrated that a greater than average Himalayan winter snow cover is followed by a less than average summer monsoon rainfall over India (and vice versa). Guo and Wang (1986) pointed out that the summer monsoon over the Tibet Plateau was relatively weak in heavy snow years. The analysis of Zhou

(1992), using monthly data from 1956 to 1987, showed that days with snow cover over the south-eastern Tibet Plateau in November and December can be taken as an important indicator of the flow rate of the upper-middle basin of the Yangtze River in August and September of the following year. The positive correlation is very significant. These relationships may be used in long-range forecasts of monsoons and monsoon rainfall.

On the basis of statistical analyses, using the Arctic sea-ice extent and meteorological data from the Northern Hemisphere from $20°$ to $90°$N over 300 months (1953–1977), Walsh and Johnson (1979) found that there is a statistically significant correlation between the dominant modes of atmosphere and sea-ice values, with atmospheric lags of up to two months and ice lays of up to four months. The surface temperature field has the strongest relationship to the sea-ice variations. The strongest correlations between ice extent anomalies and subsequent atmospheric variation can be found in the autumn months. The ice-atmospheric correlation has also been found in the mid-latitudes of the North Atlantic. For the years of extreme ice extent, statistically significant pressure differences reach up to 10–15 hPa, surface temperature differences up to $8°$–$9°$K, 700-hPa height differences up to 16-18 decametres and the anomaly centres tend to migrate seasonally with the ice edge. The statistical predictability of large-scale sea-ice variations decays to nil at a forecast interval of five to six months.

According to the synoptic-statistical analyses, based on the monthly data analysis for the period 1953–1984, Wang et al. (1992) found the influence of the Arctic sea-ice area on some atmospheric activity centres and thus on the hydro-meteorological regime of the Yangtze River basin. They suggested that, if the sea-ice area over the Bering Sea during March–May and the Kara Sea and the Okhotsk Sea during October–December, was larger than normal in the previous year, then the centre of the polar vortex over the Eastern Hemisphere would be located at a more westerly position than its average position; stronger zonal circulation intensity would result in lower 500-hPa potential height over India and Burma; and the northern Pacific subtropical high would appear weaker than normal. In these cases, the precipitation in the rainy season for the upper and middle reaches of the Yangtze River Basin, and the flow of these parts of the river, would be less (and vice versa). The time lag of the precipitation and the flow behind the sea-ice variation were about four months for the Bering Sea, while about nine months for the Kara Sea and the Okhotsk Sea. According to Fig. 7.3, the influence of the Arctic sea-ice on the area of Arctic polar vortex over Asia was more obvious at the atmospheric lag of about four seasons. According to Fig. 7.4, and our further analysis, the influence of the Antarctic sea-ice area on the polar vortex over Asia was more frequent at the atmospheric lag of about two seasons. The influence of the sea-ice area on the polar vortex over northern America was more frequent at the atmospheric lag of about four seasons. Using monthly data of sea-ice area for all eight sectors of the Antarctica during the periods of 1973–1990, Peng et al. (1996b) calculated correlation fields between the ice areas and 500 hPa heights of the Northern Hemisphere with time lags of the heights behind the ice areas from one to four seasons. When the correlation coefficient reaches 0.60, the confidence level is 0.01, and the analysis results indicated that, in many cases, there were significant influences of the ice cover for different sectors on the 500 hPa height fields with time lags for one to four seasons, depending on the locations of different atmospheric centres. For instance, there was an obvious correlation between the sea-ice area cover over the sector of 115°E–160°E of the Antarctica in July to September and the 500 hPa heights in April to June of the following year for several continental regions. The time lag is three seasons. According to Fig. 7.14, the most significant and biggest negative correlation region ($r = -0.839$) could be found over Eurasia. Another negative correlation region ($r = -0.709$) was located over northern Canada, while a positive correlation region ($r = 0.691$) could be seen over eastern USA. In each of the three cases, the confidence level exceeds 0.01. Obviously, the time lag analysis of the correlation fields, concerning the influence of the Arctic and the Antarctic sea-ice cover mentioned above are very helpful for long-range hydro-meteorological predictions.

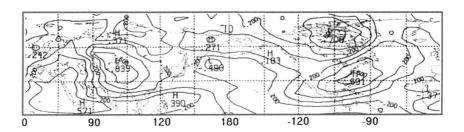

Fig. 7.14. The correlation field between the sea-ice area cover over the sector of 115°E–160°E of the Antarctica in July to September, and the 500 hPa heights in April to June of next year (after Peng et al., 1996b).

Finally, it should be pointed out that ice-snow cover can influence not only meteorological and hydrological conditions, but also sea surface temperature, ocean current features, as well as the vegetation and soil conditions, so it can also be used as an indicator to predict these environmental components.

8. Modelling and Prediction of the Terrestrial Biosphere

Catherine Ciret

8.1 Introduction and Background

The modelling and prediction of the vegetation of the Earth have long been the focus of studies in agricultural, ecological and environmental sciences. These studies allowed us to predict crop yields, improve forest managements, understand the dynamics of ecosystems population, and evaluate the impacts of land-use on soil erosion and water supply. The ecological models used in these studies were often run in a stand alone mode, i.e., they were not run in conjunction with other models. The scope of ecological modelling has, however, changed in the last decade to become part of an integrated modelling effort which aims at predicting the change of our global environment. Human activities are, indeed, producing measurable alterations of the Earth's major system components (e.g. the atmosphere, the biosphere and the ocean). The prospect of "global change" (that is, the anthropogenic changes of our natural environment) has prompted the scientific community to try to understand better the interactions of the Earth system components and to integrate the Earth and biological sciences into interdisciplinary and international scientific projects (e.g. the International Geosphere-Biosphere Program, IGBP, 1994).

The climate system has received particular attention because of the threat of climate change due to the increase emissions of greenhouse gases. The components of the climate system, i.e. the atmosphere, the hydrosphere, the cryosphere, the land and its biomass, are strongly coupled and most of these components are now included in current climate models (Houghton et al., 1996). However, the interactions that take place between the ice, the atmosphere, the ocean and the land are represented in numerical models with various degrees of details, and, until recently, the land biomass was described in climate models in a very simplified manner (Sellers et al., 1997). The study of biosphere-atmosphere interactions is becoming, however, a growing area of research. There is increasing evidence that vegetation feedbacks have played an important role in the genesis of the climates of the past and that it can play a critical role in a greenhouse-warmed climate (Melillo et al., 1996). Hence a better representation of the land biomass and a better understanding of the interactions between vegetation and atmosphere have become a high priority in climate system modelling. The interactions between the vegetation and the atmosphere refer to the exchanges of heat, moisture, trace gases, aerosols and momentum between land surfaces and the overlying atmosphere (Pielke, et al., 1998). The presence of vegetation canopy affects the radiation absorption at the surface, modifies the momentum transfer between the surface and the planetary boundary layer airflow and, when the vegetation canopy is dense, insulates the soil surface from the overlying airflow. Vegetation is particularly important for the moisture exchanges since more than half of the flux of water from the land to the atmosphere moves through vegetation (Molz, 1981). Vegetation affects the exchange of water between the land and the atmosphere through three mechanisms: biophysical control of evapotranspiration, interception of precipitation and control of soil moisture availability by the plant root system (see Sellers et al., 1997, for a comprehensive review).

The biogeochemical exchanges between vegetation and atmosphere are also important. The terrestrial biosphere is a primary sink and source of atmospheric carbon and the role played by the terrestrial ecosystems in the current atmospheric carbon budget is critical. The exchange of atmospheric carbon dioxide between the land biomass and the atmosphere is by means of processes of photosynthesis, autotropic respiration (i.e. plant respiration) and heterotrophic respiration (i.e. microbial respiration involved in the decay of litter and soil). The annual gross flux of carbon between the land and the atmosphere represents 15 to 20% of the atmosphere's carbon content (Houghton et al., 1990) and this annual gross flux is roughly equivalent to that of the world's oceans (Aber, 1992). In the calculation of the global carbon cycle undertaken in the 1980s, it appeared, however, that the atmosphere should have gained more CO_2 than it has been actually monitored (Post, 1993). The annual average emissions of $7.0 \pm 1.1 \, \text{PgC}$ from combustion of fossil fuels, changes in land use and cement manufacturing are greater than the sum of the annual uptake by the ocean ($2.0 \pm 0.8 \, \text{PgC yr}^{-1}$) and the annual accumulation of carbon in the atmosphere ($3.4 \pm 0.2 \, \text{PgC yr}^{-1}$) (Watson et al., 1990). An additional sink (or "missing sink") of $1.6 \, \text{PgC yr}^{-1}$ is required to balance the carbon budget. This "missing sink" was thought to be the terrestrial ecosystems (Post, 1993) and recent studies based on atmospheric and oceanic data, and also on forest, inventories have shown that carbon is indeed accumulating in terrestrial ecosystems (see Houghton et al., 1998, for a review). Mechanisms which could explain the terrestrial accumulation of carbon include CO_2 fertilisation and its impacts on photosynthesis and water use efficiency, nitrogen deposition, warming-enhanced nitrogen mineralisation and forest regrowth following disturbances (Houghton et al., 1998).

Modelling the biophysical and biogeochemical exchanges between the vegetation and its surrounding environment (i.e. soil, atmosphere) requires that different types of ecological models be used. The literature related to ecological modelling is abundant and this chapter does not attempt to present a comprehensive review of ecological modelling. The aim of this chapter is to present the ecological modelling in relation to other components of the Earth system and to describe the different approaches to link climate, plant physiology, biogeochemistry and biogeography in a same integrated framework. The different kinds of ecological models, their limitations and the issues related to their validation are presented in Sect. 8.2. Section 8.3 focuses on the modelling of the interactions between terrestrial ecosystems and climate system and an example of an interactive vegetation model is described in Sect. 8.4.

8.2 Terrestrial Ecosystem Models

8.2.1 Model Overview

Schematically, an ecological model consists of four main components: the external variables, or forcing functions, which influence the state of the ecosystem (e.g. climate variables, input of pollutants or fertilisers); the state variables which describe the state of the ecosystem; the mathematical equations which link the forcing functions to the state variables and represent the biological, chemical and physical processes in the ecosystem, and, finally, the coefficients, or parameters, which are needed for the mathematical representations of processes in the ecosystem (Jorgensen, 1988). In addition, most models also contain universal constants such as gas or radiation constants. Unlike the physical models (e.g. atmospheric or ocean models) which are mainly built on first principles of the functioning of the physical environment, many ecological models are empirically based. Establishing the empirical relationships between the environment and the components of the ecosystems require the access to a large range of data sources. The values of the parameters contained in the mathematical representation of processes are only known within limits, and the calibration of ecological models is often required to determine the values of the other parameters.

There are many types of ecological models and their objectives, as well as their spatial and temporal scales, vary greatly. Several model classifications and reviews can be found in the literature (e.g. Malanson, 1993, Dale and Rauscher, 1994, Hurtt et al., 1998). The ecological models can be conveniently classified as a function of the spatial scales and time steps at which they operate (Heal et al., 1993) and the grouping of these models along the two scales are shown in Fig. 8.1.

Most of the models described in this section have been coupled to other environmental models, in particular climate models.

The **leaf** models and **crop** models occupy the lower end of the spatial/temporal scales. The time steps of these models are typically hourly to daily, and their spatial scales span from less than a centimetre for the leaf model, to about a few square metres for the crop models. The **leaf** models describe in detail the processes of photosynthesis. Photosynthesis involves light-

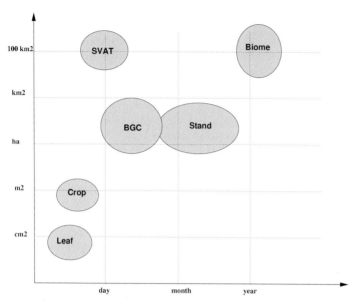

Fig. 8.1. Classification of vegetation models in relation to their temporal and spatial scales (modified from Heal et al., 1993).

driven photochemical processes, enzymatic processes without light ("dark" reactions) and exchanges of CO_2 and O_2 between the chloroplasts and the atmosphere (diffusion processes) (Larcher, 1995). Once the photosynthetically utilisable radiation and the atmospheric CO_2 are captured by the chloroplasts, the process of carboxylation takes place to convert the CO_2 into carbohydrates. This process is catalysed by the enzyme ribulose bisphosphate (RuBP) carboxylase (Rubisco). The primary means of gas exchanges between the leaf and the surrounding air is via the stomata and as CO_2 diffuses into the leaf, water vapor from the leaf is lost to the atmosphere due to the diffusive conductance of leaf surfaces to water vapour (i.e. stomatal conductance g_s). Stomata serve the conflicting role of permitting CO_2 uptake and, at the same time, restricting the leaf transpiration in order to reach an appropriate balance between photosynthesis and water loss (Collatz et al., 1991).

The first product of CO_2 fixation is a 3-carbon molecule and this process is also called C_3 pathway of CO_2 assimilation. About 10% of plant species have a different CO_2 assimilation pathway (i.e. dicarboxylic acid pathway with 4 carbon atoms or C_4 pathway) (Larcher, 1995). The distinction between these two types of carbon fixation biochemistry is important for the modelling purpose, because the primary productivity of plants depends on their metabolic types, i.e. C_3 or C_4. The C_4 plants are mainly distributed in tropical and subtropical latitudes, and arid and semi-arid environments and include herbaceous vegetation such as savannas, and crops such as maize, sorghum, millets and sugarcane. The C_3 plants represents the majority of the world's vegetation cover and include temperate grasses, all trees, and most temperate crops.

In the biochemical model of leaf photosynthesis proposed by Farquhar and von Caemmerer (1982) the photosynthetic rate A of C_3 leaves is determined as a function of absorbed light, leaf temperature, CO_2 concentration within the leaf and at the leaf surface. A is also a function of the capacity for photosynthesis of the Rubisco enzyme. Collatz et al. (1991) have added a stomatal conductance sub-model to describe the response of leaf stomatal conductance g_s to the photosynthetic rate A, and the concentrations of CO_2 and water vapour in the air. Leaf models such as the ones described by Farquhar and von Caemmerer (1982) and Collatz et al. (1991) are now widely used in larger scale vegetation models.

The **crop** growth models predict grain yields and allow the evaluation of management strategies such as the choice of sowing dates or the use of fertilisers. Crop models are often specific to the ecosystem for which they have been developed and usually their results cannot be extrapolated to other ecosystems or to a wider range of environmental conditions (White et al., 1993). Crop models are now being used in conjunction with other models, particularly GCMs, to estimate the impact of climate change on agriculture (see review in Goudriaan, 1996).

The **stand** models operate at the scale within which interactions among individual plants become important with respect to the dynamic of the species (Malanson, 1993). A stand is a patch of vegetation of relatively homogeneous species composition (Burrows, 1990) and the size of a stand is typically 0.1 ha. Most models of this type focus on forest stands (e.g. FOREST model, Shugart and West, 1977). These models, also called forest succession models, gap models, or forest growth models in treat each tree as an individual and use species-specific information on, for instance, growth rates, fecundity, mortality and seed dispersal distances. The stand models can produce realistic simulations of gap-phase dynamics (e.g. sequential replacement of plants due to the death of a big tree) (Prentice et al., 1993a) and simulate the growth, recruitment, mortality (including mortality due to disturbances), changes of diameters of individual trees and the crowding from other trees. Stand models have also been used in conjunction with GCMs output for climate impact assessment (e.g. Prentice et al., 1993b).

The **biogeochemical** (BGC) models predict the primary productivity (i.e. yield of the dry matter production) of the terrestrial biosphere. The carbon flux into the biosphere represents the gross primary productivity (GPP), that is, the carbon fixed in photosynthesis by plants (Heimann et al., 1998). The net primary productivity (NPP) is the profit remaining from the photosynthesis of the plants, that is, the GPP minus the autotropic respiration. The rate of plant respiration, as a percentage of GPP, is estimated to range between 20% to 80%, that is, 20–40% for annual plants, 40–60% for temperate successional forests and higher respiration rates can be obtained in tropical forests (Lieth and Whittaker, 1975). NPP is the sum of the net primary productions by all individual plants in a unit area. The net primary production of plants is used for building up organic matter, part of which is lost as litter over the course of the year or is grazed by consumers (Larcher, 1995). The losses include the shedding of leaves, flowers, fruit and branches and the death of tissues. The increase of plant mass in an area is the net community growth, or net ecosystem production (NEP). NEP represents the net exchange between the terrestrial biosphere and the atmosphere and is equal to NPP minus the heterotropic respiration, that is, the release of carbon by the soil microorganisms. In a mature plant community, NEP should be closed to zero, whereas in a young forest, for instance, a fraction of NPP can accumulate from one year to the next as NEP (Lieth and Whittaker, 1975). The aim of biogeochemical models is to quantify the metabolism of terrestrial ecosystems and to simulate the different aspects of the terrestrial ecosystem carbon cycle. The recognition of the terrestrial's biosphere key role in the global carbon cycle has prompted the development of an increasing number of biogeochemical models (Heimann et al., 1998). Three main types of BGC models can be identified: regression based models, remote-sensing based model, and processed based models. Regression-based BGC models, or statistical models, use empirically derived relationship between climate and primary productivity. In this type of model, the primary productivity is related in a correlative manner to, for instance, temperature, precipitation and evapotranspiration values derived from publications (e.g. MIAMI model, Lieth, 1975). Remote-sensing based models (e.g. Ruimy et al., 1994) rely on satellite observations of spectral vegetation indices to estimate the biomass net primary productivity. These models are essential derived from the Monteith model (Monteith, 1982) which assumes that the biomass primary productivity can be related to the fraction of the incoming solar energy stored into organic dry matter (see Sect. 8.4). Regression-based models and satellite-based models can only be used in a diagnostic mode since we cannot extrapolate their results in the future when climate, CO_2 and other factors (e.g. soil nutrients) may be different (Tian et al., 1998).

Process based BGC models rely on first principles of ecosystem processes. Ideally, a process based BGC model should explicitly describe the physiological processes such as photosynthesis, autotropic respiration, decomposition and nutrient cycles. Process based models depend on detailed information about the environment and biologic characteristics. However, the lack of parameters for most of the world ecosystems is one of the problem attached to this type of model when they are applied on a global scale (Ruimy et al., 1994). Examples of processed based BGC models are the Terrestrial Ecosystem Model (TEM) (Melillo et al., 1993), which is used to simulate the equilibrium response of natural ecosystem to increased atmospheric CO_2; CENTURY (Parton et al., 1988) which simulates the primary productivity and the dynamics of soil organic matter, carbon, nitrogen, sulfur and phosphorus in natural or cultivated systems; and CARAIB (Warnant et al., 1994) which extrapolates photosynthesis rates from the leaf to ecosystem level using mechanistic model of leaf photosynthesis (e.g. Farquhar and Von Caemmerer, 1982). Remote-sensing based models and mechanistic models are converging and some models (e.g. CASA, Potter et al., 1993) represent a merger between both approaches. Many differences exist between the BGC models in terms of structure, parameterisation and hence results (Tian et al., 1988). So far, the performance of the BGC models has not been thoroughly investigated, mainly because of the absence of adequate validation datasets (Heimann et al., 1998) (see Sect. 8.2.2 for a discussion about the validation of BGC models).

The **biome** models such as BIOME-1 (Prentice et al., 1992) operate on coarse spatial resolution and can be applied on a global scale. The vegetation units predicted are usually biome or plant functional type. These models predict the geographical distributions of natural ecosystems (e.g. tropical rainforest, tundra, grassland). The underlying assumption in these models is the existence of climatic climax (i.e. equilibrium state between mature vegetation and the local climate). These models are useful tools to investigate the response of the climate system to alteration of the vegetation distribution. However, they cannot simulate the dynamics of vegetation change (i.e. transient changes of vegetation cover in response to environmental conditions), hence they have little value for projecting future vegetation distributions (Woodward and Beeling, 1997).

The **biophysical** models, or Soil–Vegetation–Atmosphere Transfer (SVAT) schemes focus on the biophysical interactions between the atmosphere and the land surface (see Chap. 5 for details). SVATs are generally intended for incorporation in GCMs and numerical weather forecast models, and their main task is to determine the radiative energy balance of the land surface and its partitioning into latent and sensible heat fluxes. SVATs have evolved considerably since the first generation of schemes from the early 1970s which were based on simple aerodynamic bulk transfer and uniform prescription of surface parameters such as as albedo, aerodynamic roughness and soil moisture (Sellers et al., 1997). The second generation of SVATs took into account the effects of vegetation in the calculation of the surface energy balance by including the key vegetation characteristics, that is, the leaf area index (LAI) and the seasonality of leaf area, the roughness length, the albedo, the root structure and the stomatal resistance (Aber, 1992). The representation of these vegetation characteristics, particularly the stomatal functioning, remained however very simplified.

The "greening" of SVATs is now attracting renewed attention (Hutjes et al., 1998) and the current trend is to include more realistic plant physiology, in particular more mechanistic treatments of photosynthesis and stomatal functioning. Two main methods are currently used to represent the biological control of transpiration, (see review by Niyogi et al., 1998). Many SVATs still apply the approach proposed by Jarvis (1976) in which the leaf stomatal conductance g_s is estimated empirically as a function of environmental factors such as soil moisture, air temperature, humidity and solar radiation. The physiological processes involving stomatal aperture and photosynthesis are therefore not explicitly described. The scaling up from the leaf to the canopy is made by multiplying g_s by prescribed values of leaf area index. One of the limitations inherent to this approach is that it does not account for explicit interactions or secondary feedback processes in the system dynamics (Niyogi et al., 1998). In addition, if the LAI needs to be prescribed, rapid changes in the vegetation cover associated with climatic events such as drought will be ignored (Calvet et al., 1998). In studies that consider the impacts of CO_2 changes, the lack of feedback between g_s and the atmospheric CO_2 in the Jarvis-type approach represents a serious limitation. A few SVATs (Bonan, 1995; Sellers et al., 1996c; Calvet et al., 1998; Dickinson et al., 1998) employ a more physiological approach to estimate the evapotranspiration rates and link the photosynthetic rate A to the leaf stomatal conductance.

Although the stomatal aperture is allowed to vary in time in response to environmental conditions, it should be noted that other vegetation parameters (vegetation height and structure, LAI, albedo, rooting depth) are, in most SVATs, still prescribed. The vegetation parameters and their seasonal variation are, usually, estimated from remote-sensing databases (Sellers et al., 1996b) or from literature surveys. In earlier versions of SVATs (e.g. BATS, Dickinson et al., 1993, SECHIBA, Ducoudré et al., 1993), a parameter such as LAI varied between a minimum and a maximum value as a function of simulated soil temperatures, ignoring the effect of soil water stress on leaf display. However this issue is currently being addressed and several attempts are being made to represent the plant phenology in a more detailed and interactive manner (Dickinson et al., 1998; Calvet et al., 1998; Ciret et al., 1999)

The distinction between biophysical models (SVATs), biogeochemical models (BGCs), and biogeographical (biome) models is rapidly becoming obsolete (Hurtt et al., 1998). New types of models are now emerging in which the direct physical interactions between the atmosphere and the land surface, the air-land carbon balance, and the variations of vegetation distributions are all integrated in one framework (see Sect. 8.3).

8.2.2 Model Validation

A critical need in terrestrial ecosystem modelling is model validation. The data required for the validation of terrestrial ecosystem models depends on the type of model and the timescale and spatial scale at which the model is being applied. The different types of datasets most commonly used to validate terrestrial ecosystem models are vegetation and soil maps from inventories, satellite measurements of Spectral Vegetation Indices (SVIs) and ground observations of ecosystem parameters. The land cover maps from inventories (e.g. Olson et al., 1983) have limited applications, particularly because, as a result of human land-use activities, land cover is changing rapidly (Cramer and Fischer, 1996).

Satellite data are now being increasingly used to test and validate terrestrial ecosystem models. Satellite remote sensing have been identified by various international committees as a unique tool to repetitively acquire data at spatial, temporal and spectral resolutions appropriate to global environmental studies (Verstraete et al., 1996). The Advanced Very High Resolution Radiometer (AVHRR) embarked on satellites allows the daily measurements of spectral vegetation indices such as Normalised Difference Vegetation Index (NDVI). NDVI is defined as the combination of red (R) and near-infrared (NIR) surface reflectances of the plant canopy ($NDVI = (NIR - R)/(NIR + R)$). The underlying assumption behind the NIR/R algorithm is that the chrolophyll pigments in green leaves absorb radiation in the red wavelength, whereas most of the near-infrared radiation is reflected or transmitted by the leaves (Running et al., 1989). Since the red reflectance is inversely related to the quantity of chlorophyll present in the canopy, and the infrared reflectance is influenced by the multiple layers of leaves, it follows that the NDVI can be related to the leafiness of the vegetation. NDVI observations are used to estimate global LAI (Myneni et al., 1997), fraction of incident photosynthetically active radiation absorbed by plants (FPAR) (e.g. Sellers, 1996b), and plant phenology (e.g. shooting dates and foliation period) (Ludeke et al., 1996).

Although spectral vegetation indices provides valuable and irreplaceable source of information, this type of data has its limitations and caveats. The accuracy of NDVI is greatly influenced by the way the atmosphere and ground corrections are undertaken (Tian et al., 1998). Some corrections need to be applied to NDVI to account for various sources of anomalies (e.g. persistent cloud cover in the tropics, variations in solar zenith angle, soil background) (Sellers et al., 1996b). In addition, the algorithms devised to connect the radiance measurements to surface properties (e.g. LAI) need to be carefully tested because they are not always satisfactory. There are, indeed, some uncertainties in the relationships between NDVI and the key vegetation parameters LAI and FPAR. The relationship between NDVI and LAI is nonlinear, is affected by spatial heterogeneity and exhibits considerable variation among the cover types (Myneni et al., 1997). The NDVI-FPAR relationship is, on the other hand, quasi linear, however, some errors can be expected in the estimation of FPAR in regions with bright soil backgrounds and small vegetation cover (i.e. low NDVI) (Ruimy et al., 1994).

Ground measurements of, for instance, biomass, LAI, gas fluxes are also used to test and validate ecosystem models. In the last decade there have been several programs whose aim was to collect, in an artificial or in a natural environment, a large number of data providing information about carbon and nutrient cycles, atmosphere-biosphere trace gas exchanges, land surface characteristics (albedo, roughness), and ecosystem structure and composition (e.g. IGBP Terrestrial Transects, Koch et al., 1995). However, such datasets are scarce, and limited to a few regions and ecosystems.

Validating the ecosystem models raises a number of issues. The validation of local scale models (i.e. "leaf", "stand" and "crop" models) is usually not problematic, providing the models are tested at a scale, both spatial and temporal, at which they were intended to be applied (Rastetter, 1996). It is more difficult, however, to adequately validate larger scale models (i.e. GBC and "biome" models). One of the issues is the scarcity, or the lack, of global ecosystem datasets. The dataset used to validate a model should, for obvious reasons, be independent from the dataset used for its calibration (Cramer and Fischer, 1996). This requirement cannot always be met and it is often difficult to validate models such as the "biome" models with an independent dataset (Heal et al., 1993).

For GBC models, the lack of relevant global scale datasets is even more critical and the unequivocal and rigorous test of such models appears, at present, impossible (Oreskes et al., 1994). There are, however, various ways by which the performance of GBC models can be evaluated: test against short-term data; space for time substitution, that is, space (location of the site) is substituted for time (e.g. time since glacial retreat); reconstruction of past responses to climate change and comparison with other models (Rastetter, 1996). In addition, field-based measurements of primary productivity can be used to evaluate GBC models at specific sites, and the aggregated carbon fluxes simulated by the BGC models can also be compared to seasonal CO_2 flux fields generated by a three dimensional tracer transport model (Heimann et al., 1998).

If a crucial issue in model validation is the lack of adequate datasets, the choice of the validation procedure is probably as critical. Using synthetic timeseries from a hypothetical watershed and two simple models, Kirchner et al. (1996) demonstrated that common methods for comparing model predictions against data can be singularly ineffective at revealing problems with models. In their study, Kirchner et al. (1996) showed that, although one of the two tested models was fundamentally flawed, the conventional visual comparisons of predicted and observed timeseries and scatter plots did not allow a discrimination between these models. It is only when the relationship of interest was statistically extracted from the other confounding factors, and when the relevant test was carried out, that discrepancies between models appeared clearly. Rastetter (1996) also pointed out the inadequacy of testing a model at a temporal scale other than at which it was intended to be applied. He tested the leaf model from Farquhar and von Caemmerer (1982) at three different timescales (milliseconds, seconds and weeks) and showed that, unless the time scale of the data (i.e. in this case, seconds to minutes) against which the model is tested matches the mechanisms represented in the model, the validation of the model with other datasets was useless. In particular, this study demonstrates that validating a model against short-term data does not justify confidence in longer term projections, for instance, the long term responses of ecosystem to CO_2-induced climate change.

Although a universally applicable validation procedure is difficult to prescribe considering the diversity of modelling approaches, the methods employed to test the performance of ecosystem models should, nevertheless, be put under scrutiny.

8.2.3 Model Limitations and Research Needs

There are numerous advantages in using computer-based, mathematical models: models are useful instruments with which to survey complex systems such as ecological systems ; they can be used to reveal system properties; they help to expose the weakness in our knowledge and thus help define research priorities; and, finally, models are useful to test scientific hypotheses which cannot be tested in a laboratory (Jorgrnsen, 1988).

There are, however, limitations in the predictive ability of current ecological models. Some of these limitations are due to the insufficient understanding of ecological processes and of the interactions between the different components of the ecosystem (e.g. soil, microorganisms, plants). Many processes are not sufficiently well understood to model them based on first principles (e.g. plant and microbial competition for nitrogen in a CO_2-enhanced environment, Houghton et al., 1998; plant resource allocation, Hurtt et al., 1998). Hence fundamental assumptions have to be made to model these processes. However, several widely accepted assumptions need challenging: for instance, the assumption that the long term leaf photosynthesis responses will be comparable to initial responses and that increase CO_2 will induce increase LAI in the long term; the assumption that increase temperatures will induce increase tissue respiratory rates, or that the majority of mineralised nitrogen is taken up by plants and will lead to an increase terrestrial carbon storage (Körner, 1996; Houghton et al., 1998). The underlying assumptions in the models must therefore be scrutinised before extrapolating their results to future scenarios.

Another issue in ecological modelling is the difficulty to scale from the physiological performance of cells or leaves to the long term performance of whole plants, plant populations and whole ecosystems (Hurtt et al., 1998). Scaling up implies that we must understand how information is transferred from the fine scales to the broad scales (Levin, 1992). For instance, we need to know how to extrapolate the dynamics of individual plants to the entire terrestrial biosphere. The solution is, however, probably not to simply multiply stand results by an appropriate area-based weighting factor. Some studies have shown that the responses of plants to increase CO_2 differ whether we consider individual species or plant communities (i.e. biome) ; the variations found at the species level being by far larger than the variations at the ecosystem level (Körner, 1996). These results highlight the need to account for the effects of biodiversity in the response of ecosystems to environmental changes (Hurtt et al., 1998) and indicate that a whole ecosystem approach is, in some cases, more appropriate than the individual-based approach (Tian et al., 1998).

The question is to know how much detail should be included in the ecological models to account for the complexity of the ecological systems, and, at the same time, how much detail can be ignored without producing results that contradict observations (Levin, 1992). One trend in the development of coupled biosphere model is towards increasing physiological details of the models at small scales (Sellers et al., 1997). At the same time, there has been a tendency in ecological studies to adopt a reductionist approach, that is, to reduce the complexity of the ecological systems by focusing on subsystems (Brown, 1994). However, it has been argued that the breakdown of system complexity is unpractical and is unlikely to lead to more accurate predictions of the overall system response (Odum, 1997; Körner, 1996). For instance, predicting plant growth by simulating the photosynthetic processes at the tissue level is an approach which does not take into account the other growth determinants, such as plant interactions, plant-animal interactions, plant age, and, therefore, does not allow a reliable prediction of ecosystems responses to enhanced CO_2 and temperature (Körner, 1996). It seems that one way to progress is by formulating models at intermediate scales of biological detail (Hurtt et al., 1998) and to a develop more holistic approaches that confront the system complexity directly (Brown, 1994).

Finally, a critical limitation in the current biosphere models' predictions is due to the difficulty to account for changes in land use (i.e. the management regime humans impose on a site such as plantations or agroforestry; Dale, 1997) and the impacts of ecological disturbances (e.g. insect outbreaks, fire, savanna grazing). The prediction and characterisation of land use and ecological disturbances represent, indeed, crucial but challenging areas of research. The prediction of land use is particularly critical in the evaluation of the role played by terrestrial ecosystems in the global carbon balance. If the terrestrial ecosystems, and in particular the forests, can serve as carbon sinks, the spatial extents of these forests, and the proportion of forests cleared should be known (Houghton et al., 1998).

The next section will focus on the modelling of the interactions between the terrestrial ecosystems and the climate system. As noted in the introduction, there has been a dramatic increase in the development and use of vegetation models in the last decades to study the biosphere–atmosphere feedbacks. The next section presents an overview of the numerical experiments undertaken to simulate those feedbacks, and describes the different approaches employed to model interactively the biosphere and the climate system.

8.3 Modelling Biosphere–Atmosphere Interactions

8.3.1 Biosphere–Atmosphere Feedbacks: Overview of Numerical Experiments

The first attempt to demonstrate the existence of biogeophysical feedbacks following land cover change was made by Charney (1975). Charney (1975) suggested that Sahel desertification could be explained by a feedback mechanism in which decreasing vegetation cover is leading to a decrease of precipitation, hence allowing the desert to feed back upon itself. Since Charney's seminal work on Sahel desertification, the mechanisms of biogeophysical feedbacks have been investigated in a large number of numerical experiments. A wide range of land cover changes have been considered: desertification (Laval and Picon, 1986; Xue and Shukla, 1993), tropical deforestation (Dickinson and Henderson-Sellers, 1988; Polcher and Laval, 1994), boreal deforestation (Bonan et al., 1992; Thomas and Rowntree, 1992; Douville and Royer, 1997), vegetation change in response to global warming (Betts et al., 1997), and vegetation change in response to past climatic regimes, such as the Holocene Epoch (Foley et al., 1994), or the last glaciation (De Noblet et al., 1996).

The biogeophysical feedbacks simulated by the climate models are dominated, in many experiments, by the modification of the surface albedo resulting from the land cover change. The effect of surface albedo change on the simulated global climate is particularly pronounced when this change occurs in boreal regions. Several numerical experiments have investigated the influence of vegetation masking on snow albedo and its effects on regional and global climate . (Thomas and Rowntree, 1992; Bonan et al., 1992; Douville and Royer, 1997). The primary perturbation in these experiments is to the surface energy balance. Results show that the presence of boreal forests, by decreasing land surface albedo in the winter and therefore changing the surface energy balance, significantly warms both winter and summer temperatures of the northern hemisphere and increases latent heat flux and atmospheric moisture (Bonan et al., 1992). Conversely, climate model experiments indicate that the boreal forest removal leads to a surface cooling of the mid and high latitudes of the northern hemisphere (Douville and Royer, 1997), a decrease of latent heat flux and a systematic decrease of precipitation in the deforested areas (Thomas and Rowntree, 1992). The presence or absence of boreal forests seem to affect the atmospheric circulation well beyond the high and middle latitudes (Thomas and Rowntree, 1992; Douville and Royer, 1997).

The investigation of possible interactions between past climates and vegetation cover suggests that large feedbacks between climate and boreal forests may have taken place. Foley et al. (1994) conducted a numerical experiment to evaluate the relative effects of orbitally-induced insolation variations and of the northward extension of boreal forests on the mid-Holocene climate. Foley et al. (1994) found that vegetation forcing lead to a significant temperature increase (about 1.8°C annual average) in the high latitudes in addition to the warming caused by change in orbital variations. A numerical experiment undertaken by De Noblet et al. (1996) indicated that orbital forcing alone was not sufficient to initiate a glaciation in their climate model. Their results suggested that the southward migration of boreal forests, by creating favourable conditions for continental ice-sheet growth, might have played a role in triggering the last glaciation.

The other type of vegetation–atmosphere feedbacks which has been widely investigated in numerical models is tropical deforestation. Tropical deforestation experiments have been carried out since the early 1980s. The key surface parameters in the deforestation experiments are the surface albedo and the roughness length (Sellers, 1992). The model simulations have mostly focussed on the world's largest rain forest, the Amazon tropical forest. The coupling between convection, large-scale atmospheric dynamics and land processes is very strong in this region (Zeng, 1998), and the complexity of the land–atmosphere interactions is illustrated by the substantial differences in the responses of the models to the vegetation forcing, particularly with regard to precipitation and moisture convergence (Polcher and Laval, 1994; Hahmann and Dickinson, 1997). Most studies find, however, that the large scale removal of the Amazon tropical forest seems to lead to warmer and drier climate conditions in this region (Hahmann and Dickinson, 1997). The key processes determining the response of the climate system to tropical deforestation are not yet fully identified and understood (Zeng, 1998) and several questions regarding the impact of deforestation on the global climate and the role played by the ocean and cloud feedbacks remained unanswered (Hahmann and Dickinson, 1997).

The prospect of global warming due to increase greenhouse gases and its potential impact on land cover have been studied in numerical experiments since the early 1990s. The numerical experiments investigating the relationships between vegetation feedbacks and CO_2-induced climate change have demonstrated that vegetation feedbacks are potentially significant and must be included in future climate change assessments. These studies show that the nature of vegetation feedbacks span a wide range of processes, including biophysical, biogeochemical, physiological, structural and ecological (Neilson and Drapek, 1998). Early greenhouse climate simulations (e.g. Henderson-Sellers and McGuffie, 1994) focussed on the biophysical feedback mechanisms between the vegetation and the atmosphere, ignoring the effects of altered plant physiology resulting from the increase of atmospheric CO_2. In these preliminary studies, the vegetation distribution was modified according to CO_2-induced changed climate conditions and the resulting new vegetation characteristics (rooting factors, stomatal resistances, canopy albedo, roughness length) were imposed as boundary conditions to the climate model. The impact of the changed vegetation cover on global scale temperature appeared to be small, however some more significant differences in temperature were found at the regional scale (i.e. in some high latitude regions, in the Tibetan Plateau and the Arabian Gulf) (Henderson-Sellers and McGuffie, 1994). The shift from short to taller vegetation in these regions and the resulting decrease in surface albedo could be the cause of these regional temperature changes.

The physiological effect of increase atmospheric CO_2 on plant metabolism and its impact on climate has later been considered (Henderson-Sellers et al., 1995; Pollard and Thompson, 1995; Sellers et al., 1996a; Betts et al., 1997; Levis et al., 1999; Cox et al., 1999). The current state of research shows that the short-term exposure of most plants (i.e. C_3) to increase atmospheric CO_2 leads to a stimulation of photosynthesis and a decrease of stomatal conductance (Bounoua et al., 1999). The long-term responses of natural ecosystems to increase atmospheric CO_2 and climate change is still, however, uncertain (Houghton et al., 1998; Hurtt et al., 1998). The potential effect of reduced stomatal conductance on climate was first investigated by Henderson-Sellers et al. (1995) and Pollard and Thompson (1995). In these experiments, the stomatal conductance was simply decreased by 50%. In more recent experiments, the relationship between photosynthesis and stomatal conductance was, however, explicitly accounted for (Sellers et al., 1996a; Cox et al., 1999; Levis et al., 1999). The results show that the combined effect of increase CO_2 and stomatal resistance could lead to regional decrease in evapotranspiration and increase in air temperatures, either in boreal regions (Henderson-Sellers et al., 1995), or in tropical regions (Sellers et al., 1996a). The physiological response of terrestrial vegetation to increase CO_2 seem to amplify the changes resulting from atmospheric radiative effects alone. One study (Betts et al., 1997) also considered the changes in vegetation structure (i.e. density of vegetation cover, leaf display and LAI) resulting from increase atmospheric CO_2 and global warming. In this study, the changed environmental conditions lead to a widespread increase in plant productivity and water use efficiency. The structural change (i.e. increase LAI) is found to partially offset the effects of the physiological changes (i.e. reductions in stomatal conductance) in the long term. However it should be noted that this study does not account for the dynamical changes in both vegetation physiology and structure. At present, it is still unclear whether global warming will lead to an accumulation (negative feedback) or a loss of terrestrial carbon (positive feedback) (Houghton et al., 1998), and, overall, the uncertainties in the magnitude of the biological feedbacks on climate remain large (Hurtt et al., 1998).

Tropical deforestation, desertification, boreal deforestation or vegetation change following global warming represent large-scale alterations of the land cover to which the climate system is likely to be sensitive. The sensitivity of the climate system to less radical alterations of the vegetation cover has also been investigated. Chase et al. (1996) have evaluated the sensitivity of the climate system to the geographical and temporal distribution of LAI and fractional vegetation cover. In their experiment, the maximum values of LAI prescribed in the GCM were allowed to vary. They found that large differences in the climate simulations occurred in the high northern latitude in January despite the fact that LAI differences were non-existent in these regions at this time of year. Chase et al. (1996) hypothesised that LAI forcing in the tropics could have long ranging effects in the winter of the northern hemisphere. A study undertaken by Xue et al. (1996a) also demonstrates that the accurate prescription of vegetation parameters is critical to assess the climatic impact of land surface degradation. Xue et al. (1996a) validated their land surface scheme using data from the Anglo Brazilian Amazonian Climate Observation Study (ABRACOS) over a forest clearing site. Compared to previous deforestation experiments, the authors found that the new vegetation dataset produces significantly different latent heat flux (i.e. 25 W m^{-2} lower) and surface temperatures (i.e 2 K higher) in GCM simulations and they found that one of the most important parameters was LAI. The importance of the accurate representation of the seasonality of fractional vegetation cover, LAI and associated canopy albedo for the climate simulation was also stressed by Xue et al. (1996b). Xue et al. (1996b) found that the simulated positive temperature bias in central US was caused by the inadequate representation of the vegetation canopy properties (i.e. vegetation cover, leaf area index and green leaf fraction) and in particular the inadequate prescription of the seasonality of crop in central US. Xue et al. (1996b) modified the seasonality of vegetation canopy properties and the soil properties and found that the major GCM biases (positive surface temperature biases in summer and negative precipitation biases) were reduced. These studies demonstrate that different prescriptions of the geographical and temporal distribution of vegetation canopy properties can have a significant impact on the GCM output.

The numerical experiments described in this section have demonstrated that vegetation-atmosphere interactions are likely to have a significant impact on climate. However, these studies show that if the potential feedbacks on climate are large, the uncertainties in the direction and magnitude of those feedbacks are also large. Predicting the biophysical and biogeochemical feedbacks on climate depends critically on simulating the relevant ecological processes. Hence one of the prerequisites to conduct these experiments is to employ a realistic, fully interactive, and integrated, biosphere model.

8.3.2 Different Approaches Toward the Development of an Interactive Biosphere Model

One way coupling and asynchronous coupling. The approach traditionally employed to simulate land cover changes in numerical experiments consists of imposing once a new state of vegetation and land surface as boundary conditions to the climate model. The vegetation characteristics most commonly altered are albedo, roughness length, stomatal resistance, leaf area index and fractional vegetation cover. The aim is to evaluate the sensitivity of the climate system to a single perturbation of the land surface. Another approach consists of coupling iteratively the climate model to an ecosystem model which simulates the equilibrium state of potential vegetation (i.e. "biome" models, see Sect. 8.2). In this type of experiment (Henderson-Sellers, 1993; Claussen, 1994; Betts et al., 1997), the models are coupled asynchronously, that is, the models do not communicate at every time step, but instead they communicate with each other at the end of an integration period of varying lengths, typically 10 model years (Claussen, 1993; Ciret and Henderson-Sellers, 1995). The iterative coupling between the climate model and the ecosystem model allows each model to provide new boundary conditions to the other, until an equilibrium state is reached. One outcome of this type of experiment is to evaluate the sensitivity of the climate models to sudden and asynchronous changes of the boundary conditions and to assess the effects of those changes on the stability of the climate model. These experiments also produce valuable information about the potential feedbacks of vegetation in the genesis of paleoclimates (e.g. Claussen and Gayler, 1997). Although these experiments contribute to our understanding of the sensitivity of the climate system to vegetation feedbacks, the type of approach has one serious limitation, the lack of representation of vegetation dynamics. Vegetation is, indeed, a dynamical system that is continuously responding to variations of the environment and consequently alternative modelling approaches are required to simulate the full range of vegetation-atmosphere feedbacks.

Dynamic biosphere model. Two temporal scales of vegetation dynamics can be distinguished for modelling purposes: long timescale (decades to centuries) and short timescale (hours to years). Representing the long timescale dynamics of vegetation consists of simulating the transient changes (e.g. migration, invasion, dieback) which affect vegetation composition and structure in response to changed environmental conditions. Representing the short timescale dynamics of vegetation characteristics consists of simulating the diurnal exchanges of water, energy and carbon between vegetation and atmosphere, the seasonal variations of vegetation in relation to plant phenological phases (e.g. leaf emergence, growth, and shedding), and the interannual fluctuations of vegetation in response to short-term climate variability (e.g. El-Nino events) and ecological disturbances (e.g. fire).

Developing an interactive biosphere model which fully represents the different temporal scales of vegetation dynamics is a challenging task, and this activity is still in its infancy. A fundamental problem in linking climate and vegetation models is to choose the appropriate scales, both temporal and spatial (Aber, 1992), and to understand how information is transferred across those scales (Levin, 1992). The mismatch in spatial scales is a particularly critical issue, since general circulation models operate on spatial scales many order of magnitude greater than the scale at which most ecological studies are carried out. Another issue to consider is the accuracy of the climate simulation if the vegetation model is used in a prognostic mode, that is, if the vegetation distribution is predicted and not prescribed (Ciret and Henderson-Sellers, 1997). Although vegetation models are usually validated using observed climate variables, these models need to be also carefully tested using GCM output. Biases in the GCM variables may compensate each other and lead to the spurious simulation of apparently realistic vegetation cover (Ciret and Henderson-Sellers, 1998). The compatibility between vegetation and climate models and the need for an interface are also critical issues which need to be accounted for. Many vegetation models are developed independently from the climate model to which they will be coupled. Vegetation models require information about soil moisture, and often this variable is computed by the vegetation model separately. If a critical environmental variable such as soil moisture is estimated in a different manner in the two models, the representation of vegetation–atmosphere interactions may contain physical inconsistencies. Finally, the interactive biosphere model needs to be applied on a large enough scale to allow feedback processes to take place at a scale relevant for climate models. Hence the interactive biosphere model needs to be sufficiently robust to be applied to various regions of the globe.

One approach to model coupling consists of improving the representation of vegetation dynamics in the existing coupled SVATs-climate models (i.e. "greening" of SVATs, see Sect. 8.2). As noted in Sect. 8.2, several SVATs are now accounting for the effect of atmospheric CO_2 on plant stomatal aperture (Bonan, 1995; Sellers et al., 1996c). A few land surface schemes have included a plant growth model to represent the seasonal variations of vegetation density and leaf display (i.e. Calvet et al, 1998; Dickinson et al., 1998 and see also Sect. 8.4). Calvet et al. (1998) have modified the land surface scheme ISBA in order to describe, in a more realistic way, the canopy stomatal conductance and to allow a climate-derived prediction of leaf area index. The new version of ISBA simulates the leaf net assimilation of CO_2 and leaf conductance and, in addition, estimates the plant growth and mortality, hence allowing the LAI to be predicted consistently with the simulated climate and CO_2 concentrations. Dickinson et al. (1998) have also improved the representation of vegetation in the land surface scheme BATS by allowing the leaves stomatal conductance to be consistent with the leaves assimilation of CO_2 and by simulating the growth and death of the green foliage and LAI. In addition, BATS includes simplified representations of plant phenology and soil carbon cycles.

Another approach to biosphere–atmosphere model coupling consists of developing so-called Dynamic Global Vegetation Models (DGVMs). The aim of a DGVM is to simulate, on a global scale, both the short term and long term responses of vegetation to changes in climate, atmospheric CO_2 concentrations and soil nutrients. Several research groups around the world are actively involved in the development of DGVMs. The design of a DGVM for use with global climate models has been a key focus of the Global Change and Terrestrial Ecosystems (GCTE) of the IGBP for many years (Steffen et al., 1992). Initially, several approaches were considered, the "bottom-up" approach, that is, the scaling up of "stand" models, and the "top-down" approach, that is, using "biome" models in which the vegetation types would be assigned various parameters relating to rates of vegetation change (e.g. growth, mortality and dispersal) (Smith et al., 1993). A DGVM was developed at the University of Sheffield in collaboration with other research groups (Woodward, 1996). The Sheffield DGVM (SDGVM) is designed to process the short term and long term vegetation dynamics in two separate submodels. The first submodel, described in Woodward et al. (1995), simulates the short term vegetation response to climate changes, CO_2 concentration and soil nutrient and predicts global scale distributions of LAI and NPP for natural vegetation types. The vegetation types are grouped into the three major life-forms, that is, trees, shrubs and grasses. The processes of photosynthesis and the stomatal functioning are simulated in a mechanistic manner. The objective of the second submodel is to used the predictions of LAI and NPP to determine the potential changes in structure of the three major life-forms (Woodward, 1996). The rate of change of the three life-forms is simulated by a matrix compartment model (Woodward and Lee, 1995). Each cell contains a distribution of life-forms and a percentage of bare ground area and can be affected by ecological disturbances. One important feature of the SDGVM is that no vegetation initialisation is required. Hence the current vegetation data and remote-sensed vegetation indexes have not been already used to calibrate the model and therefore can be used with confidence to validate the model. The first submodel was tested using data from a catchment of boreal vegetation which has been exposed to elevated CO_2 and temperature (Beerling et al., 1997). In that study, the SDGVM was coupled to the biogeochemistry-decomposition model CENTURY (Parton et al., 1988) in order to account for the effects of soil nutrients on NPP.

The Climate, People and Environmental Program (CPEP), from the University of Wisconsin, is another research group which has developed a DGVM. Their DGVM called Integrated BIosphere Simulator (IBIS1.1) is a model in which land surface processes, plant physiology, biogeochemistry and biogeography are combined in a same framework (Foley et al., 1996). One interesting feature of IBIS is that it was designed to be coupled directly to a GCM. The treatment of land surface processes and ecophysiological processes is entirely carried out by IBIS1.1, and the only processes that the GCM are simulating are the physics and general circulation of the atmosphere. IBIS1.1 has a modular structure: it contains a land surface module which simulates the energy, water, carbon and momentum balance of the soil–vegetation–atmosphere interface with a 30 minutes timestep (Foley et al., 1998); a vegetation phenology module; a carbon balance module which estimates the annual carbon balance after simulating the gross photosynthesis, maintenance and growth respiration of each of the nine functional types; and finally a vegetation competition module which determines the changes in vegetation structure. Many of IBIS1.1. features and algorithms need improvement: the phenology module uses very simple temperature and productivity criteria to control leaf display in plants ; the processes of vegetation change are also very simplified and the ecological disturbances and interannual climatic variations are not taken into account (Foley et al., 1996; Foley et al., 1998). To date, however, IBIS1.1. is the only DGVM which has been fully coupled with a GCM (Foley et al., 1998; Levis et al., 1999) and this model is currently probably the most advanced and integrated DGVM.

Improving the representation of vegetation dynamics in SVATs is an important step toward the implementation of a fully dynamic biosphere-climate model. However, this approach ignores the long timescale of vegetation dynamics: each land grid cell is assigned one or several vegetation types and the changes in vegetation structure resulting from climate and environmental changes are, therefore, not predicted. The development of a DGVM represents a more ambitious approach. The question is, however, to know which degree of confidence we can have in current DGVMs' predictions, considering the uncertainties in our ability to simulate various ecological processes, such as vegetation phenology, changes in vegetation structure, impacts of ecological disturbances, and long term physiological responses of the system plant-soil to variations of, for instance, atmospheric CO_2. In addition, a major issue with DGVMs is the difficulty to validate this type of model (Woodward and Beerling, 1997). Hence, at present, the "greening" of SVATs can be seen as a pragmatic way to increase our understanding of vegetation-atmosphere interactions. An example of SVAT "greening" is given in the next section.

8.4 Simulation of an Interactive Vegetation Canopy in a Climate Model: Example of the Plant Production and Phenology (PPP) Model

8.4.1 Description of the Plant Production and Phenology Model

The aim of this section is to present an approach to simulate the seasonal variations of vegetation cover, including those due to ecological perturbations, in coupled SVAT-climate models. As discussed above, there have been, to date, very few attempts to simulate the plant phenology in the current coupled SVAT-climate models, and the representation of ecological disturbances such as fires has been usually neglected. A plant production and phenology model (hereinafter referred to as the PPP model) was developed and tested using the LMD (the Laboratoire de Météorologie Dynamique) GCM output. The objective of the PPP model is initially to allow the general circulation model LMD and land surface scheme SECHIBA (Ducoudré et al., 1993) to update daily LAI values in the regions covered by tropical savannas. The choice of the modelling approach is motivated by the desire to employ a simple model that contains a small number of parameters and is sufficiently robust to be applied to various regions without major modifications.

8.4 Simulation of an Interactive Vegetation Canopy in a Climate Model

The PPP model, represented schematically in Fig. 8.2 is fully described in Ciret et al. (1999). The PPP model has evolved from the savanna function model PEPSEE–Grass developed by Le Roux (2000a). PEPSEE–Grass was developed to simulate the primary productivity and phenology of savanna and the site chosen for its implementation, calibration and testing was the Lamto Scientific Station, Ivory Coast, Western Africa (Le Roux, 2000b).

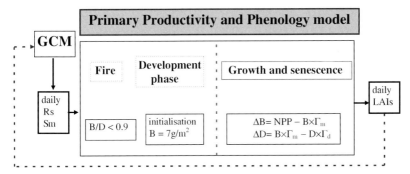

Fig. 8.2. Schematic representation of the Plant Prodection and Phenology (PPP) model. R_s is incoming shortwave radiation received at the surface, S_m is the soil moisture, B is the green biomass, D is the dead biomass, Γ_m is the mortality rate, Γ_d is the rate of dead biomass disappearance, NPP is the net primary productivity, LAIs are the green and dead leaf area indexes(Ciret et al., 1999).

Simulation of plant production. The modeling of primary production is based on the parametric model proposed by Monteith (1972). Monteith (1972) have demonstrated that there is a relationship between the net primary productivity of agricultural crops and the photosynthetically active radiation (PAR) absorbed by plants (i.e., APAR). This relationship has been shown to also hold for forests (Jarvis and Leverenz, 1983). It is thus possible to predict the vegetation production by evaluating the biomass/energy quotient, or "efficiency", of the conversion of APAR to biomass:

$$\text{NPP} = \int_t \varepsilon \, FPAR \, \varepsilon_c \, R_s \, dt \tag{8.1}$$

where NPP is the net aboveground primary productivity (g m^{-2}); R_s is the total downward solar radiation (MJ d^{-1}m^{-2}); ε_c is the climatic efficiency (i.e., PAR/R$_s$), ε is the conversion efficiency or dry aboveground matter yield of APAR (i.e., NPP/APAR) (g MJ^{-1} APAR); and, finally, FPAR, called absorption efficiency, is the fraction of incident PAR absorbed by the canopy (i.e., APAR/PAR).

Since about 45% of the incoming solar radiation is in the PAR wave band (i.e., 0.4–0.7 μm) (Larcher, 1995), the climatic efficiency ε_c usually ranges from 0.45 to 0.50. At Lamto, ε_c can be considered as a constant and is equal to 0.48 (Le Roux, 1995). The absorption efficiency FPAR is related to the green LAI as follows:

$$\text{FPAR} = 1 - e^{-k \times LAI} \tag{8.2}$$

where k is the canopy extinction coefficient and is equal to 0.48 according to Le Roux and Gauthier (1997).

In the parametric model applied by Monteith (1972) to crops from temperate regions, the conversion efficiency ε was presented as a conservative quantity. Several studies provide some physiological interpretation of the conservative nature of ε (e.g., Dewar, 1996). However, empirical values of ε for natural ecosystems are scarce (Ruimy et al., 1994), and there are some uncertainties about the values found in the literature because of differences in the method employed to calculate ε and errors in the estimation of plant production (Hanan et al., 1995). Reviews (e.g., Prince, 1991) show that although ε is far from being a constant, its values occupy a relatively narrow range. For natural ecosystems the conversion efficiency of APAR into aboveground dry matter (DM) is estimated to range from about 1 to 2 g DM MJ^{-1} APAR (see, e.g., Landsber et al., 1996). According to observations made at Lamto (Le Roux et al., 1997), the maximum value (i.e., unstressed value) of the aboveground conversion efficiency (ε_{max}) is equal to 1.15 g DM MJ^{-1} APAR. In the PPP model the conversion efficiency ε is parameterised to account for the effect of water stress on plant production.

Simulation of plant phenology. In savanna regions the vegetative cycle of grass starts at the end of the dry season, after the possible passage of fire. The transfer of organic matter from the roots to the shoots after the passage of fire is not simulated. Instead, a certain amount (7 g m^{-2}) of aboveground biomass is allowed to remain immediately after the occurrence of fire. The daily variation of aboveground green biomass B and dead biomass D (g m^{-2}) is simulated as

$$\Delta B = \text{NPP} - \Gamma_M B \tag{8.3}$$

$$\Delta D = \Gamma_M B - \Gamma_d D \tag{8.4}$$

where Γ_M (g biomass g^{-1} biomass d^{-1}) is the rate of biomass mortality and Γ_d (g dead biomass g^{-1} dead biomass d^{-1}) is the rate of dead biomass disappearance rate through decomposition, grazing, and allocation to living parts. Γ_M is a function of soil water content and Γ_d is equal to a constant (i.e. 0.015 g dead biomass g^{-1} dead biomass d^{-1} as estimated from in situ measurements by Le Roux, 2000a, for Lamto savanna).

Green and dead leaf areas (LAI$_g$ and LAI$_d$) are calculated according to the seasonal courses of green and dead specific leaf areas measured by Le Roux (2000a) for the Lamto grass layer.

$$\text{LAI}_g = [128 - 62(1 - e^{(0.0102\,B)}]B \times 10^{-4} \tag{8.5}$$

$$\text{LAI}_d = 144\,N \times 10^{-4} \tag{8.6}$$

8.4 Simulation of an Interactive Vegetation Canopy in a Climate Model

Simulation of fire occurrence. Fire is an essential component of savanna ecosystems. An estimated 8680 Tg DM of biomass is consumed annually and the burning of savanna grassland represents 42% of this amount (Levine, 1991). The majority of savanna fires are set by humans. Burning is usually used to control pests, weeds, tree seedlings, and litter accumulation in order to prepare the fields for cultivation or for grazing. Important parameters to determine fire intensity and frequency are fuel abundance and curing, wind speed, relative humidity, and air temperatures (Tapper et al., 1993). The task of the PPP model is not, however, to simulate the detail of fire regimes, such as fire intensity, spread, and patterns, but, instead, to predict when the conditions are considered favourable for fire occurrence. Simulating the occurrence of fire as a function of climatic conditions is, indeed, a difficult task when dealing with simulated climate variables. Given the fact that green and dead biomass B and D integrate the effects of environmental conditions throughout the year, it is proposed to use these biomass variables to simulate fire occurrences.

Observation of green and dead biomass at Lamto during a 7-year period indicates that the ratio of green biomass to dead biomass is about 0.9 at the date of fire. Hence it can be assumed that fire occurrence can be predicted (see Fig. 8.3) whenever the ratio B/D is lower than 0.9. Since the influence of human management on fire is clearly beyond the scope of this study, it is assumed here that, whatever the causes of fire ignition (anthropogenic or natural), fire takes place as soon as $B/D < 0.9$.

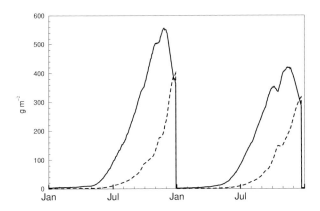

Fig. 8.3. Seasons variation of green biomass B (bold line) and dead biomass D (dashed line) simulated during 2 consecutive years in the African grid cell. The occurrence of fire is simulated when $B/D < 0.9$ and the biomass is reduced to a very small value (i.e. $7\,\mathrm{gm}^2$).

8.4.2 Application of the PPP Model

The performance of the PPP model was assessed by comparing the simulated biomass with observed values. The aim was to evaluate the realism of the seasonal variations of the simulated biomass in the two regions of savanna selected to test the PPP model. These two regions are the region of Lamto, Ivory Coast and the region of Victoria River District, North Australia. Biomass and climatological data are available for both sites. The campaign of biomass measurement in Lamto was conducted from 1981 to 1993 (Le Roux, 2000b). In Victoria River District, the period of field measurements took place in 1993–96 (Dyer et al., 1997). Lamto is classified as a region of humid savanna, whereas the Victoria River district is classified as a region of semiarid rangeland. A comparison of water index, growth index, and temperature constraints between Lamto savanna and the " Monsoon Tallgrass" savanna of northern Australia (Mott et al., 1984) indicates that the region of Lamto is equivalent to the wettest areas of the Australian monsoon region. Australian soils are particularly deficient in phosphorus and are generally of lower fertility than African soils, and as a result, the primary productivity is lower in Australian savannas than African savannas (Braithwaite, 1991). Climate and vegetation characteristics differ markedly in these two regions, offering adequate conditions to test the robustness of the PPP model. Several LMD GCM grid cells corresponding to the two selected regions were chosen. The north Australian sites cover three grid cells of the GCM, whereas one grid cell is sufficient to cover the west African site.

The PPP model was forced with 11-year time series of climate variables simulated by version 6 of the LMD GCM. The 10-year time series of simulated LAI over the grid cell Lamto is displayed in Fig. 8.4a. Fig. 8.4b represents these time series averaged over the last 10 years of the simulation. The simulated LAI exibits a strong seasonal variability, enhanced by the passage of fire. The PPP model simulates fire occurrence in the grid cell Lamto in December or early January, with a frequency of one year (Fig. 8.4a). These results are realistic since savanna fires in the region of Lamto occur every year, generally in January. The LAI maxima simulated by the PPP model using LMD6 GCM climate data range between 2.0 and 4.0, with an average of 2.8. The LAI maxima observed during the 4 years of biomass measurements range between 2.2 and about 4.0, and the average is 2.9. Hence, as far as LAI amplitude is concerned, the PPP model forced with simulated climate generates realistic results. However, the amount of simulated biomass and LAI is insufficient during the first part of the year. As shown in Fig. 8.4b, the start of the growing season is delayed by about 1 to 2 months in comparison to observations.

The results of the application of the PPP model on the Northern Australian sites are shown in Figs. 8.5 and 8.6. The values of simulated phytomass in the three grid cells remain approximatively within the range of observed yields (Fig. 8.5) and, as in the Lamto grid cell, the interannual variability of

8.4 Simulation of an Interactive Vegetation Canopy in a Climate Model 341

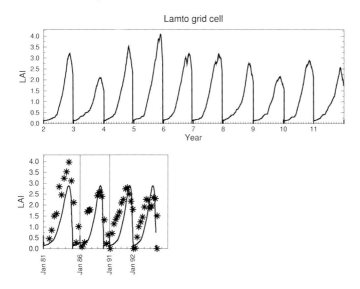

Fig. 8.4. (a) Top panel: Time series of green LAI simulated by the PPP model in the grid cell Lamto using LMD6 GCM climate variables. (b) Bottom panel: Green LAI averaged over the last 10 years of the simulation (line) and LAI observed at Lamto in 1981, 1986, 1991 and 1992 (stars). (Note that the averaged green LAI represented by the bold line is plotted twice to allow comparison with observations in 1991 and 1992).

predicted LAI is relatively high (Fig. 8.6). In contrast with Lamto, however, there is no lag in the simulated growth at the beginning of the rainy season. In the three North Australian grid cells, fire occurence is predicted to occur every two to three years, during winter (i.e. from June to September) (Fig. 8.6). In both regions, the simulation of fire occurence is consistent with the observations of fire frequency and timing. The frequency of fire is, indeed, generally high in regions of savanna with high productivity (i.e., humid or mesic savanna) and decreases in drier savanna with lower fuel availability (Menaut et al., 1991). The results show that the predicted fire frequency is approximatively proportional to the amount of biomass, that is, high fire frequency in the region of Lamto, and low fire frequency in Northern Australia.

The PPP model represents a promising tool to investigate the biophysical feedback mechanisms, including those due to fire disturbances, in tropical regions. In particular, the model can be used to evaluate the importance of vegetation feedbacks during specific climatic events, such as El-Nino-related droughts.

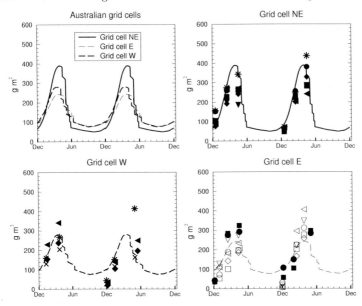

Fig. 8.5. Top left panel: simulated phytomass (i.e., green plus dead biomass) averaged over the last 10 years of the simulations in the 3 Australian grid cess. Other panels: Comparison of simulated phytomass (line) with yield of standing phytomass plus litter measured at different sites (symbols) (Dyer et al., 1997). (Note each symbol type corresponds to one set of biomass measurements and the simulated average phytomass is pollted twice to allow a comparison with the time series of observations).

8.5 Summary

As shown in this chapter, there are many different types of ecological models currently used to simulate the terrestrial ecosystems (see Table 8.1). The growing knowledge that the interactions between the terrestrial biosphere and the global environment are critical has lead, however, to a recent trend towards an integrative approach to simulate the terrestrial biosphere. As a result, there has been a dramatic increase in the development and use of so-called "biosphere" models. The interactions between the terrestrial ecosystems and the climate system have received, in the last decade, considerable attention and it was decided in this chapter to focus on this area of research. This research area represents a whole new field in ecological modelling, and has been described as "squarely between ecology and meteorology/climatology" (Field and Avissar, 1998). The implementation of the biosphere models depend on interdisciplinary approaches since these models must include some elements of plant physiology, biogeochemistry, biogeography, and must also simulate the short term (seasonal) and the long term (decadal) dynamics of vegetation. A number of issues need to be addressed in

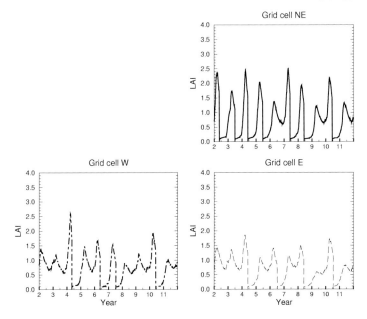

Fig. 8.6. Time series of green LAI (10 years) simulated by the PPP model in the 3 Australian grid cells using LMD6 GCM climate variables.

the development of these models. In particular, it is important to overcome the problem of scaling and to determine which are the relevant ecological and land surface processes. In addition, it is critical to find some ways of testing and validating the global scale predictions of these models.

Since the biosphere models must be linked to a climate model, it seems essential to think about the different methods of coupling at the time the biosphere models are being developed. That implies that a close collaboration with the climate modelling community is required. The role of the ocean needs also to be accounted for in the modelling of interactions between the biosphere and the global environment. Experiments which include fully dynamic ocean model show different biophysical feedbacks, for instance a positive feedback in the case of boreal deforestation (Bonan et al., 1992), or different biosphere responses in enhanced greenhouse gases experiments (Neilson and Drapek, 1998).

Simulating the full range of interactions between the terrestrial biosphere and the other components of the Earth system is a very challenging task, and one which will require the development of many types of models and many types of coupling. Although this research area is still in its infancy, considerable progress has been made in the last recent years and we are starting now to understand better the complex biosphere and climate responses to multiple physical, biophysical and biogeochemical feedbacks.

Table 8.1. Frequently used acronyms in ecological modelling.

APAR	Absorbed photosynthetically active radiation
AVHRR	Advance Very High Resolution Radiometre
BATS	Biosphere–Atmosphere Transfer Scheme
BGC	Biogeochemical Cycles
CPEP	Climate, People and Environmental Programm
DGVM	Dynamic Global Vegetation Model
FPAR	Absorption efficienty
GCM	General Circulation Model
GCTE	Global Change and Terrestrial Ecosystems
IBIS	Integrated BIosphere Simulator
IGBP	International Geosphere-Biosphere Program
ISBA	Interactions Soil Biosphere Atmosphere
LAI	Leaf Area Index
LMD	Laboratoire de Météorologie Dynamique
NDVI	Normalised Difference Vegetation Index
NPP	Net primary Productivity
SECHIBA	Schématisation des Echanges Hydriques à l'Interface entre la Biosphre et l'Atmosphère
SVAT	Soil–Vegetation–Atmosphere Transfer scheme
SVI	Spectral Vegetation Index
PPP	Plant Production and Phenology model

9. Theoretical Basis of Biological Models in Environmental Simulation

Yunhu Tan

9.1 Introduction

Biological reactions govern many environmental processes. Models of environmental processes must therefore incorporate mathematical descriptions of biological processes as well as physical and chemical processes. The most widely used model for the description of enzyme and/or bacteria catalysed reactions, cell growth, and prey-predator relation is the so-called Michaelis-Menton or Monod equation. Although more complex models exist for the description of those processes, the Monod-type equations are the predominant equations used to describe enzyme and/or bacteria catalysed reactions, growth kinetics and prey-predator relationship in the environment. This is partially due to low substrate concentrations, and slow growth and reaction rates, under which conditions, the Monod-type equations perform satisfactorily, therefore not warranting added complexity. In addition, the Monod-type equations can be readily incorporated into the macroscopic transport equations that are often used to describe environmental processes.

The Monod-type equations have been used in a number of different fields. For example, it has been used in describing reactions that are catalysed by microorganisms and/or enzymes, carbon and nutrient diagenesis in both freshwater and ocean systems (e.g., Boudreau, 1996), degradation of organics and other pollutants in the environment (Borden and Bedient, 1986; Chen et al., 1992; Tan and Bond, 1994), microbially catalysed oxidation of pyrite (Liu et al., 1987; Tan, 1996), and photosynthetic CO_2 assimilation in leaves (e.g., Farquhar et al., 1980). The Monod type equations have been treated as universally accepted models for the description of bacteria growth limited by single and multi-substrates (Tsao and Hanson, 1975; Tan and Bond, 1994; Tan and Wang, 1997), nutrient and light limitation of phytoplankton growth (Chen and Orlob, 1975; Sherman and Webster, 1994), and prey-predator relationships. Nutrient uptake by plant roots have also been reported to follow the Monod equation (Barber, 1981).

Modified forms of the Monod equation such as the Moser equation (Moser, 1958, 1985), the multiple Monod equations and the Tsao-Hanson equation also exist for the description of cell growth and enzyme and/or bacteria catalytic reactions to account for environmental effects. The Moser equation introduces another parameter, the multiple-Monod equation is the sum of two or more single Monod equations, and the Tsao-Hanson equation takes into consideration multiple substrates (both essential and growth-promoting substrates). It is important to note that none of the above mentioned equations, i.e., the Monod equation, the Moser equation, the multiple Monod equation and the Tsao-Hanson equation can be used to describe transformations that are inhibited by reactants, a common phenomenon in enzyme or bacteria catalytic reactions. A number of empirical models exist for the description of substrate inhibition of enzyme or bacteria catalytic reactions. The most commonly used is a homologous equation proposed for uncompetitive inhibition of enzymes by Heldane and later applied to substrate-inhibited microbial reaction kinetics by Andrews (Haldane, 1930; Andrews, 1968). The Haldane-Andrews equation is based on the assumption of an enzyme forming an inactive enzyme-substrate complex involving two substrate molecules per enzyme. In addition, the acidity or alkalinity as indicated by pH is a factor that profoundly affects all organisms and enzyme or bacteria catalysed reactions in the environment. The existing equation for the description of pH effects is the so-called symmetry pH equation, which has been successfully used to describe symmetrical pH curves (Bailey and Ollis, 1986).

Despite the fact that the Monod equation has found applications in many fields such as plant physiology, microbiology, biotechnology, geochemical cycling, and bioremediation, to date there has been no rigorous mathematical derivation given and the parameters appearing in this equation are strictly empirical. Despite the wide acceptance of the Tsao-Hanson equation in describing microbially-catalysed reactions involving multiple substrates, no rigorous mathematical derivation existed and the parameters appearing in this equation do not have a well-defined physical meaning probably due to the complexity of the metabolism of microbial cells under multiple substrate limitation. The same can be said about the existing pH model and the substrate inhibition model (Heldane equation). It should be noted that similar equations exist in enzyme kinetics studies and some equations are adapted from enzyme kinetics. It is well established that relationships exist between enzyme reactions and bacteria catalytic reactions. For example, the Monod model is a mathematical homologue of the Michaelis-Menten equation (Michaelis and Menten, 1913) while Moser equation derives from that of Hill (1910). Equations developed for enzyme kinetics should be applicable to the kinetics of cell growth and bacteria catalytic reactions, the behaviour of which depends on the integrated action of a multitude of synthetic processes catalysed by enzymes (Roels, 1983). The equations adapted from enzyme kinetics are seldom derived specifically for bacteria catalysed reactions in the natural environment although the relationship between enzyme kinetics and bacteria catalysed reactions is recognised and the adapted models were sometimes given explanations and interpretations in relation to enzyme activities.

Tan and Bond (1994) derived a general approach using a statistical thermodynamic method to describe microbial growth kinetics under the limitation of a single substrate. This approach provided a unifying theoretical basis for the empirical models such as the Monod, Moser and multiple Monod equations and was shown to be widely applicable in representing experimental data. This general approach was later extended to microbial growth under the limitation of multiple substrates and the Tsao-Hanson equation was derived using this general approach (Tan, 1996). This approach has been successfully extended to microbial growth as influenced by substrate inhibition and pH values (Tan et al., 1996, 1998). The object of this chapter is to describe the basic theory of this general approach and to extend this theory to the description of enzyme and/or bacteria catalytic reactions. Four subject areas will be covered: single substrate limitation, multiple substrate limitation, substrate inhibition and pH effects.

9.2 Single Reactant or Substrate

Reactions catalysed by microorganisms are often considered to result from a sequence of enzymatic reactions which have a large overall free enthalpy change and in which one of the reactions is rate limiting (Dabes et al., 1972; Roels, 1983). In a system of solution containing a large number of reactants or substrates, the enzyme catalysed reaction of the rate limiting step satisfies fast equilibrium conditions if the reaction is much slower than the binding of reactants or substrates to the reaction sites. This solution can be regarded as a quasi-equilibrium thermodynamic system (i.e., the equilibrium state is maintained for a very short time interval). Substrate molecules need to bind with enzymes or bacteria cells to be degraded. We assume that microorganisms or enzymes contain a number of basic identical functional units, part of which are substrate receptors, enzymes, or enzyme complexes. Each basic functional unit has n identical binding and reactive sites for a reactant or a limiting substrate. In the case of living cells, each functional unit can contain a complete set of catabolic functions required to catalyse both the anabolic and the catabolic reactions of cell growth. If the interactions between functional units are very weak, substrates or reactants bound to different functional units would be practically independent from each other. Such a functional unit can then be regarded as a system in a grand canonical ensemble in a reservoir of free substrates (Hill, 1985).

Following the theory of a grand canonical ensemble in statistical thermodynamics, the probability, $P_{|j(i)>}$, that a functional unit is in the quantum state $|j(i)>$ with energy $E_{j(i)}$ ($j = 1, 2, ..., i = 0, 1, ..., n$) and combines with exactly i molecules of the substrate is given by (Hill, 1985; Tan et al., 1994)

348   9. Theoretical Basis of Biological Models in Environmental Simulation

$$P_{|j(i)>} = \frac{1}{Q}\exp\{[-E_{j(i)} + i\phi]/k_BT\} \quad (9.1)$$

k_B is the Boltzmann constant, T is the absolute temperature, ϕ is the chemical potential of substrate molecules bound on enzymes or microbial cells, and Q is the grand partition function expressed as

$$Q = \sum_{i=0}^{n}\sum_{|j(i)>} \exp\{[-E_{j(i)} + i\phi]/k_BT\} \quad (9.2)$$

If the total concentration of enzymes or cell mass, C, at a given time can be made to represent the total concentration of the functional units of the biotic phase, the concentration of the functional units in the quantum state $|j(i)>$ with energy $E_{j(i)}$ combining with exactly i substrate molecules becomes $P_{|j(i)>}C$. It is noted that the change in C is small within a short time interval. If the binding of substrates to enzymes or microbial cells is much faster than the subsequent reactions, i.e., formation of reaction products, the initial rate of enzyme catalytic reactions, v, can be expressed as

$$v = \frac{\sum_{i=0}^{n}\sum_{|j(i)>} k_{j(i)}C\exp\{[-E_{j(i)} + i\phi]/k_BT\}}{\sum_{i=0}^{n}\sum_{|j(i)>} \exp\{[-E_{j(i)} + i\phi]/k_BT\}} \quad (9.3)$$

where $k_{j(i)}$ is the initial reaction rate constant for enzyme-substrate complexes in the quantum state $|j(i)>$. Since the system is considered to be in thermodynamic equilibrium, the chemical potential (ϕ) of a substrate in solution would be equal to that of the substrate bound on enzymes or cells. In dilute solution, the chemical potential is expressed as

$$\phi = \phi^0 + k_BT\ln[S] \quad (9.4)$$

where ϕ^0 is the standard chemical potential and $[S]$ the concentration of the substrate. Introducing (9.4), the initial reaction rate becomes

$$v = \frac{\sum_{i=0}^{n}\sum_{|j(i)>} k_{j(i)}C\exp\{[-E_{j(i)} + i\phi^0]/k_BT\}[S]^i}{\sum_{i=0}^{n}\sum_{|j(i)>} \exp\{[-E_{j(i)} + i\phi^0]/k_BT\}[S]^i} \quad (9.5)$$

This equation can be further reduced to

$$v = \frac{\sum_{i=0}^{n}\alpha_i[S]^i}{\sum_{i=0}^{n}\beta_i[S]^i} \quad (9.6)$$

where

$$\alpha_i = \sum_{|j(i)>} k_{j(i)} C \exp\{[-E_{j(i)} + i\phi^0]/k_B T\}$$

$$\beta_i = \sum_{|j(i)>} \exp\{[-E_{j(i)} + i\phi^0]/k_B T\}$$

Equation (9.6) is the general equation for reactions that are catalysed by microorganisms and/or enzymes under the influence of a single substrate. The theoretical and practical importance of this general equation lies in the fact that it provides a unifying theoretical basis for the commonly used Monod, Moser or Hill equation, Heldane-Andrews, multiple Monod equations of empirical nature and it has also been used to describe experimental data as demonstrated by Tan et al. (1994).

The Monod model considers a simplified case where each functional unit has n identical and non-interactive binding sites for the growth limiting substrate. Tan et al. (1994) showed that the average number of substrate molecules bound per functional unit can be obtained by differentiating Q with respect to ϕ and multiplying the result by $k_B T/Q$, i.e.,

$$\bar{n} = k_B \frac{1}{Q} \frac{\partial Q}{\partial \phi} = k_B T \frac{\partial \ln Q}{\partial \phi} \tag{9.7}$$

Substitution of (9.4) into (9.7) and further reduction yield

$$\bar{n} = \frac{n[S]}{1/\{\exp\left[(\varepsilon_0 + \phi^0)/k_B T\right] + [S]\}} \tag{9.8}$$

where ε_0 is the energy level decrease when each functional unit combines with an additional molecule of substrate. If the rate constant is the same for all binding sites of a functional unit, the initial reaction rate v divided by the maximum reaction rate $v_{\max}$ would be equal to the average number of occupied sites ($\bar{n}$) divided by the total number of binding sites (n), i.e., $v/v_{\max} = \bar{n}/n$ and the Monod equation results

$$v = \frac{v_{\max}[S]}{K_s + [S]} \tag{9.9}$$

where K_s is the saturation constant and

$$K_s = \frac{1}{\exp\left[(\varepsilon_0 + \phi^0)/k_B T\right]}$$

The Moser or Hill equation derives from (9.6) if we assume that the n binding sites exist in two states: that is, all binding sites are either occupied or vacant, i.e.,

$$v = \frac{v_{\max}[S]^n}{K_s + [S]^n} \tag{9.10}$$

Similarly the multiple Monod equation can be derived from (9.6) (Tan et al., 1994)

$$v = \sum_{i=0}^{n} \frac{v_{\max_i}[S]}{K_{s_i} + [S]} \tag{9.11}$$

where $v_{\max_i}$ and K_{s_i} are the maximum specific growth rate and the saturation constant related to the i-th binding site.

9.3 Multiple Reactants or Substrates

Microbes or enzymes may simultaneously act upon more than one substrate. For systems with m kinds of substrates in the solution phase and ns number of binding sites for s-th type substrate molecules ($s = 1, 2, ..., m$), the corresponding probability can be found. If i_s expresses the number of s-th kind of substrate molecules bound on bacterial cells or enzymes, then the probability that a functional unit is in quantum state $|j(i_1, i_2, ..., i_m) >$ and combines with exactly $i_1, i_2, ..., i_m$ substrates is

$$P_{|j(i_1,i_2,...,i_m)>} = \frac{1}{Q}\exp\left[\left(-E_{j(i_1,i_2,...,i_m)} + \sum_{s=1}^{m} i_s\phi_s\right)/k_BT\right] \tag{9.12}$$

and the grand partition function is written as

$$Q = \sum_{i_1}\sum_{i_2}\cdots\sum_{i_m}\sum_{|j(i_1,i_2,...,i_m)>} \exp\left\{\left[-E_{j(i_1,i_2,...,i_m)} + \sum_{s=1}^{m} i_s\phi_s\right]/k_BT\right\} \tag{9.13}$$

where $E_{j(i_1,i_2,...,i_m)}$ is the energy level of functional units in the quantum state $|j(i_1, i_2, ..., i_m) >$ combining with $i_1, i_2, ..., i_m$ number of substrate molecules (M L^{2}t^{-2}), k_B is the Boltzmann constant (M L^{2}t^{-2}T^{-1}), and ϕ_s ($s = 1, 2, ..., m$) is the chemical potential of s-th kind of substrate bound on binding sites (M L^{2}t^{-2}).

In dilute solution, the chemical potential can be similarly expressed as

$$\phi_s = \phi_s^0 + k_BT\ln[S_s] \tag{9.14}$$

where ϕ_s^0 is the standard chemical potential and $[S_s]$ the concentration of s-th type substrate bound.

The concentration of enzymes or cell mass in the quantum state $|j(i_1,i_2,...,i_m)>$ with energy level $E_{j(i_1,i_2,...,i_m)}$ combining with exactly $i_1, i_2, ..., i_m$ substrate molecules becomes $P_{|j(i_1,i_2,...,i_m)>}C$. Assuming that the binding process is much faster than the step leading to the formation of reaction products, the initial rate (v) of enzyme catalytic reactions in the presence of multiple limiting substrates (M L^{-3}t^{-1}) can be expressed as

$$v = \frac{\sum_{i_1}\cdots\sum_{i_m}\sum_{|j(i_1,...,i_m)} k_{j(i_1,...,i_m)} C \exp\{[-E_{j(i_1,...,i_m)} + \sum_{s=1}^{m} i_s\phi_s]/k_BT\}}{\sum_{i_1}\cdots\sum_{i_m}\sum_{|j(i_1,...,i_m)>} \exp\{[-E_{j(i_1,...,i_m)} + \sum_{s=1}^{m} i_s\phi_s]/k_BT\}} \quad (9.15)$$

where $k_{j(i_1,...,i_m)}$ is the initial rate constant for reaction from the enzyme-substrate complexes in the quantum state $|j(i_1,...,i_m)>$ leading to the degradation of substrate molecules (t^{-1}).

For the case of one essential substrate and two growth-promoting substrates, the Tsao-Hanson equation is written as (Tsao and Hanson, 1975)

$$v = \left(k_0 + \frac{k_1[S_1]}{K_1 + [S_1]} + \frac{K_2[S_2]}{k_2 + [S_2]}\right)\left(\frac{[S_e]}{K + [S_e]}\right) C \quad (9.16)$$

where k_0 is a constant indicating the maximum reaction rate in the absence of growth-promoting substrates (t^{-1}); k_1 and k_2 are empirical constants related to growth-promoting substrate 1 and 2 (t^{-1}); K_e, K_1 and K_2 are empirical constants related to the essential substrate and growth-promoting substrate 1 and 2 (M L^{-3}); $[S_e]$, $[S_1]$ and $[S_2]$ are concentrations of the essential substrate and growth-promoting substrate 1 and 2 respectively (M L^{-3}).

From (9.15), the reaction rate in the presence of one essential substrate and two growth-promoting substrates, a case considered by Tsao and Hanson (1975) can be expressed as

$$v = \frac{1}{Q}\sum_{i_1}\sum_{i_2}\sum_{i_E}\sum_{|j(i_1,i_2,i_E)>} k_{j(i_1,i_2,i_E)} C \exp\left[\left(-E_{j(i_1,i_2,i_E)} + i_1\phi_1 + i_2\phi_2 + i_E\phi_E\right)/K_BT\right] \quad (9.17)$$

where

$$Q = \sum_{i_1}\sum_{i_2}\sum_{i_E}\sum_{|j(i_1,i_2,i_E)>} \exp\left[\left(-E_{j(i_1,i_2,i_E)} + i_1\phi_1 + i_2\phi_2 + i_E\phi_E\right)/K_BT\right]$$

and $k_{j(i_1,i_2,i_E)}$ is the reaction rate constant in the presence of growth-promoting substrate 1 and 2 and the essential substrate (t^{-1}), and ϕ_1, ϕ_2 and ϕ_E are the chemical potential of growth-promoting substrate 1, 2 and the essential substrate respectively (M L^{2}t^{-2}).

Assumptions are made that the energy level is zero when no substrates are bound and decreases by ε_E when each additional molecule of essential substrate combines with the functional unit and by ε_1 or ε_2 (M L^{2}T^{-2}) when a functional unit combines with each additional molecule of growth-promoting substrate 1 or 2. It is noted that when all the binding sites are occupied or saturated, there will not be further decrease in energy levels. This occurs when the limiting substrate concentration is very high. If there exist one binding site for the essential substrate, and n_1 and n_2 binding sites for the growth-promoting substrate 1 and 2, the grand partition function (Q) is given by

$$Q = \sum_{i_1=0}^{n_1} \sum_{i_2=0}^{n_2} \sum_{i_E=0}^{1} \binom{n_1}{i_1} \binom{n_2}{i_2} \binom{1}{i_E} \exp\left[(i_1\varepsilon_1 + i_2\varepsilon_2 + i_E\varepsilon_E \right.$$
$$\left. + i_1\phi_1 + i_2\phi_2 + i_E\phi_E)/k_B T\right]$$
$$= \left\{\sum_{i_1=0}^{n_1} \binom{n_1}{i_1} \exp\left[i_1(\varepsilon_1 + \phi_1)/k_B T\right]\right\}\left\{\sum_{i_2=0}^{n_2} \binom{n_2}{i_2} \exp\left[i_2(\varepsilon_2 + \phi_2)/k_B T\right]\right\}$$
$$\{1 + \exp\left[(\varepsilon_E + \phi_E)/k_B T\right]\}$$
$$= \{1 + \exp\left[(\varepsilon_1 + \phi_1)/k_B T\right]\}^{n_1} \{1 + \exp\left[(\varepsilon_2 + \phi_2)/k_B T\right]\}^{n_2}$$
$$\{1 + \exp\left[(\varepsilon_E + \phi_E)/k_B T\right]\} \quad (9.18)$$

Substituting (9.14) into (9.18) yields

$$Q = (1 + [S_1]/K_1)^{n_1} (1 + [S_2]/K_2)^{n_2} (1 + [S_E]/K_E) \quad (9.19)$$

where $K_1 = \exp\left[-(\varepsilon_1 + \phi_1^0)/k_B T\right]$, $K_2 = \exp\left[-(\varepsilon_2 + \phi_2^0)/k_B T\right]$, and $K_E = \exp\left[-(\varepsilon_E + \phi_E^0)/k_B T\right]$, ϕ_1^0, ϕ_2^0 and ϕ_E^0 are the standard chemical potentials of growth-promoting substrate 1 and 2, and the essential substrate respectively (M L^{2}t^{-2}), and $[S_E]$, $[S_1]$ and $[S_2]$ are concentrations of the essential substrate and growth-promoting substrate 1 and 2 respectively (M L^{-3}).

Further assumptions are made that the reaction rate constant is k_0 when only the essential substrate is bound, $k_0 + i_1 k_1'$ when the functional units bind with the essential substrate combine with i_1 number of growth-promoting substrate 1, $k_0 + i_2 k_2'$ when the functional units bind with the essential substrate combine with i_2 number of growth-enhancing substrate 2, and $k_0 + i_1 k_1' + i_2 k_2'$ when the functional units bind with the essential substrate combine with i_1 number of substrate 1 and i_2 number of substrate 2. Because C is not expected to change significantly within the short time interval of consideration, the reaction rate or microbial growth rate can be written as

$$v = \frac{1}{Q}\left\{\sum_{i_1} \sum_{i_2} k \binom{n_1}{i_1} \binom{n_2}{i_2} \exp\left[i_1(\varepsilon_1 + \phi_1)/K_B T\right] \exp\left[i_2(\varepsilon_2 + \phi_2)/K_B T\right]\right\}$$
$$\exp\left[(\varepsilon_E + \phi_E)/K_B T\right] C$$
$$= \frac{1}{Q}\left\{k_0 \left[1 + \frac{[S_1]}{K_1}\right]^{n_1}\left[1 + \frac{[S_2]}{K_2}\right]^{n_2} + n_1 k_1' \frac{[S_1]}{K_1}\left[1 + \frac{[S_1]}{K_1}\right]^{n_1-1}\left[1 + \frac{[S_2]}{K_2}\right]^{n_2}\right.$$
$$\left. + n_2 k_2' \frac{[S_2]}{K_2}\left[1 + \frac{[S_1]}{K_1}\right]^{n_1}\left[1 + \frac{[S_2]}{K_2}\right]^{n_2-1}\right\}\frac{[S_E]}{K_E} C \quad (9.20)$$

where $k = k_0 + i_1 k_1' + i_2 k_2'$. Substitution of (9.19) into (9.20) yields the Tsao-Hanson equation for two growth-enhancing substrates and one essential substrate if we set $k_1 = n_1 k_1'$ and $k_2 = n_2 k_2'$.

Similarly, a more general expression can be derived for p growth-promoting substrates and q essential substrates

$$v = \left\{1 + \sum_{i=1}^{p} \frac{k_i[S_i]}{K_i + [S_i]}\right\} \left\{\prod_{j=1}^{q} \frac{k_{E_j}[S_{E_j}]}{K_{E_j} + [S_{E_j}]}\right\} C \tag{9.21}$$

where k_i is an energy constant related to the maximum specific growth rate for growth-promoting substrate i, k_{E_j} is an energy constant related to the maximum specific growth rate for essential substrate j, K_i is the Monod saturation constant related to growth-promoting substrate i, K_{E_j} is the Monod saturation constant related to essential substrate j, $[S_i]$ is the concentration of growth-promoting substrate i, and $[S_{E_j}]$ is the concentration of essential substrate j.

9.4 Substrate Inhibition

If each functional unit considered in the case of single substrate has h inhibition sites, the occupation of any of which results in inhibition of reactions, the substrate molecules occupy at least one of the h inhibition sites when $n - h + 1$ substrate molecules combine with a functional unit. This means that the functional unit bound with i ($i > n - h$) substrates is no longer able to combine with any more substrates (i.e., the initial reaction rate constant for the binding reaction is zero for this functional unit). When $i > n - h$, the coefficients α_i before the term $[S]^i$ are zero and therefore the degree of $[S]$ in the numerator of (9.6) is at most $n - h$. Equation (9.6) can then be written as

$$v = \frac{\alpha_1[S] + \alpha_2[S]^2 + \cdots + \alpha_{n-h}[S]^{n-h}}{\beta_0 + \beta_1[S] + \beta_2[S]^2 + \cdots + \beta_n[S]^n} \tag{9.22}$$

This is the general equation for the description of substrate inhibition.

Various equations can be obtained for different numbers of binding sites (n) and/or inhibition sites (h). For the case of $n = 2$ and $h = 1$, the general equation reduces to

$$v = \frac{A_1[S]}{1 + B_1[S] + B_2[S]^2} \tag{9.23}$$

where $A_1 = \alpha_1/\beta_0$, $B_1 = \beta_1/\beta_0$, and $B_2 = \beta_2/\beta_0$. If we assume $A_1 = v_{\max}/K_s$, $B_1 = 1/K_s$, and $B_2 = 1/K_s K_i$, Equation (9.23) becomes the Haldane-Andrews equation, i.e.,

$$v = \frac{v_{\max}[S]}{K_s + [S] + [S]^2/K_i} \tag{9.24}$$

where K_s and K_i are empirical constants. Note that $v_{\max}$ is not attainable in the presence of inhibition.

9.5 pH Effects

The effects of pH on enzymes is due to changes in the state of ionization of the components of the system as the pH changes (Dixon and Webb, 1979). Since enzymes contain ionizable groups, which exist in a whole series of different states of ionization, and the distribution of the total enzyme among the various ionic forms depends on the pH and ionization constants of the various groups. The so-called active centre in enzymes often contain ionizable groups, which must be in proper ionic forms to maintain the conformation of the active centre, bind substrates and catalyze reactions (Segel, 1975). In addition, the active components of microbial cells, which are often enzymes, enzyme complexes, or other ionizable substrate receptors, must also be in proper ionic forms to bind substrates, which leads to reactions. In particular, microbial catalytic reactions and cell growth are often considered to result from a sequence of enzymatic reactions (Roels, 1983). It is therefore also appropriate to explain the effects of pH on microbially-catalysed reactions and microbial growth in terms of the ionization of the active components of microbial cells.

We need to consider the effects of pH on the enzyme catalytic reactions when the system contains hydrogen ions (H$^+$) besides limiting substrate molecules and enzymes or microorganisms and the functional units have ionisable groups for H$^+$ binding. We consider the case where each functional unit has n ionizable groups for H$^+$ binding and one binding site for a limiting substrate. We assume one binding site for the limiting substrate only because the main purpose of this section is to discuss the effects of H+ on reaction kinetics. Following Hill (1985), Tan et al. (1994) and Tan and Wang (1997) the probability that a functional unit is in the quantum state $|j(l,i)>$ with energy level $E_{j(l,i)}$ and combines with exactly l number of hydrogen ions ($l = 0, 1, ..., n$) and i substrate molecules ($i = 0, 1$) can be expressed as

$$P_{|j(l,i)>} = \frac{1}{Q}\exp\left[(-E_{j(l,i)} + l\phi_H + i\phi)/k_B T\right] \tag{9.25}$$

where

$$Q = \sum_l \sum_i \sum_{|(l,i)>} \exp\left[(-E_{j(l,i)} + l\phi_H + i\phi)/k_B T\right]$$

is the grand partition function, ϕ_H and ϕ are the chemical potentials of the hydrogen ions and bound substrate molecules. Assuming that the binding process is much faster than the formation of reaction products, the initial rate of enzyme catalytic reactions can be written as

$$v = \frac{\sum_l \sum_{|(l,1)>} k_{j(l,1)} C \exp\left[(-E_{j(l,1)} + l\phi_H + \phi)/k_B T\right]}{\sum_l \sum_i \sum_{|(l,i)>} \exp\left[(-E_{j(l,i)} + l\phi_H + i\phi)/k_B T\right]} \qquad (9.26)$$

where $k_{j(l,1)}$ is the turn-over number for substrate reaction for the functional unit bound with l hydrogen ions. Since the system is considered to be in thermodynamic equilibrium, the chemical potential (ϕ_H) of the hydrogen ions in solution would be equal to that of the hydrogen ions bound. In dilute solution, the chemical potential of hydrogen ions is given by

$$\phi_H = \phi_H^0 + k_B T \ln[H] \qquad (9.27)$$

where ϕ_H^0 is the standard chemical potential of H^+ in solution, and $[H]$ is the concentration of H^+ in solution. Substitution of (9.27) into (9.26) yields

$$v = \frac{\sum_l \alpha_l [H]^l}{\sum_l \beta_l [H]^l} \qquad (9.28)$$

where

$$\alpha_l = \sum_{|j(l,1)>} k_{j(l,1)} C \exp\left[(-E_{j(l,1)} + l\phi_H^0 + \phi^0)/k_B T\right][S]$$

and

$$\beta_l = \sum_{|j(l,0)>} \exp\left[(-E_{j(l,0)} + l\phi_H^0)/k_B T\right]$$
$$+ \sum_{|j(l,1)>} \exp\left[(-E_{j(l,1)} + l\phi_H^0 + \phi^0)/k_B T\right][S]$$

Note that α_l and β_l can be regarded as constants when the change in substrate concentration due to consumption and the increase in the concentration of cell mass are small.

As explained above, a large number of ionizable groups exist in various states of ionization depending on the pH of the surrounding solution and the ionization constants of the various groups. The ionization of some ionizable groups will affect enzyme catalytic reactions and cell growth while that of others may be irrelevant. We divide the ionizable groups into f functional groups and $(n-f)$ non-functional groups. The ionization of the functional groups affects enzyme catalytic reactions whereas that of other ionizable groups $(n-f)$ has little or no effect on the processes of enzyme catalytic reactions. If no interactions exist between the f functional groups and other $n-f$ groups and the enzymes of the components of microbial cells are catalytically active in some ionic forms only in which t of the f functional ionizable groups must be in protonated state and r of the f functional ionizable groups in deprotonated state, (9.28) can be written as

$$v = \frac{\alpha_t[H]^t + \alpha_{t+1}[H]^{t+1} + \cdots + \alpha_{f-r-1}[H]^{f-r-1} + \alpha_{f-r}[H]^{f-r}}{\beta_0 + \beta_1[H] + \beta_2[H]^2 + \cdots + \beta_f[H]^f} \quad (9.29)$$

where f is the number of functional ionizable groups in a basic unit. It is seen that the highest power of $[H]$ in the numerator is $(f - r)$ and the smallest is t. Besides $(r + t)$ essential groups, the functional groups may contain several non-essential groups, which exert their effects by interacting with the essential groups but differ from the essential groups in that there is finite activity even if they are in protonated or deprotonated states. As enzyme catalytic reactions and cell growth are usually confined to a relatively small pH range it is likely that the ionization of only a small number of functional groups influences the reactions, i.e., f is a small integer number.

By defining $k_l = \alpha_l/\beta_l$, $i = t, ..., f - r$, and $K_j = \beta_j - 1/\beta_j$, $j = 1, ..., f$, $j \geq 1$, we can rewrite the rate equation as

$$v = \frac{\frac{k_t}{K_1 K_2 \cdots K_t}[H]^t + \frac{k_{t+1}}{K_1 K_2 \cdots K_t K_{t+1}}[H]^{t+1} + \cdots + \frac{k_{f-r}}{K_1 K_2 \cdots K_{f-r}}[H]^{f-r}}{1 + \frac{1}{K_1}[H] + \frac{1}{K_1 K_2}[H]^2 + \cdots + \frac{1}{K_1 K_2 \cdots K_f}[H]^f} \quad (9.30)$$

It should be noted that $f \geq (r + t)$ and that when $t = 0$, the first term in the numerator is defined as k_0. Equation (9.29) can be further simplified if there are only a few functional ionizable groups.

In the case of two functional ionizable groups ($f = 2$), (9.30) can be expressed as

$$v = \frac{k_0 + \frac{k_1}{K_1}[H] + \frac{k_2}{K_1 K_2}[H]^2}{1 + \frac{1}{K_1}[H] + \frac{1}{K_1 K_2}[H]^2} \quad (9.31)$$

The constants k_0, k_1 and k_2 determine the shapes of the $v - $ pH curves. For example, v approaches zero at both low and high pH values symmetrically when both k_0 and k_2 equal zero, i.e., one of the two functional ionizable groups is in protonated state and the other in deprotonated state ($r = 1, t = 1$) for enzymes to be catalytically active. Equation (9.31) reduces to the existing pH model

$$v = \frac{v_{\max}}{1 + [H]/K_H + K_{OH}/[H]} \quad (9.32)$$

if we assign $k_0 = 0$, $k_2 = 0$, $k_1 = v_{\max}$, $K_{OH} = K_1$ and $K_H = K_2$. Equation (9.31) can be further reduced if either the term $[H]/K_H$ or $K_{OH}/[H]$ can be eliminated.

9.6 Comparison of Models with Observations

We include one example as an illustration of the capability of the biological models to describe experimental data. We use the experimental data of Kistner et al. (1979) which investigated the effects of pH on the specific growth rate of a rumen bacterium, Butyrivibrio fibrisolvens. The growth curve is asymetrical as a function of pH. The least complex equation that can describe asymetrical curves is (9.30) with $f = 3$, $t = 1$, and $r = 2$. The same set of experimental data is fitted to (9.32) for comparison. The results using both (9.30) and (9.32) are presented in Fig. 9.1 together with experimental data. It is easily seen that (9.30) describes this set of experimental data much better than (9.32).

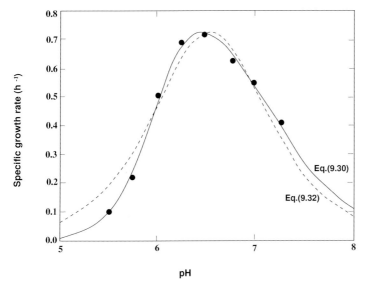

Fig. 9.1. Specific growth rate of B. fibrisolvens as a function of pH and the best fit to experimental data of Equation (9.30) with $k_2 = 0$ and Equation (9.32). Data of Kistner et al. (1979).

Few experimental studies have been carried out to test both specific mathematical models and the conceptual approaches summarised above. Of those studies reported, most have found that specific models provide a good description of the experimental results, but very few models have been validated using independently measured parameters. Similarly, few models have been tested beyond their initial development or applied to other scenarios. Much experimental work remains to be done before models of biological systems can be used with confidence to make a priori environmental predictions. Nevertheless, existing experimental studies have provided valuable information for the study and evaluation of biological processes in environmental simulation.

9.7 Discussion and Conclusions

This chapter discusses the theoretical basis of widely used biological models such as the Monod equation and explores the linkages between biological transformations in different disciplines of environmental science. Models such as the widely used Monod equation, the Tsao-Hanson equation, the Heldane-Andrews equation and the symmetry pH equation are shown to have a common theoretical basis. They can all be derived from the general approach shown above. The linkage is the binding of influencing agents to reaction sites associated with enzymes or bacteria that catalyse reactions. As the binding process is usually faster than the formation of reaction products, the rate of initial enzyme or bacteria catalytic reactions can be related to the concentration of influencing agents. The influencing agents considered in this chapter include a single limiting substrate, multiple limiting substrates, inhibitory substrates, and hydrogen ions.

Equation (9.6) is the most general expression for the description of enzyme or bacteria catalytic reactions . Firstly, no assumptions are made regarding the absolute number of functional units in an enzyme or cell. Each enzyme or cell can have any number of functional units and each functional unit can be a single enzyme or enzyme-complexes with a number of enzymes or other substrate receptors. For example, the functional unit can be a CO_2 receptor or a FeS_2 receptor. Secondly, this general approach can be used for the description of enzyme or bacteria catalytic reactions under substrate limitation and the same equation can also be used to describe the influences of inhibitory substrates and pH depending on the nature of the binding sites and the interactions between the sites. Thirdly, no specific assumptions are made in deriving this equation with relation to the number and the equivalence of the binding sites. From the derivation of the Monod equation, the Tsao-Hanson equation, the Heldane-Andrews equation and the symmetry pH equation, it can be seen that none of them represents the most general approach to the kinetics of enzyme or bacteria catalytic reactions because various specific assumptions are made regarding the number and the equivalence of the binding sites and the interactions between them. For example, various specific assumptions are incorporated into the Tsao-Hanson equation regarding the number and equivalence of the binding sites and the interactions between them. The binding sites of a functional unit are assumed to be independent of each other, that is, no interactions exist between the binding substrate molecules. The reaction rate constants are assumed to be additive regardless of whether the substrates are essential or growth-promoting.

All the empirical parameters contained in the Monod equation, the Tsao-Hanson equation, the Heldane-Andrews equation and the symmetry pH equation are physically well-defined as demonstrated by the theoretical derivation. The initial reaction rate constant, $v_{\max}$ is a function of the binding energy and/or the initial reaction rate constant. The number of bindings sites or ionizable groups f, h, n, r, and t should all be integers. The K parameters are functions of the binding energy having positive values and the k parameters are also functions of binding energy related to the maximum reaction rates. All parameters should have positive values although negative values may provide a good match with experimental observations. It is possible to determine all energy-related parameters (such as the Monod saturation constants) theoretically as an alternative method to curve fitting. This relies on the quantification of the energy levels of the functional units. If the energy levels are known from quantum mechanics, the values of all energy-related parameters can be evaluated theoretically using the expressions presented. It is important to note that some of those constants cannot be directly measured experimentally without making assumptions. For example, the K values or the pK values of individual ionizable groups are of great importance to structural and mechanistic studies. Even for the simplest dibasic acids system, the pK values obtained are relatively complex system constants rather than those associated with a simple group. Assumptions have to be made to assign pK values to the individual ionizable groups involved.

Environmental modelling studies have often adopted the Monod equation to describe biological transformations because of its mathematical simplicity and its success in matching experimental observations under some conditions. The Monod equation gives very simplified interpretations of the complexities of enzyme or bacteria catalysed reactions under the limitation of one single substrate or nutrient. The symmetry pH equation and the Heldane equation are also known for their mathematical simplicities when describing the influences of hydrogen concentrations and inhibitory substrates. They all provide convenient expressions to fit experimental data and have been successfully used to fit some observations. Tan et al. (1994, 1996, 1998) demonstrated that the general equation always provides comparable or better representations of microbial growth curves. Further work needs to be conducted to validate the general approach in its ability to describe light limitation, plant nutrient uptake, photosynthetic CO_2 assimilation and prey-predator relationships. It is expected that the general approach will always provide comparable or better representations of experimental observations as the empirical models are all special cases of the general approach.

10. Modelling of Photochemical Smog

Hiep Duc, Vo Anh and Merched Azzi

This chapter is concerned with air chemistry and its applications in air-quality modelling. Photochemical smog, i.e., the formation of high ground-level ozone concentrations, has been one of the main topics of air quality research in the last three decades. The chemistry of ground-level ozone is different from that of stratospheric ozone. Photochemical smog is a anthropogenic air pollution problem in many urban areas around the world, which is related to population increases and the reliance on motor vehicles for transport.

There are gaseous and particle pollutants, both of which have health effects. Many studies have been carried out on the effects of carbon monoxide, sulfur dioxides, nitrogen oxides, ozone and fine particles on mortality rates, respiratory and cardiac symptoms such as asthma, bronchitis and angina attacks. Apart from fine particles, smog formation during summer can produce high levels of ozone, hence degrading the air quality in the environment. The photochemistry of ozone formation is complex. In some urban airshed models there are hundred of equations involving many chemical species.

The mechanism of smog formation has been studied by the Commonwealth Scientific and Industrial Research Organisation of Australia (CSIRO) in smog chamber experiments. A simple semi-empirical model has been formulated to reduce the complexity of the photochemical reaction, which involves nitrogen oxides and reactive organic compounds under strong sunlight. This model is called the Integrated Empirical Rate (IER) model. In addition, a set of reaction equations has been proposed to simplify the explanation of the smog formation. This Generic Reaction Set (GRS) only involves seven reaction equations. In this chapter, we describe both the IER model and GRS mechanism and their application to air quality modelling.

10.1 Air Chemistry of Smog Formation

10.1.1 Chemistry of Ground Level Ozone Formation

Ozone is a reactive and unstable form of oxygen, often found at relatively high concentration during hot and hazy summer weather. Such haze may build up over a period of days into a "photochemical smog". The yellow or brownish colour of the smog is due to fine particles, rather than ozone which is invisible.

Unlike most other air pollutants, ozone is not directly emitted, but forms in air when strong sunlight acts on a mixture of hydrocarbons and nitrogen oxides (ozone precursors). High concentrations of ozone can irritate eyes and throat, cause headaches, bronchial irritation and shortness of breath, and produce an acrid taste and smell. Ozone also damages plants, weakens rubber and attacks metals and painted surfaces.

Ozone production in the troposphere requires the photolysis of nitrogen dioxide NO_2, in turn, NO_2 is produced by reaction of NO with either ozone or peroxy radicals. Ozone can accumulate only when sufficient peroxy radicals are present. NO_x is removed from the NO-to-NO_2-to-NO cycle by the reaction of NO_2 with radicals to produce stable gaseous nitrogen and stable non-gaseous nitrogen. Since radical reactions results in both ozone formation and conversion of NO_x to its reaction products, a correlation of ozone with NO_x should occur whether or not NO_x is the limiting precursor.

The chemistry of ozone in the atmosphere is highly complex and is difficult to simulate successfully with computer models, which are the main tools used to devise control strategies. This poses a ongoing challenge to air quality managers in their quest for the most efficient means to target pollution sources which contribute to ozone.

Ozone formation was extensively studied at the Division of Coal and Energy of the CSIRO using environmental chamber experiments. The smog chamber results enabled the rates of smog producing processes to be elucidated and provided the missing key for quantifying the atmospheric processes. This CSIRO innovation was called the Integrated Empirical Rate (IER) approach and the Generic Reaction Set (GRS) model.

By identifying the key property of reactive organic compounds (ROC) to be the photolytic rate coefficient for photochemical smog oxidant formation (R_{smog}), the quantitative description of photochemical smog was made tractable.

Ozone pollution is greatly influenced by weather conditions. A combination of light winds and strong sunshine, persisting with little change over several days, allows ozone precursor compounds to build up with little dilution. Under these conditions, ozone concentrations often rise progressively day by day as long as the warm, stable weather lasts. Because of the influence of weather, ozone pollution is highly variable from one year to the next. In warm summers, high ozone concentrations may be experienced repeatedly, whereas during more unsettled seasons, virtually no high values may be recorded.

10.1.2 Generic Reaction Set (GRS) Mechanism

The Generic Reaction Set (GRS) resulting from the smog chamber work is a type of photochemical reaction mechanism that utilises a set of only seven generic equations, but allows for the speciation of the individual reactive compound (ROC) precursors.

Traditionally the quantitative description of the chemical processes which give rise to photochemical smog has been approached via lengthy and detailed chemical mechanisms. These types of models require extensive computing resources and are difficult to implement and verify. In contrast, the GRS mechanism has the advantages of rapid computer execution and that, for the atmosphere, the principal rate coefficient can be measured in real time (by the Airtrak air monitoring system) thus providing a straightforward and practical means for the validation of airshed models which incorporate the GRS approach.

The GRS model consists of the following seven reactions

$$ROC + h\nu \xrightarrow{O_2} RP + ROC$$
$$RP + NO \longrightarrow NO_2$$
$$NO_2 + h\nu \longrightarrow NO + O_3$$
$$NO + O_3 \longrightarrow NO_2$$
$$Rp + RP \longrightarrow RP$$
$$RP + NO_2 \longrightarrow SGN$$
$$RP + NO_2 \longrightarrow SNGN$$

where: ROC = Reactive Organic Compound, RP = Radical Pool, SGN = Stable gaseous nitrogen products, and SNGN = stable non gaseous nitrogen products.

This can be compared with other mechanisms, such as Carbon Bond IV (CB-IV) mechanism containing 70 chemical reactions and 28 species or the California Institute of Technology (CIT) lumped molecule model containing 52 reactions and 32 species (Hess et al., 1992).

GRS mechanism has been applied in a number of studies (Venkatram et al., 1997; MAQSP, 1996). The GRS mechanism also have been included in the airshed modeling (MAQSP, 1996), in which a modified CIT airshed model with the LCC chemical mechanism was replaced by GRS mechanism. The results from the two model outputs are comparable.

The Generic Reaction Set (GRS) model is based on the representation of all processes that lead to ozone formation. Firstly, the chemical mechanism of NO_2 and O_3 production is started. The sunlight induces photochemical reactions in the ROC species, producing strongly oxidising peroxy species and these oxidise NO to NO_2,

$$\text{NO} \xrightarrow[\text{oxygen}]{\substack{\text{sunlight} \\ \text{ROC}}} \text{NO}_2 \qquad (1)$$

Then NO_2 can itself be photolysed, producing O_3 and regenerating NO

$$\text{NO}_2 \xrightarrow[\text{oxygen}]{\text{sunlight}} \text{NO}_2 + \text{O}_3 \qquad (2)$$

The NO regenerated is then free to again participate in another oxidation cycle. This process can continue producing ozone until there is insufficient NO available to sustain reaction (1). This occurs when competing reactions consume the nitrogen oxides and produce stable products such as nitric acid and nitrogenous aerosol particles,

$$\text{NO}_2 \xrightarrow[\text{oxygen}]{\text{sunlight}} \text{stable nitrogen species} \qquad (3)$$

A further complication which needs to be considered is the reaction of NO with O_3, which is rapid and produces NO_2,

$$\text{NO} + \text{O}_3 \longrightarrow \text{NO}_2 \qquad (4)$$

Thus there are two means by which NO can be oxidised to NO_2, namely by reactions (1) and (4). When NO is emitted into air which already contains ozone, reaction (4) is important and NO immediately reacts with O_3 to form NO_2 until either the NO or O_3 is consumed. On the other hand, if the air has little O_3 content, the rate of NO_2 production will be governed by the rate of reaction (1), that is by the concentration of ROC and the intensity of sunlight. The rate of reaction (1) is also temperature sensitive, the rate increasing with increasing temperature. Thus the same factors (i.e. sunlight intensity, ROC concentration and temperature) dictate the production in the air of both NO_2 and O_3 [via reactions (1) and (2)].

10.1.3 Integrated Empirical Rate (EIR) Model

Photochemical smog formation is a complex process. To reduce the complexities and still have the ability to access fairly accurately and interpret air quality data, a semi-empirical model, resulted from smog chamber studies, has been formulated. This model is called Integrated Empirical Rate (IER) model. The method is based on quantifying photochemical smog in terms of NO oxidation.

The IER model defines Smog Produced (SP) as the quantity of NO consumed by photochemical processes plus the quantity of O_3 produced. A key feature of the IER model is that SP increases approximately linearly with respect to cumulative sunlight exposure during a light-limited regime, until the available NO_x are consumed by reaction, then the NO_x-limited regime occurs and SP production ceases.

The current concentration of SP compared to the SP concentration that would be present if the NO_x-limited regime existed is indicative of how far toward attaining the NO_x-limited regime the photochemical reactions have progressed. During the light-limited regime, SP is calculated from the expression

$$SP(t) = R_{smog}(t) \int J_{NO_2} f(T(t)) dt \qquad (10.1)$$

where R_{smog} is the photolytic rate coefficient (measured directly from the CSIRO Airtrak system), J_{NO_2} is the rate coefficient for NO_2 photolysis (a measure of sunlight intensity), $f(T)$ is a function of temperature.

$$f(T) = \exp\left[-1000\gamma \left(\frac{1}{T} - \frac{1}{360}\right)\right] \qquad (10.2)$$

$\gamma = 4.7$ from the smog chamber studies, T is given in °K.

For the NO_x-limited regime, where there is no new smog production, the concentration of SP is at its maximum and is proportional to the NOx previously emitted in the air

$$SP_{max}(t) = \beta NO_x^0(t) \qquad (10.3)$$

$\beta = 4.1$ from the smog chamber studies.

The ratio $E(t) = \dfrac{SP(t)}{SP_{max}}$ is then defined as the extent of smog produced.

When $E = 1$, smog formation is in the NO_x-limited regime and the NO, NO_2 concentration approach zero. When $E < 1$, smog production is in the light-limited regime.

Model description. The IER model defines the formation of photochemical oxidants in terms of "Smog Produced" (SP) where SP represents the concentration of NO consumed by photochemical processes plus the concentration of O_3 produced, viz.,

$$[SP]_0^t = [NO]_0^t - [NO]^t + [O_3]^t - [O_3]_0^t \qquad (10.4)$$

where $[NO]_0^t$ and $[O_3]_0^t$ denote the NO and O_3 concentrations that would exist in the absence of atmospheric chemical reactions occurring after time $t = 0$ and $[NO]^t$ and $[O_3]^t$ are the NO and O_3 concentrations existing at time t. $[SP]_0^t$ denotes the concentration of smog produced by chemical reactions occurring during time $t = 0$ to time $t = t$. The use of the parameter SP to describe smog production removes the complicating influences of the following competing chemical reactions (i) and (ii):

$$NO_2 + O_2 \xrightarrow{h\nu} NO + O_3 \quad \text{(i)}$$
$$NO + O_3 \longrightarrow NO_2 + O_2 \quad \text{(ii)}$$

The IER model provides an alternative concept of smog description by treating smog production as a function of the cumulative exposure of the reactants to sunlight rather than as a function of time. From this representation it was shown there are two regimes for photochemical smog. Firstly smog produced increases approximately linearly with respect to the cumulative sunlight. This stage is called "light-limited regime". This is followed by the NO_x-limited regime where the concentrations of NO and NO_2 decrease to zero. During the NO_x-limited regime there is no new smog production and the concentration of SP was found to be proportional to the concentration of NO_x previously emitted into the air

$$[SP]_{max} = \beta[NO_x]_0^t \tag{10.5}$$

where, from smog chamber data, the β coefficient for urban air can be assigned the value of 4.1. In this regime we can derive an expression for the ozone concentration as follows:

$$[O_3]^t = (\beta - F)[NO_x]_0^t \tag{10.6}$$

where the coefficient F is the proportion of NO_x emitted into the air in the form of NO; usually $F \cong 0.9$.

For the light-limited regime the concentration of smog produced, SP, at a given time t, can be written as:

$$[SP]^t = \int_0^t R_{smog}^t J_{NO_2}^t F(T^t) dt \tag{10.7}$$

where R_{smog} is the photolytic rate coefficient for smog production and J_{NO_2} is the rate coefficient for photolysis of NO_2, a measure of sunlight intensity. $F(T)$ is the temperature function:

$$F(T) = \exp\left[-1000\gamma(1/T - 1/316)\right] \tag{10.8}$$

where γ is a temperature coefficient determined from smog chamber studies and has the value 4.7; T is given in °K.

The ratio of the current concentration of SP to the concentration that would be present if the NO_x-limited regime existed is defined as the parameter "Extent" of smog production (E) and is given by:

$$E^t = [SP]^t / [SP]_{max} \tag{10.9}$$

When $E = 1$, smog production is in the NO_x-limited regime and the NO_2 concentration approaches zero. When $E < 1$, smog production is in the light-limited regime.

The IER approach also enables SP and $[NO_x]_0^t$ to be determined from ambient measurements of ozone and nitrogen oxides. For the light-limited regime, the smog production can be calculated as:

$$[SP]^t = ([O_3]^t + [NO_y]^t - [NO]^t - (1-F)[NO_y]^t)/(1-FP) \quad (10.10)$$

where $[NO_y]^t$ is the concentration of oxidised nitrogen conventionally measured by nitrogen oxides analysers and P is a coefficient for the loss of NO_x into species and forms not detected as NO_y. For urban air an appropriate value of P is 0.122.

For the NO_x-limited regime:

$$[SP]^t = \beta[O_3]^t/(\beta - F) \quad (10.11)$$

The value of $[NO_x]_0^t$ for ambient air can also be determined from monitoring data.

For the light-limited regime:

$$[NO_x]_0^t = ([NO_y]^t + P([O_3]^t - [NO]^t))/(1-FP) \quad (10.12)$$

and for the NO_x-limited regime:

$$[NO_x]_0^t = [O_3]^t/(\beta - F) \quad (10.13)$$

In addition to the nitrogen oxides and ozone concentration of the air it is also necessary to know the photolytic rate at which new smog will be produced [see (10.7)]. The key parameters, which determine this rate, are the sunlight intensity and the value of R_{smog}, a photolytic rate coefficient. The values of R_{smog} for ambient air are related to the emissions of ROC and R_{smog} values can be routinely measured with the Airtrak system, which was especially developed for this purpose (Blanchard et al., 1995). Airtrak also simultaneously determines the nitrogen oxides and ozone concentrations of the air.

Time series data from a monitoring station represents the concentrations of different air parcels as they reach the monitor. With suitably located monitoring stations use of such data removes the need to model or guess the composition of the ambient air with which the NO_x plume is expected to mix. This also has the advantage that the intrinsic variability in atmospheric composition can be taken into account.

The IER chemistry formulation shows that the chemical rate dependent processes are significant only to the light-limited regime and that during this regime the rate of SP production is independent of the atmospheric NO_x concentration. Thus, when the ambient air is in the light-limited regime, the SP formation is unchanged by the presence of additional NO_x from the chimney plume. During the NO_x-limited regime, mixing of additional NO_x from the plume into the surrounding air can cause resumption of SP formation. For NO_x-plumes these findings enable the IER-reactive plume model to be formulated so that the plume dispersion calculations can be carried out independently of the photochemical reaction modelling.

Application of the IER model has been used in a number of studies (Hess et al., 1992; Pryor, 1998; Blanchard et al., 1995). The simplicity of the IER numerical calculations and the ability to execute the model on the basis of air quality monitoring data are its most attractive uses. It gives a technique that is particularly useful for environmental impact assessments of NO_x sources that have moderate emissions rates and for experiments which can not support the cost of a full airshed study.

10.1.4 Application

An application of the IER model is to access the smog event at a particular day with Airtrak measurements. Observations were collected during three summer days (10–12 February, 1994) at Liverpool station on Sydney. These successive days cover three different categories of high, medium and low photochemical reactive air, respectively. Six minute averages of Airtrak data recorded during these three days are presented in Figs. 10.1a and b.

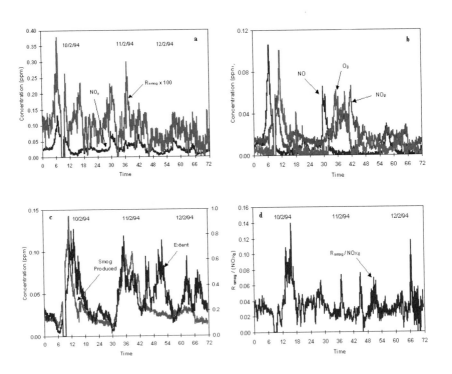

Fig. 10.1. (a) and (b) Airtrak measurements at Liverpool station in southwest of Sydney during February 10 to 12, 1994. (c) Calculated IER parameters using the model of Johnson and Azzi (1992). (d) the (R_{smog}/NO_x) ratio.

The selected three days of Airtrak data were in good agreement with the simultaneous measurements taken by conventional instrumentation. On the 10th of Feb between 0830 hr and 0940 hr the Airtrak recorded data is missing. On the 10th of Feb between around 0940 hr and 1200 hr the ozone concentration is greater than 0.08 ppm. At other times the concentration is less than 0.06 ppm. On the 10th and 11th between 0500 hr and 0800 hr Fig. 10.1a indicates a plume of NO$_x$ emissions and another plume of reactive organic compounds. On the 12th these plumes are not strongly pronounced.

The concentrations of NO, NO$_y$, O$_3$, and R$_{smog}$ determined by Airtrak were used to calculate the various IER parameters according to version 2.2 of the model (Johnson and Azzi, 1992). The modeling results are given in Fig. 10.1c where the variation of smog production and extent are plotted in separate axis against time. As well, in Fig. 10.1d, the (R$_{smog}$/NO$_x$) ratio is plotted against time showing how sensitive the photochemical smog formation is to changes in anthropogenic hydrocarbon and NO$_x$ emissions. The IER modeling results show that the light-limited regime was dominating over the three days period where the extent parameter values were always less than 1 (Fig. 10.1d).

On the 10th, polluted air parcels were detected between 0930 hr and 1200 hr, and on the 11th polluted air parcels were detected between 1100 hr and 1500 hr. These parcels are chararterised by a smog produced concentration greater than 0.8 ppm. In the morning before sunrise and after 2000 hr, the air sampled by Airtrak during the pervious two days had the characteristics of clean or background air where SP and extent values were less than 0.02 ppm and 0.2 respectively. On the 12th the concentration of smog produced was less than 0.03 ppm all day indicating that there was no pollution episodes detected on this day.

Age of precursor emissions. The IER parameters, which describe the photochemical characteristic of an air parcel, can be used to approximate the parcel's age. The difference between the cumulative sunlight at the moment of sampling and the corresponding SP/R$_{smog}$ ratio, gives insight on how much the air parcel has been exposed to sunlight. The deduced value will allow a determination of the approximate time of precursor emissions.

A better estimation for the age of precursor emissions would be made by subtracting the existing background smog produced concentration allowing only the anthropogenic emissions to be evaluated. This can be done by using (SP − SP$_{back}$)/R$_{smog}$ where SP$_{back}$ is the time series of smog produced for background air. In the absence of any knowledge about the SP$_{back}$ profile for the Sydney airshed and wishing to proceed with the selected Airtrak data analysis, we have adopted the following methodology: The following two values of SP$_{back}$ 0.015 ppm and 0.030 ppm were used in this study. The modeling results, which are given in Fig. 10.2, show for every selected day four different plots designated by A, B, C, and D. The first, second and third were obtained by using SP/R$_{smog}$, (SP − 0.015)/R$_{smog}$, and (SP − 0.03)/R$_{smog}$ respectively and the fourth plot represents the cumulative sunlight profile for the selected day and its previous day.

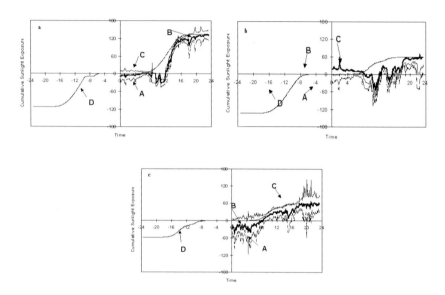

Fig. 10.2. Age of precursor emissions. (**a**) 10 February, 1994; (**b**) 11 February, 1994; (**c**) 12 February, 1994.

The ordinate serves to indicate how much cumulative sunlight an air parcel has seen. The time increments are quantified on the abscissa, which indicate how much time has elapsed before or after midnight of the selected day. The negative values indicate that the air parcel has been emitted from the previous day. For example, let us consider the plot A at 11am parcel on the 10th, the calculations indicate that the average time of precursor emissions was some 11 hours before midnight e.g. around 13pm from the previous day. The plots A, B, and C indicate the sensitivity of the method to the accuracy of the SP$_{back}$ concentration. Since plot A does not have any SP$_{back}$ correction, predictions of the age of the selected air parcels will be overestimated.

For a given SP$_{back}$, if the calculated value at a given time is higher than the corresponding cumulative sunlight value, then the used SP$_{back}$ is not appropriate for that time. This can be seen on the following cases: on the 10th and 12th, between midnight and around 730 hr and between around 1800 hr and 2400 hr the use of plot C was not appropriate. On the 10th, the plot C was not appropriate between midnight and around 800 hr.

The relationships represented in Fig. 10.2 show that what ever the value used for SP$_{back}$, on the 10th, the air parcels sampled between around 930 hr and 1200 hr would be emitted in the afternoon of the previous day. On the 11th and in the range of the variability of the selected SP$_{back}$ the three plots show that the polluted episodes, ranged between 1000 hr and 1200 hr, originated from the previous day. On the 12th and in the accepted SP$_{back}$ range of confidence, all plots can be fitted to have the shape of the corresponding cumulative sunlight profile without showing any time delay. The findings allow us to say that the 12th was a clean day.

The last two days illustrate that the air parcels passing the Airtrak have a low extent, indicating that they have either not received sufficient light flux to generate substantial photochemical products, or the dispersion of pollutants was high. These assumptions can be verified if we refer to an available windfield for these selected days.

The data sets from three selected Airtrak in the Liverpool area were implemented in the IER model to assess the air quality of the area. These Airtrak data sets show typical levels for NO, NO$_y$, NO$_2$, O$_3$ and R$_{smog}$. Airtrak measurements were in general agreement with measurements taken with conventional instruments. The IER parameters have produced information about ambient air quality including the age of the photochemical episodes precursors. This analysis provides valuable and effective tools for assessing ozone precursor control strategies. In addition, the Airtrak data analysis has shown that it is necessary to predict the background SP concentration to allow higher accuracy for an effective control strategy to take place.

10.2 Air Chemistry in Air Quality Modelling

10.2.1 Review of Air Quality Modelling

There are a significant number of air quality models, which have been used, applied and described in the literature. The models range from simple to large and highly complex mathematical modelling of the physical phenomena. The different kind of models can also be classified as empirical/statistically-based or physically-based models. In general, physically-based models are too large and complex to be used on a personal computer to predict and forecast the behaviour of the physical system under consideration.

The progress of air quality modelling is associated with the advance in the computing technology. In the past 3 decades, air quality models has been developed from simple box models and Gaussian models in the early 1970s to the advanced Eulerian and Lagrangian models currently used. Reviews of these models can be found in Collett and Oduyemi (1997) and Zanetti (1990).

Eulerian models use a stationary frame of reference, in which the concentration field relative to the domain is derived. In these models, the equation for the mass conservation of a species c is

$$\frac{\partial c_i}{\partial t} = -\overline{U} \cdot \nabla c_i - \nabla \cdot c'_i U' + D \nabla^2 c_i + S_i \tag{10.14}$$

where the wind field vector $U = \overline{U} + U'$ and the concentration of species i, $c_i = \overline{c_i} + c'_i$ consist of an average component ($\overline{U}$ or $\overline{c_i}$) and a fluctuating component (U' or c'_i).

The first 3 terms on the right hand side of the above equation represent the rate of advection by wind motion, rate of turbulent diffusion and rate of molecular diffusion of pollutant species. The last term represents the contribution from sources or sinks (e.g. deposition) of the species. In most cases, the rate of molecular diffusion can be neglected and the rate of turbulent diffusion in the second term can be assumed to follow a "first-order" equation in which the diffusion rate is linearly proportional to the local concentration. This method is also called the K theory of diffusion.

Lagrangian models use a reference system which follows the prevailing vector of atmospheric motion. The general atmospheric dispersion equation for a species is (Collett and Oduyemi, 1997)

$$c(\mathbf{r}, t) = \int_{-\infty}^{t} \int p(\mathbf{r}, t | \mathbf{r}', t') S(\mathbf{r}', t') d\mathbf{r}' dt' \qquad (10.15)$$

where $c(\mathbf{r}, t)$ is the average concentration of the species at location $\mathbf{r}$ at time t, $S(\mathbf{r}', t')$ is the source term and $p(\mathbf{r}, t | \mathbf{r}', t')$ is the probability density function that an air parcel is moving from $\mathbf{r}'$ at time t' to $\mathbf{r}$ at time t.

The Lagragian model requires more intensive computing resources compared to the Eulerian model. One of the difficulties with Lagrangian models is that their results can not be easily compared with observed measurements (Collett and Oduyemi, 1997).

The newest breed of models are the Lagrangian particle models. These models combine both Eulerian and Lagrangian approaches in which the conceptual view of "inert particles" being transported through a domain as a result of force applying on them. Various methods have been used and applied in the majority of current models using this approach.

These atmospheric dispersion models are strongly influenced by meteorological conditions. Due to the lack (or the impracticality) of meteorological observations in the modelled domain, many models (especially the later advanced models) incorporate a full meteorological modelling component.

Many air quality and dispersion models are developed to estimate the near-field impacts from a variety of industrial point sources and are used for regulatory purposes in local and national environment agencies. In the USA, regulatory models such as the Industrial Source Complex Short-Term (ISCST2) (and its recently updated AERMOD model), Buoyant Line and Point Source (BLP), RAM, COMPLEX I, MESOPUFF, Rough Terrain Dispersion Model (RTDM), Shoreline Dispersion Model (SDM) are used by the US EPA and local environment authorities. Some of these models have been adapted and used in Australia.

These models mostly use the classic Gaussian dispersion method and involve a limited number of non-reactive pollutant species from industrial point

10.2 Air Chemistry in Air Quality Modelling   373

sources. Such models are usually less complex than those models used for research and planning of air quality control strategies in large urban domains, which have many point sources and a variety of diffuse sources. Large complex photochemical smog models are used for modelling air pollution events or simulations of different control scenarios in an urban airshed. These models contain many reactive species. Examples of such models are the Urban Airshed Model (UAM) developed by the US EPA or the one developed by the California Institute of Technology (CIT) for an large urban area such as the Los Angeles airshed.

In Europe, long distance photochemical smog models such as the EUMAC model used in EUROTRAC project or various models used in different countries (such as LOTOS in the Netherlands, PHOXA in Germany) are used to study the effects of Volatile Organic Compound (VOC) and NO_x control at regional (European) scale on ozone levels in Europe (Fowler et al., 1999).

Depending on the domain under consideration and its grid size, air quality models can also be classified into regional or long distance (global) transport models. Long distance transport models involve inter-regional transport of air pollutants. Examples of such models are the EMEP model used for SO_2 or ozone transport across Europe or the Regional Acidic Deposition Model (RADM) used in the USA for modelling SO_2 deposition across the Eastern USA.

Air quality models are also designed for modelling different pollutants: gaseous pollutants (such as sulfur dioxide, ozone or photochemical smog) and particles. The most complex of these is the photochemical smog model. As the species are reactive, the air chemistry component in the model, reflecting the complex mechanism of ozone formation in the atmosphere under sunlight with many air pollutants, is substantial.

10.2.2 Air Pollution Forecasting Using GRS and Reactive State-space Models

There is a need for a simple system to predict and forecast the level of ozone in an operational network of monitoring stations in an urban area. Many of the existing operational forecasting schemes, used by many environment authorities, are based on the knowledge of meteorological conditions. The GRS model is implemented in the multivariate model of three main species of ozone, nitrogen oxide and nitrogen dioxides in the form of state-space equations. Because of the nature of reaction between the three variables, this model is called the reactive state-space model.

The model is developed and reduced to a Kalman filtering system suitable for implementation in a fast algorithm on the computer. The advantage of this model is that it is simple, accurate and fast enough to be used on an operational daily basis. In contrast, the mechanism for photochemical smog formation as used in the airshed models (such as the Carbon Bond IV (CB-IV) with 78 reactants and 170 reactions (Gerry et al., 1988; Morris et al., 1990) as used in the US EPA Urban Airshed Model (UAM), or in the California Institute of Technology (CIT) Model) is too complex and CPU intensive to be used routinely.

10.2.3 The Extended Space-time Model

The concentrations $c_i(x, y, z, t)$, $i = 1, ..., n$ of n chemically reactive species are assumed to satisfy the mass conservation equations:

$$\frac{\partial c_i}{\partial t} + Lc_i = r_i + s_i + d_i = f_i \tag{10.16}$$

where r_i is the rate of formation of species i by chemical reaction, s_i is the rate of formation of species i from the sources and d_i is the rate of deposition of species i.

Under the assumptions of the K theory, the operator L takes the form

$$Lc_i = \overline{u}\frac{\partial c_i}{\partial x} + \overline{v}\frac{\partial c_i}{\partial y} + \overline{w}\frac{\partial c_i}{\partial z} - K_H\frac{\partial^2 c_i}{\partial x^2} - K_H\frac{\partial^2 c_i}{\partial y^2} - \frac{\partial}{\partial z}\left(K_V\frac{\partial c_i}{\partial z}\right) \tag{10.17}$$

where K_H and K_V are horizontal and vertical diffusivity constants and $\overline{u}, \overline{v}, \overline{w}$ are the deterministic wind velocity components in the x, y and z direction respectively.

10.2.4 The GRS Mechanism

The chemistry component of f_i of the above equation is based on the GRS mechanism, which consists of the following reactions:

ROC+hv $\longrightarrow$ RP+ROC
RP+NO $\longrightarrow$ NO$_2$
NO$_2$+hv $\longrightarrow$ NO+O$_3$
NO+O $\longrightarrow$ NO$_2$
RP+RP $\longrightarrow$ RP
RP+NO$_2$ $\longrightarrow$ SGN
RP+NO$_2$ $\longrightarrow$ SNGN

where ROC = reactive organic compound (NMHC and oxygenated products); RP = radical pool (lumped radical species); SGN = stable gaseous nitrogen product; SNGN = stable non-gaseous nitrogen product.

Writing $[O_3](t)$ to mean the concentration of O_3 at time t, etc., the reaction rates for the above seven equations are given by

$R_1 = k_1[\text{ROC}]$
$R_2 = k_2[\text{RP}][\text{NO}]$
$R_3 = k_3[\text{NO}_2]$

$R_4 = k_4[\text{NO}][\text{O}_3]$
$R_5 = k_5[\text{RP}][\text{RP}]$
$R_6 = k_6[\text{RP}][\text{NO}_2]$
$R_7 = k_7[\text{RP}][\text{NO}_2]$

with k_i as the reactive constant of each of the above seven chemical reactions.

The chemical species that we interest in are ozone (O_3), nitrogen oxide (NO) and nitrogen dioxides (NO_2). By putting $c_1 = [O_3]$, $c_2 = [\text{NO}]$, $c_3 = [NO_2]$, the extended space-time model which combines advection, diffusion and GRS is

$$\frac{\partial c_1}{\partial t} = Lc_1 + k_3 c_3(t) - k_4 c_2(t) c_1(t) \tag{10.18}$$

$$\frac{\partial c_2}{\partial t} = Lc_2 + k_3 c_3(t) - k_4 c_2(t) c_1(t) - k_2[\text{RP}](t) c_2(t) + s_2 + d_2 \tag{10.19}$$

$$\frac{\partial c_3}{\partial t} = Lc_3 + k_2[\text{RP}](t) c_2(t) + k_4 c_2(t) c_1(t) - k_3 c_3(t) -$$
$$- k_6[\text{RP}](t) c_3(t) - k_7[\text{RP}](t) c_3(t) + s_3 + d_3 \tag{10.20}$$

with boundary conditions $K_H \dfrac{\partial c_i}{\partial z} = 0$, for $z = 0$ and $H(x, y, t) = 0$, $i = 1, 2, 3$, where $H(x, y, t)$ is the height of the inversion layer for the airshed under consideration.

In the above system, the next value of $[\text{RP}](t)$ is obtained from the current values of $[O_3]$, [NO] and $[NO_2]$. Hence, the system must be solved recursively.

10.2.5 The State-space Form

For a discrete approximation to the above continuous system, we may consider forward time difference to approximate time derivatives, second-order centred finite difference to integrate the advection and horizontal diffusion terms as well as the Crank-Nicholson method to integrate the vertical diffusion term. The resulting system can then be written in the state-space form (Anh et al., 1998)

$$\begin{cases} c(t+1) = Ac(t) + B + \varepsilon(t) \\ y(t) = Hc(t) + u(t) \end{cases} \tag{10.21}$$

where $y(t)$ is the vector of concentrations observed at the sites, H is a matrix whose elements are either 0 or 1, the value 1 corresponding to a grid point coinciding with a site. The matrix A is determined by the diffusion and advection characteristics (diffusion constants and wind data). The matrix B contains by the chemistry and emission or deposition characteristics.

The covariance matrix $Q(t)$ of the state equation is taken to have the form

$$Q_{ij}(t) = E(\varepsilon_i(t)\varepsilon_j(t)) = \frac{r_{ij}}{2\alpha}K_1(\alpha r_{ij}) \tag{10.22}$$

where r_{ij} is the distance between site i and site j on the grid, and K_1 is the modified Bessel function of the second kind, order 1. The above form is deduced from the atmospheric diffusion equation of the K theory (Anh et al., 1997). Nonlinear least squares fitting to the cumulative semivariogram of monthly averaged ozone data at 18 monitoring stations of the Sydney region yields that $\alpha = 0.148$ for the period under study (January 1994). This form seems adequate in representing the spatial variability of a homogeneous and isotropic concentration field (Anh et al., 1997).

The elements of the covariance matrix $R(t)$ of the observation equation are set at the values

$$R_{ij}(t) = \begin{cases} 0.1 & i = j \\ 0 & i \neq j \end{cases}$$

to reflect moderate random errors in the measurements.

The system is derived for a species c_j. The complete system for c_1, c_2 and c_3 can now be written in the state-space form above with

$$c = [c_1\ c_2\ c_3], \quad B = [B_1\ B_2\ B_3]^T, \quad \varepsilon = [\varepsilon_1\ \varepsilon_2\ \varepsilon_3]^T, \quad u = [u_1\ u_2\ u_3]^T$$

$$A = \begin{bmatrix} A_1 & 0 & 0 \\ 0 & A_2 & 0 \\ 0 & 0 & A_3 \end{bmatrix} \quad H = \begin{bmatrix} H_1 & 0 & 0 \\ 0 & H_2 & 0 \\ 0 & 0 & H_3 \end{bmatrix}$$

T denoting the transpose matrix. Since there is no change in the notation, we shall continue to use (10.21) to denote this complete system.

The Kalman algorithm for filtering and prediction of system (10.21) is

$$c(t|t) = c(t|t-1) + G(t)(y(t) - Hc(t|t-1))$$
$$c(t+1|t) = Ac(t|t) + B$$
$$G(t) = P(t|t-1)H'(HP(t|t-1)H' + R(t))^{-1}$$
$$P(t|t) = (I - G(t)H)P(t|t-1)$$
$$P(t+1|t) = AP(t|t)A' + Q(t+1)$$

where $c(t|t)$ is the estimation of $c(t)$ based on the new data $y(t)$, $c(t+1|t)$ is the prediction of $c(t+1)$ made at time t, $G(t)$ is the Kalman gain, which gives a correction to the previous forecast $c(t|t-1)$, and $P(t|t-1)$ is the covariance matrix of the prediction error, which is recursively computed through the last two equations of the algorithm.

The system (10.21) for the three species c_1, c_2 and c_3 is much more complex than those considered in Bankoff and Hanzevack (1975), Fronza et al. (1979) and Hernandez et al. (1991).

In these latter studies, the state-space form was obtained for a single species, while our system (10.21) allows for reactive relationships between the three species. These relationships connect and describe the reactive dynamics of the individual equations of the system. The Kalman algorithm also offers a convenient framework for determining the influence of different effects on the generation of c_1, c_2 and c_3. For example, under suitable conditions such as a calm day with strong sunlight, the chemistry component [mainly the matrix B in (10.21)] will be dominant in the algorithm. On the other hand, under different meteorological conditions such as strong wind, advective and diffusive transport will take effect while the chemical reaction rates become close to zero rendering the term B insignificant in the algorithm (apart from some adjustments for emissions and deposition rates).

10.2.6 Application

The state-space model, which accommodates the GRS photochemical mechanism, advection, diffusion and emissions, is applied to an irregular grid of monitoring network stations in Sydney, Australia. The location of these stations in the Sydney network (of 18 monitoring stations) is given in Fig. 10.3.

Two configurations of the Sydney monitoring network are considered:

- Grid A consisting of six stations: St Marys, Blacktown, Westmead, Lidcombe, Liverpool and Bringelly;
- Grid B consisting of seven stations: St Marys, Blacktown, Westmead, Lidcombe, Liverpool, Bringelly and Campbelltown.

The domain under consideration is urban. Two levels, namely the ground and the mixing height (maximum 2000 m above ground level) are considered since there are no observations currently available at the intermediate levels. Hence Grid A is considered as a rectangular grid with L = 3, M = 2, N = 2. January 9th 1994 is characterised by moderate concentrations of ozone in south west Sydney and is used as a test case. The data set consists of ten-minute observed concentrations of ozone, NO, NO_2, wind speed, wind direction, temperature, humidity and solar radiation at the monitoring stations. As an example, the time series of ozone concentrations for the six sites of Grid A are displayed in Fig. 10.4.

The diffusivity coefficients are obtained from the CSIRO Lagrangian atmospheric dispersion model (LADM), which is an air pollution dispersion model simulating the transport and diffusion of emissions of pollutants from discrete sources. It has a prognostic windfield component and a Lagrangian particle dispersion component (Physick et al., 1994).

The mixing height h is estimated using the diagnostic equations. For the period under study, neutral and stable conditions prevailed at night according to the wind speed classification, while unstable conditions are observed during the day.

378   10. Modelling of Photochemical Smog

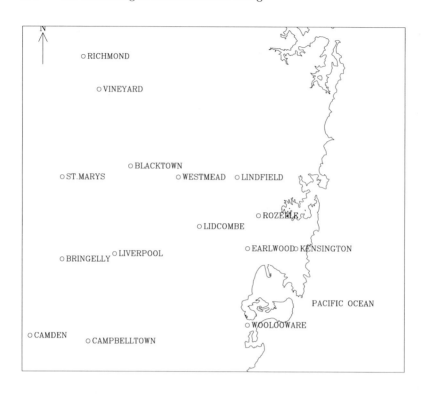

Fig. 10.3. Air quality monitoring stations in Sydney.

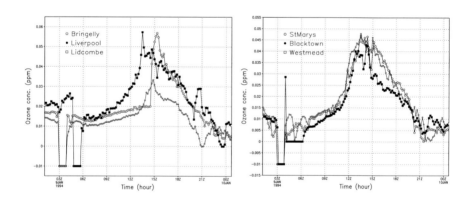

Fig. 10.4. Time series of ozone concentrations for the 6 sites.

10.2 Air Chemistry in Air Quality Modelling

The rate coefficients, as well as the observed values of a number of ROC mixtures, are obtained using the data from the CSIRO smog chamber experiments. These values are assumed for ambient Sydney suburban hydrocarbons. This is considered to be space and time-invariant and is consistent with the information on ROC from the existing NSWEPA emission inventory.

Due to lack of reliable data on emission and deposition rates, we follow Fronza et al. (1979), Melli et al. (1981) by heuristically correcting these values through an a posteriori pollutant mass balance, which is calculated from a comparison between filtered and previously predicted overall mass of pollutants at ground level. The filtering and prediction of O_3, NO, NO_2 and RP are carried out in an iterative scheme. The algorithm is applied to Grid A, and one-step ahead forecasts are derived for O_3, NO, NO_2 at each station.

As an example, the predicted and observed values of O_3, NO and NO_2 are presented in Fig. 10.5 for Blacktown, and in Fig. 10.6 for Liverpool. Calibration was performed between 3 and 4 am, hence data was not available in this period and was set at -0.01. Due to this artificial setting, the Kalman filter would not trace out these values.

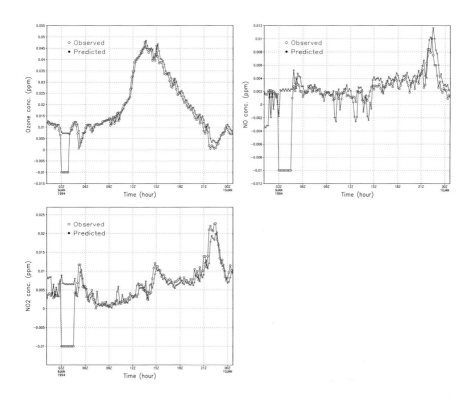

Fig. 10.5. Predicted and observed values of O_3, NO and NO_2 for Blacktown.

380   10. Modelling of Photochemical Smog

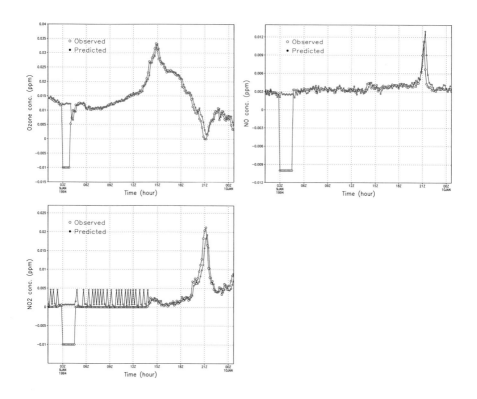

Fig. 10.6. Predicted and observed values of O_3, NO and NO_2 for Liverpool.

The exercise is repeated for Grid B, where Campbelltown is added to Grid A; but in this exercise, we assumed that there were no measurements at Campbelltown. The Kalman filter is then activated based on information at the other six sites of the grid and produces one-step ahead forecasts for Campbelltown. The forecasts for ozone at Campbelltown are displayed in Fig. 10.7. The forecasts for NO and NO_2 at Campbelltown were also obtained; but since there are no observed values for them at this station, the results are not displayed here.

From Figs. 10.5 – 10.7, it is seen that the model works quite well, tracing out the pollution episodes closely in each case.

The space-time model which accommodates the main elements of air chemistry, advection, diffusion and emissions is put into the state-space form using appropriate stable numerical schemes which allow for irregular grids, a common feature of monitoring networks. The covariance structure of the noise term of the state equation reflects the spatial variability of the Sydney airshed established in a previous study.

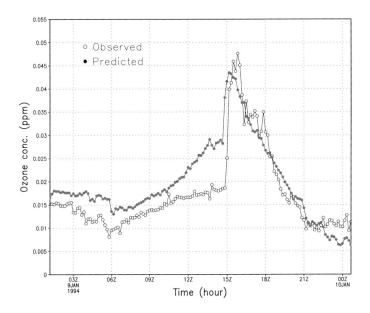

Fig. 10.7. Ozone forecast for Campbelltown.

Another significant contribution of this model is the modelling of the reactive dynamics of three key species (O_3, NO, NO_2) of photochemical smog within a stochastic state-space framework. It should be noted that previous state-space models concentrated on a single species or several non-reactive species, hence were not concerned with reactive relationships between the species, which is an important aspect of photochemical smog production.

Due to its reasonably compact size and the fast Kalman algorithm, the model offers a tool for experimentation and scenario analyses, particularly suited to investigating the effect of different meteorological conditions on the generation of O_3, NO, NO_2. Numerical results on a grid of seven stations indicate that the performance of the model is quite credable in terms of producing forecasts for an unobserved location, which is a difficult but important task of airshed modelling.

The photochemical smog formation is represented by the GRS mechanism. Since the solution of the gas phase chemistry is typically the largest consumer of CPU time in air quality models, a simplified chemical reaction set, such as the GRS mechanism, is highly desirable. However, the mechanism lacks representation of inorganic processes, which may be important for the description of photochemical smog production in the NO_x-rich region, i.e. for low VOC-to-NO_x ratios of emissions.

In this case, it would be appropriate to use a more detailed model such as the condensed version of the non-linear Lurmann-Carter-Coyner (LCC) mechanism (Lurmann et al., 1987; Harley et al., 1993). This version includes 26 differential and 9 steady-state chemical species. In addition to eight lumped organic classes, the chemical mechanism explicitly includes the chemistry of methane, methanol, ethanol, methyl tert-butyl ether (MTBE), isoprene, hydrogen peroxide and sulfur dioxide. The chemistry of isoprene is relevant to the modelling of contributions of emissions from biogenic/natural sources to the urban photochemical smog development. This chemistry is used in the model as a surrogate for all biogenic hydrocarbon emissions. The oxygenated species (methanol, ethanol, and MTBE) are of interest because they are the main ingredients in alternative or reformulated fuels, which are introduced gradually in many states in the USA. However, MTBE is being withdrawn due to problems with the contamination of ground water.

11. Applications of Integrated Environmental Modelling

Yaping Shao and Sixiong Zhao

11.1 Recent Developments in Integrated Environmental Modelling

As mentioned in Chap. 2, environmental prediction involves the estimation of future environmental states for given current and/or past conditions. It consists of three categories, namely:

- The prediction of key variables that quantify the state of the individual components, such as wind, air temperature, humidity and precipitation for the atmosphere; flow speed, temperature and salinity for the ocean; or mass and area distribution of ice and snow, etc;
- The prediction of the exchange of mass, momentum and energy between the components, in association with major cycles in the environmental system, e.g., the cycles of water, energy, CO_2 and aerosol; and
- The prediction of environmental events and phenomena, such as severe weather, air- and water-pollution episodes, ozone concentration, dust storms, salinisation, etc.

From the time-scale perspective, environmental predictions can be divided into short-range (up to a few days), medium-range (several days to several weeks) and long-range predictions (weeks to decades). From the perspective of the spatial domain, they can be classified as local, regional or global.

The methods of environmental prediction include deterministic numerical modelling, statistical and dynamic-stochastic methods, or a combination of these (e.g., ensemble prediction). The choice of which method to use depends on the time scale of the problem concerned. Deterministic models are the most popular for environmental predictions at all temporal and spatial scales, although the general view is that they are most appropriate for short- to medium-range predictions. For medium- to long-range predictions, statistical and dynamic-stochastic methods are alternatives, as deterministic models become less reliable.

As the environment is inherently dominated by non-linear processes and interactions, there are no simple relations between its responses to external forcing. This is true for both global and local events. It is increasingly understood that quantitative environmental predictions depend on integrated environmental modelling systems, which couple various numerical modules for the components of the environment, with high quality data.

Integrated environmental modelling embraces two basic streams of scientific development. The first is the coupling of dynamic models initially designed for individual components of the environment, i.e., those for the five spheres. The second is the coupling of dynamic models with spatial data. In recent years, exciting progress has been made in information technology for the collection, manipulation and exchange of large quantities of environmental data. This includes the development of geographic information systems (GIS), the global positioning system (GPS), remote sensing techniques, and the Internet. This progress has been achieved in parallel with the improvement of dynamic models and has created the possibility of bridging the gap between dynamic models and spatial data.

The driving force behind integrated environmental modelling is the effort devoted to the prediction of global climate change. A climate model is a coupled system of global circulation models (GCMs) for the atmosphere and the ocean, together with modules for the land surface, ice and snow, and vegetation, dealing with physical, chemical and biogeochemical processes. A sample of the climate models is given in Table 2.2. Accompanying the development of climate models are major international research programs, such as the Global Energy and Water Cycle Experiment (GEWEX), initiated in 1988 by the World Climate Research Programme for the observation and modelling of energy and water cycles through the atmosphere, land surface and the upper oceans. It currently supports 25 projects and is an integrated programme of research, observation and scientific activity leading to the improved prediction of global and regional climate change.

According to the assessment of Bengtsson (1998), much has been achieved in modelling the environment as a dynamic system. The achievements in modelling atmospheric circulation are most obvious. This is not surprising, as atmospheric general circulation models (AGCMs) have been under development for over 40 years, as the basis of numerical weather prediction models, and atmospheric observations are the most comprehensive and best analysed. Over the past 10 years, around 30 groups have carried out long-term simulations of the atmospheric general circulation forced by the same sea-surface temperature data (Gates, 1992). These numerical experiments have produced accurate predictions, when compared with the averaged observed atmospheric circulation and the characteristic natural variability on different time scales, as shown in Table 11.1.

Integrated environmental modelling systems have also been developed for environmental predictions on smaller scales, e.g., regional atmospheric predictions and catchment to continental scale hydrological predictions. Before the 1970s, short-term atmospheric processes were considered to be adiabatic,

Table 11.1. Correlation between observed (ECMWF) and simulated pattern of variability by the ECHAM3_T42 for the Northern Hemisphere (30-year average) (from Perlwitz, 1997).

ECMWF obs.	stationary pattern			high frequency (2.5 – 6 days)			low frequency (10 – 90 days)		
SEASON	1000 hPa	500 hPa	200 hPa	1000 hPa	500 hPa	200 hPa	1000 hPa	500 hPa	200 hPa
DJF	0.90	0.91	0.93	0.97	0.95	0.94	0.97	0.93	0.90
MAM	0.84	0.87	0.89	0.96	0.96	0.95	0.96	0.96	0.98
JJA	0.88	0.75	0.69	0.92	0.92	0.95	0.94	0.96	0.95
SON	0.92	0.95	0.95	0.98	0.98	0.99	0.95	0.94	0.96

and the interactions between the atmosphere and the underlying surface were not carefully treated. Today, these processes are considered to be diabatic, and the atmosphere-land and atmosphere-ocean interactions are taken into consideration. Land-surface schemes have shown marked improvements in the simulation of surface hydrological flows, and the simulation of biomass, which evolves slowly with time and dynamically interacts with atmospheric and hydrological processes. Improvements in land-surface models, and the coupling of atmospheric and land-surface models, have resulted in much improved atmospheric predictions. In addition, this coupling has enabled the prediction of a wide range of land-surface processes not thought possible before, such as the prediction of soil moisture, plant growth, and soil erosion by wind and water (Shao and Leslie, 1997; Lu and Shao, 2000). The development in the coupling of atmospheric and land-surface models has also created better input data for hydrological models, e.g., increased accuracy in the estimate of precipitation and evapotranspiration over large areas (Shao et al., 1997).

A large number of integrated environmental modelling systems have been developed for a wide range of specific environmental problems on local scales. They are used to simulate certain environmental problems in real time and to produce short-time predictions or scenario studies. The prediction of air quality in urban areas is a typical example. In advanced air-quality modelling systems, a local atmospheric flow model is commonly coupled with an air-chemistry model and a database for pollution sources, such as traffic-related and industrial pollution sources.

In short, various integrated environmental modelling systems have been developed recently and these models already have a substantial capacity for the predictions of complex environmental problems on scales ranging from local to global. They are increasingly applied to study interactions within the environment, the cycles of energy and water, as well as a range of specific environmental events. Several examples are given in this chapter.

11.2 Atmosphere-Ocean Interactions

The coupling of atmosphere and ocean models depends on the time scale of the problem which we are interested in. For problems that have short time scales, i.e., several days, the temporal variations in deep ocean temperature and flow speed can be considered to be small. In these cases, a simple model for the mixed layer of the ocean (top few metres), often known as the "slab" model, can be employed. The "slab" ocean model is commonly used in numerical weather prediction models for limited areas, or alternatively, the observed ocean surface temperature is given as input data. For the predictions of phenomena on longer time scales, e.g., seasonal to decades, the thermodynamic behaviour of the entire ocean plays an important, or even a dominating, role. In these cases, the atmosphere-ocean interactions must be modelled to include the deep sea. A common approach is to couple an atmospheric prediction model, usually an AGCM, with an OGCM.

Sea-level variations are important to the understanding and prediction of ocean dynamics, especially the large-scale and low-frequency variabilities of the ocean circulation, and hence, to the behaviour of the global environment. For example, the gradients of sea level can be used to compute the variations of equatorial currents. The sea level is also a good indicator of vertical displacements of the thermocline, as it is closely related to the dynamic height, heat content and thermocline depth. Consequently, the sea level can be used to determine water movement above the thermocline, and the water-mass budget. Sea-level observations are also valuable in analysing and monitoring large-scale climatic anomalies, including the ENSO phenomena (e.g., Wyrtki, 1985).

ENSO arises from atmosphere-ocean interactions in the tropical Pacific region. Studies have shown that ENSO has a profound influence on the global environment. The following observations can be made:

- Under normal conditions, the Walker circulation is strong, with the main cumulus convections occurring near 130°E; During El-Nino years it is weak, with the main cumulus convections occurring near 180°E;
- The Inter Tropical Convective Zone (ITCZ) over the tropical Pacific is located closer to the equator during El-Nino years than during La-Nina years. The location of ITCZ is closely related to the occurrence of hurricanes over the western Pacific;
- The subtropical high is located further to the south during El-Nino years;
- During El-Nino, precipitation is greatly increased in South America and decreased over large areas of China, Australia, Indonesia and southeast Africa. There is a possible increase in storm activities in North America;
- The number of hurricanes over the western Pacific decreases during El-Nino years (about 21). In addition, more hurricanes make landfall in Japan and fewer in China. During La-Nina years, the number of hurricanes increases (about 26).

Attempts have been made to estimate the changes in the sea level and to predict the El-Nino phenomena using a variety of models and techniques. Examples of relatively simple models can be found in Busalacchi and O'Brien (1981). More sophisticated OGCMs have also been used to hindcast sea-level variations, and the thermal and current structure of the tropical Pacific Ocean, in response to observed atmospheric forcing. Philander and Seigel (1985) and Philander et al. (1987) have been able to simulate both the general patterns and the evolution of the temperature structure and current system in the upper tropical Pacific Ocean for the 1982–1983 El-Nino and seasonal cycle. With the establishment of the Tropical Ocean Global Atmosphere (TOGA) sea-level network, and the availability of satellite sensed sea-surface topography, spatial sea-level measurements and sea-level time series are increasingly used, such as for the validation of numerical results and data assimilation of OGCMs.

In early OGCMs, the rigid-lid approach was commonly used (e.g., Bryan, 1969). Consequently, these OGCMs are unable to explicitly predict the sea-surface elevation, although the sea-surface variation can manifest as dynamic height, thermocline depth or heat content. The rigid-lid approach also leads to difficulties in comparing simulated and observed sea levels. Since the rigid-lid approach removes divergence of the vertical motion on all scales, including those of the large-scale barotropic Rossby waves, it artificially excludes the available potential energy arising from sea-level variation and its conversion to kinetic energy. Thus, the rigid-lid approach distorts the energy conversion and energy cycle in the model, and introduces errors in the computation of surface currents, the propagation of very long waves, and the variation of gigantic gyres.

More recent OGCMs employ the free-surface approach (e.g., Zeng et al., 1991). Consider the ocean in spherical coordinates, λ, θ and z, where λ is longitude, θ is latitude, and z is vertical height, and define

$$p = \bar{p}(z) + \rho_m p' \tag{11.1}$$

$$\rho = \bar{\rho}(z) + \rho' \tag{11.2}$$

$$T = \bar{T}(z) + T' \tag{11.3}$$

$$S = \bar{S}(z) + S' \tag{11.4}$$

where p is pressure, ρ is density, T is temperature, S is salinity, and ρ_m is the mean density of the seawater. For OGCMs, the baroclinic primitive equations governing ocean dynamics are often written in the σ coordinate with σ being defined as

$$\sigma = \frac{z - z_o}{h_s + z_o}$$

where h_s is the depth of the ocean and z_o is sea surface. Obviously, σ varies between 0 at $z = z_o$ and -1 at $z = -h_s$. Between the σ and z coordinate systems, the following relationships can be established

$$H_o = gz_o \quad \Phi = \sqrt{g(h_s + z_o)}$$
$$U = \Phi u \quad V = \Phi v \quad W = \Phi w$$
$$\Theta_T = \Phi T' \quad \Theta_S = \Phi S'$$

where (u, v, w) are velocity components in the (λ, θ, z) coordinate system. The baroclinic primitive equations can be written in the σ coordinate system as

$$\frac{\partial \mathbf{V}}{\partial t} = -L(\mathbf{V}) - f\mathbf{k} \times \mathbf{V}_\Phi \nabla p'_\Phi \frac{\rho'}{\rho_m}(\nabla H_o + 2\Phi\sigma\nabla\Phi) + \Phi\mathbf{F} \quad (11.5)$$

$$\frac{\partial \Theta_T}{\partial t} = -L(\Theta_T) - \Phi\frac{\Gamma_T}{g}\left\{\delta(1+\sigma)\frac{\partial H_o}{\partial t} + \Phi W + \right.$$
$$\left. + \frac{1}{\Phi}(\delta\mathbf{V}\cdot\nabla H_o + 2\Phi\sigma\mathbf{V}\cdot\nabla\Phi)\right\} + \Phi F_T \quad (11.6)$$

$$\frac{\partial \Theta_S}{\partial t} = -L(\Theta_S) - \Phi\frac{\Gamma_S}{g}\left\{\delta(1+\sigma)\frac{\partial H_o}{\partial t} + \Phi W + \right.$$
$$\left. + \frac{1}{\Phi}(\delta\mathbf{V}\cdot\nabla H_o + 2\Phi\sigma\mathbf{V}\cdot\nabla\Phi)\right\} + \Phi F_S \quad (11.7)$$

$$\delta\frac{\partial H_o}{\partial t} = -\frac{1}{a\sin\theta}\left(\frac{\partial \Phi U}{\partial \lambda} + \frac{\partial \Phi V \sin\theta}{\partial \theta}\right) - \frac{\partial \Phi W}{\partial \sigma} \quad (11.8)$$

$$\frac{\partial p'}{\partial \sigma} = -\frac{\Phi^2}{\rho_m}\rho' \quad (11.9)$$

where δ is an indicator introduced to distinguish the treatment of the sea surface as right-lid, or free surface, in the model and $L(f)$ is an advective operator defined as

$$L(f) = \frac{1}{2a\sin\theta}\left[\left(\frac{\partial fu}{\partial \lambda} + u\frac{\partial f}{\partial \lambda}\right)\right] + \left[\left(\frac{\partial fv\sin\theta}{\partial \theta} + v\sin\theta\frac{\partial f}{\partial \theta}\right)\right] +$$
$$+ \frac{1}{2}\left(\frac{\partial f\dot\sigma}{\partial \sigma} + \dot\sigma\frac{\partial f}{\partial \sigma}\right) \quad (11.10)$$

$\mathbf{F}$, F_T and F_S are the turbulent terms for momentum, temperature and salinity, respectively. These terms can be written as

$$\mathbf{F} = \nu_{mh}\left(\nabla^2\mathbf{v} + \frac{1-\cot^2\theta}{a^2}\mathbf{v} + \frac{2\cos\theta}{a^2\sin^2\theta}\mathbf{k}\times\frac{\partial\mathbf{v}}{\partial\lambda}\right) +$$
$$+ \frac{\partial}{\partial z}\left(\nu_{mv}\frac{\partial\mathbf{v}}{\partial z}\right) \quad (11.11)$$

$$F_T = \nu_{Th}\nabla^2 T + \frac{\partial}{\partial z}\left(\frac{\nu_{Tv}}{\delta_c}\frac{\partial T}{\partial z}\right) \qquad (11.12)$$

$$F_S = \nu_{Sh}\nabla^2 S + \frac{\partial}{\partial z}\left(\frac{\nu_{Sv}}{\delta_c}\frac{\partial S}{\partial z}\right) \qquad (11.13)$$

where ν_{mh} and ν_{mv} are the horizontal and vertical eddy viscosities, respectively, ν_{Th} and ν_{Sh} are the horizontal eddy diffusivities for temperature and salinity, respectively, ν_{Tv} and ν_{Sv} are the vertical ones, and $\delta_c = 0$ or 1, with the case $\delta_c = 0$ representing instantaneous convective overturning that restores a neutral stratification whenever a vertical unstable stratification develops.

The lateral boundary conditions along the coast are $u = v = 0$ and $\partial T/\partial n = \partial S/\partial n = 0$ and the surface and bottom boundary conditions are:

Surface
$\sigma = 0$
$\dot{\sigma} = 0$
$\dfrac{g\rho_m \nu_{mv}}{\Phi^2}\left(\dfrac{\partial u}{\partial \sigma}, \dfrac{\partial u}{\partial \sigma}\right) = (\tau_{xo}, \tau_{yo})$
$\dfrac{g\rho_m C_p \nu_{Tv}}{\Phi^2}\dfrac{\partial T}{\partial \sigma} = Q_T$
$\dfrac{g\rho_m \nu_{Sv}}{\Phi^2}\dfrac{\partial S}{\partial \sigma} = S_s Q_w$
$p' = \dfrac{p'_a}{\rho_m} + H_o$

Bottom
$\sigma = -1$
$\dot{\sigma} = 0$
$\dfrac{g\rho_m \nu_{mv}}{\Phi^2}\left(\dfrac{\partial u}{\partial \sigma}, \dfrac{\partial u}{\partial \sigma}\right) = (\tau_{xb}, \tau_{yb})$
$\dfrac{g\rho_m C_p \nu_{Tv}}{\Phi^2}\dfrac{\partial T}{\partial \sigma} = 0$
$\dfrac{g\rho_m \nu_{Sv}}{\Phi^2}\partial T \partial \sigma = 0$

where $\mathbf{n}$ is the unit vector normal to the lateral boundary; C_p is the heat capacity of seawater; p'_a is the atmospheric pressure departure at the sea surface; (τ_{xo}, τ_{yo}) and (τ_{xb}, τ_{yb}) are the wind shear stress at the ocean surface and the shear stress at the ocean bottom, respectively; S_s is the sea surface salinity; Q_T and Q_w are heat and water fluxes at the sea surface, respectively. More details of the model can be found in Zhang and Endoh (1992).

The coupling of the atmospheric model with the ocean model is achieved through the calculation of the energy and momentum fluxes occurring at the atmosphere–ocean interface, i.e., (τ_{xo}, τ_{yo}), Q_T and Q_w. Methods for determining these fluxes have been described in detail by Kraus and Businger (1994). As outlined in Chapter 2, one possible method for estimating Q_T is to use the bulk transfer method, namely,

$$Q_T = -c_p \rho C_H U(T_a - T_o) \qquad (11.14)$$

where C_H is the bulk transfer coefficient for sensible heat, U is the relative speed between air and ocean flows, T_a is the air temperature at the reference level, and T_o is the temperature of the ocean surface. The estimate of C_H can be made using the Monin–Obukhov similarity theory (Liu et al., 1979;

Bradley et al., 1991). The water flux at the sea surface, Q_w, is determined in general by precipitation, P_r, and evaporation, E (near the coast, fresh water from the continental hydrological system also plays a role). It therefore follows that

$$Q_w = P_r - E \tag{11.15}$$

where E can also be estimated using the bulk transfer method

$$E = -\rho C_E U (q_a - q_o) \tag{11.16}$$

where q_a and q_o are the specific humidity at the reference level and the sea surface, respectively, and C_E is the bulk transfer coefficient for water vapour. The behaviour of Q_T and E are constrained by the energy-balance equation at the ocean surface, namely, Equation (2.36). In that equation, the net radiation at the ocean surface must be determined from the atmospheric model and the temperature at the ocean surface [see Equation (2.37)]. At the ocean surface, we also have

$$\rho_a u_*^2 = \sqrt{\tau_{x0}^2 + \tau_{y0}^2} \tag{11.17}$$

where the friction velocity, u_*, at the ocean surface is given by

$$u_*^2 = C_D (U_a - U_o)^2 \tag{11.18}$$

where $U_o = \sqrt{u_o^2 + v_o^2}$ and U_a is the atmospheric flow speed at the reference level (commonly set to 10 m). C_D is the drag coefficient determined by the properties of the atmospheric turbulence. Methods for calculating C_D can be found in Kraus and Businger (1994).

As an example, we present the AGCM-OGCM coupled predictions for the 1997/1998 El-Nino and 1999 La-Nina events. In this example, a number of ENSO predictions up to 24 months have been made using the IAP (Institute for Atmospheric Physics, Chinese Academy of Sciences) TP-CGCM, which is a coupled system of a tropical Pacific Ocean model and an AGCM. The atmospheric component is the IAP two-level AGCM (Zeng et al., 1990). The oceanic component is the OGCM as described by Zhang and Endoh (1992). The domain of the ocean extends from 121°E to 69°W and between 30°S to 30°N, with a horizontal grid size of 1° in latitude and 2° in longitude. In the vertical direction, it has 14 layers with 9 in the upper 240 m. Both the AGCM and the OGCM can reproduce reasonable climatology and interannual variabilities, when forced by observed fluxes (Zeng et al., 1990; Zhang and Endoh, 1994). The model components are coupled synchronously, i.e., the flux exchange takes place daily at the atmosphere–ocean interface, in which the so-called "linear statistical correction" technique is applied to correct the variables at this interface. The artificially induced disturbances are limited as little as possible in the correction, under the premise that "climate drift" is controlled. The 100-year control run of this coupled system indicated that not only the climatology, including annual mean and seasonal variation, but also the ENSO like inter-annual oscillation, can be simulated successfully (Zhou et al., 1999).

For the initialization of the coupled system the observed sea-surface temperature anomaly (SSTA) is used. Starting from 1981, the AGCM is integrated continuously for 18 years, forced by the mean SST superposed with observed SSTA from November 1981 to December 1997. This is followed by running the OGCM for the same period, forced by the output of the AGCM, e.g., the wind stress and heat fluxes. For the end of every month of 1996, the coupled system outputs the instantaneous values, which are then used for an ensemble of predictions of the 1997/1998 El-Nino event. Figure 11.1 shows the simulated results of the system. The coupled system successfully predicted the 1997/1998 El-Nino event, and indicated that this event would reach its maximum at the end of 1997, then become weaker from the beginning through to the summer of 1998. The predictions for the 1999 La-Nina event also agree well with the observations (Fig. 11.2).

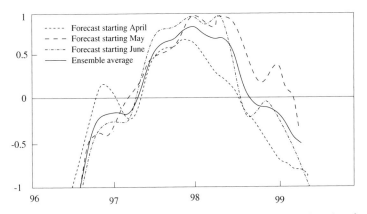

Fig. 11.1. The prediction of the 1998 El-Nino event, its signal represented by the SSTA (sea-surface temperature anomaly) in the Pacific Ocean, using the IAP-AGCM coupled with the tropical Pacific OGCM.

As another example, we consider the long-range predictions (i.e., monthly, seasonal and annual) of climatic anomalies. These anomalies are of great significance for social and economic activities. For instance, in the monsoon regions of Asian countries, such as China, India and Japan, precipitation anomalies strongly influence agricultural activities. Traditionally, long-range forecasts of monsoon anomalies in these regions have been mostly empirical. The recent developments in coupled systems of AGCM and OGCM have made long-range numerical atmospheric predictions possible. The capacity for carrying out these forecasts has been further enhanced by the rapid progress in computing power, data collection, data visualization and manipulation.

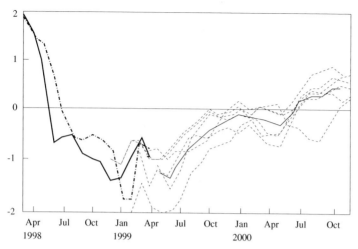

Fig. 11.2. The prediction of the 1999 La-Nina event, its signal represented by the SSTA (sea-surface temperature anomaly) in the Pacific Ocean, using the IAP-AGCM coupled with the tropical Pacific OGCM.

For these reasons, recent years have seen rapid developments in long-range atmospheric predictions, although a number of problems are yet to be solved (e.g., Zeng et al., 1990; Krishnamurti, 1991; and Shukla, 1991). For long-range atmospheric forecasts, an AGCM is usually coupled with an OGCM and the operation is supported by a wide range of numerical processes, including the initialisation of the AGCM and OGCM schemes for anomaly prediction, ensemble prediction, and correction and verification of the numerical outputs.

Figure 11.3 shows the observed precipitation anomalies (in %) for June, July and August 1991 and the summer season. From June to August 1991, there was very severe flooding, with a precipitation anomaly exceeding 100%, in the Huaihe River and Yangtze River region. At the same time, a severe drought with a negative precipitation anomaly exceeding -50% occurred in north and south China, accompanied by severe-to-moderate flooding in northeast China. Figure 11.4 shows the predicted anomalies using a coupled system of AGCM and OGCM with the initial fields for 15 February 1991. As shown, the numerical system has correctly predicted the very severe flood event in the Huaihe River and Yangtze River region, although it has not predicted the drought in south China. The predicted maximum positive anomaly is large (over +50%), but falls short of the observed value, which is over +100%, except for July. Nevertheless, the results using the coupled system of AGCM and OGCM are encouraging. The operational long-range forecasts of disastrous climate events, such as severe flood and drought, have been qualitatively successful in most cases, although in a quantitative sense these predictions often differ from the observations. In order to carry out operational long-range predictions successfully, significant improvements of the coupled system are needed, especially in the OGCM, the coupling procedure, the four-dimensional data assimilation and the derivation of initial fields. It appears that long-range predictions are always contaminated by systematic errors. It is desirable to remove these systematic errors, possibly through the introduction of statistical corrections to the numerical output. Such corrections are currently under examination.

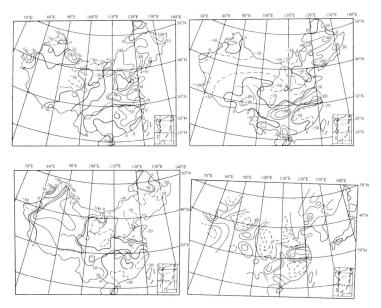

Fig. 11.3. (a) Observed monthly mean precipitation anomaly (in %) for June 1991. The contour interval is 50%. (b) as (a), but for July 1991; (c) as (a), but for August 1991; and (d) observed seasonal mean precipitation anomaly (in %) for summer (JJA) 1991. The contours are $0 \pm 20\%$, $\pm 50\%$ and $\pm 100\%$.

11.3 Atmosphere-Land Interactions

Land-surface models have been developed as a component of atmospheric models in the first instance, but they also have a wide range of applications in hydrological and ecological models. The implementation of land-surface models in atmospheric models has resulted in improved atmospheric predictions on scales ranging from local to global. It is also important to realize that when coupled with atmospheric, hydrological and ecological models, land-surface models also enable the simulation and prediction of a number of land-surface quantities, such as soil moisture, plant growth, soil erosion and salinisation, etc., never before thought possible.

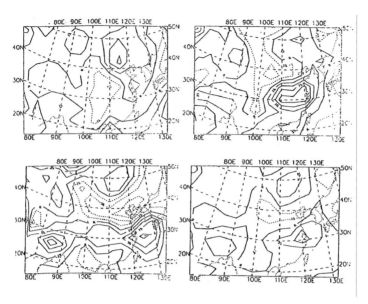

Fig. 11.4. Predicted precipitation anomalies over China for the same periods as for Fig. 11.3.

The implementation of land-surface models for regional to global modelling purposes has three major difficulties. Firstly, the performance of a land-surface scheme is limited by the empirical and/or semi-empirical nature of the parameterizations. Numerical tests using various schemes with prescribed atmospheric forcing data and land-surface parameters for given sites have demonstrated that, while most land-surface schemes generally perform well (e.g., Shao and Henderson-Sellers, 1996; Wood et al., 1999), significant uncertainties exist. Secondly, land-surface schemes require a large number of input parameters to describe soil hydrological, surface aerodynamic and vegetation properties. At present, these data sets are still difficult to obtain. Lastly, atmosphere and land-surface interactions involve complex feedback mechanisms, which are not yet fully understood or adequately represented in land-surface models. Despite these difficulties, much has been achieved in applying land-surface models to environmental modelling and prediction.

11.3.1 Soil Moisture Simulation

Soil moisture plays an important role in atmospheric and surface hydrological processes, since it influences the partitioning of surface net radiation into sensible and latent heat fluxes, and the partitioning of precipitation into evapotranspiration and runoff. It is also an important variable for ecological processes, as its availability is critical to plant growth. There are two approaches to the simulation of soil moisture. The first approach is to develop an atmosphere-land-surface coupled model and implement such a model over large areas. The second approach is to use remote sensing data combined with a surface-energy balance model to derive soil moisture (e.g., McVicar and Jupp, 1999). Both approaches are still at an early stage of development. We shall concentrate on the first one only.

Soil moisture simulation requires atmospheric forcing data, which are either analysed from meteorological observations or predictions using regional or global atmospheric models. Obviously, the availability and reliability of this data limits soil moisture prediction. For this reason, current studies on soil moisture are confined mainly to either soil moisture hindcast or real-time simulation. In a typical hindcast or real-time simulation, the land-surface model is forced continuously by the atmospheric model, which functions essentially as a tool for dynamic interpolation of atmospheric data in space and time.

Accurate initializations of soil moisture and soil temperature for large areas are virtually impossible. Instead, they are normally specified as best guesses. Because of the inaccuracies in the initializations, a long spin up time (up to 2 years) is normally required for coupled atmosphere-land-surface modelling to produce meaningful soil moisture output. During the spin up time, the nudging technique can be applied to control the simulation. Suppose a selected set (often not the full set) of atmospheric variables, A_o, e.g., air temperature, specific humidity and wind speed, are observed for the simulation time period. The observations are interpolated onto the atmospheric model grid for the time step t to obtain $A_o(i,j,k,t)$, using techniques such as the optimal analysis. Suppose also the same set of atmospheric variables are estimated using the coupled atmosphere-land-surface model to obtain $A_m(i,j,k,t)$, which are affected for the early period of the simulation by the incorrect initialization of the land surface. Then, applying the nudging technique, we would obtain

$$A_n(i,j,k,t) = W_o(i,j,k,t)A_o(i,j,k,t) + (1 - W_o)A_m(i,j,k,t) \quad (11.19)$$

where W_o is a weighting function depending on space and time. The new atmospheric quantities, $A_n(i,j,k,t)$, are then used to force the land-surface model for the $t + 1$ time step. There is no unified design of W_o, but its dependency on space and time is as illustrated in Fig. 11.5. At the boundary of the atmospheric simulation domain, W_o is always unity, but decreases towards the interior. As time progresses, W_o in the interior gradually decreases to zero.

Shao et al. (1997) have carried out soil moisture simulations for the entire Australian continent. This study can be used to illustrate simulations with coupled atmospheric and land-surface models. Those authors have developed an integrated system that couples a limited-area atmospheric model, a land-surface scheme and a database of land-surface parameters derived from a geographic information database. The basics of the land-surface scheme have been described in Chap. 5 and its performance has been examined by Irannejad and Shao (1998). The atmospheric model is as described by Leslie and

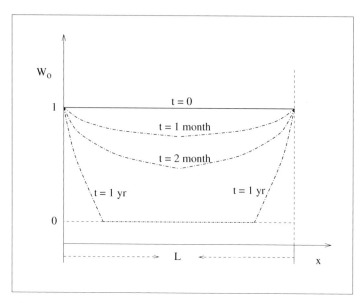

Fig. 11.5. A schematic illustration of the spatial and temporal changes of the nudging function used for soil moisture simulation.

Purser (1995). The horizontal resolution of the atmospheric model is 20 km, with 31 levels in the vertical. The initial and boundary conditions for the atmospheric model are derived from an AGCM.

The geographic information database available for the Australian continent includes data for soil type, vegetation and land use. The parameters required by the land-surface schemes are listed in Table 11.2. According to the database, Australian soils are classified into 28 soil classes, with 21% shallow permeable sandy soil, 17% deep massive earths, 11.2% cracking clay soils, etc. For each soil class, there is a qualitative description of the soil properties and associated land forms. Soil hydrological parameters are derived from these qualitative data, corresponding to the 12 USDA soil-texture classes ranging from sand to heavy clay.

The vegetation data set provides a range of parameters such as vegetation height, fractional vegetation cover, leaf area index (LAI), minimum vegetation stomatal resistance, vegetation albedo and root distribution. In the data set, vegetation has been divided into 35 classes according to the height, density and number of canopy layers, with 31.5% open-tall-sparse shrub lands, 27% low woodlands and low open woodlands, and 22% medium and short vegetation. From the data, an estimation can be made for quantities such as vegetation height, fractional vegetation cover, vegetation albedo, minimum vegetation stomatal resistance, and plant root systems.

The estimation of LAI draws on the remotely sensed Normalized Difference Vegetation Index (NDVI) data. A composite of satellite images over a

11.3 Atmosphere-Land Interactions

Table 11.2. A tentative list of parameters required by a land-surface scheme.

Symbol	Physical meaning	Dimension
K_s	Saturated hydraulic conductivity	m s^{-1}
θ_s	Saturated volumetric water content	m^3m^{-3}
θ_r	Air dry volumetric water content	m^3m^{-3}
λ_s	Macroscopic capillary length scale	m
b	Soil hydraulic characteristic parameter	-
θ_f	Volumetric water content at field capacity	m^3m^{-3}
θ_w	Volumetric water content at wilting point	m^3m^{-3}
D_{hs}	Heat diffusivity for dry soil	m^2s^{-1}
α_s	Soil-surface albedo	-
f_v	Fraction of vegetation cover	-
h_v	Height of vegetation	m
LAI	Leaf-area index	-
$R_{st,min}$	Minimum vegetation stomatal resistance	s m^{-1}
$p_r(z)$	Root fraction in different soils	-
α_v	Vegetation albedo	-

two-week period is usually necessary to eliminate the effects of clouds. For different vegetation types, LAI can be derived from NDVI using empirical relationships. For instance, from *in situ* observations McVicar et al. (1996) found the relationship between LAI and NDVI for the dominant vegetation type in the Murray Darling Basin in Australia to be

$$\text{LAI} = -4.65 + 4.24 \frac{1 + \text{NDVI}}{1 - \text{NDVI}} \qquad (11.20)$$

An example of the predicted soil moisture pattern is shown in Fig. 11.6, where the soil moisture of layer 1 (0 – 0.05 m), layer 2 (0.05 – 0.2 m) and layer 4 (0.5 – 1 m) are illustrated, together with the total soil water in the top 1 m, for 15 February 1996. The basic soil moisture pattern is typical for the Australian continent in summer. In large areas in the north-western part of Australia, including the Great Sandy Desert, Gibson Desert, Great Victoria Desert, and Nullabor Plain, soil moisture is low for the time of the year (late summer in the Southern Hemisphere). There is a slight increase in soil moisture toward deeper layers, for a considerable soil depth; the soil moisture falls in the range 0.05 and 0.15 m^3m^{-3}, with the total soil water in the top 1 m of soil being around 100 mm. From the desert areas, soil moisture gradually increases both toward the east and west coasts. Over large areas of the "Channel Country" and the Murray Darling Basin, typical values of soil moisture are around 0.25 m^3m^{-3} in layers 1 and 2, and 0.3 m^3m^{-3} in layer 4. Total soil water in the top 1 m of soil is around 250 mm. Further towards the east coast, soil moisture is over $0.3\,\mathrm{m}^3\mathrm{m}^{-3}$, under the influence of rainfall that occurred during the simulation period.

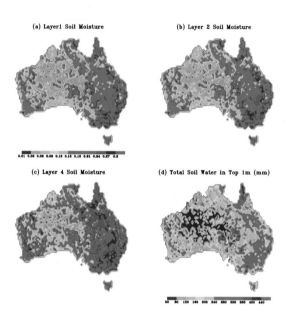

Fig. 11.6. Soil moisture distribution over the Australian continent for 15 February 1996. (a) Predicted soil moisture in $\mathrm{m}^3\mathrm{m}^{-3}$ for layer 0 – 0.05 m; (b) for layer 0.05 – 0.20 m; (c) for layer 0.5 – 1.0 m, and (d) total soil water in mm for the top 1 m of soil.

The spatial patterns of soil moisture are closely related to soil hydraulic properties. For instance, the low soil moisture in the desert areas is characteristic of the predominant sandy soils in the region. It is known that this region has little precipitation, and as the sandy soils have a high (saturated) hydraulic conductivity and low air-dry soil moisture content, soil moisture is low for most of the time. The exception is immediately after rainfall. In the desert areas, soil moisture is rapidly lost through drainage or evaporation. Soil moisture is significantly higher towards the east coast, apart from patches of dry areas in Queensland. For a large area in the eastern parts of Australia, the soil moisture is around $0.3\,\mathrm{m}^3\mathrm{m}^{-3}$, as the soils in this region are predominantly sandy clay or silty clay with high values of θ_r. For some of these areas, although the absolute soil moisture is quite high when compared with that of sandy soils, the available soil moisture is not necessarily much higher. The strong similarity between soil moisture patterns and soil type patterns supports the notion that, in summer, the Australian continent is generally under water stress.

Although there is not yet a comparison of the predictions through independent studies, apart from single point evaluations with observational data, the results are as expected. The simulations show that over the Australian continent in summer, the soil-moisture pattern is closely related to the distribution of soil types. This implies that, apart from isolated areas and at certain times under the influence of precipitation, drying factors such as evapotranspiration and drainage are primarily responsible for soil-moisture status. This phenomenon forms an interesting contrast to the distribution patterns of soil temperature, which in the top soil layers is more closely linked to weather patterns, and exhibits a south-north gradient in deep soil layers. This reflects the gradient in the incoming solar radiation received at the surface.

11.3.2 Influence of Land-surface Processes on Weather and Climate

The impact of land-surface processes on weather and climate can be profound. The impact can be examined using coupled atmosphere and land-surface models, following similar procedures as described above. Figure 11.7 shows an example of simulated surface energy fluxes over the Australian continent during a summer day, together with precipitation in the previous six hours, and the near-surface flow and temperature patterns. The patterns of energy fluxes clearly reflect the interactions between the atmosphere and the land surface. In the cooler region behind the front, for example, the sensible heat fluxes are generally larger than those in the warmer region before the front. This situation is accompanied by the somewhat smaller latent heat fluxes and negative ground heat fluxes behind the front, and the larger and positive values before the front.

Figure 11.8 shows a sensitivity test of weather to land-surface conditions. In case (a), the initial soil temperature and moisture are respectively set to 283° and field capacity over the entire Australian continent, while in case (b), the initial soil temperature and moisture are the 'correct' values. After 216 hours of simulation of the coupled system with identical initial and boundary conditions, significant differences in the weather patterns can be identified. These are reflected in the near-surface flow and temperature fields, as well as the cumulative precipitation starting from the beginning of the simulation. The wetter soil surface in case (a) is primarily responsible for the cooler atmosphere and high precipitation.

GCM simulations also show that land-surface processes play a significant role in the global climatic system. Figure 11.9 shows a comparison of the simulated climatological precipitation patterns for July using the IAP-AGCM with and without a land-surface scheme. As Figure 11.9c reveals, the differences between the two simulations are profound, especially in the Asia Pacific region. In Fig. 11.9b, the rain band is located over the South China

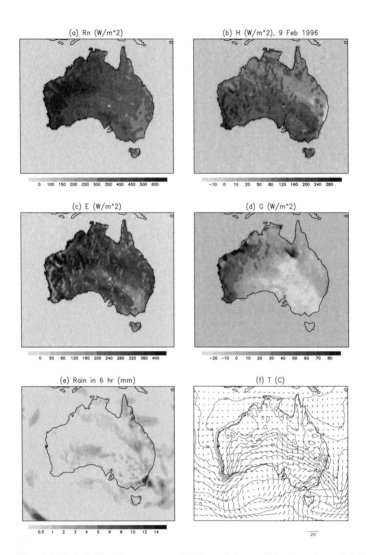

Fig. 11.7. (a) Simulated net radiation; **(b)** sensible heat flux; **(c)** latent heat flux; **(d)** ground heat flux; **(e)** precipitation in the previous six hours; and **(f)** near-surface wind and temperature over the Australian continent for a summer day. The synoptic situation is characterised by the northeastward moving cold front.

11.3 Atmosphere-Land Interactions 401

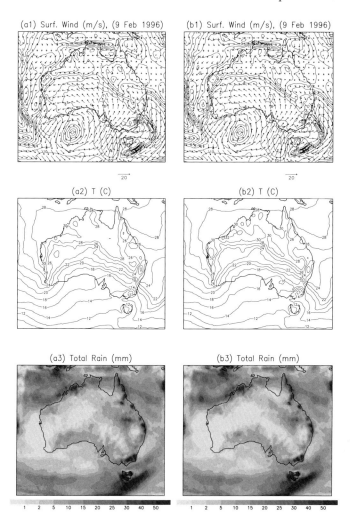

Fig. 11.8. A sensitivity test of weather to land-surface conditions. In (**a**), soil temperature and soil moisture are initially set to 283° and field capacity, respectively, for the entire Australian continent. In (**b**), the initial soil temperature and soil moisture are the "correct" values. Near-surface wind (a1, b1), temperature (a2, b2) and cumulative precipitation over nine days (a3, b3) are shown.

Sea in July, in disagreement with the known climatology for the region. In Fig. 11.9a, the main rain band is located along the Yangtze River, in much better agreement with observations. The predictions of total rainfall in the south and east Asian regions have also been improved significantly through the implementation of the land-surface model.

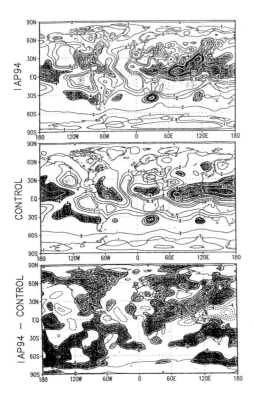

Fig. 11.9. Climatological distribution of the simulated precipitation using the IAP-AGCM with an improved land-surface scheme (**a**), and with a poor land-surface scheme (**b**). The differences between the two are shown in (**c**).

11.4 Atmosphere-Ocean-Ice Interactions

Ice (including snow) also plays an important role in the environmental system. Variations in the total amount of ice is small when compared with the fluctuations occurring in the atmosphere and ocean. Hence, these variations are good indicators for long-term climate changes. The impact of ice on the environment is reflected in three aspects: 1) the presence of ice increases surface albedo and enhances the reflection of short-wave solar radiation back into space; 2) ice absorbs heat from (releases heat to) the atmosphere and ocean during melting (formation) and hence moderates the changes in atmospheric and ocean temperatures; and 3) ice is a source of fresh water, the melting (formation) of ice reduces (increases) the salt concentration in ocean currents and hence affects ocean circulation.

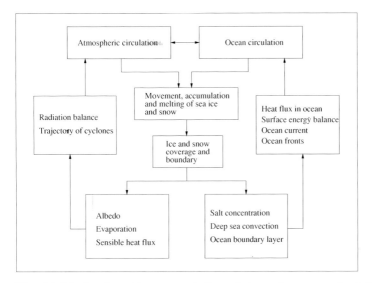

Fig. 11.10. A schematic representation of atmosphere-ocean-ice interactions.

The modelling of atmosphere-ocean-ice interactions requires the coupling of modules for at least the three components involved, namely, the atmosphere, ocean, and ice. A schematic illustration of the atmosphere-ocean-ice interactions is given in Fig. 11.10. Conceptually, the role of each component is as follows.

- Ocean: 1) evaporation from the ocean surface is the main source of atmospheric water vapour; ocean circulation and convection influence cloud formation, the distribution of precipitation and the global energy balance; 2) the ocean has a large heat capacity and consequently, the sea-surface temperature experiences smaller temporal variations than the land and ice surfaces; 3) ocean circulation causes heat transfer from the equatorial region to the polar region, influencing the global energy balance; 4) the distribution of sea temperature is closely related to that of ice. The salt

concentration of sea water influences the melting point and hence impacts on the formation of sea ice; 5) the ocean absorbs CO_2 from the atmosphere.
- Atmosphere: 1) shear stress at the ocean surface is the main driving force for surface ocean circulation and the motion of sea ice; 2) the energy exchange between the atmosphere and the ice surface is determined for the growth and decay of ice; 3) the exchange of energy and mass between the atmosphere and ocean influences the ocean current and the thermal structure of the upper ocean layers.
- Ice: 1) the high albedo of the ice surface increases the reflection of short-wave solar radiation back to the universe and reduces the absorption of energy at the surface; 2) increased ice cover over the ocean reduces global evaporation and consequently reduces cloud formation and precipitation; 3) the melting of ice in summer releases latent heat, and the formation of ice in winter absorbs sensible heat from the atmosphere and ocean. This process reduces the temperature gradient in the latitudinal direction. Accompanying the processes of ice formation and melting are the release and absorption of fresh water, which have an impact on the ocean effective temperature.

In sea-ice models, the impact of the atmosphere and the ocean on ice are realized through modelling τ_a and τ_O in (2.17), and modelling G_a and G_o in (2.18). The air and water stresses can be estimated as follows,

$$\tau_a = C_{Da}(\mathbf{V_g} \cos \varphi_a + \mathbf{k} \times \mathbf{V_g} \sin \varphi_o) \tag{11.21}$$
$$\tau_O = C_{Do}[(\mathbf{V_O} - \mathbf{v}_{ice}) \cos \varphi_o + \mathbf{k} \times (\mathbf{V_O} - \mathbf{v} \sin \varphi_o)] \tag{11.22}$$

where $\mathbf{V_g}$ is the geostrophic wind, $\mathbf{V_O}$ the geostrophic ocean current, C_{Da} and C_{Do} are air- and water-drag coefficients, and φ_a and φ_o air- and water-tuning angles. The geostrophic wind and the geostrophic ocean current can be derived from the atmospheric and ocean models. Two common formulations for C_{Da} and C_{Do} can be used, namely, the linear formulation in which C_{Da} and C_{Do} are taken to be constant, and the quadratic formulation in which they are expressed as

$$C_{Da} = \rho_a c_a |\mathbf{V_g}| \tag{11.23}$$
$$C_{Do} = \rho_o c_o |\mathbf{V_w} - \mathbf{v}_{ice}| \tag{11.24}$$

where c_a and c_o are dimensionless drag coefficients, often chosen as 0.0012 and 0.0055, respectively (McPhee, 1980).

A possible scheme for estimating G_a and G_o is illustrated in Fig. 11.11. This is a two-layer ice growth model in conjunction with an ice-surface energy budget and an ocean mixed layer. As depicted, the ice/snow system comprises a snow layer of thickness H_s and an ice layer of thickness H_i. Below the ice layer, an ocean mixed layer exists and underneath that is the pycnocline.

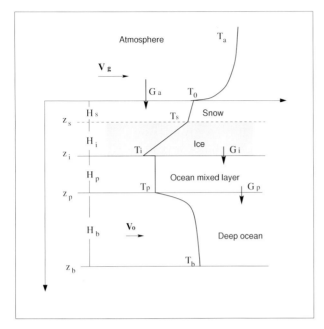

Fig. 11.11. An illustration for a possible atmosphere-ocean-ice coupling scheme based on the energy balance.

Neglecting the heat capacity of the ice and snow, and assuming no snow melting at the snow-ice interface, we have

$$G_a = -(T_s - T_0)\frac{\kappa_s}{H_s} = -(T_i - T_s)\frac{\kappa_i}{H_i} \tag{11.25}$$

where κ_s and κ_i are the thermal conductivities of snow and ice. The above expression can be formulated as

$$G_a = -\gamma(T_i - T_0) \tag{11.26}$$

with

$$\gamma = \frac{\kappa_i \kappa_s}{\kappa_s H_i + \kappa_i H_s} \tag{11.27}$$

The calculation of G_a involves the surface radiative temperature, T_0, which can be determined iteratively from the energy-balance equation at the atmosphere-ice interface, namely,

$$R_n - H - \lambda E - G_a = 0 \tag{11.28}$$

where the net radiation, R_n, sensible heat flux, H, and latent heat flux, E, are given by

$$R_n = (1-\alpha)R_s + R_l - \epsilon\sigma T_0^4 \tag{11.29}$$
$$H = -\rho_a C_p C_H \mid V_a \mid (T_0 - T_a) \tag{11.30}$$
$$E = -\rho_a L_v C_E \mid V_a \mid [q_s(T_0) - q_a] \tag{11.31}$$

where q_s is the saturation specific humidity at T_0, ϵ is emissivity, σ is the Stefan-Boltzman constant, ρ_a is air density, C_p is specific heat at constant pressure, L_v is latent heat of sublimation, C_H and C_E are aerodynamic coefficients for sensible and latent heat respectively.

The heat exchange between the ice and ocean can be expressed simply as

$$G_i = -\rho_o c_o C_{Do} \mid \mathbf{V_o} - \mathbf{v}_{ice} \mid (T_p - T_i)$$

where ρ_o and c_o are the density of sea water and heat capacity, respectively, both depending on salt concentration. The determination of G_i also involves the determination of T_p. This can be treated as follows. Suppose temperature and salinity have the following profiles

$$T(z) = T_p \qquad z_i < z \leq z_p \tag{11.32}$$
$$S(z) = S_p \qquad z_i < z \leq z_p \tag{11.33}$$
$$T(z) = T_\infty + (T_p - T_\infty)\exp[(z - z_p)/d_T] \qquad z_p < z < z_b \tag{11.34}$$
$$S(z) = S_\infty + (S_p - S_\infty)\exp[(z - z_p)/d_S] \qquad z_p < z < z_b \tag{11.35}$$

where d_T and d_S are e-folding depths of the thermocline and halocline. The magnitude of z_b is chosen to be large, say about 3000 m. In this case, we have $T_b \approx T_\infty$ and $S_b \approx S_\infty$. The temporal changes of T_p and S_p are given by

$$\frac{\partial T_p}{\partial t} = -\frac{G_p - G_i}{H_p} \tag{11.36}$$
$$\frac{\partial S_p}{\partial t} = -\frac{G_{sp} - G_{si}}{H_p} \tag{11.37}$$

Following Lemke et al. (1990), we can write

$$G_p = -(T^* - T_p)w_e \tag{11.38}$$
$$G_{sp} = -(S^* - S_p)w_e \tag{11.39}$$

where

$$T^* = \frac{1}{\delta}\int_{z_p}^{z_p+\delta} T(z)dz \tag{11.40}$$
$$S^* = \frac{1}{\delta}\int_{z_p}^{z_p+\delta} S(z)dz \tag{11.41}$$

with δ being a constant of about 10 m. An evaluation of the above integrations gives

$$T^* - T_p = a_T(T_b - T_p) \qquad (11.42)$$
$$S^* - S_p = a_S(S_b - S_p) \qquad (11.43)$$

where

$$a_{T,S} = 1 + d_{T,S}[\exp(-\frac{\delta}{d_{T,S}}) - 1]/\delta$$

Also the temporal variations of d_S, d_T and H_p are determined by

$$\frac{\partial d_S}{\partial t} = \frac{d_S}{S_b - S_p}\frac{\partial S_p}{\partial t} + \left[\frac{S^* - S_p}{S_b - S_p} - 1\right]w_e \qquad (11.44)$$

$$\frac{\partial d_T}{\partial t} = \frac{d_T}{T_b - T_p}\frac{\partial T_p}{\partial t} + \left[\frac{T^* - T_p}{T_b - T_p} - 1\right]w_e \qquad (11.45)$$

$$\frac{\partial H_p}{\partial t} = w_e - \frac{\partial H_i}{\partial t} \qquad (11.46)$$

The final closure of the scheme is achieved by specifying the entrainment velocity at the pycnocline, which depends on the intensity of ocean turbulence. In earlier studies of atmosphere-ocean-ice interactions, the numerical models for the atmosphere, ocean and ice were mostly partially coupled. The first simulation of the seasonal cycle of ice extent in the Weddell Sea only considered the local, thermodynamic response of sea ice to atmospheric forcing with a prescribed oceanic heat flux (Washington et al., 1976). In a more recent study, Lemke et al. (1990) have developed a sea ice-mixed layer pycnocline model, considering the atmospheric effect as the driving force. The model has been applied successfully to reproduce the large scale flow features of the ocean and sea ice in the Weddell Sea (Fig. 11.12). Fully coupled atmosphere-ocean-ice models have been developed more recently to simulate global climate changes (e.g., Manabe et al., 1990, 1991).

11.5 Prediction of Environmental Cycles

A number of cycles occur in the environment, including those of energy, water, CO_2, and aerosol. Integrated environmental modelling provides a powerful tool for studying these cycles. Two examples are given in this section.

11.5.1 Energy and Water Cycle

The hydrological cycle is closely related to the energy cycle. A schematic illustration of the former is as shown in Fig. 11.13. Evaporation (including

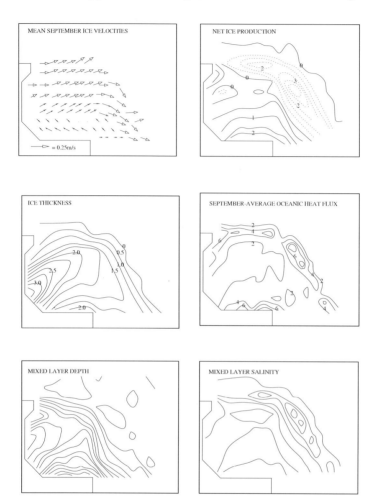

Fig. 11.12. Numerical modelling results for the Weddell Sea (adapted from Lemke et al., 1990).

transpiration) takes place over the ocean, land, and ice surfaces. Then, water vapour is convected from the surface to higher levels in the atmosphere and advected by atmospheric flows. Condensation, formation of clouds and eventually precipitation return liquid water back to the surface as rain and/or snow. The distribution of water on the continental surface is affected by surface runoff and groundwater flows.

One outstanding issue in many parts of the world is the management of water resources, including response strategies to flooding in major river valleys and the rationalization of limited available water during dry years. Until recently, the modelling of hydrological problems had been conducted more or less independently from the modelling of the atmosphere and the land

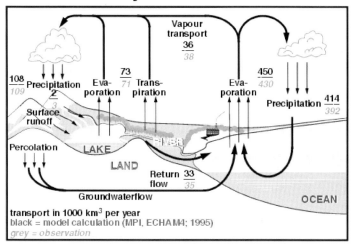

Fig. 11.13. Global annual mean hydrological cycle for the marine and continental areas, respectively. Upper figures show model calculations with the ECHAM4 model and lower figures are empirical estimates (averages result from Baumgartner and Reichel, 1975, and Chahine, 1992). Snowfall according to Bromwich (1990). Units are given in 1000 km^3 y^{-1} (adapted from Bengtsson, 1998).

surface. The most crucial driving variables, such as precipitation and evaporation, were estimated crudely, which resulted in inadequacies in hydrological modelling. Hydrological predictions are not truly meaningful unless hydrological models are coupled with atmospheric and land-surface models.

Budyko (1956) first recognised the need for coupling the atmospheric and hydrological models. This recognition has led to the recent development of land-surface schemes. However, so far, these schemes have been designed mainly for AGCMs and numerical weather prediction models. With the development of land-surface schemes, the coupling of atmospheric and hydrological models is now possible. In recent major international projects, such as GEWEX, integrated modelling of continental scale hydrology has gained much in terms of further development.

An integrated system for the modelling and prediction of continental hydrological problems consists of four major components: (1) a three-dimensional ground-water model; (2) a land-surface scheme, modelling surface soil hydrology in unsaturated soils; (3) an atmospheric model; and (4) a geographic information database and observed data. A schematic illustration for such a system is given in Fig. 11.14.

Atmosphere-land-surface coupled models can be used to estimate the exchange of energy and water at the interface between the atmosphere and land. Figures 11.15 and 11.16 show surface sensible and latent heat fluxes over the Australian continent over a five-day period. The diurnal variation

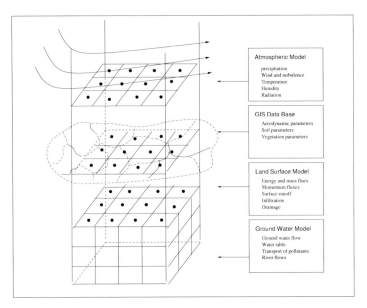

Fig. 11.14. An illustration of the atmospheric, land-surface and ground-water models with GIS data.

and the spatial distributions of these energy fluxes can be clearly seen. This type of model output is essential for an understanding of energy and water cycles in the environment.

Recent studies have shown that the hydrological cycle can now be reasonably well simulated. Figure 11.13 shows the simulated result of the global water cycle using the ECHAM4 model developed by the Max Planck Institute for Meteorology in Hamburg (Bengtsson, 1998). The model has been forced by observed sea-surface temperature data for the period 1979–1988. The agreement between the simulated results and the best available empirical estimates is within the accuracy of the latter. The model has also been able to accurately reproduce the secondary water circulation over land areas. Even for limited areas, the models can reasonably well reproduce the large-scale water cycle features. Figure 11.17 shows the simulated hydrological cycle for the same period using the same modelling system for the Baltic Sea region. The simulated data are in broad agreement with observational estimates, although a detailed evaluation indicates deviations in the annual cycle and it is found that the model produced too much precipitation during winter, and too little during summer.

11.5.2 Dust Cycle

Aerosols influence the atmospheric radiation balance both directly, through scattering and absorbing various radiation components, and indirectly, through

11.5 Prediction of Environmental Cycles   411

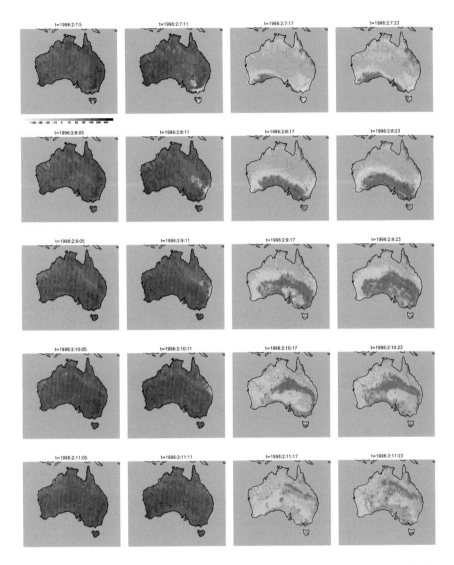

Fig. 11.15. Calculated sensible heat fluxes over the Australian continent for five days in February 1996.

modifying the optical properties and lifetime of clouds. Mineral aerosols are by far the most important aerosols in the atmosphere. The global total dust emission of 3000 Mt is more than twice the second largest aerosol source, the sea salt, which is about 1300 Mt. The global mean column dust load is approximately 65 mg m^{-2}, more than 9 times that of sea salt (about 7 mg m^{-2}). Many contaminants that pose significant risks to human health and the environment are found, or associated, with dust, including metal, pesticides,

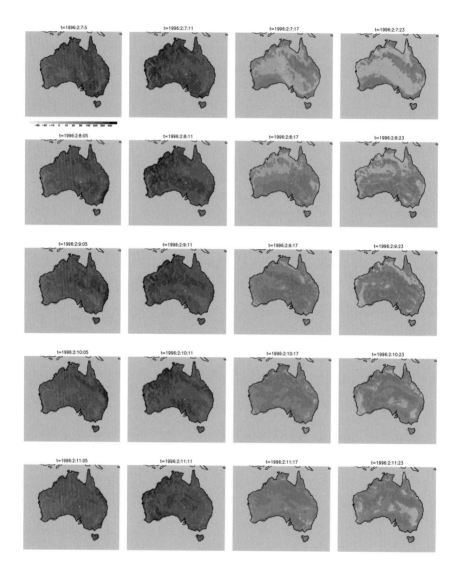

Fig. 11.16. Calculated latent heat fluxes over the Australian continent for five days in February 1996.

dioxins and radionuclides. For these reasons, it is important to determine areas and intensities of dust emission for both atmospheric modelling and air quality studies (e.g., Westphal et al., 1988; Tegen and Fung, 1994, 1995).

The dust cycle involves dust emission by wind erosion, dust transport by atmospheric turbulence and flow fields, and dust deposition through turbulent and molecular motion (dry deposition) and through precipitation (wet deposition). An important approach to modelling the dust cycle is to de-

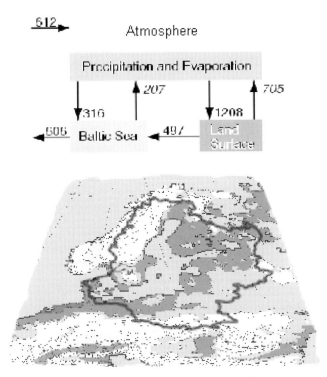

Fig. 11.17. Simulated hydrological cycle in the Baltic Sea drainage basin with the ECHAM4 model (10-year average). Precipitation over land and sea is indicated by downward arrows and the evaporation by upward arrows. The river runoff from land, and the net runoff into the North Sea are shown by arrows pointing to the right. The net flow of water vapour into the drainage basin is shown by an arrow pointing to the left. Units are given in km^{3}y^{-1} (adapted from Bengtsson, 1998).

velop an integrated wind-erosion modelling system that has the capacity for modelling the entrainment, transport and deposition processes. This is a formidable task because these processes are governed by a wide range of factors involving atmospheric conditions, soil state, and surface properties.

A fully-integrated wind-erosion modelling system couples the following three major components: 1) an atmospheric prediction model, together with a land-surface model; 2) a wind-erosion scheme; and 3) a geographic information database. The framework of an integrated wind-erosion modelling system is illustrated in Fig. 11.18. The atmospheric model, either a GCM, a weather-prediction model or a local flow model, provides the necessary input data for the wind-erosion scheme, including wind speed and precipitation. In the context of wind-erosion modelling, the land-surface scheme produces soil moisture as an important output. The wind-erosion scheme obtains friction velocity from the atmospheric model, soil moisture from the land-surface scheme and other spatially distributed parameters from the GIS database. The wind-erosion scheme predicts the streamwise saltation flux and the dust emission rate for different particle-size groups. The transport and deposition model obtains flow velocity, turbulence data and precipitation from the atmospheric model, and dust-emission rate and particle-size information from the wind-erosion scheme. The geographic information database provides spatially distributed parameters, such as soil type and vegetation coverage, for the atmospheric, land-surface and wind-erosion models. Figure 11.18 also illustrates a possible computational procedure, although this may differ for different models. In this example, the atmospheric model is first run after initialisation for atmospheric dynamics and atmospheric physics. This is followed by running the land-surface scheme and the wind-erosion scheme. Finally, the calculation of dust transport and deposition is carried out.

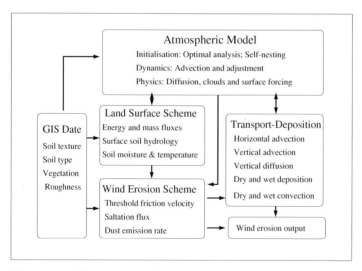

Fig. 11.18. The structure of an integrated wind-erosion modelling system consisting of an atmospheric prediction model, a land-surface model, wind-erosion model, a transport and deposition scheme, and a GIS database.

Land-surface parameters are required to quantify the land-surface properties that influence wind erosion. These parameters can be divided into three categories. The first category includes parameters relating to soil properties, including soil texture, soil-salt concentration, and parameters used to specify the soil binding strength. The second category includes parameters relating to surface roughness and drag partitioning, such as frontal area index (total area of roughness elements projected in the direction of wind per unit ground area), the overall roughness length of the surface, that of the underlying erodible surface, and the erodible fraction of the surface. The third category includes parameters relating to the soil thermal and hydraulic properties required by land-surface modelling.

There are obvious difficulties in quantitative wind-erosion modelling. The dust emission rate is sensitive to input data, such as soil moisture and frontal-area index, which are difficult to determine accurately. Nevertheless, wind-erosion models developed recently have produced estimates of wind-erosion intensity and patterns that are in reasonable agreement with observations (Marticorena and Bergametti, 1995; Shao et al., 1996; Shao and Leslie, 1997; and Lu and Shao, 2000). In the following, we present an example application of a wind-erosion modelling system for the prediction of dust storm events of 8 - 11 February 1996, over the Australian continent.

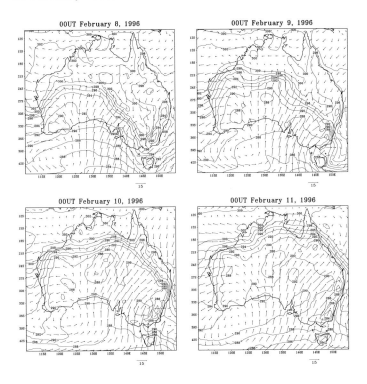

Fig. 11.19. Numerical predictions of near-surface wind and temperature fields associated with the cold front (which can be identified from the temperature gradient) for the 8 - 11 February 1996 wind-erosion event in Australia (from Shao and Leslie, 1997).

In the early 1990s, the central and eastern parts of Australia experienced a severe drought. Consequently, vegetation cover was significantly destroyed in some areas. With such a climatic background, wind-erosion risk in inland Australia is high. In the southern Hemisphere summer, cold fronts follow on from gusty warm airflow, but little rainfall may result in a dry top soil and wind erosion. During summer 1996, frequent wind-erosion activities were observed in Australia. In New South Wales, for example, there were 10 reported dust events in February 1996, but the 8 – 11 February events were the most severe.

The weather pattern that produced the dust storms of 8 – 11 February 1996 is common in Australia. A deep low pressure system was crossing the Southern Ocean to the south of the continent and an associated cold front produced strong north to northwesterly winds ahead of the front. This was followed by a vigorous, cool to cold, south to southeasterly air stream that persisted for several days after the passage of the cold front. The predictions of the near-surface wind speed and air temperature fields, using an atmospheric prediction model, are shown in Fig. 11.19. The location of the front can be seen from the narrow regions with a sharp temperature gradient. By 9 February 1996 strong winds reached up to 20 ms^{-1} in central Australia. Over the next two days, the low pressure system moved further east, and by 10 February the cold front had moved offshore of the Australian east coast. By 11 February, the Australian continent was dominated by the high pressure system with wind speed easing. During this period, dust storms were reported over large parts of the Australian continent, from the west and southwest coasts and adjacent regions, and across much of the interior of the continent. The dust storms were most pronounced on 9 and 10 February.

Wind erosion occurred on 20 days in February at various locations in Australia. The predicted daily average dust emission rates are shown in Fig. 11.20 for these days. For 8 February 1996, the system predicted strong wind erosion in the Simpson Desert (central Australia) and scattered, weak erosion activity in the adjacent areas to the west and south. The areas affected by erosion were in good agreement with the regions under the influence of the cold front. By 9 February 1996, the intensity and extent of wind-erosion activities had increased over the Australian continent, especially in the Simpson Desert and surrounding areas, as near-surface winds in these regions increased (Fig. 11.19b). Over the four-day period, wind erosion was strongest on 9 February 1996. By 10 February, while erosion remained severe in central Australia and extended further towards the northeast, it was reduced in the western parts of Australia. As the frontal system moved further east and the wind speed decreased over the continent, erosion activities were significantly weaker on 11 February 1996, and virtually reduced to zero late that day. The highest six-hourly averaged near-surface dust concentration, streamwise saltation flux and vertical dust emission rate were estimated to be 9528 μg m^{-3}, 1890 mg m^{-1}s^{-1}, and 0.65 mg m^{-2}s^{-1}, respectively, at

11.5 Prediction of Environmental Cycles   417

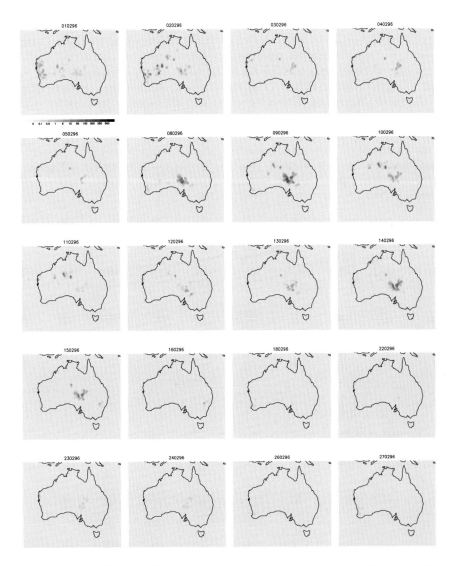

Fig. 11.20. Predicted daily averages of dust emission rate over the Australian continent, in $g\,m^{-2}s^{-1}$ (the values are multiplied by 10^4 for plotting), for 20 selected days of February 1996. The dust-storm event that occurred between 8 and 11 February 1996 was the most severe (from Lu and Shao, 2000).

(136.8°E, -27.9°S) during the time period between 5 hours and 11 hours on 2 February 1996. From the model results, the total dust emission and that for each particle-size group can be estimated and used for the calculation of dust transport.

Overall, wind erosion was most active in the first half of February 1996 and the major erosion areas were located in Western Australia and the Simpson Desert. Three dust storms were particularly severe, one occurring at the beginning of February, one between 8 and 11 February, and one between 14 and 15 of the month. Figure 11.21 shows the simulated temporal evolution of the total amount of suspended dust, and the amount of dust for each particle-size group for February 1996. The figure also shows that while a large amount of the giant sized particles were emitted from the surface, only a small amount of these large particles remained suspended in air for a long time, due to their large settling velocity. As expected, the residence time of suspended dust increases with decreasing particle size.

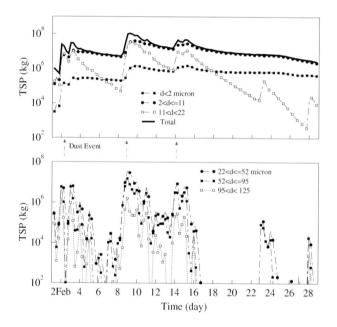

Fig. 11.21. Simulated temporal evolution of the total amount of suspended dust, and the amount of suspended dust for each particle-size group for February 1996.

11.6 Specific Events: Air Quality

Integrated environmental modelling systems have also been widely applied to the simulation and prediction of specific environmental events. Among numerous examples is the modelling of air quality. In recent years, many different versions of the air-quality modelling system have been developed with a similar framework and structure. Figure 11.22 shows the structure of the European Air Pollution Dispersion Model System (EURAD), developed by Ebel et al. (1997). The system consists of three major components, namely, the Meteorological Meso-scale Model (MM5), the EURAD Emission Model (EEM) and the Chemical Transport Model (CTM). These three major components are supported by many more modules, designed for model input, output, and interfacing between the various components.

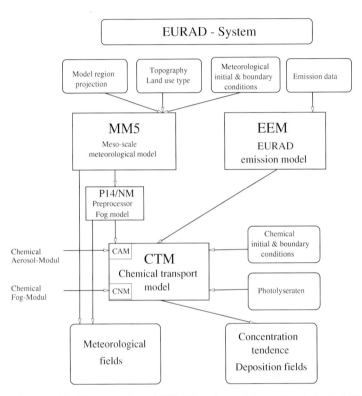

Fig. 11.22. Framework of EURAD (adapted from Ebel et al., 1997).

The functioning of the three major components is closely related. The prerequisite for the simulation and prediction of the transport, chemical changes and deposition of air pollutants is the adequate forecast of the meteorological variables, such as wind speed, air temperature, air humidity and turbulence intensity. In EURAD, this is achieved through the implementation of MM5 (Grell et al., 1994). The concentration of air pollution at a given location is influenced by several factors, including pollution emission, its transport through advection and turbulent diffusion, transformation through chemical reactions, and deposition. These processes are considered in the chemical transport model. The emission rates of various pollutants are estimated using

EEM, which is mainly used to establish a database for emission distribution and intensity. Such a model not only requires raw data relating to emission sources, such as traffic- and industry-generated emissions, but also the social-economic activities under a changing environment, as well as the distribution of population. For the determination of emission rates, it is also necessary to apply spatial and temporal interpolation to account for missing data.

Like other environmental models, the determination of initial and boundary conditions for CTMs is of vital importance. For air-pollution modelling, these conditions pose great practical difficulties, as they are often unknown for the particular air-pollution episode we wish to model and predict. Chang et al. (1987) have proposed a method for the initialisation of the air-pollution fields. In this method, within the framework of the chemical mechanisms, a self-consistent data set is generated. Such a data set is produced through the use of climatological values of air pollution, then forced for a time period (e.g., two days) with the actual meteorological and emissions data. These air-pollution "climate" data are mostly based on measurements (often profiles of pollution species, such as NO_x and O_3) and their spatial interpolation. At the boundaries of the simulation domain, either the pollution concentration or its gradient must be specified.

As in meteorological and ocean simulations, data assimilation using optimal interpolations and variational data assimilation techniques are useful for improved accuracy of the initial air pollution field and forecast. A detailed discussion on optimal interpolation is given by Daley (1991). It involves the interpolation of measurements obtained at irregularly distributed locations over a regular grid of the numerical model, e.g., MM5 and CTM. The optimal interpolation is not only important for the preparation of initial and boundary conditions from observed data, but also the diagnoses of the spatial distribution of air pollutants. The model simulation for a certain time is combined in an optimal way to give the spatial distribution of the concentration field of a particular air pollution species. Figure 11.23 shows an example of the assimilated field of the surface ozone concentration for 00 UT 26 July 1994. Figure 11.23a shows the measured and background field of ozone in μg m^{-3}, and Figure 11.23b shows the assimilated concentration field of ozone.

Variational data assimilation can also be implemented to improve the estimate of the initial concentration field of air pollutants. The method of variational data assimilation has been increasingly used in meteorological and oceanic modelling. The purpose of this method is to optimize the estimate of the initial condition for the model by taking into consideration the observations for a certain period of time. The variational data assimilation is achieved through minimizing the function

$$J = \int_{t_0}^{t_1} (\mathbf{x_t} - \mathbf{x_t}^{obs})^T W_t (\mathbf{x_t} - \mathbf{x_t}^{obs}) dt \tag{11.47}$$

11.6 Specific Events: Air Quality   421

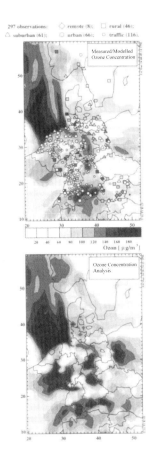

Fig. 11.23. Optimal analysis combining the measured and modelled fields of surface ozone concentration for 00 UT 26 July 1994, central-west Europe (from Ebel et al., 1997).

where $\mathbf{x}_t$ is the modelled vector in the phase space for time t, $\mathbf{x}_t^{obs}$ are the corresponding measured one and W_t is the error-covariance matrix, which is a description of the quality of the measurements. For the efficient minimization of J, the gradient of J at $\mathbf{x}_0$, $\nabla J|\mathbf{x}_0$, is required. Suppose the model equations are given by

$$\frac{\mathrm{d}\mathbf{x}}{\mathrm{d}t} = \mathbf{M}(\mathbf{x}) \tag{11.48}$$

where $\mathbf{M}(\mathbf{x})$ represents the model equations at time t. The corresponding equation for the adjunct model is then

$$\frac{\mathrm{d}\delta'\mathbf{x}}{\mathrm{d}t} = -\mathbf{M}'^{*}_{\mathrm{x}(t)}\delta'\mathbf{x} \qquad (11.49)$$

where $\mathbf{M}'^{*}_{\mathrm{x}(t)}$ is the adjunct of the Jacobi Matrix $\mathbf{M}'_{\mathrm{x}(t)}$. The backwards integration of the above equation, taking the measurements into account, gives the vector $\delta'\mathbf{x}$, which is the required gradient.

As an example of integrated air-quality modelling, we use the 1990 summer smog episode during the time period between 31 July 1990 and 5 August 1990. During this episode, ozone concentrations in central-west Europe exceeded 120 ppbv. Figure 11.24 shows the near-surface meteorological situation. On 31 July 1990, central Europe was under the influence of a strengthening high-pressure zone. In the regions dominated by the westerly, the advection of relatively clean air prevailed. During the following days, the high-pressure system moved slowly eastwards, with an easterly stream dominating its southern flank. This was followed by a southwesterly stream over central and western Europe, bringing increasingly warm and humid air into the region. On 5 August 1990, a cold front moved through central Europe, significantly reducing the concentration level of ozone.

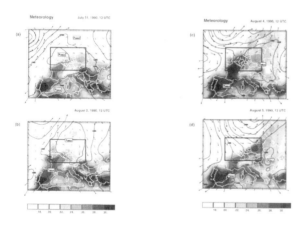

Fig. 11.24. Weather simulation for the 1990 summer smog episode in Europe (from Ebel et al., 1997).

The simulated near-surface concentration of ozone is shown in Fig. 11.25. On 31 July 1990, the ozone concentration in western and central Europe was mostly below 100 ppbv, although in a few coastal cities in the Mediterranean it was above that value. On 2 August, the ozone concentration was over 100 ppbv in large areas of western and central Europe. Part of the ozone generated over the continent reached the Golf of Biskaya, where the ozone concentration was higher than over the Atlantic. In Great Britain, under a prevailing humid and warm air mass, ozone values reached their highest value on 2 and 3 August 1990, before they were reduced by the passage of a cold front on 4 August 1990. On that day, central Europe was still under the influence of a warm and humid air mass, when the highest ozone values during the entire episode occurred.

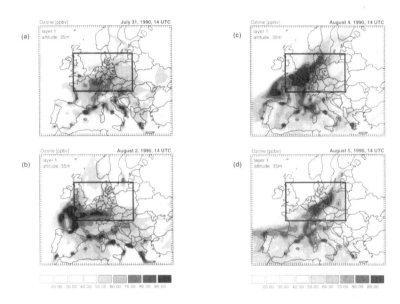

Fig. 11.25. Ozone prediction for the 1990 summer smog episode in Europe (from Ebel et al., 1997).

11.7 Performance of Integrated Environmental Modelling Systems

The performance of the various integrated environmental modelling systems is extremely difficult to assess. Often, this is because we do not have sufficient and accurate data for the validation of the very complex systems. Weather and climate models are probably the best evaluated models so far. In particular, substantial international efforts have been devoted to the evaluation of climate models. Gates et al. (1996) have compared the modelled outputs, mainly surface parameters, of the existing climate models with climatological observations. After examining a range of parameters, they have concluded that the large-scale features of the current climate are well simulated, on average, by the current coupled models. Different coupled models simulate the current climate with varying degrees of success. It has been found that atmospheric models generally provide a realistic portrait of the phase and amplitude of the seasonal march of the large-scale distribution of pressure, temperature and circulation. The largest discrepancies in the seasonal cycle of sea-level pressure and surface-air temperature occur in the higher latitudes, while the largest discrepancies in the seasonal cycle of precipitation are found in the tropics. Land-surface models generally perform well if site specific parameters and atmospheric conditions are specified well. Hence, the large uncertainties in land-surface modelling occur due to the inaccuracies of land-surface parameters. Ocean GCMs are also found to realistically simulate the large-scale structure of oceanic gyres and the gross features of thermohaline circulation.

References

1. Abdulla F. A. and Lettenmaier D. P. (1997) Application of regional parameter estimation schemes to simulate the water balance of a large continent river. J. Hydrol. 197:258–285.
2. Aber J. D. (1992) Terrestrial ecosystems. In: Trenberth K. E. (Ed.) Climate System Modeling, Cambridge University Press, 173–200.
3. Abramopoulos F., Rosenzweig C. and Choudhury B. (1988) Improved ground hydrology calculations for Global Climate Models (GCMs): Soil moisture movement and evapotranspiration. J. Climate 1:921–941.
4. Advisory Council of the National Aeronautics and Space Administration (1998) Earth system science (a close view). University Corporation for Atmospheric Research, Boulder, USA.
5. Allen J. S. (1973) Upwelling and coastal jets in a continuously stratified ocean. J. Phys. Oceanogr. 3:245–257.
6. Allen J. S. (1980) Models of wind-driven currents on the continental shelf. Ann. Rev. Fluid Mech. 12:389–433.
7. Allen J. S., Newberger P. A and Federiuk J. (1995) Upwelling circulation on the Oregon continental shelf. Part I: Response to idealized forcing. Phys. Oceanogr. 25:1,843–1,866.
8. Alley W. M., Dawdy D. R. and Schaake J. C. (1980) Parametric deterministic urban watershed model. Proc. of ASCE, J. Hydraul. Div. 106(HY5):679–690.
9. Ambroise B., Beven K. and Freer J. (1996) Toward a generalization of the TOPMODEL concepts: Topographic indices of hydrological similarity. Water Resour. Res. 32(7):2,135–2,145.
10. Ambrose R. B. Jr., Wool T. A., Connolly J. P. and Schanz R. W. (1987) WASP4, a Hydrodynamic and Water Quality Model: Model Theory, User's Manual, and Programmer's Guide (revision for WASP4.3x). Environmental Research Laboratory, Office of Research and Development, U.S. Environmental Protection Agency, Athens, GA.
11. André J. C., Goutorbe J. P. and Perrier A. (1986) HAPEX-MOBILHY: a hydrologic atmospheric experiment for the study of water budget and evaporation flux at the climate scale. B. Am. Meteorol. Soc. 67:138–144.
12. Andrews J. F. (1968) A mathematical model for the continuous culture of microorganisms utilizing inhibitory substrates. Biotechnol. Bioeng. 10:707–723.
13. Anh V., Duc H. and Shannon I. (1997) Spatial variability of Sydney air quality by cumulative semivariogram. Atmos. Environ. 31:4,073–4,080.
14. Anh V., Azzi M., Duc H., Johnson G. and Tieng Q. (1998) A reactive state-space model for prediction of urban air pollution. Environ. Modell. Softw. 13:239–246.
15. Anthes R. A. (1986) The general question of predictability. Mesoscale meteorology and forecasting. In: Ray P. S. (Ed.) Mesoscale meteorology and forecasting. American Meterology Scoiety. Boston, MA, USA.

16. Anthes R. A. and Kuo Y. H. (1986) The influence of soil moisture on circulations over North America on short-time scales. In: Jerome Namias Commemorative Volume, Namias Symp., La Jolla, Scripps Inst. Oceanogr. Ref. Ser., 86–17, Univ. Calif., 132–147.
17. Arola A. and Lettenmaier D. P. (1996) Effects of subgrid spatial heterogeneity on GCM-Scale land surface energy and moisture fluxes. J. Climate 9(6):1,339–1,349.
18. Ashraf M., Loftis J.C. and Hubbard K.G. (1997) Application of geostatisticals to evaluate partial weather station network. Agri. Forest Meteorol. 84:255–271.
19. Austin J. A. and Barth J. A. (2001) Variation in the position of the upwelling front on the Oregon shelf. J. Geophys. Res. (submitted).
20. Avissar R. (1992) Conceptual aspects of a statistical–dynamical approach to represent landscape subgrid-scale heterogeneities in atmospheric models. J. Geophys. Res. 97:2,729–2,742.
21. Avissar R. and Pielke R. A. (1989) A parameterization of heterogeneous land surface for atmospheric numerical models and its impact on regional meteorology. Mon. Weather Rev. 117:2,113–2,136.
22. Avissar R. and Schmidt T. (1998) An evaluation of the scale at which ground-surface heat flux patchiness affects the convective boundary layer using large-eddy simulation model. J. Atmos. Sci. 55:2,666–2,689.
23. Azzi M. and Johnson G. (1992) An introduction to the generic reaction set photochemical smog mechanism. Proceedings of the 11th International Clean Air Conference, Brisbane, 451–462.
24. Bai S. and Wu H. (1998) Numerical sea ice forecast for the Bohai Sea. Acta Meteorological Sinica 56(2):139–153.
25. Bailey J. E. and Ollis D. F. (1986) Biochemical Engineering Fundamentals. 2nd edition, McGraw-Hill, New York.
26. Baldwin M., Treadon R. and Contorno S. (1994) Precipitation type prediction using a decision tree approach with NMC's mesoscale eta model. Preprints, 10th Conf. on Numerical Weather Prediction, Amer. Meteorol. Soc., Portland, Oregon, 30–31.
27. Bales J. D. and Giorgino M. J. (1998) Dynamic modeling of water-supply reservoir physical and chemical processes. Proceedings of the First Federal Interagency Hydrological Modeling Conference, Las Vegas, U.S., April 19–23, 1998. 2–61.
28. Bankoff S. C. and Hanzevack E. L. (1975) The adaptive-filtering transport model for prediction and control of pollutant concentration in an urban airshed. Atmos. Environ. 9:793–808.
29. Barber S. A. (1981) Soil chemistry and the availability of plant nutrients. In: Baker D. E. et al. (Eds.) Chemistry in the Soil Environment. ASA Special Publication No. 40. SSSA. Madison, Wisconsin 53711 USA.
30. Bardsley W. and Campbell D. I. (1994) A new method for measuring near-surface moisture budget in hydrological systems. J. Hydrol. 154:245–254.
31. Barnier B. (1998) Forcing the ocean. In: Chassignet E. P. and Verron J. (Eds.) Ocean Modelling and Parameterization. Kluwer Academic Pub., Netherlands, 45–80.
32. Barnier B., Siefried L. and Marchesiello P. (1995) Thermal forcing for a global ocean circulation model Using a 3-year climatology of ECMWF analyses. Mar. Syst. 6:363–380.
33. Béranger K., Siefried L., Barnier B., Garnier E. and Roquet H. (1999) Evaluation of operational ECMWF surface Freshwater fluxes over oceans during 1991–1997. J. Mar. Sys. 22:13–36.

34. Barry R. G. (1987) The cryosphere-neglected component of the climate system. Toward understanding climate change. Westview Press, Boulder and London.
35. Baumgartner A. and Reichel E. (1975) The World Water Balance. Elsevier, New York.
36. Baumann D. D., Boland J. J., and Hanemann W. M. (1998) Urban Water Demand Management and Planning. McGraw-Hill, New York.
37. Beerling D. J., Woodward F. I., Lomas M. and Jenkins A. J. (1997) Testing the response of a dynamic global vegetation model to environmental change – A comparison of observations and predictions. Global Ecol. Biogeogr. 6:439–450.
38. Beljaars A. C. M. and Bosveld F. C. (1997) Cabauw data for the validation of land surface parameterization schemes. J. Climate 10:1,172–1,193.
39. Bengtsson L. (1998) Climate modelling and prediction – achievements and challenges. Proceedings of the International CLIVAR Conference Paris, December 2–4, 1998.
40. Bennett A. F. (1992) Inverse Methods in Physical Oceanography. Cambridge Univ. Press, New York.
41. Bennett A. F., Leslie L. M., Hagelberg C. R. and Powers P. E. (1993) Tropical cyclone prediction using a barotropic model initialized by a generalized inverse method. Mon. Weather Rev. 121:1,714–1,729.
42. Bentamy A., Grima N., Quilfen Y., Harscoat V., Maroni C. and Pouliquens S. (1997) An Atlas of Surface Wind from ERS-1 Scatteromoter Measurements 1991–1996. IFREMER, Brest.
43. Bergstrom S. and Graham L. P. (1998) On the scale problem in hydrological modelling. J. Hydrol. 211:253–265.
44. Bertalanffy L. Von. (1968) General System Theory. G. Braziller.
45. Betts A. K. and Ball J. H. (1998) FIFE surface climate and site-average dataset: 1987–1989. J. Atmos Sci. 55:1,091–1,108.
46. Betts R., Cox P., Lee S. and Woodward F. (1997) Contrasting physiological and structural vegetation feedbacks in climate change simulations. Nature 387:796–799.
47. Beven K. J. and Kirkby M. J. (1979) A physically-based, variable contributing area model of basin hydrology. Hydrol. Sci. Bull. 24(1):43–69.
48. Bhumralkar C. M. (1975) Numerical experiments on the computation of ground surface temperature in atmospheric general circulation models. J. Appl. Meteorol. 14:1,246–1,258.
49. Biddiscombe E. F., Rogers A. L., Allison H. and Litchfield R. (1985) Response of ground water levels to rainfall and to leaf growth of farm plantation near salt seeps. J. Hydrol. 78:19–34.
50. Biftu G. F. and Gan T. Y. (1999) Retrieving near-surface soil moisture from RADARSAT SAR Data. Water Resour. Res. 35(5):1,569–1,580.
51. Biswas A. K. and Qu G. (Ed.)(1987) Environmental Impact Assessment for Developing Countries' Natural Resources and the Environment Series, Volume 19, London.
52. Black T. (1994) The new NMC mesoscale eta model: description and forecast Examples. Weather Forecasting 9:265–278.
53. Blackadar A. K. (1976) Modeling the nocturnal boundary layer. In: Proceedings of the Third Symposium on Atmospheric Turbulence, Diffusion and Air Quality. Am. Meteorol. Soc., Boston, Mass, 46–49.
54. Blanchard C., Roberts P., Chinkin L. and Roth P. (1995) Application to the Smog Production (SP) Algorithm to the TNRCC COAST Data. 88th Annual Air and Waste management Association Meeting, Paper 95-TP15P.04.

55. Bleck R. and Boudra D. B. (1986) Wind-driven spin-up in eddy-resolving ocean models formulated in isopycnic and isobaric coordinates. J. Geophys. Res. 91:7,611–7,621.
56. Bleck R., Rooth C., Hu D. and Smith L. (1992) Salinity-driven thermohaline transients in a wind- and thermohaline-forced isopycnic coordinate model of the North Atlantic. Phys. Oceanogr. 22:1,486–1,505.
57. Bleck R., Sun S. and Dean S. (1997) Global ocean simulations with an isopycnic coordinate model. In: Bhanot G., Chen S. and Seiden P. (Eds.) Some New Directions in Science on Computers. World Scientific, Singapore, 297–317.
58. Blumberg A. F. and Kantha L. H. (1985) Open boundary conditions for circulation models. Hydr. Eng. 111:237–255.
59. Blumberg A. F. and Mellor G. L. (1987) A description of a three-dimensional coastal ocean circulation model. In: Heaps N. (Ed.) Three-dimensional Coastal Ocean Models. Amer. Geophys. Union, 1–16.
60. Bolle H. J., Andre J. C., Arrue J. L., Barth H. K., Bessemoulin P., Brasa A., de Bruin H. A. R., Dugdale G., Engman E. T., Evans D. L., Fantechi R., Fiedler F., van de Griend A., Imeson A. C., Jochum A., Kabat P., Kratzsch T., Lagouarde J. P., Langer I., Llamas R., Lopes-Baeza E., Meli Miralles J., Muniosguren L. S., Nerry F., Noilhan J., Oliver H. R., Roth R., Sanchez Diaz J., de Santa Alalla M., Shuttleworth W. J., Sogaard H., Stricker H., Thornes J., Vauclin M. and Wickland D. (1993) European field experiment in a desertification threatened area. Annals Geophysicae 11:173–189.
61. Bonan G. B. (1995) Land atmosphere CO_2 exchanges simulated by a land surface process model coupled to an atmospheric general circulation model. J. Geophys. Res. 100:2,817–2,831.
62. Bonan G. B., Pollard D. and Thompson S. L. (1992) Effects of boreal forest vegetation on global climate. Nature 359:716–718.
63. Boning C. W., Holland W. R., Bryan F. O., Danabasoglu G. and McWilliams J. C. (1995) An overlooked problem in model simulations of the thermohaline circulation and heat transport in the Atlantic Ocean. Clim. 8:515–523.
64. Boning C. W., Bryan F. O., Holland W. R. and Doscher R. (1996) Deep-water formation and meridional overturning in a high-resolution model of the North Atlantic. J. Phys. Oceanogr. 26:1,142–1,164.
65. Borden R. C. and Bedient P. B. (1986) Transport of dissolved hydrocarbons influenced by oxygen-limited biodegradation: 1. Theoretical development. Water Resour. Res. 22:1,973–1,982.
66. Boudreau B. P. (1996) A method-of-line code for carbon and nutrient diagenesis in aquatic sediments. Comput. Geosci. 22:479–496.
67. Bounoua L., Collatz G. J., Sellers P. J., Randall D. A., Dazlich D. A., Los S. O., Berry J. A., Fung I., Tucker C. J., Field C. B. and Jensen T. G. (1999) Interactions between vegetation and climate: radiative and physiological effects of doubled CO_2. J. Climate 12:309–324.
68. Bourao F., Vachaud G., Haverkamp R. and Normand B. (1997) A distributed physical approach for surface-subsurface water transport modelling in agricultural watersheds. J. Hydrol. 203:79–92.
69. Box E. O. (1988) Estimating the seasonal carbon source-sink geography of a natural, steady-state terrestrial biosphere. J. Appl. Meteorol. 27:1,109–1,124.
70. Bradley E. F., Coppin P. A. and Godfrey J. S. (1991) Measurements of sensible and latent heat flux in the Western Equatorial Pacific Ocean. J. Geophys. Res. 96:3,375–3,389.
71. Braithwaite R. (1991) Australia's unique biota: Implications for ecological processes. In: Werner P. (Ed.) Savanna ecology and management: Australian per-

spectives and intercontinental comparisons, 3–11. Blackwell Scientific Publications, Oxford.
72. Bras R. L. (1990) Hydrology: An introduction to hydrological sciences. Addison-Wesley Publishing Company. Reeding, Mass.
73. Bromwich D. H. (1990) Estimates of antarctica precipitation. Nature 343:627–629.
74. Brooks R. H. and Corey A. T. (1964) Hydraulic properties of porous media. Hydrology Paper 3, Colorado State University, Fort Collins.
75. Brooks H. E., Stensrud D. J. and Tracton M. S. (1996) Short-range ensemble forecasting pilot project: a status report. Preprints, 11th Conf. on Numerical Weather Prediction, Amer. Meteor. Soc., Norfolk, Virginia.
76. Brooks H. E., Tracton M. S., Stensrud D. J., DiMego G. and Toth Z. (1995) Short-range ensemble forecasting: Report from a workshop (25–27 July 1994). Bull. Amer. Meteor. Soc. 76:617–624.
77. Brown M. E. and Arnold D. L. (1998) Land-surface–atmosphere interactions associated with deep convection in Illinois. Int. J. Climate 18:1,637–1,653.
78. Brown J. H. (1994) The ecology of coexistence (book review). Science 263:995–996.
79. Brubaker K. L., Entekhabi D. and Eagleson P. S. (1993) Estimation of continental precipitation recycling. J. Climate 6:1,077–1,089.
80. Brutsaert W. (1982) Evaporation into the Atmosphere. D.Reidel, Dordrecht.
81. Bryan K. (1969) A numerical method for the study of the circulation of the world ocean. J. Comp. Phys. 4:347–376.
82. Bryan K. (1984) Accelerating the convergence to equilibrium of ocean-climate models. J. Phys. Oceanogr. 14:616–673.
83. Bryan F. (1987) Parameter sensitivity of primitive equation in ocean general circulation models. J. Phys. Oceanogr. 17:970–985.
84. Bryan K. and Cox M. D. (1967) A numerical investigation of the oceanic general circulation. Tellus 19:54–80.
85. Bryan K. and Cox M. D. (1968) A nonlinear model of an ocean driven by wind and differential heating: Part I. Description of the three-dimensional velocity and density fields. J. Atmos. Sci. 25:945–967.
86. Bryan K. and Lewis L. J. (1979) A water mass model of the world ocean. J. Geophys. Res. 84:2,503–2,517.
87. Bryant E. (1997) Climate Process and Change. Cambridge University Press, Melbourne.
88. Buckley B. W and Leslie L. M (1998) High resolution numerical of Tropical Cyclone Drena undergoing extra-tropical transition. Meteorology and Atmospheric Physics 65:207–222.
89. Budyko M. I. (1956) Teplovoi Balans Zemnol Poverkhnosti. Gidrometeoizdat, Leningrad; Translated by Stepanova, N. A., 1958, The Heat Balance of the Earth's Surface, U.S. Weather Bur..
90. Budyko M. I. (1969) The effect of solar radiations on the climate of earth. Tellus 21:611–619.
91. Budyko M. I. (1974) Climate and Life. English edition Miller D. H. (Ed.) Int. Geophys. Series, Vol.18, Academic Press.
92. Buishand T. A. (1982) Some methods for testing homogeneity of rainfall records. J. Hydrol. 58:11–27.
93. Burrows C. J. (1990) Processes of Vegetation Change. Unwin Hyman, London.
94. Busalacchi A. J. and O'Brien J. J. (1981) Interannual variability of the equatorial Pacific in the 1960s. J. Geophys. Res. 86:10,901–10,907.

95. Claussen M. (1993) Shift of biome patterns due to simulated climate variability and climate change. Tech. Rep. 115, Max-Planck-Institut für Meterologie, Hamburg.
96. Cai W. and Godfrey J. S. (1995) Surface heat flux parameterizations and the variability of thermohaline circulation. J. Geophys. Res. 100:10,679–10,692.
97. Calder J. R., Harding R. Z. and Rosier P. T. W. (1983) An objective assessment of soil-moisture deficit models. J. Hydrol. 60:329–355.
98. Calvet J. C., Noilhan J., Roujean J. L., Bessemoulin P., Cabelguenne M., Olioso A. and Wigneron J. P. (1998) An interactive SVAT model tested against data from six contrasting sites. Agr. Forest Meteorol. 92:73–95.
99. Camillo P. J., Gurney R. J. and Schmugge T. J. (1983) A soil and atmospheric boundary layer model for evapotranspiration and soil moisture studies. Water Resour. Res. 19:371–380.
100. Camillo P. J. and Gurney R. J. (1986) A resistance parameter for bare soil evaporation models. Soil Sci. 141:95–105.
101. Campbell G. S. (1985) Soil Physics with Basic. Elsevier, New York, 73–97.
102. Campolo M., Andreussi P. and Soldati A. (1999) River flood forecasting with a neural network model. Water Resour. Res. 35(4):1,191–1,198.
103. Cane M. A. and Sarachik E. S. (1977) Forced baroclinic ocean motions, II, the linear equatorial bounded case. J. Mar. Res. 35:395–432.
104. Cane M. A. (1992) Tropical pacific ENSO models: ENSO as a mode of the coupled system. In: Trenberth K. E. (Ed.) Climate System Modeling, Cambridge University Press, 583–614.
105. Canter L. W. (1996) Environmental Impact Assessment. Second Edition, McGraw-Hill Publishing Company, Inc., New York.
106. Caporali E., Entekhabi D. and Castelli F. (1996) Rainstorm statistics conditional on soil moisture index: temporal and spatial characteristics. Meccanica 31:103–116.
107. Celia M. A., Bouloutas E. T. and Zebra R. L. (1990) A general mass-conservative numerical solution for the unsaturated flow equation. Water Resour. Res. 26:1,483–1,496.
108. Ciret C. and Henderson-Sellers A. (1995) "Static" vegetation and dynamic global climate: Preliminary analysis of the issues of time steps and time scales. J. Biogeogr. 22:843–856.
109. Chahine M. (1992) The hydrological cycle and its influence on climate. Nature 359:373–380.
110. Chang, J. S., Brost J. A., Isaksen S. A., Madronich S., Middleton P., Stockwell W. R. and Walcek C. J. (1987) A three dimensional Eulerian acid deposition model: physical concepts and formulation. J. Geophys. Res. 92:14,681–14,700.
111. Chao J. and Chen Y. (1980) The effects ice caps albedo feedback on the global climate in two-dimensional energy balance model. Sci. Sinica 12:1,198–1,207.
112. Chapman D. C. (1985) Numerical treatment of cross-shelf open boundaries in a barotropic coastal ocean model. J. Phys. Oceanogr. 15:1,060–1,075.
113. Chapman D. C. and Lentz S. J. (1997) Adjustments of stratified flow over a sloping bottom. J. Phys. Oceanogr. 27:340–356.
114. Charney J. G., Fjortoft R. and von Neumann J. (1950) Numerical integration of the barotropic vorticity equation. Tellus 2, 237–254.
115. Charney J. G. (1975) Dynamics of deserts and droughts in the Sahel. Q. J. Roy. Meteor. Soc. 101:193–202.
116. Charney J. G., Quirk W. J., Chow S. H. and Kornfield J. (1977) A comprative study of the effects of albedo change on drought in semi-arid regions. J. Atmos. Sci. 34:1,366–1,385.

117. Charney J. G. (1975) Dynamics of deserts and drought in the Sahel. Q. J. Roy. Meteor. Soc. 101:193–202.
118. Chase T., Pielke R., Kittel T., Nemani R. and Running S. (1996) Sensitivity of a general circulation model to global changes in leaf area index. J. Geophys. Res. 101:7393–7408.
119. Chehbouni A., Goodrich D. C., Moran M. S., Watts C. J., Kerr Y. H., Dedieu G., Kepner W. G., Shuttleworth W. J. and Sorooshian S. (2000) A preliminary synthesis of major scientific results during the SALSA program. Agri. For. Meteor. 105:311–323.
120. Chen S. P. (1998) Urgent tasks of citinization problems and citi-geoinformation system in China. Geo-informations 2:20–25. (in Chinese).
121. Chen S. P., Lu J. and Zhou C. H. (1999) An Introduction to Geographical Information System. Science Press. (in Chinese).
122. Chen F. and Avissar R. (1994) The impact of land-surface wetness heterogeneity on mesoscale heat fluxes. J. Appl. Meteor. 33:1,323–1,340.
123. Chen Y. M., Abriola L. M, Alvarez P. J. J., Anid P. J. and Vogel T. M. (1992) Modeling transport and biodegradation of Benzene and Toluene in sandy aquifer material: comparisons with experimental measurements. Water Resou. Res. 28:1,833–1,847.
124. Chen L., Wang Y. and Miao Q. (1996) The numerical experiments of the influence of Antarctic sea ice variety on the general circulation. In Zhou X., Lu L. (Eds.) The study on the interaction between Antarctic and global climate. China Meteorol. Press, 36–42.
125. Chen T. H., Henderson-Sellers A., Milly P. C. D., Pitman A. J., Beljaars A. C. M., Polcher J., Abramopoulos F., Boone A., Chang S., Chen F., Dai Y., Desborough C. E., Dickinson R. E., Dumenil L., Ek M., Garratt J. R., Gedney N., Gusev Y. M., Kim J., Koster R., Kowalczyk E. A., Laval K., Lean J., Lettenmaier D., Liang X., Mahfouf J. F., Megelkamp H. T., Mitchel K., Nasonova O. N., Noilhan J., Polcher J., Robock A., Rosenzweig C., Schaake J., Schlosser C. A., Schulz J. P., Shmakin A. B., Verseghy D. L., Wetzel P., Wood E. F., Xue Y., Yang Z. L. and Zeng Q. (1997) Cabauw experimental results from the project for intercomparison of landsurface parameterization schemes. J. Climate 10:1,144–1,215.
126. Chen X. and Wheeler A. (1998) Enhancement of three-dimensional reservoir quality modelling by Geographic Information Systems. Geographical Environmental Modelling 2(2):125–139.
127. Chen L. and Miao Q. (1996) Numerical experiments of Antarctic sea ice cover and ocean conditions on climate changes. In: Zhou X. and Lu L. (Eds.) The Interactions between Antarct and Global Climate. China Meteorol. Press, 51–66.
128. Chen C. W. and Orlob G. T. (1975) Ecology simulation for aquatic environments. In: Patten B. C. (Ed.) System Analysis and Simulation in Ecology. Academic Press, New York, 476–587.
129. Cheng J. C. and Li Q. (1998) Study on informatics technique of flood forecast and protection of the Yangtze River. Geo-informatics 3:17–20. (in Chinese).
130. Cheng A. and Preller R. H. (1994) An ice-ocean coupled model for the Northern Hemisphere. In: Yu Z. et al. (Eds.) Sea Ice Observation and Modelling. China Ocean Press, Beijing, 78–87.
131. Chou J. F. (1995) Long-range Numerical Weather Prediction. Meteorol. Press. (in Chinese).
132. Christensen F. T. and Lu Q. (1994) Numerical sea ice modeling for industrial application. In: Yu Z. et al. (Eds.) Sea Ice Observation and Modelling. China Ocean Press, Beijing, 165–182.

133. Chu J. and Xu H. (1992) The numerical simulation of water quality in fluvial network. J. Hohai Univ. 1:16–22. (in Chinese).
134. Ciret C. and Henderson-Sellers A. (1997) Sensitivity of global vegetation models to present-day climates simulated by global climate models. Global Biogeochem. Cy. 11:415–434.
135. Ciret C., Polcher J. and Le Roux X. (1999) An approach to simulate the phenology of savanna in the Laboratoire de Météorologie Dynamique GCM. Global Biogeochem. Cy. 13:603–622.
136. Ciret C. and Henderson-Sellers A. (1998) Sensitivity of ecosystem models to the spatial resolution of the NCAR Community Climate Model CCM2. Clim. Dynam. 14:409–429.
137. Clapp R. B.and Hornberger G. M. (1978) Empirical equations for some soil hydraulic properties. Water Resour. Res. 14:601–604.
138. Clark C. A. and Arritt R. W. (1995) Numerical simulations of the effects of soil moisture and vegetation cover on the development of deep convention. J. Appl. Meteorol. 34:2,029–2,045.
139. Clarke C. C. and Munasinghe M. (1994) Economic aspect of disasters and sustainable development: An introduction. In: Munasinghe M. and Clarke C. C. (Eds.) Disaster prevention for sustainable development: Economic and policy issues, a report from the Yokohama world conference on natural disaster reduction, May 23–27, 1994, 1–9.
140. Claussen M. (1994) On coupling global biome models with climate models. Climate Res. 4:203–221.
141. Claussen M. and Gayler V. (1997) The greening of Sahara during the Mid-Holocene: results of an interactive atmosphere-biome model. Global Ecol. Biogeogr. 6:369–377.
142. Cohn S. E., da Silva A., Guo J., Sienkiewicz M. and Lamich D. (1998) Assessing the effects of data selection with the DAO physical–space statistical analysis system. Mon. Weather Rev. 126:2,913–2,926.
143. Cole T. M. and Buchak E. M. (1995) CE-QUAL-W2: A two-dimensional, laterally averaged, hydrodynamic and water-quality model, version 2.0, user's manual: Vicksburg, Mississippi, Instruction Report EL-95-1, U.S. Army Engineer Waterways Experiment Station.
144. Collatz G. T., Ball J. T., Grivet C. and Berry J. A. (1991) Physiological and environmental regulation of stomatal conductance, photosynthetis and transpiration: A model that includes a laminar boundary layer. Agr. Forest Meteorol. 53:107–136.
145. Collett R. S. and Oduyemi K. (1997) Air quality modelling: A technical review of mathematical approaches. Meteorol. Appl. 4:235–246.
146. Cox M. D. (1975) A baroclinic numerical model of the world ocean: preliminary results. In: Contribution to Numerical Models of Ocean Circulation (Proceedings of the Symposium held at Durham), New Hampshire, 17–20 October, 1972. 107–120.
147. Cox M. D. (1984) A primitive equation, three-dimensional model of the ocean. GFDL Ocean Group Tech Rep No.1.
148. Cox M. D. (1987) Isopycnal diffusion in a z-coordinate ocean model. Ocean Modelling 74:1–5. (Unpublished manuscript).
149. Cox P. M., Betts R. A., Bunton C. B., Essery R. L., Rowntree P. R. and Smith J. (1999) The impact of new land surface physics on the GCM simulation of climate and climate sensitivity. Clim. Dynam. 15:183–203.
150. Cramer W. and Fischer A. (1996) Data requirements for ecosystem modelling. In: Walker B. and Steffen W. (Eds.) Global Change and Terrestrial Ecosystems. Cambridge University Press, Cambridge, IGBP Book Series, 529–565.

151. Cuenca, R. H., Ek, M. and Mahrt, L. (1996) Impact of soil water property parameterization on atmospheric boundary layer simulation. J. Geophys. Res. - Atmospheres 101(D3):7,269–7,277.
152. Dabes J. N., Finn R. K. and Wilke C. R. (1973) Equations of substrate-limited growth: the case for Blackman kinetics. Biotechnol. Bioeng. 15:1,159–1,177.
153. Dale V. H. (1997) The relationship between land-use change and climate change. Ecol. Appl. 7:753–769.
154. Daley R. (1991) Atmospheric Data Analysis. Cambridge University Press, Cambridge.
155. Darnell W. L., Staylor W. F., Ritchey N. A., Gupta S. K. and Wilber A. C. (1996) Surface Radiation Budget: A Long Term Global Dataset of Shortwave and Longwave Fluxes. EOS Transactions, Amer. Geophs. Union.
156. Darocha H. R., Nobre C. A., Bonatti J. P., Wright I. R. and Sellers P. J. (1996) A vegetation–atmosphere interaction study for Amazonia deforestation using field data and a single column model. Q. J. Roy. Meteor. Soc. 122:567–594.
157. Da Silva A., Young C. and Levitus S. (1994) Atlas of Surface Marine Data 1994, Volume 4: Anomalies of Freshwater Fluxes. NOAA atlas NESDIS 9, U.S. Department of Commerce, NOAA, NESDIS.
158. Deardorff J. W. (1977) A parameterization of ground-surface moisture content for use in atmosphere prediction models. J. Appl. Meteorol. 16:1,182–1,185.
159. Deardorff J. W. (1978) Efficient prediction of ground surface temperature and moisture with inclusion of a layer of vegetation. J. Geophys. Res. 83:1,889–1,903.
160. de Cuevas B. A. (1992) The Main Runs and Datasets of the Fine Resolution Antarctic Model Project (FRAM). Part I: The Coarse Resolution Runs. Institute of Oceanographic Sciences, Internal Document No.315.
161. de Cuevas B. A. (1993) The Main Runs and Datasets of the Fine Resolution Antarctic Model Project (FRAM). Part II: The Data Extraction Routines. Institute of Oceanographic Sciences, Internal Document No.319.
162. de Cuevas B. A., Webb D. J., Coward A. C., Richmond C. S. and Rourke E. (1998) The UK Ocean Circulation and Advanced Modelling Project (OCCAM). In: Allan R. J., Guest M. F., Simpson A. D., Henty D. S. and Nicole D. A. (Eds.) High Performance Computing, Proceedings of HPCI Conference 1998, Manchester, 12–14 January, 1998. Plenum Press, 325–335.
163. Delecluse P., Davey M. K., Kitamura Y., Philander S. G. H., Suarez M. and Bengtsson L. (1998) Coupled general circulation modeling of the Tropical Pacific. J. Geophys. Res. 103(C7):14,357–14,373.
164. Delworth T. L. and Manabe S. (1988) The influence of potential evaporation on the variability of simulated soil wetness and climate. J. Climate 1:523–547.
165. Delworth T. L. and Manabe S. (1989) The influence of soil wetness on near-surface atmospheric variability. J. Climate 2:1,447–1,462.
166. De Noblet N., Prentice C., Joussaume S., Texier D., Botta A. and Haxeltine A. (1996) Possible role of atmosphere-biosphere interactions in triggering the last glaciation. Geophys. Res. Lett. 23:3191–3194.
167. Desborough C. E. and Pitman A. J. (1998) The BASE land surface model. Global Planet Change 18:3–18.
168. DePinto J. V., Lin H., Guan W., Atkinson J. F., Densham P. J., Calkins H. W. and Rodgers P. W. (1994) Development of GEO-WAMS: An approach to the integration of GIS and great lakes watershed analysis models. Special issue of Microcomputers in Civil Engineering, 9:251–262.
169. De Troch F. P., Troch P. A., Su Z. and Lin D. S. (1996) Application of remote sensing for hydrological modelling. In: Abbort M. B. and Refsgaard J. C. (Eds.) Distributed Hydrological Modelling. Kluwer Academic Publishers, Printed in the Netherlands 165–191.

170. de Vries D. A. (1975) Heat transfer in soils. In: de Vries D. A. and Afghan N. H. (Eds.) Heat and Mass Transfer in the Biosphere. Part 1: Transfer Processes in the Plant Environment. John Willy, 4–28.
171. Dewar R. (1996) The correlation between plant growth and intercepted radiation: An interpretation in terms of optimal plant nitrogen content. Ann. Bot.-LONDON 78:125–136.
172. Dey B. and Kumar O. B. (1982) An apparent relationship between Eurasian spring snow cover and the advance period of Indian summer monsoon. J. Appl. Meteorol. 21(12):1,929–1,923.
173. Dey B. and Kumar O. B.(1983) Himalayan winter snow cover area and summer monsoon rainfall over India. J. Geophys. Res. 88:5,471–5,474.
174. Dickinson R. E. (1983) Land surface processes and climate - surface albedo and energy balance. In: Saltzman (Ed.) Theory of Climate. Advances in Geophysics. Academic Press, New York, 25:305–353.
175. Dickinson R. E. (1986) Biosphere–atmosphere scheme (BATS) for NCAR community climate model. NCAR, Boulder. Co. Tech. Note/TN-275+STR.
176. Dickinson R. E. (1992) Land surface. In: Trenberth K. E. (Ed.) Climate System Modeling. Cambridge University Press, 149–171.
177. Dickinson R. E., Henderson-Sellers A., Kennedy P. and Wilson M. F. (1986) Biosphere–atmosphere Transfer Scheme (BATS) for NCAR Community Climate Model. Tech Note TN–275+STR, National Centre for Atmospheric Research, Boulder, Col.
178. Dickinson R. E., Henderson-Sellers A. and Kennedy P. (1993) Biosphere–Atmosphere Transfer Scheme (BATS) Version 1e as Coupled to the NCAR Community Climate Model. Tech Note TN–387+STR, National Centre for Atmospheric Research, Boulder, Col.
179. Dickinson R. E., Meleshko V., Randall D., Sarachik E., Silva-Dias P. and Slingo A. (1996) Climate processes. In: Houghton J. T., Meira Filho L. G., Callander B. A., Harris N., Kattenberg A. and Maskell K. (Eds.) Climate Change 1995, the Science and Climate Change, 193–227.
180. Dickinson R. E., Shaikh M., Bryant R. and Graumlich L. (1998) Interactive canopies for a climate model. J. Climate 11:2,823–2,836.
181. Dickinson R. E. and Henderson-Sellers A. (1988) Modelling tropical deforestation: a study of GCM land-surface parametrizations. Q. J. Roy. Meteor. Soc. 114:439–462.
182. Dietachmayer G. S. (1990) Comments on "Noninterpolating Semi-Lagrangian Advection Scheme with Minimized Dissipation and Dispersion Errors". Monthly Weather Review 118(10):2,252–2,252.
183. Dietrich D. E. (1998) Application of a modified Arakawa 'a' grid ocean model having reduced numerical dispersion to the Gulf of Mexico circulation. Dynamics of Atmospheres and Oceans 27(1-4):201–217.
184. Dirmeyer P. A. and Shukla J. (1996) The effect on regional and global climate of expansion of the worlds deserts. Q. J. Roy. Meteor. Soc. 122:451–482.
185. Dixon M. and Webb E. C. (1979) Enzymes. Longman, London.
186. Domenico P. A. and Schwartz F. W. (1997) Physical and Chemical Hydrogeology. John Wiley and Sons, New York.
187. Dong Y., Scharffe D., Lobert J. M., Crutzen P. J. and Sanhueza E. (1998) Fluxes of CO_2, CH_4 and N_2O from a temperate forest soil: the effect of leaves and humus layers. Tellus 50(B): 243–252.
188. Dong Y., Scharffe D., Qi Y. and Peng G. (2000) Nitrous oxide emissions from cultuvated soils in North China plain. Tellus 53(B):1–9.
189. Dooge J. C. L. (1960) The routing of groundwater recharge through typical elements of linear storage. ISHA Publ. (52):286–300.

190. Dooge J. C. I. (1992) Hydrological models and climate change. J. Geophys. Res. 97(D3):2,677–2,688.
191. Douville H. and Royer J. (1997) Influence of the temperate and boreal forests on the northern hemisphere climate in the METEO-FRANCE climate model. Clim. Dynam. 13:57–74.
192. Dale V. and Rauscher M. (1994) Assessing impacts of climate change on forests: The state of biological modeling. Climatic Change 28:65–90.
193. Dregne H. E. (1977) Desertification of arid land. Econ. Geogr. 53:322–331.
194. Droegemeier K. K, Smith J. D., Businger S., Doswell C. III, Doyle J., Duffy C., Foufoula-Georgio E., Grazino T., James L. D., Krajewski V., Lemone M., Lettenmaier D., Mass C., Pielke R. Sr., Ray P., Rutledge S., Schaake J., and Zipser E. (2000) Meeting summary: hydrological aspects of weather prediction and flood warnings: Report of the ninth prospect development team of the U.S. weather research Program. Bull. Am. Meteorol. Soc. 81(11):2,665–2,680.
195. Ducoudré N. I., Laval K.and Perrier A. (1993) SECHIBA, A new set of parameterizations of the hydrologic exchanges at the land–atmosphere interface within LMD atmospheric general circulation model. J. Climate 6:248–273.
196. Ducoudré N., Laval K. and Perrier A. (1993) SECHIBA, a new set of parameterizations of the hydrologic exchanges at the land-atmosphere interface within the LMD Atmospheric General Circulation Model. J. Climate 6:248–273.
197. Dukowicz J. K. and Smith R. D. (1994) Implicit free-surface method for the Bryan–Cox–Semtner ocean model. J. Geophys. Res. 99:7,991–8,014.
198. Dümenil L. and Todini E. (1992) A rainall-runoff scheme for use in the Hamburg Climate Model. In: J. P. O'Kane (Ed.) Advances in Theoretical Hydrology, A Tribute to James Dooge. Eur. Geophys. Soc. Ser. Hydrol. Sci. 1, Elsevier, Amsterdam, 129–157.
199. Durran D. R. (1999) Numerical Methods for Wave Equations in Geophysical Fluid Dynamics. Springer, New York.
200. Dyer R., Cobiac M., Cafe L. and Stockwell T. (1997) Developing sustainable pasture management practices for the semi-arid tropics of the Northern Territory. Tech. Rep. MRC Project NTA 022, Northern Territory Department of Primary Industry and Fisheries, Katherine, Australia.
201. DYNAMO Group (1997) DYNAMO: Dynamics of North Atlantic Models: Simulation and Assimilation with High Resolution Models. Report Nr. 294, Institut fur Meereskunde, Kiel, Germany.
202. Ebel, A., Elbern H., Feldmann H., Jakobs H. J., Kessler C., Memmesheimer M., Oberreuter A. and Peikorz G. (1997) Air pollution studies with the EURAD model system (3): EURAD - European Air Pollution Dispersion Model System. Mitteilungen aus dem Institut für Geophysik und Meteorologie der Universität zu Köln, Heft 120.
203. Eby M. and Holloway G. (1994) Sensitivity of a large-scale ocean Mmdel to a parameterisation of topographic stress. J. Phys. Oceanogr. 24:2,577–2,588.
204. Ek M. and Cuenca R. H. (1994) Variation in soil parameters: implications for modeling surface fluxes and atmospheric boundary-Layer development. Bound. Lay. Meteorol. 70:369–383.
205. Eltahir E. A. B. and Bras R. L. (1994) Precipitation recycling in the amazon basin. Q. J. Roy. Meteor. Soc. 120:861–880.
206. England M. H. (1992) On the formation of Antarctic intermediate and bottom water in ocean ceneral circulation models. J. Phys. Oceanogr. 22:918–926.
207. England M. H. (1993) Representing the global-scale water masses in ocean general circulation models. Phys. Oceanogr. 23:1,523–1,552.
208. England M. H. (1995) Using chlorofluorocarbons to assess ocean climate models. Geophys. Res. Lett. 22:3,051–3,054.

209. England M. H., Godfrey J. S., Hirst A. C. and Tomczak M. (1993) The mechanism for Antarctic intermediate water renewal in a world ocean model. J. Phys. Oceanogr. 23:1,553–1,560.
210. England M. H., Garçon V. C. and Minster J. F. (1994) Chlorofluorocarbon uptake in a world ocean model; 1. Sensitivity to the surface gas forcing. J. Geophys. Res. 99:25,215–25,233.
211. England M. H. and Garçon V. C. (1994) South Atlantic Circulation in a World Ocean Model. Annales Geophysicae (Special edition on the South Atlantic Ocean) 12:812–825.
212. England M. H. and Hirst A. C. (1997) Chlorofluorocarbon uptake in a world ocean model, 2. Sensitivity to surface thermohaline forcing and subsurface mixing parameterisation. J. Geophys. Res. 102:15,709–15,731.
213. Errico R. M. (1997) What is an adjoint model. B. Am. Meteorol. Soc. 78:2,577–2,591.
214. Esbenson S. K. and Kushnir Y. (1981) The Heat Budget of the Global Ocean: An Atlas Based on Estimates from Surface Marine Observations. Report No. 29, Climate Research Institute, Oregon State University.
215. Ezer T. and Mellor G. L. (1997) Simulations of the atlantic ocean with a free surface sigma coordinate ocean model. J. Geophys. Res. 102(C7):15,647–15,657.
216. Famiglietti J. S. and Wood E. F. (1995) Effects of spatial variability and scale on areally averaged evapotranspiration. Water Resour. Res. 31:699–712.
217. Farquhar G. D., von Caemmerer S. and Berry J. A. (1980) A biochemical model of photosynthenic CO_2 assimilation in leaves of C_3 species. Planta 149:78–90.
218. Farquhar G. D. and Von Caemmerer S. (1982) Modelling of photosynthetic response to environmental conditions. In: Lange O., Nobel P., Osmond C. B. and Ziegler H. (Eds.) Encyclopedia of Plant Physiology, Physiological Plant Ecology, Springer-Verlag, Berlin, 549–587.
219. Feddes R. A. (1981) Water use models for assessing root zone modification. In: Arkin G. F. and Taylor H. M. (Eds.) Modifying the Root Environment to Reduce Crop Stress. ASAE Monograph, St. Joseph, Michigan, 4:347–390.
220. Fichefet T., Maqueda M., Planto S. and Bellevaux C. (1994) A global sea-ice-upper-ocean model: some preliminary results. In: Yu Z. et al. (Eds.) Sea Ice Observation and Modelling. China Ocean Press, Beijing, 64–77.
221. Field C. and Avissar R. (1998) Bidirectional interactions between the biosphere and the atmosphere – Introduction. Global Change Biology 4:459–460.
222. Findell K. L.and Eltahir E. A. B. (1997) An analysis of the soil moisture-rainfall feedback, based on direct observations from illinois. Water Resour. Res. 33:725–735.
223. Flather R. A. (1976) A tidal model of the Northwest European Continental Shelf. Mem. Soc. R. Sci. Leige Ser. 6, 10:141–164.
224. Flato G. M. (1991) Numerical investigation of the dynamic of a variable thickness of the Arctic ice cover. Ph. D. thesis, Thayer School of Engineering, Dartmouth College, Hanover, NH.
225. Flato G. M. and Hibler W. D. (1992) Modeling pack ice as a cavitating fluid. J. Phys. Oceanogr. 22:626–651.
226. Foley J. A. (1994) Net primary productivity in the terrestrial biosphere: the Application on a global model. J. Geophys. Res. 99:20,773–20,783.
227. Foley J. A., Levis S., Prentice I. C., Pollard D. and Thompson S. L. (1998) Coupling dynamic models of climate and vegetation. Glob. Change Biol. 4:561–580.

228. Foley J. A., Prentice I. C., Ramankutty N., Levis S., Pollard D., Sitch S. and Haxeltine A. (1996) An integrated biosphere model of land surface processes, terrestrial carbon balance and vegetation dynamics. Global Biogeochem. Cy. 10:603–628.
229. Foley J. A., Kutzbach J., Coe M. T. and Levis S. (1994) Feedbacks between the climate and boreal forests during the Holocene Epoch. Nature 371:52–54.
230. Fong D. A. and Geyer W. R. (2001) Response of a river plume during an upwelling favourable wind event. J. Geophys. Res. 106:1,067–1,084.
231. Foster I. D. L. and Charlesworth S. M. (1996) Heavy metals in the hydrological cycle: Trends and explanation. Hydrological Processes 10:227–261.
232. Fowler D., Cape J., Coyle M., Smith R., Hjellbrekke A., Simpson D., Derwent R. and Johnson C. (1999) Modelling photochemical oxidant formation, transport, deposition and exposure of terrestrial ecosystems. Environ. Pollut. 100:43–55.
233. Fraedrich K. and Leslie L. M. (1988) Real time short-term forecasting of precipitation at an Australian tropical station. Wea. Forec. 3:104–114.
234. Fraedrich, K. (1988) El-Nino/Southern Oscillation predictability. Mon. Wea. Rev. 116:1,001–1,012.
235. Frankenberger J. R., Brooks E. S., Walter M. T., Walter M. F. and Steenhuis T. S. (1999) A GIS-based variable source area hydrology model. Hydrological Processes 13:805–822.
236. Franks S., Gineste W. P., Beven K. J. and Merot P. (1998) On constraining the prediction of a distributed model: the incorporation of fuzzy estimates of saturated areas into the calibration processes. Water Resour. Res. 34(4):787–798.
237. Fronza G., Spirito A. and Tonielli A. (1979) Real-time forecast of air pollution episodes in the venetian region. Part 2: The Kalman predictor. Appl. Math. Model. 3:409–415.
238. Fung I., Prentice K., Matthews E., Lerner J. and Rusel G. (1983) Three-dimensional tracer model study of atmospheric CO_2 response to seasonal exchanges with the terrestrial biosphere. J. Geophys. Res. 88:1,281–1,294.
239. Gan J. and Allen J. S. (2001) A modeling study of shelf circulation off northern California in the region of the Coastal Ocean Dynamics Experiment. Part 1, response to relaxation of upwelling winds. J. Geophys. Res. (submitted).
240. Gardiner V. and Herrington P. (1986) Water demand forecasting. Proceedings of a workshop sponsored by the Economic and Social Research Council, Short Run Press Ltd., Exrter.
241. Garnier E., Barnier B., Siefridt L. and Béranger K. (2000) Investigating the 15 years air-sea flux climatology from the ECMWF re-analysis project as a surface boundary condition for ocean models. Int. J. Climatol. 20:1,653–1,673.
242. Garratt J. R. (1978) Transfer characteristics for a heterogeneous surface of large aerodynamic roughness. Q. J. Ror. Meteorol. Soc. 104:491–502.
243. Garratt J. R. (1992) The Atmospheric Boundary Layer. Cambridge University Press.
244. Gash J. H. C. and Nobre C. A. (1997) Climatic effects of Amazonian deforestation: some results from ABRACOS. Bull. Amer. Meteorol. Soc. 78(5):823–830.
245. Gates W. L. (1976) Modeling the ice-age climate. Science 191:1,138–1,144.
246. Gates W. L. (1992) AMIP: The atmospheric model intercomparison project. B. Am. Meteorol. Soc., 73:1,962–1,970.
247. Gates W. L., Henderson-Sellers A., Boer G. J., Folland C. K., Kitoh A., McAvaney B. J., Semazzi F., Smith N., Weaver A. J. and Zeng Q. C. (1995) Climate models - Evaluation. In: Houghton J. T., Meira Filho L. G., Callander B. A.,

Harris N., Kattenberg A. and Maskell K. (Eds.) Climate Change 1995, the Science and Climate Change. 233–276.
248. Gates W. L., Henderson-Sellers A., Boer G. J., Folland C. K., Kitoh A., McAvaney B. J., Semazzi F., Smith N., Weaver A. J. and Zeng Q. C. (1996) Climate model - Evaluation. In: Climate Change 1995, the Science of Climate Change. Cambridge University Press.
249. Gauntlett D. J. and Leslie L. M. (1975) Numerical methods for predicting precipitation in catchment hydrology. In: Chapman T. G. and Dunin F. X. (Eds.) Prediction in Catchment Hydrology. The Griffin Press, Australia Acaemy of Sciences 33–45.
250. Geleyn, Jean-Francois (1987) Use of a modified Richardson number for parameterising the effect of shallow convection in short and medium range numerical weather prediction. Meteor. Soc. Japan. Tokyo, Japan.
251. Gent P. R., Willebrand J., McDougall T. J. and McWilliams J. C. (1995) Parameterizing eddy-induced tracer transports in ocean circulation models. J. Phys. Oceanogr. 25:463–474.
252. Gent P. R. and McWilliams J. C. (1990) Isopycnal mixing in ocean circulation models. J. Phys. Oceanogr. 20:150–155.
253. Gerry M. W., Whitten G. Z. and Killus J. P. (1988) Development and Testing of the CBM-IV for Urban and Regional Modelling. EPA/600/3-88/012, U.S. EPA, Research Triangle Park, NC.
254. Ghan S. J., Liljegren J. C., Shaw W. J., Hubbe J. H. and Doran J. C. (1997) Influence of subgrid variability on surface hydrology. J. Climate 10:3,157–3,166.
255. Gibbs M. T., Marchesiello P. and Middleton J. H. (1997) Nutrient enrichment of Jervis Bay, during the Massive 1992 Coccolithophorid Bloom. Mar. Freshwater Res. 48:473–478.
256. Gibson J. K., Kallberg P., Uppala S., Hernandez A., Nomura A. and Serrano E. (1997) ECMWF Re-analysis Project, 1. ERA Description. Project Report Series, ECMWF report, July 1997.
257. Gill A. E. (1980) Some simple solutions for heat induced tropical circulation. Q. J. Roy. Meteor. Soc. 106:447–462.
258. Gill A. E. (1982) Atmosphere–Ocean Dynamics. International Geophysics Series. Academic Press.
259. Giorgi S. and Mearns L. O. (1991) Approaches to the simulation of regional climate change: a review. Rev. Geophys. 29:191–216.
260. Giorgi F. (1997) An approach for the representation of surface heterogeneity in land surface models. Part I: theoretical framework. Mon. Weather Rev. 125:1,885–1,899.
261. Giorgi F. and Avissar R. (1997) Representation of heterogeneity effects in earth system modeling: experience from land surface modeling. Rev. Geophys. 35:413–437.
262. Gjevik B., Nost E. and Straume T. (1994) Model simulation of the tides in the Barents Sea. J. Geophys. Res. 99:3,337–3,350.
263. Glansdorff P. and Prigogine I. (1971) Thermodynamic Theory of Structure, Stability and Fluctuation. John Wiley and Sons.
264. Gloersen P., Cavalieri D. J., Comiso J. C., Parkinson C. L. and Zwally H. J. (1992) Arctic and Antarctic Sea Ice. 1,978–1,987.
265. Godfrey J. S. and Schiller A. (1997) Tests of Mixed-layer Schemes and Surface Boundary Conditions in an Ocean General Circulation Model, Using the IMET Flux Data Set. CSIRO Marine Laboratories Report No. 231.
266. Golding B. W. and Leslie L. M. (1993) The impact of resolution and formulation on model simulations of an east coast low. Aust. Met. Mag., 42:105–116.

267. Gordon A. L. (1986) Inter-ocean exchange of thermocline water. J. Geophys. Res. 91:5,037–5,046.
268. Goudriaan J. (1996) Predicting crop yields under global change. In: Walker B. and Steffen W. (Eds.) Global Change and Terrestrial Ecosystems. Cambridge University Press, Cambridge, 260–274.
269. Goutorbe J. P., Lebel T., Tinga A., Bessemoulin P., Bouwer J., Dolman A. J., Engman E. T., Gash J. H. C., Hoepffner M., Kabat P., Kerr Y. H., Monteny B., Prince S. D., Sad F., Sellers P. and Wallace J. S. (1994) Hapex-Sahel: a large scale study of land-surface interactions in the semi-arid tropics. Ann. Geophysicae 12:53–64.
270. Gravel S. and Staniforth A. (1994) A mass conserving semi-Lagrangian scheme for the shallow-water equations. Monthly Weather Review 122:243–244.
271. Grell, G. A., Dudhia J. and Stauffer D. R. (1994) A description of the fifth-generation PENN State/NCAR Mesoscale model (MM5). NCAR Technical Note, NCAR/TN-398-STR.
272. Griffies S., Gananadesikan A., Pacanowski R., Larichev V., Dukowicz J. and Smith R. (1998) Isoneutral diffusion in a z-coordinate ocean model. J. Phys. Oceanogr. 28(5):805–830.
273. Griffies S. M., Boning C., Bryan F. O., Chassignet E. P., Gerdes R., Hasumi H., Hirst A. C., Treguier A. M. and Webb D. (2000) Developments in ocean climate modelling. Ocean Modelling 2:123–192.
274. Guo B. R., Jiang J. M., Fan X. G., Zhang H. L. and Chou J. F. (1996) Nonlinear characteristics of climate system and their prediction theory. Meteorological Press. (in Chinese).
275. Guo B. R., Jiang J. M., Fan X. G., Chang H. N. and Qiao J. F. (1996) Nonlinear characteristics and prediction theory of the climatic system. Meteorol. Publishers (in Chinese), Beijing. (in Chinese).
276. Guo Q. and Wang J. (1986) Snow accumulation over the Tibet Plateau and its influence on eastern Asia monsoon. Plateau Meteor. 5(2). (in Chinese).
277. Hahmann A. N. and Dickinson R. E. (1997) RCCM2-BATS model over tropical South America: Applications to tropical deforestation. J. Climate 10:1,944–1,964.
278. Haidvogel D. B., Wilkin J. L. and Young R. (1991) A semi-spectral primitive equation model using vertical sigma and orthogonal curvilinear horizontal coordinates. J. Comp. Phys. 94:151–185.
279. Haken H. (1983) Synergetics. Springer-Verlag, Berlin.
280. Haldane J. B. S. (1930) Enzymes. Longmans, London.
281. Haltiner G. J. and Williams R. T. (1980) Numerical Predition and Dynamical Meteorology. Wiley Publication, New York.
282. Hanan N. P., Prince S. D. and Bégué A. (1995) Estimation of absorbed photosynthetically active radiation and vegetation net production efficiency using satellite data. Agr. Forest Meteorol. 76:259–276.
283. Haney R. L. (1971) Surface thermal boundary condition for ocean circulation models. J. Phys. Oceanogr. 1:241–248.
284. Hanks R. J. (1985) Soil water modelling. In: Aderson M.G. and Burt T. B. (Eds). Hydrological Forecasting. Chichester, New York: Wiley.
285. Hansen J. and Lebedeff S. (1988) Global surface temperatures: update through 1987. Geophys. Res. Lett. 15:323–326.
286. Harley R. A., Russell A. G., McRae G. J., Cass G. R. and Seinfeld J. H. (1993) Photochemical modelling of the Southern California air quality study. Environ. Sci. Technol. 27:378–388.

287. Hasumi H. and Suginohara N. (1999) Effects of locally enhanced vertical diffusivity over rough bathymetry on the world ocean circulation. J. Geophys. Res. 104:23,367–23,374.
288. Haverkamp R., Vauclin M., Tauma J., Wierenga P. J. and Vachaud G. (1977) A comparison of numerical simulation models for one-dimensional infiltration. Soil Sci. Soc. Am. J. 41:285–294.
289. Hydrology Bureau of Changjiang Water Resources Commission (1999) Operational flood forecasting for middle Reaches from Yichang to Luoshan of the Yangtze River. Special Issue for the national Hydrological Forecasting Techniques Competition, Supplement, August, 1999, Hydrol. (in Chinese).
290. Heal O. W., Menaut J. C. and Steffen W. L. (1993) Toward a global terrestrial observing system (GTOS), detecting and monitoring change in terrestrial ecosystems. Tech. Rep. MAB Digest 14 and IGBP Report 26, UNESCO-IGBP, Paris.
291. Heimann M. et al. (1998) Evaluation of terrestrial carbon cycle models through simulations of the seasonal cycles of atmospheric CO_2: First results of a model comparison study. Global Biogeochem. Cy. 12:1–24.
292. Heller M. and Wang Q. (1996) Improving potable water demand forecasts with neural networks. In: Proceedings of UCOWR 1996, San Antonio, TX.
293. Hellerman S. and Rosenstein M. (1983) Normal monthly wind stress over the world ocean with error estimates. J. Phys. Oceanogr. 13:1,093–1,104.
294. Henderson-Sellers A., Yang Z. L. and Dickinson R. E. (1993) The Project for Intercomparison of Land-Surface Schemes (PILPS). Bull. Amer. Meteor. Soc. 74:1,335–1,349.
295. Henderson-Sellers A., Pitman A. J, Love P. K., Irannejad P. and Chen T. H. (1995) The Project for Intercomparison of Land-Surface Schemes (PILPS): Phases 2 & 3. Bull. Amer. Meteor. Soc. 73:1,962–1,970.
296. Henderson-Sellers A. and Gornitz V. (1984) Possible climatic impacts of land cover transformations with particular emphasis on tropical deforestation. Climatic Change 6:231–257.
297. Henderson-Sellers A. (1993) Continental vegetation as a dynamic component of a global climate model: a preliminary assessment. Climatic Change 23:337–377.
298. Henderson-Sellers A. and McGuffie K. (1994) Land-surface characterization in greenhouse climate simulations. Int. J. Climatol. 14:1,065–1,094.
299. Henderson-Sellers A., McGuffie K. and Gross C. (1995) Sensitivity of global climate model simulations to increased stomatal resistance and CO_2 increases. J. Climate 8:1,738–1,756.
300. Herman G. F. and Johnson W. T. (1978) The sensitivity of the general circulation to arctic sea ice boundaries. Mon. Weather Rev. 106:1,649–1,664.
301. Hernandez E., Martin F., Valero F. (1991) State-space modelling for atmospheric pollution. J. Appl. Meteorol. 30:793–811.
302. Hess G., Carnovale F., Cope M. and Johnson G. (1992) The evaluation of some photochemical smog reaction mechanisms – I. Temperature and initial composition effects. Atmos. Environ. 26A(4):625–651.
303. Hibler W. D. (1979) A dynamic thermodynamic sea ice model. J. Phys. Oceanogr. 9:815–846.
304. Hibler W. D. (1980) Modeling a variable thickness sea ice cover. Mon. Weather Rev. 108:1,942–1,973.
305. Hibler W. D. and Bryan K. (1987) A diagnostic ice-ocean model. J. Phys. Oceanogr. 17:987–101.
306. Hibler W. D. and Flato G. M. (1992) Sea ice models. In: Trenberth K. E. (Ed.) Climate System Modelling. Cambridge University Press, Cambridge, 413–436.

307. Hicks B. B. and Wesely M. L. (1981) Heat and momentum characteristics of adjacent fields of soybean and maize. Bound. Lay. Meteorol. 20:175–185.
308. Hill A. V. (1910) The possible effects of the aggregation of the molecules of hemoglobin on its dissociation curves. J. Physiol. 40:4–7.
309. Hill T. L. (1985) Cooperativity Theory in Biochemistry: Steady-State and Equilibrium Systems. Springer-Verlag, New York, 1–15.
310. Hillel D. (1980) Foundamentals of Soil Physics. Academic Press.
311. Hino M. (1970) Runoff forecasts by linear predictive filter. Proc. ASCE. J. Hydraul. Div. 96(Hy3):681–701.
312. Hirst A. C. and Cai W. (1994) Sensitivity of a world ocean GCM to changes in subsurface mixing parameterization. J. Phys. Oceanogr. 24:1,256–1,279.
313. Hirst A. C. and McDougall T. J. (1996) Deep water properties and surface buoyancy flux as simulated by a z-coordinate model including eddy-induced advection. J. Phys. Oceanogr. 26:1,320–1,343.
314. Hoffman R. N. and Kalnay E. (1983) Lagged average forecasting: an alternative to Monte Carlo forecasting. Tellus 35:100–118.
315. Holland D. M., Mysak L. A. and Oberhuber J. M. (1991) Simulation of the seasonal Arctic sea ice cover with a dynamic thermodynamic sea ice model. Res. Rep. 17, Cent. for Clim. and Global Change Res. McGill. Univ., Montreal, Quebec.
316. Holland D. M., Ingram G. and Myssak L. A. (1995) A numerical simulation of the sea ice cover in the northern Greenland Sea. J. Geography. Res. 100(C3):4,751–4,760.
317. Hollinger S. E. and Issard S. A. (1994) A soil moisture climatology of Illinois. J. Climate 7:822–833.
318. Holloway P. E. (1984) On the semi-diurnal internal tide at a shelf-break region on the Australian North West Shelf. J. Phys. Oceanogr. 14:1,778–1,790.
319. Holloway G. (1992) Representing topographic stress for large-scale ocean models. J. Phys. Oceanogr. 22:1,033–1,046.
320. Holton James R. (1972) An Introduction to Dynamical Meteorology. Academic Press, New York.
321. Horton C. M. C., Cole D., Schnitz J. and Kantha L. (1992) Operational modeling: Semi-enclosed basin modeling at the Naval Oceanographic Office. Oceanogr. 5:69–72.
322. Houghton J. T., Callander B. A. and Varney S. K. (1992) Climate change. The Supplementary Report to the IPCC Scientific Assessment, Meteorology Office, Bracknell.
323. Houghton J. T., Filho L. G. M., Callander B. A., Harris N., Kattenberg A. and Maskell K. (Eds.) (1996) Climate Change 1995: the Science of Climate Change; The Second Assessment Report of the Intergovernmental Panel on Climate Change: Contribution of Working Group I. Cambridge University Press, Cambridge.
324. Houghton J. T., Jenkins G. J. and Ephraums J. J. (Eds.) (1990) Climate change: the Intergovernmental Panel on Climate Change Scientific Assessment. Cambridge University Press, Cambridge.
325. Houghton R. A., Davidson E. A. and Woodwell G. M. (1998) Missing sinks, feedbacks, and understanding the role of terrestrial ecosystems in the global carbon balance. Global Biogeochem. Cy. 12:25–34.
326. Houtekamer P. L. (1995) The construction of optimal perturbations. Mon. Weather Rev. 123:2,888–2,898.
327. Hollinger S. E. and Isard S. A. (1994) A soil moisture climatology of Illinois. J. Climate 7:822–833.

328. Huang P. (1997) The perspective of the study and application of catchment distributed hydrological model. Hydrol. 5:5–9. (in Chinese).
329. Huang J., van den Dool H. M. and Georgakakos K. P. (1996) Analysis of model-calculated soil moisture over the United States (1931–1993) and applications to long-range temperature forecasts. J. Climate 9:1,350–1,362.
330. Huo S. (1999) The variation of historic flood in the middle reach of Yellow River and the method for long-term hydrological forecasting. Report of the project of flood control of Yellow River. 97E01:37–38. (in Chinese).
331. Hurni H. (1998) A multi-level stakeholder approach to sustainable land management. In: Blume H. P., Eger H., Fleischhauer E., Hebel A., Reij C. and Steiner K. G. (Eds.) Towards Sustainable Land Use. Advances in Geoecology, 31:827–836.
332. Hurtt G. C., Moorcroft P. R., Pacala S. W. and Levin S. A. (1998) Terrestrial models and global change: Challenges for the future. Glob. Change Biol. 4:581–590.
333. Hutjes R. W. A. et al. (1998) Biospheric aspects of the hydrological cycle: Preface. J. Hydrol. 212:1–21.
334. Hydrological Engineering Center (1990) HEC-1 Flood hydrograph package, User's Manual, computer program document No. 1A.
335. Idso S. B., Jackson R. D., Reginato R. J., Kimball B. A. and Nakayama F. S. (1975) The dependence of bare soil albedo on soil water content. J. Appl. Meteor. 14:109–113.
336. IGBP Global Change Report (1994) IGBP in Action: Work Plan 1994-1998. NO. 28, Stockholm.
337. Intera Environmental technologies Ltd, Regional ice model implementation guide. Calgary, Canada, Vol. 1.
338. IPCC (1995) The Science of Climate Change (Contribution of Working Group I to the Second Assessment Report of the Intergovernmental Panel on Climate Change). In: Houghton J. T. et al. (Eds.) Climate Change, 1995. Cambridge University Press.
339. Irannejad P. and Shao Y. (1998) Description and validation of the Atmosphere–Land–Surface Interaction Scheme (ALSIS) with HAPEX and Cabauw data. Global Planet. Change 19:87–114.
340. Irannejad P. and Shao Y. (2000) Impacts of soil moisture modelling on the simulation of furface energy fluxes by a land surface scheme. Global Planet. Change. (submitted).
341. Jorgensen S. E. (1988) Fundamentals of Ecological Modelling. Elsevier, Amsterdam.
342. Jackson T. J., Schmugge T. J. and O'Connell P. O. (1983) Remote sensing of soil moisture from an aircraft platform using passive microwave sensors. In: Hydrological Applications of Remote Sensing and Remote Data Transmission. Proceedings of the Hamburg Symposium, August, 1983. IAHS Publ. 145:529–539.
343. Jacobson M. Z. (1999) Fundamentals of Atmospheric Modeling. Cambridge University Press, Cambridge.
344. Jaeger L. (1976) Monatskarten des Niederschilages fur die ganze Erde. Ber D. Wetterdienstes 139:1–38.
345. James A. (1993) An Introduction to Water quality models. 2nd Edition. Chichester; New York: Wiley.
346. Jarvis P. G. (1976) The interpretation of the variations in leaf-water potential and stomatal conductance found in canopies in the field. Philos. T. Roy. Soc. London B 723:517–529.

347. Jarvis P. (1976) The interpretation of leaf water potential and stomatal conductance found in canopies in the field. Philos. Trans. Roy. Soc., London 273B:593–610.
348. Jarvis P. and Leverenz J. W. (1983) Productivity of temperate, deciduous and evergreen forests. In: Lange O., Nobel P., Osmomd C. B. and Ziegler H. (Eds.) Encyclopaedia of Plant Physiology. New Series, Vol. 12. Springer-Verlag, New York.
349. Jenkins G. (1982) Some practical aspects of forecasting in organization. J. Forecasting 1(1):3–22.
350. Jiang Y (2000) The contuntive utilization of ground water and surface water by WRDSS. Ph.D dissertation to Institute of Geography, Chinese Academy of Sciences, Beijing.
351. Jiang H. et al. (1999) Study on application of conceptual soil moisture content model in Huoquan irrigation District. Hydrol. 6:12–16. (in Chinese).
352. Jin Z. and Han N. (1998) A new water quality model for plain rivers system: A combined units water quality model. Adv. Water Sci. 9(1):35–40. (in Chinese).
353. Johnson G. (1984) A Simple Model for Prediction Ozone Concentration of Ambient Air. Proceedings of the 8th International Clean Air Conference, Melbourne, 751–731.
354. Johnson K. D., Entekhabi D. and Eagleson P. (1993) The implementation and validation of improved land-surface hydrology in an Atmospheric General Circulation Model. J. Climate 6:1,009–1,026.
355. Jonch-Claussen T. (1979) Systeme hydrologique European: short description. SHE Report 1: Danish Hydraulic Institute, Hosholm, Denmark.
356. Jones P. D. C (1988) Hemispheric surface air temperature variations: recent trends and an update to 1987. J. Climate 1:654–660.
357. Josey S., Kent C. E. and Taylor P. K. (1999) New insights into the ocean heat budget closure problem from analysis of the SOC air-sea flux climatology. J. Climate 12:2,856–2,880.
358. Kabat P. R., Hutjes R. W. A. and Feddes R. A. (1997) The scaling characteristics of soil parameters: From plot scale heterogeneity to subgrid parameterization. J. Hydrol. 190:363–396.
359. Kalnay Eugenia, Seon Ki Park, Zhao-Xia Pu and Jidong Gao (2000) Application of the quasi-inverse method to data assimilation. Monthly Weather Review 128(3):864–875.
360. Kalnay E., Kanamistu M., Kistler R., Collins W., Deaven D., Gandin L., Iredell M., Saha S., White G., Woolen J., Zhu Y., Cheliah M., Ebisuzaki W., Higgins W., Janowiak J., Mo C. K., Ropelewski C., Leetma A., Reynolds R. and Jenne R. (1996) The NCEP/NCAR Reanalysis Project. Bull. Amer. Meteor. Soc. 77:437–471.
361. Kantha L. H. and Clayson C. A. (1994) An improved mixed layer model for geophysical applications. J. Geophys. Res. 99:25,235–25,266.
362. Killworth P. D., Stainforth D., Webb D. J. and Patterson S. M. (1991) The development of a free-surface Bryan–Cox–Semtner ocean model. J. Phys. Ocean. 21:1,333–1,348.
363. Kitanidis P. K. and Bras R. L. (1980) Real-time forecasting with a conceptual hydrological model, 1: Analysis of uncertainty; 2: Applications and results. Water Resour. Res. 16(6):232–240.
364. Kim J. and Ek M. (1995) A simulation of the surface energy budget and soil water content over the hydrologic atmospheric pilot experiments: Modélissation du Bilan Hydrique Forest Site. J. Geophys. Res. 100(D10):20,845–20,854.

365. Kim K., Jackman A. P. and Triska F. J. (1990) Modelling transient storage and nitrate uptake kinetics in a flume containing a natural periphyton community. Water Resour. Res. 26(3):505–515.
366. Kirchner J. W., Hooper R. P., Kendall C., Neal C. and Leavesley C. (1996) Testing and validating environmental models. Sci. Total. Environ. 183:33–47.
367. Kirkham D. and Powers W. L. (1972) Advanced Soil Physics. Wiley-Interscience, New York.
368. Kirkland M. R., Hills R. G. and Wierenga P. J. (1992) Algorithms for solving Richards' Equation for variably aaturated soils. Water Resour. Res. 28:2,049–2,058.
369. Kistner A., Therion J., Kornelius J. H. and Hugo A. (1979) Effect of pH on specific growth rates of rumen bacteria. Ann. Rech. Vet. 10:268–270.
370. Kivman G. A. (1997) Weak constraint data assimilation for tides in the arctic ocean. Prog. Oceanogr. 40:179–196.
371. Knorr W. and Heimann M. (1995) Impact of drought stress and other factors on seasonal land biosphere CO_2 exchange studied through an atmospheric tracer transport model. Tellus 47B:471–489.
372. Koch G. W., Vitousek P., Steffen W. L. and Walker B. H. (1995) Terrestrial transects for global change research. Vegetation 121:53–65.
373. Kondratyev K. Y. and Nikolsky G. A. (1970) Solar radiation and solar activity. Q. J. Roy. Meteor. Soc. 96:509–522.
374. Körner C. (1996) The responses of complexes multispecies systems to elevated CO_2. In: Walker B. and Steffen W. (Eds.) Global Change and Terrestrial Ecosystems. Cambridge University Press, Cambridge, IGBP Book Series, 20–42.
375. Kosjin S. E. and Pakrofskia T. V. (1957) Climatology. China Higher Education Press. (in Chinese translated from Russian).
376. Koster R. D. and Suarez M. J. (1992) Modeling the land surface boundary in climate models as a composite of independent vegetation stands. J. Geophys. Res. 97(D3):2,697–2,715.
377. Kotliakof V. M. and Krenke A. N. (1989) Role of continental snow cover of the Southern Hemisphere. In: Moisture Redistributions. Reports of the Academy of Sciences of the Soviet Union, 304(5):1,221–1,226. (in Russian).
378. Kowalczyk E. A., Garratt J. R. and Krummel P. B. (1991) A soil-Canopy Scheme for Use in a Numerical Model of the Atmosphere – 1D Stand-alone Model. CSIRO, DAR, Technical Paper.
379. Kowalik Z. and Proshutinsky A. Y. (1995) Topographic enhancement of tidal motion in the Western Barents Sea. J. Geophys. Res. 100:2,613–2,637.
380. Kraijenhoff D. A. and Moll J. R. (1986) River Flow Modelling and Forecasting. D. Reidel Publication (Holland).
381. Kraus E. B. and Businger J. A. (1994) Atmosphere–Ocean Interactions. Oxford University Press, New York.
382. Krishnamurti T. N. and Bounoua L. (1996) An Introduction to Numerical Weather Prediction Techniques. CRC press.
383. Krishnamurti T. N. (1991) Seasonal monsoon integrations with the FSU model. Simulation of interannual and intraseasonal monsoon variability, WCRP-68, WMO/TD-No. 470.
384. Krishnamurti T. N., Bedi H. S. and Hardiker V. M. (1998)An Introduction to Global Spectral Modeling. Oxford University Press, New York.
385. Kurapov A. L. and Kivman G. A. (1999) Data assimilation in a finite element model of M2 tides in the Barents Sea. Oceanology 39:306–313.
386. Lacis A. A. and Hansen J. E. (1974) A paramterisation for the absorption of solar radiation in the earth's atmosphere. J. Atmos. Sci. 31:118–133.

387. Landsber J. J., Prince S. D., Jarvis P. G., McMurtrie R. E., Luxmoore R. and Medlyn B. E. (1996) Energy conversion and use in forests: An analysis of forest production in terms of radiation utilisation efficiency. In: Gholz H. L., Nakane K. and Shimoda H. (Eds.) The use of remote sensing in the modeling of forest productivity. Kluwer, Dordrecht, 273–298.
388. Larcher W. (1995) Physiological Plant Ecology. Springer-Verlag, Berlin.
389. Lau N. C. (1985) Modeling the seasonal dependence of the atmosphere response to observed El-Nino in 1962–76. Mon. Wea. Rev. 113:1,970–1,996.
390. Laurenson E. M. and Mein R. G. (1993) Unusual applications of the RORB program. Engineering for hydrology and water respources Conf. I.E. Aust. Nat. Conf. Publ. 93/14, 125–131.
391. Laval K. and Picon L. (1986) Effect of a change of the surface albedo of the Sahel on climate. J. Atmos. Sci. 43:2,418–2,429.
392. Lean J. and Rowntree P. R. (1997) Understanding the sensitivity of a GCM simulation of Amazonian Deforestation to the specification of vegetation and soil Characteristics. J. Climate 10:1,216–1,239.
393. Ledwell J. R., Watson A. J. and Law C. S. (1993) Evidence for slow mixing across the pycnocline from an open-ocean tracer-release experiment. Nature 364:701–703.
394. Ledwell J. R., Watson A. J. and Law C. S. (1998) Mixing of a tracer in the pycnocline. J. Geophys. Res. 103:21,499–21,529.
395. Lei Z., Yang S. and Xie S. (1988) Soil Water Dynamics. Tsinghua Press. Beijing. (in Chinese).
396. Leith C. E. (1974) Theoretical skill of Monte Carlo forecasts. Mon. Weather Rev. 102:409–418.
397. LeMarshall J. F. and Leslie L. M. (1998) Tropical cyclone track prediction using very high resolution satellite data. Aust. Met. Mag. 47:261–266.
398. Lemke P. (1990) A coupled sea ice-mixed layer-phcnoline model for the Weddell Sea. J. Geophys. Res. 95:9,513–9,523.
399. Lemke P., Owens W. B. and Hibler W. D. (1990) A coupled sea ice-mixed layer-pycnocline model for the Weddell Sea. J. Geophys. Res. 95:9,513–9,525.
400. Lentz S. J. (1995) Sensitivity of inner-shelf circulation to the form of the eddy viscosity profile. J. Phys. Oceanogr. 25:19–28.
401. Le Roux X. (1995) Etude et modélisation des échanges d'eau et d'énergie sol-végétation-atmosphère dans une savane humide (Lamto, Cote d'Ivoire). Ph.D. thesis, Université Pierre et Marie Curie, Paris.
402. Le Roux X. (2000a) Modelling the savanna radiation balance, water balance and primary production. In: Menaut J. C., Abbadie L. and Lepage M. (Eds.) Lamto, a savanna ecosystem. Thirty years of ecological studies in West Africa. Springer Verlag, Ecological Studies. (in press).
403. Le Roux X. (2000b) Soil-Plant-Atmosphere Exchanges. In: Menaut J. C., Abbadie L. and Lepage M. (Eds.) Lamto, a savanna ecosystem. Thirty years of ecological studies in West Africa. Springer Verlag, Ecological Studies. (in press).
404. Le Roux X., Gauthier H., Bégué A. and Sinoquet H. (1997) Radiation absorption and use by a humid savanna grass canopy, and assessment by remotely sensed data. Agr. Forest Meteorol. 85:117–132.
405. Leslie L. M. and Fraedrich K. (1997) A new general circulation model: Formulation and preliminary results. Climate Dynamics 13:35–43.
406. Leslie L. M. and Holland G. J. (1993) Data assimilation techniques for tropical cyclone track prediction. In: Lighthill M. J. (Ed.) Tropical Cyclones. Beijing University Press, 92–103.

407. Leslie L. M., LeMarshall J. F., Spinoso C., Purser R. J. and Morison P. R. (1998) Prediction of Atlantic hurricanes in the 1995 season. Monthly Weather Review 125:1,248–1,257.
408. Leslie L. M., Mills G. A., Logan L. W., Gauntlett D. J., Kelly G. A., Manton M. J., McGregor J. L. and Sardie J. M. (1985) A high resolution primitive equations NWP model for operations and research. Aust. Met. Mag. 33:11–35.
409. Leslie L. M. and Purser R. J. (1991) High-order numerics in a unstaggered three-dimensional time-split semi-Lagrangian forecast model. Monthly Weather Review 119:1,612–1,623.
410. Leslie L. M. and Purser R. J. (1995) Three-dimensional mass-conserving semi-Lagrangian scheme employing forward trajectories. Monthly Weather Review 123:2,551–2,566.
411. Leslie L. M. and Skinner T. C. L. (1994) Real-time forecasting of the Western Australian summertime trough: Evaluation of a new regional model. Weather and Forecasting 9:371–383.
412. Leslie L. M. and Speer M. S. (1998) Short range ensemble forecasting of explosive Australian east coast cyclogenesis. Weather and Forecasting 12:822–832.
413. Leuning R. (1995) A critical appraisal of a combined stomatal-photosynthesis model for C3 plants. Plant, Cell and Environment 18:339–355.
414. Levin S. A. (1992) The problem of patterns and scale in ecology. Ecology 73:1,943–1,967.
415. Levine J. S. (1991) Global Biomass Burning. MIT Press, Cambridge.
416. Levis L., Foley J. A. and Pollard D. (1999) Potential high-latitude vegetation feedbacks on CO_2-induced climate change. Geophys. Res. Lett. 26:747–750.
417. Levitus S. (1982) Climatological Atlas of the World Ocean. NOAA Prof. Paper No.13, US Govt. Printing Office, Washington DC.
418. Levitus S., Boyer T. P. and Antonov J. (1994) World Ocean Atlas 1994, 1–4. NOAA Atlas NESDIS 1–4, National Oceanographic Data Center, Washington DC.
419. Li P. (1990) Snow amount variation of China for the last 30 years. Acta Meteorol. Sinica, 48(4).
420. Li C. Y. (1995) An Introduction to Climate Dynamics. Meteorol. Press. (in Chinese).
421. Li B. and Avissar R. (1994) The impact of spatial variability of land-surface characteristics on land-surface heat fluxes. J. Climate 7:527–537.
422. Li Q., Peng G. and Qi L. (1996) A theory for the interaction between ice-snow cover and climate. In: Zhou X. and Lu L. (Eds.) The Interaction Between Antarctic and Global Climate. China Meteorology Press, 263–266.
423. Li Z., Kong X. and Zhang C. (1998a) Improving Xin'anjiang Model. Hydrol. 4:19–23. (in Chinese).
424. Li Z., Kong X., Zhu Z. and Li J. (1998b) Half-self adaptive updating Kalman filter model of channel flow routing. Adv. Water Sci. 9(4):367–373. (in Chinese).
425. Liang G. C. (1995) Mathematical Methods for River Flow Forecasting. Lecture notes for the Advanced Workshop/Course on River Flow Forecasting in 1995. Part II.
426. Liang X., Lettenmaier D. P., Wood E. F. and Burges S. J. (1994) A simple hydrologically based model of land surface water and energy fluxes for General Circulation Models. J. Geophys. Res. 99(D7):14,415–14,428.
427. Li D., Zhang J., Quan J. and Zhang K.(1998c) A study on the feature and cause of runoff in the upper reaches of Yellow River. Adv. Water Sci. 9(1):22–28. (in Chinese).

428. Lieth H. (1975) Modelling the primary productivity of the world. In: Lieth H. and Whittaker R. H. (Eds.) Primary Productivity of the Biosphere. Springer Verlag, New York, 237–263.
429. Lieth H. and Whittaker R. H. (1975) Primary Productivity of the Biosphere. Springer-Verlag, New York.
430. Lighthill J. (Ed.) (1992) Tropical Cyclones. Beijing University Press.
431. Liu C., Cheng B., Cheng X. and Zhang S. (1998) A case study on simulating runoff with BATS model. Hydrol. 1:8–13. (in Chinese).
432. Liu G. and Wang J. (1997) A study on hydrologic cycle of land–atmosphere system of China. Adv. Water Sci. 8(2):99–107. (in Chinese).
433. Liu S. (1994) Parameter estimation of multi-constituent water quality model for the Liangxi River by Marquardt method. J. Chinese Geogr. 4(1/2):110–168.
434. Liu W. T., Katsaros K. B. and Businger J. A. (1979) Bulk parameterization of air–sea exchanges of heat and water vapour including the molecular constraints at the interface. J. Atmos. Sci. 36:1,722–1,735.
435. Liu M. S., Branion R. M. R. and Duncan D. W. (1987) Oxygen transfer to thiobacillus cultures. In: Norris P. R. and Kelly D. P. (Eds.) Biohydrometallurgy, Proceedings of the International Symposium, University of Warwick, Warwick, 12–16 July 1987, 375–384.
436. Liu S., Mo X., Li H., Peng G. and Robock A. (2001) The spatial variation of soil moisture in China: Geostatistical characterization. J. Meteorol. Soc. Jap. GAME Special Issue 79 (1B):555–574.
437. Liu S. and Liu C. (1994) Concept of representative elementary length and its case applications. Yangtze River 8:19–22. (in Chinese).
438. Liu S. (1997a) Development of a water transfer equation for a groundwater/surface water interface and use of it to forecast floods in the Yanghe Reservoir Basin. In: Gibert J. et al. (Eds.) Groundwater/Surface Water Ecotones: Biological and Hydrological Interactions and Management Options. International Hydrology Series. Cambridge University Press, Paris, 135–139.
439. Liu X. (1997b) Multi-scale hydrologic modelling. J. Hohai Univ. 25(3):7–14. (in Chinese).
440. Lorenz E. N. (1963) Deterministic non-periodic flow. J. Atmos. Sci. 20:130–141.
441. Lorenz E. (1975) Climate predictability. The physical basis of climate modelling. WMO GARP Publication Series, 16:132–136.
442. Lorenz E. N. (1984) Irregularity: a fundamental property of the atmosphere. Tellus 36(A):98–110.
443. Loth B., Graf H. F. and Oberhuber J. M. (1993) Snow cover model for global climate simulations. J. Geophys. Res. 98(D6):10,451–10,464.
444. Louis J. F. (1979) A parametric model of the vertical eddy fluxes in the Atmosphere. Bound. Layer Meteor. 17:187–202.
445. Lu H. and Shao Y. (2000) Toward quantitative prediction of dust storms: an integrated wind erosion modelling system and its applications. Environ. Modelling and Software. (in press).
446. Ludeke M., Ramge P. H. and Kohlmaier G. H. (1996) The use of satellite NDVI data for the validation of global vegetation phenology models: application to the Frankfurt Biosphere Model. Ecol. Model. 91:255–270.
447. Lurmann F. W., Carter W. P. L. and Coyner L. A. (1987) A Surrogate Species Chemical Reaction Mechanism for Urban-scale Air Quality Simulation Models. Volumes I and II. Report to the U.S. Environmental Protection Agency under Contract 68-02-4104. ERT Inc., Newbury Park, CA, and Statewide Air Pollution Research Centre, University of California, Riverside, CA.

448. Lynn B. H. Tao W. K. and Wetzel P. J. (1998) A study of landscape-generated deep moist convection. Mon. Weather Rev. 126:928–942.
449. Ma W. and Zhang C. (1998) The numerical simulation of water quality based on GIS – a case study in Suzhou River of Shanghai City. Acta Geographica Sinica, 53 (supplement):67–75. (in Chinese).
450. Mahfouf J. F. and Noilhan J. (1991) Comparative study of various formulations of evaporation from bare soil using in situ data. J. Appl. Meteor. 30:1,354–1,365.
451. Mahfouf J. F., Richard E. and Mascart P. (1987) The influence of soil and vegetation on the development of mesoscale circulations. J. Climate Appl. Meteor. 26:1,483–1,495.
452. Mahrt L.and Ek M. (1993) Spatial variability of turbulent fluxes and roughness lengths in HAPEX-MOBILHY. Bound. Lay. Meteorol. 65:381–400.
453. Mahrt L. and Pan H. (1994) A two-layer model of soil hydrology. Bound. Lay. Meteorol. 29:1–20.
454. Mahrt L., Sun, J., Vickers D., MacPherson J. I., Pederson J. R. and Desjardins R. L. (1992) Observations of fluxes and inland breezes over a heterogeneous surface. J. Atmos. Sci. 51:2,165–2,178.
455. Mahrt L. and Sun J. L. (1995) Dependence of surface exchange coefficients on averaging scale and grid size. Q. J. Roy. Meteor. Soc. 121:1,835–1,852.
456. Maidment D. R. (1993) Developing a spatially unit hydrograph by using GIS. IAHS Publ. 211:181–193.
457. Malanson G. P. (1993) Comment on modeling ecological response to climatic change. Climatic Change 23:95–109.
458. Maltrud M. E., Smith R. D., Semtner A. J. and Malone R. C. (1998) Global eddy-resolving ocean simulations driven by 1985-95 atmospheric winds. J. Geophys. Res. 103:30,825–30,853.
459. Manabe S. (1969) Climate and ocean circulation: 1. The atmospheric circulation and the hydrology of the earth's surface. Mon. Weather Rev. 97:739–774.
460. Manabe S. and Ahn D. C. (1977) Simulation of the tropical climate of an ice age. J. Geophys. Res. 82:3,889–3,911.
461. Manabe S., Bryan K. and Spelman M. J. (1990) Transient response of a global Ocean–Atmosphere model to a doubling of atmospheric carbon dioxide. J. Phys. Oceanogr. 20:722–749.
462. Manabe S., Stouffer R. J., Spelman M. J. and Bryan K. (1991) Transient responses of a coupled ocean–atmosphere model to gradual changes of atmospheric CO_2. Part I: Annual mean response. J. Climate 4:785–818.
463. Manabe S. and Stouffer R. J. (1988) Two stable equilibria of a coupled ocean–atmosphere model. J. Climate 1:841–866.
464. Manabe S. and Stouffer R. J. (1993) Century-scale effects of increased atmospheric CO_2 on the ocean–atmosphere system. Nature 364:215–218.
465. Manabe S. and Stouffer R. J. (1996) Low-frequency variability of surface air temperature in a 1000-year integration of a coupled ocean–atmosphere–land surface model. J. Climate 9:376–393.
466. Manabe S. and Wetherald R. T. (1987) Large-scale changes of soil wetness induced by an increase in atmosphere carbon dioxide. J. Atmos. Sci. 44:1,211–1,235.
467. Metropolitan Air Quality Study Reports, 1996, New South Wales EPA, Australia.
468. Marticorena B. and Bergametti G. (1995) Modeling the atmospheric dust cycle: 1. Design of a soil-derived dust emission scheme. J. Geophys. Res., 100:16,415–16,430. Quality Models. Second Edition. John Wiley and Sons.

469. Marchuk G. (1982) Mathematical Modelling in Environmental Problems. Science, Moscow.
470. Marlatt W. E., Havens A. V., Willets N. A. and Brill G. D. (1961) A comparison of computed and measured soil moisture under snap beans. J. Geophys. Res. 66:535–541.
471. Marotzke J. (1997) Boundary mixing and the dynamics of three-dimensional thermohaline circulation. J. Phys. Oceanog. 27:1,713–1,728.
472. Martz L. W. and Garbrecht J. (1992) Numerical definition of drainage network and subcatchment areas from digital elevation models. Comp. Geosci. 18:747–761.
473. Martz L. W. and Garbrecht J. (1998) The treatment of flat areas and depressions in automated drainage analysis of raster digital elevation models. Hydrological Processes 12:843–855.
474. Mathieu P. P. (1998) Parameterization of mesoscale turbulence in a World Ocean Model. PhD dissertation, Universite Catholique de Louvain.
475. Mauser W. and Schadlich S. (1998) Modelling the spatial distribution of evapotranspiration on different scale using remote sensing data. J. Hydrol. 212-213:250–267.
476. May L. W. (Ed.) (1996) Water Resources Handbook. McGraw-Hill,New York.
477. McCreary J. P. (1976) Eastern tropical ocean response to changing wind systems with application to El-Niño. J. Phys. Oceanogr. 6:632–645.
478. McCarthy G. T. (1938) The unit hyrograph and flood routing. Paper presented at Conf. of the North Atlantic Div. of US Corps of Engineers, New London, Connecticut.
479. McCartney M. S. (1977) Subantarctic Mode Water. Contribution to George Deacon: 20th Anniversary Volume, Angel M. V. (Ed.) Pergamon Press, 103–119.
480. McCumber M. C. and Pielke R. A. (1981) Simulation of the effects of surface fluxes of heat and moisture in a mesoscale numerical model. J. Geophys. Res. 86(C10):9,929–9,938.
481. McDougall T. J. (1991) Water mass analysis with three conservative variables. J. Geophys. Res. 96(C5):8,687–8,693.
482. McPhee M. G. (1980) An analysis of pack ice drift in summer. In: Pritchard R. S. (Ed.) Sea Ice Processes and Models. University Washington Press, 62–75.
483. McVicar T. R. and Jupp D. L. B. (1999) Using AVHRR data and meteorological surfaces as covariates to spatially interpolate moisture availability in the Murray-Darling Basin. CSIRO Land and Water, Tech. Rep. 50/99.
484. McVicar T. R, Walker J., Jupp D. L. B., Pierce L. L., Byrne G. T. and Dallwitz R. (1996) Relating AVHRR vegetation indices to in-situ measurements of leaf area index. CSIRO, Division of Water Resour., Tech. Memo. 96.5.
485. Melillo J. M., McGuire A. D., Kicklighter D. W., Moore III B., Vorosmarty C. J. and Schloss A. L. (1993) Global change and terrestrial net primary production. Nature 363:234–240.
486. Melillo J. M., Prentice I. C., Farquhar G. D., Schulze E. D. and Salsa O. E. (1996) Terrestrial biotic responses to environmental change and feedbacks to climate. In: Houghton J., Mera Filho L., Callander B., Harris N., Kattenberg A. and Maskell K. (Eds.) Climate Change 1995: The Science of Climate Change; The Second Assessment Report of the Intergovernmental Panel on Climate Change: Contribution of Working Group I. Cambridge University Press, Cambridge.
487. Melli P., Bolzern P., Fronza G. and Spirito A. (1981) Real-time control of sulphur dioxide emissions from an industrial area. Atmos. Environ. 15:653–666.
488. Mellor G. L. (1986) Numerical simulation and analysis of the mean coastal circulation off California. Continental Shelf Res. 6:689–713.

489. Mellor G. L. (1992) User's Guide for a Three-dimensional, Primitive Equation, Numerical Ocean Model. Prog in Atmos and Ocean Sci. Princeton University.
490. Mellor G. L. (1996) User's Guide for a Three-dimensional, Primitive Equation, Numerical Ocean Model (June 1996 version). Prog in Atmos and Ocean Sci. Princeton University.
491. Mellor G. L. (1998) User's Guide for a Three-dimensional, Primitive Equation, Numerical Ocean Model. Available from Program in Atmospheric and Ocean Sciences, Princeton University, Princeton, NJ 08544–0710.
492. Mellor G. L. and Yamada T. (1982) Development of a turbulence closure model for geophysical fluid problems. Rev. Geophys. Space Phys. 20:851–875.
493. Menaut J. C., Abbadie L., Lavenu F., Loudjani P. and Podaire A. (1991) Biomass burning in West Africa. In: Levine J. S. (Ed.) Global Biomass Burning, MIT Press, Cambridge, 133–141.
494. Merz B. and Bardossy A. (1998) Effects of spatial variability on the rainfall runoff Processes in a Small Loess Catchment. J. Hydrol. 212-213:304–317.
495. Michaelis L. and Menten M. L. (1913) Kinetics of invertase action. Biochem. 49:333–369.
496. Michaud J. and Sorooshian S. (1994) Comparison of simple versus complex distributed runoff models on a midsized semiarid watershed. Water Resour. Res. 30(3):593–605.
497. Middleton J. F. and Cirano M. (1999) Wind-forced downwelling slope currents: a numerical study. J. Phys. Oceanogr. 29(8). Part 1:1,723–1,743.
498. Mihailović D. T., Rajković B., Dekić L., Pielke R. A., Lee T. L. and Ye Z. (1995) The validation of various schemes for parameterizing evaporation from bare soil for use in meteorological models: A numerical study using in situ data. Bound. Lay. Meteorol. 76:259–289.
499. Miller A. J., L. M. Leslie (1984) Short-term single-station forecasting of precipitation. Monthly Weather Review 112(6):1,198–1,205.
500. Miller R. N. (1986) Towards the application of the Kalman filter to regional open ocean modeling. J. Phys. Oceanogr. 16:72–86.
501. Milly P. C. D. (1988) Advances in the modeling of water in the unsaturated zone. Transport Porous Med. 3:491–518.
502. Milly P. C. D. (1992) Potential evaporation and soil moisture in General Circulation Models. J. Climate 5:209–226.
503. Milly P. C. D. (1997) Sensitivity of greenhouse summer dryness to changes in plant rooting characteristics. Geophys. Res. Letters 24(3):267–271.
504. Mo X. (1997) The advances in Regional Evaporation Researches. Adv. Water Sci. 7(2):180–185. (in Chinese).
505. Mölder N., Raabe A. and Tetzlaff G. (1996) A comparison of two strategies on land surface heterogeneity used in a mesoscale β meteorological model. Tellus 48A:733–749.
506. Molteni F., Buizza R., Palmer T. N. and Petroliagis T. (1996) The ECMWF ensemble system: Methodology and validation. Q. J. R. Meteorol. Soc. 122:73–119.
507. Molz F. (1981) Models of water transport in the soil-plant system: a review. Water Resour. Res. 17:1,245–1,260.
508. Monod J. (1942) Recherches sur la croissance des cultures bacteriennes. Hermann et Cie, Paris.
509. Monteith J. L. (1965) Evaporation and Environment. Proc. Symp. Soc. Exp. Biol. 205–237.
510. Monteith J. (1972) Solar radiation and productivity in tropical ecosystems. J. Appl. Ecology 2:747–766.

511. Monteni F. (1993) Predictability and finite-time instability of the northern winter circulation. Q. J. Roy. Meteor. Soc. 119:269–298.
512. Moore A. M. and Reason C. J. (1993) The response of a global ocean general circulation model to climatological surface boundary conditions for temperature and salinity. J. Phys. Oceanogr. 23:300–328.
513. Morris E. M. (1980) Forecasting flood flows in grassy and forest basins using a deterministic distributed mathematical model. In: Hydrological Forecasting. IAHS publication, 129:247–255.
514. Morris R. E., Myers T. C. and Haney J. L. (1990) User's Manual for UAM (CB-IV). Reports by Systems Applications, Inc. No. SYSAPP-90/018a.
515. Morton F. I. (1983) Operational estimates of areal evapotranspiration and their significance to the science and practice. J Hydrol. 66:1–76.
516. Moser A. (1958) The dynamics of bacterial populations maintained in the Chemostat. Ph.D. thesis, The Carnegie Institute, Washington DC.
517. Moser A. (1985) Kinetics of batch fermentations. In: Rehm H. J. and Reed G. (Eds.) Biotechnol., 2. VCH Verlagsgesellschaft mbH, Weinheim, 243–283.
518. Moteni F., Buizza R., Palmer T. N. and Petroliagis T. (1996) The ECMWF ensemble prediction system: methodology and validation. Q. J. Roy. Meteor. Soc. 122:73–119.
519. Mott J., Williams J., Andrew M. and Gillison A. (1984) Australian savanna ecosystems. In: Tothill J. C. and Mott J. J. (Eds.) Ecology and Management of the world's savanna. Australian Academy of Science, Canberra, 56–82.
520. Munasinghe M. (1990) Managing Water Resources to avoid Environmental Degradation. World Bank Environmental Paper No. 41, Washington.
521. Murray R. J. and Simmonds I. (1995) Responses of climate and cyclones to reductions in Arctic winter sea ice. J. Geophys. Res. 100:4,791–4,806.
522. Myneni R. B., Nemani R. R. and Running S. W. (1997) Estimation of global Leaf Area Index and absorbed PAR using radiative transfer models. IEEE Transactions on Geoscience and Remote Sensing 35:1,380–1,393.
523. Neelin J. D., Battisti D. S., Hirst A. C., Jin F. F., Wakata Y., Yamagata T. and Zebiak S. E. (1998) Enso theory. J. Geophys. Res. 103(C7):14,261–14,290.
524. Neilson R. P. and Drapek R. J. (1998) Potentially complex biosphere responses to transient global warming. Glob. Change Biol. 4:505–521.
525. Nemry B., Francois L., Warnant P., Robinet F. and Gerard J. C. (1996) The seasonality of the CO_2 exchange between the atmosphere and the land Biosphere – A study with a Global Mechanistic Vegetation Model. J. Geophys. Res. 101(D3):7,111–7,125.
526. Neralla V. R., Jessup R. G. and Venkatesh S. (1988) The atmospheric environmental service regional ice model (RIM) for operational applications. Marine Geodesy 12:135–153.
527. Neralla V. R., Ramsier R. O. and Gross W. J. (1993) Utility of passive microwave observations for operational sea ice modelling. Proc. 3rd Int. Offshore and Polar Eng. Conf., Singapore, 6–11 June, 2:600–606.
528. Neralla V. R., Sayed M., Serrer M. and Savage S. B. (1994) The influence of ice rheology on ice forecasting. In: Yu Z. et al. (Eds.) Sea Ice Observation and Modelling. China Ocean Press, Beijing, 155–164.
529. Newkirk G. A. (1982) The Nature of Solar Variability, Solar Variability. Weather and Climate, 33047, National Academy Press, Washington, D.C.
530. Nicholls N., Gruza G. V., Jouzel J., Karl T. R., Ogallo L. A. and Parker D. E. (1996) Observed climate variability and change. In: Houghton J. T., Filho L. G. M., Callander B. A., Harris N., Kattenberg A. and Maskell K. (Eds.) Climate Change 1995, the Science of Climate Change. 132–192.

531. Niyogi D. S., Raman S. and Alapaty K. (1998) Comparison of four different stomatal resistance schemes using FIFE data. Part II: Analysis of terrestrial biospheric-atmospheric interactions. J. Appl. Meteorol. 37:1,301–1,320.
532. Noilhan J. and Mahfouf J. F. (1996) The ISBA land surface parameterization scheme. Global Planet. Change 13:145–159.
533. Noilhan J. and Planton S. (1989) A simple parameterization of land surface processes for meteorological models. Mon. Weather Rev. 117:536–549.
534. Nunuz Vaz R. A. and Simpson J. H. (1994) Turbulence closure modeling of estuarine stratification. J. Geophys. Res. 99:20,063–20,077.
535. Oberhuber J. M. (1990) Simulation of the Atlantic circulation with a coupled sea ice-mixed layer-isophcnal general circulation model. Rep. 59, Max-Planck-Inst. Meteorol., Hamburg.
536. Oberhuber J. M. (1993a) Simulation of the Atlantic circulation with a coupled sea ice-mixed layer-isophcnal general circulation model, I, Model description. J. Phys. Oceanogr. 23:808–829.
537. Oberhuber J. M. (1993b) Simulation of the Atlantic circulation with a coupled sea ice-mixed layer-isophcnal general circulation model, II, Model experiment. J. Phys. Oceanogr. 23:830–845.
538. O'Connel P. E. and Todini E. (1977) Real-time hydrological forecasting and control. Proceedings of the 1st international workshop. July, 1977, 4–29.
539. Odum E. P. (1977) The emergence of ecology as a new integrative discipline. Science 195:1,289–1,293.
540. Oglesby R. J. (1991) Springtime soil moisture, natural climatic variability, and north American drought as simulated by NCAR Community Climate Model 1. J. Climate 4:890–897.
541. Oglesby R. J. and Erickson D. J. (1989) Soil moisture and the persistence of North American Drought. J. Climate 2:1,362–1,380.
542. Oke P. R. and Middleton J. H. (1998) A Modelling Study of Upwelling off Port Stephens, Part II: Port Stephens Regional Model. New South Wales Environmental Protection Authority internal report.
543. Oke P. R. and Middleton J. H. (2000) Topographically Induced Upwelling off Eastern Australia. J. Phys. Oceanogr. 30:512–531.
544. Oke P. R. and Middleton J. H. (2001) Nutrient Enrichment off Port Stephens: the Role of the East Australian Current. Cont. Shelf Res. 21:587–606.
545. Oke P. R., Miller R. N., Allen J. S., Egbert G. D. and Kosro P. M. (2001) Assimilation of surface velocity data into a primitive equation coastal ocean model. J. Geophys. Res. (submitted).
546. Olson J. S., Watts J. A. and Allison L. J. (1983) Carbon in Live Vegetation of Major World Ecosystems. Rep ORNL-5862, Oak Ridge Natl. Lab., Oak Ridge, Tenesse.
547. Onstad C. A. and Jamieson D. G. (1970) Modelling the effect of land use modifications on runoff. Water Resour. Res. 6(5):1,287–1,295.
548. Oreskes N., Shrader-Frechette K. and Belitz K. (1994) Verification, validation and confirmation of numerical models in the Earth Sciences. Science 263:641–646.
549. Orlanski I. (1976) A Simple Boundary Condition for Unbounded Hyperbolic Flows. US Environmental Protection Agency Report.
550. Pacanowski R. C. (1995) MOM 2 Documentation, User's Guide and Reference Manual. GFDL Ocean. Tech. Rep. No.3.
551. Pacanowski R. C., Dixon K. W. and Rosati A. (1991) The GFDL Modular Ocean Model Users Guide version 1.0. GFDL Ocean Group Tech Rep No.2.

552. Pacanowski R. C. and Philander S. C. H. (1991) Parameterization of Vertical Mixing in Numerical Models of Tropical Oceans. J. Phys. Oceanogr. 11:1,443–1,451.
553. Paduan J. D. and Rosenfield L. K. (1996) Remotely sensed surface currents in Monterey Bay from Shore-based HF Radar (Coastal Ocean Dynamics Application Radar). J. Geophys. Res. 101:20,669–20,686.
554. Palma E. D. and Matano R. P. (1997) On the implementation of passive open boundary conditions for the Princeton ocean model: The Barotropic Mode. J. Geophys. Res. 103:1,319–1,341.
555. Palmer T. N., Shutts G. J. and Swinbank R. (1986) Alleviation of a systematic westerley bias in General Circulation and Numerical Weather prediction models through an orography gravity wave drag parameterisation. Q. J. R. meteor. Soc. 112:1,001–1,040.
556. Pan H. and Mahrt L. (1987) Interaction between soil hydrology and boundar layer development. Bound. Lay. Meteorol. 38:185–202.
557. Pan L. and Wierenga P. J. (1997) Improving numerical modeling of two-dimensional water flow in variably saturated, heterogeneous porous media. Water Resour. Res. 61:335–346.
558. Pan Z. T., Takle E., Segal M. and Turner R. (1996) Influences of model parameterization schemes on the response of rainfall to soil moisture in the Central United States. Mon. Weather Rev. 124:1,786–1,802.
559. Panagoulia D. and Dimou G. (1997) Linking space-time scale in hydrological modelling with respect to global climate change. J. Hydrol. 194:15–17.
560. Parkinson C. L. and Cavalieri D. J. (1989) Arctic sea ice 1973-1987: Seasonal, regional and inter-annual variability. J. Geophys. Res. 94:14,499–14,523.
561. Parkinson C. L. and Washington W. M. (1979) A large-scale numerical model of sea ice. J. Geophys. Res. 84:311–337.
562. Parton W. J., Steward J. W. B. and Cole C. V. (1988) Dynamics of carbon, nitrogen, phosphorus and sulfur in grassland soils: a model. Biogeochemistry 5:109–131.
563. Paulson C. A. (1970) The mathematical representation of wind speed and temperature profiles in the unstable atmospheric surface layer. J. Appl. Meteor. 9:857–861.
564. Peng G. and Domroes M. (1987a) Connections of the West Pacific subtropic high and some hydro-climatic regimes in China with Antarctic ice-snow indices. Meteorol. and Atmos. Phys. 37:61–71.
565. Peng G. and Domroes M. (1987b) Statistical studies of the atmospheric circulation of the Northern Hemisphere, hydro-climatic regimes in China and Antarctic ice-snow cover. Large scale effects of seasonal snow cover. IAHS Press, 61–78.
566. Peng G. B., Li Q. and Qian B. D. (1992) Climate and ice-snow cover. Meteorol. Press (in Chinese).
567. Peng G. B. and Lu W. (1982) Development of atmospheric circulation and irregulation of earth rotation. Scientia Sinica (Ser. B) 25(5):531–546.
568. Peng G. B. and Lu W. (1983) Fourth-kind natural factors of climate. Science Press, Beijing. (in Chinese).
569. Peng G. B., Qian B. and Li Q. (1996) Possible influence of Antarctic sea-ice on the general circulation over the Northern Hemisphere. In: Zhou X. and Lu L. (Eds.) Interactions between Antarctic and Global Climate. China Meteorol. Press, 247–254. (in Chinese).
570. Peng G. B., Qian B. D. and Qi L. X. (1996) The connections between the polar sea-ice variety and global atmospheric circulation. In: Zhou X. and Lu L. (Eds.) The study on the interaction between Antarctic and global climate. China Meteorol. Press, 240–246. (in Chinese).

571. Peng G. B. and Si Y. Y. (1983) Application of geophysical factors to long range weather forecasts. Kexue Tonghao (China Science Bulletin) 28(7):930–936.
572. Peng G. B., Wang B. G. (1989) The influence of antarctic sea-ice on the northwest Pacific sub-tropical high and the background of the ocean and atmosphere circulation. China Science Bulletin 34(17):56–59. (in Chinese and English).
573. Perlwitz J. (1997) Zeitscheibenexperimente mit dem atmosphärischen Zirkulationsmodell T42-ECHAM3 füer eine verdoppelte und verdreifachte CO_2-Konzentration unter besonderer Beachtung der Änderungen der nordhemisphärischen troposphärischen Dynamik. Dissertation, University Hamburg.
574. Petts G. and Calow P. (1996) River restoration: selected extracts from the river handbook. Rivers hand book, selections. Cambridge, MA: Blackwell Science.
575. Peuquet D. I. and Marble D. F. (Eds.) (1990) Introductory readings in Geographical Information Systems. Taylor and Francis.
576. Philander S. G. H. and Pacanowski R. C. (1980) The Generation of Equatorial Currents. J. Geophys. Res. 85:1,123–1,136.
577. Philander S. G. (1990) El-Nino, La-Nina and the Southern Oscillation. Academic Press.
578. Philander S. G. H., Hurlin W. J. and Siegel A. D. (1987) Simulation of the seasonal cycle of the tropical Pacific ocean. J. Phys. Oceanogr. 17:1,986–2,002.
579. Philander S. G. H. and Siegel A. D. (1985) Simulation of El-Nino of 1982–83. In: Nihoul J. (Ed.) Coupled Ocean–Atmospheric Models. Elsevier, New York, 517–741.
580. Philip J. R. (1957–1958) The theory of infiltration, 1: the infiltration equation and its solution; 2: the profile at infinity; 3: moisture profiles and relation to experiment; 4: sorptivity and algebraic infiltration equations; 5: the influence of the initial soil moisture content; 6: effects of water depth over soil; 7: soil science. 83–84; 345–357; 435–448; 163–178; 257–265; 329–339; 278–286; 333–337.
581. Philip, J. R. (1966) Plant water relations: some physical aspects. Ann Rev. Plant Physiol. 17:245–268.
582. Philip J. R. and de Vries D. A. (1957) Moisture movement in porous materials under temperature gradients. Trans. Amer. Geophys. Union. 38:222–232.
583. Physick W., Noonan J. A., McGregor J. L., Hurley P. J., Abbs D. J. and Manings P. C. (1994) LADM: A Lagrangian Atmospheric Dispersion Model. CSIRO Div. of Atmos. Res. Tech. Paper No. 24.
584. Pielke R. A., Avissar R., Raupach M., Johannes A., Dolman A. J., Zeng X. and Denning A. S. (1998) Interactions between the atmosphere and terrestrial ecosystems: Influence on weather and climate. Glob. Change Biol. 4:461–475.
585. Pielke R. A. and Segal M. (1986) Mesoscale circulations forced by differential terrain heating. In: Ray P. (Ed.) Mesoscale Meteorology and Forecasting. American Meteotology Society, 516–548.
586. Polcher J. and Laval K. (1994) The impact of African and Amazonian deforestation on tropical climate. J. Hydrol. 155:389–405.
587. Pollard D. and Thompson S. L. (1995) Use of a land surface transfer scheme (LSX) in a global climate model: The response to doubling stomatal resistance. Global Planetary Change 286:1–32.
588. Polzin K. L., Toole J. M., Ledwell J. R. and Schmitt. R. W. (1997) Spatial Variability of Turbulent Mixing in the Abyssal Ocean. Science 276:93–96.
589. Post W. M. (1993) The terrestrial carbon cycle. In: Solomon A. M. and Shugart H. H. (Eds.) Vegetation dynamics and global change. Chapman and Hall, London.
590. Potter C. S., Randerson J. T., Field C. B., Matson P. A., Vitousek P. M., Mooney H. A. and Klooster S. A. (1993) Terrestrial ecosystem production: A

process model based on global satellite and surface data. Global Biogeochem. Cy. 7:811–841.
591. Power S. B. and Kleeman R. (1994) Surface heat flux parameterization and the response of ocean general circulation models to high latitude freshening. Tellus 46A:86–95.
592. Prather M., Derwent R., Ehhalt D., Fraser P., Sanhueza E. and Zhou X. (1995) Other trace gases and atmospheric chemistry. In: Houghton J. T., Meira Filho L. G., Bruce J., Lee H., Callandwe B. A., Haites E., Harris N. and Maskell K. (Eds.) Climate Change 1994. Cambridge University Press, 71–126.
593. Preller R. H. (1994) U.S. Navy sea ice forecasting systems: Past, present and future. In: Yu Z. et al. (Eds.) Sea Ice Observation and Modelling. China Ocean Press, Beijing, 125–133.
594. Prentice I. C., Monserud R. A., Smith T. M. and Emanuel W. R. (1993) Modeling large-scale vegetation dynamics. In: Solomon A. M. and Shugart H. H. (Eds.) Vegetation dynamics and global change. Chapman and Hall, London.
595. Prentice C., Cramer W., Harrison S., Leemans R., Monserud R. and Solomon M. (1992) A global biome model based on plant physiology and dominance, soil properties and climate. J. Biogeogr. 19:117–134.
596. Prentice I. C., Sykes M. T. and Cramer W. (1993) A simulation model for the transient effects of climate change on forest landscapes. Ecol. Model. 65:51–70.
597. Priestley A. (1993) A quasi conservative version of the semi-Lagrangian advection scheme. Monthly Weather Review 121:311–337.
598. Priestley C. H. B. and Taylor R. J. (1972) On the assessment of surface heat flux and evaporation using large-scale parameters. Monthly Weather Review 100:81–92.
599. Prince S. D. (1991) A model of regional primary production for use with coarse resolution satellite data. Int. J. Remote Sens. 12:1,313–1,330.
600. Pryor S. C. (1998) A Case study of emission changes and ozone responses. Atmos. Environ. 32(2):123–131.
601. Puri K., Dietachmayer G. S., Mills G. A., Davidson N. E., Bowen R. and Logan L. W. (1998) The new BMRC Limited Area Prediction System. LAPS. Aust. Met. Mag. 47:203–223.
602. Qi L. and Huang X. (1997) Simulation on slope runoff and soil erosion in a raining event. Acta Mechanica Sinica 29(3):344–348. (in Chinese).
603. Qian B., Fan Z., Peng G. and Zhou E. (1990) Impact of the variation of Antarctic sea-ice extent on Atmospheric Circulation and runoff for the upper Yangtze River. Adv. Water Sci. 2(2):99–105. (in Chinese).
604. Qian B., Fan Z. and Peng G. (1996) The influence of the polar sea ice variety to the heat transportation in the atmosphere. In: Zhou X. and Lu L. (Eds.) Interactions between Antarctic and Global Climate. China Meteorol. Press, 255–262. (in Chinese).
605. Qian X. S. (1994) On Geographical Sciences. Zhejiang Education Press, Hangzhou. (in Chinese).
606. Qu G. P. (1987) Development of Environmental Problems in the World. China Environ. Sci. Press. (in Chinese).
607. Rahmstorf S. and Willebrand J. (1995) The role of temperature feedback in stabilizing the thermohaline circulation. J. Phys. Oceanogr. 25:787–805.
608. Rastetter E. B. (1996) Validating models of ecosystem response to global change. J. Geophys. Res. 46:190–198.
609. Raupach M. R. (1988) Canopy transfer processes. In: Steffen W. L. and Denmead O. T. (Eds.) Flow and Transport in the Natural Environment: Advances and Applications. Springer-Verlag, Berlin, 95–127.

610. Raupach M. R. (1992) Drag and drag partition on rough surfaces. Bound.-Lay. Meteorol. 60:375–395.
611. Raupach M. R. (1994) Simplified expressions for vegetation roughness length and zero-plane displacement as functions of canopy height and area index. Bound. Lay. Meteorol. 71:211–216.
612. Raupach M. R., Antonia R. A. and Rajagopalan S. (1991) Rough-wall turbulent boundary layers. Appl. Mech. Rev. 44:1–24.
613. Redi M. H. (1982) Oceanic Isopycnal Mixing by Coordinate Rotation. J. Phys. Oceanogr. 12:1,154–1,158.
614. Redinger G. J., Campbell G. S., Saxton K. E. and Papendick R. I. (1984) Infiltration rate of slot mulches: measurement and numerical simulation. Soil Sci. Soc. Am. J. 48:982–986.
615. Refsgaard J. C. and Knudsen J. (1996) Operational validation and intercomparison of different types of hydrological models. Water Resour. Res. 32(7):2,189–2,202.
616. Ren M. E. (1999) Environmental problems induced by cut off water current of the Yellow River and their protections. Geo-information Science 1(1):4–11. (in Chinese).
617. Rhines P. B. (1982) Basic dynamics of the large-scale geostrophic circulation. Summer Study Program in Geophysical Fluid Dynamics, Woods Hole Oceanographic Institution, 1–47.
618. Richtmyer R. D. and Morton K. W. (1967) Difference Methods for Initial-value Problem, John Wiley, New York.
619. Rinaldo A., Marani A. and Bellin A. (1989) On mass response function. Water Resour. Res. 127(7):568–580.
620. Roberts M. J. and Wood R. A. (1997) Topographic sensitivity studies with a Bryan–Cox–Type Ocean Model. J. Phys. Oceanogr. 27(5):823–836.
621. Robock A. (1980) The seasonal cycle of snow cover, sea ice and surface albedo. Mon. Weather Rev. 108:267–285.
622. Robock A., Schlosser C. A., Vinnikov K. V., Liu S. and Speranskaya N. A. (1995) Validation of humidity, moisture fluxes and soil moisture in GCMs. Report of AMIP Diagnostic Subproject 11:Part 1 – Soil moisture. In: Gates W. L. (Ed.) World Climate Research Programmes, Proceedings in the first AMIP Science Conference. WCRP, WMO/TD, Geneva, 732:85–90.
623. Robock A., Vinnikov K. Y., Schlosser C. A., Speranskaya N. A. and Xue Y. (1995) Use of midlatitude soil moisture and meteorological observations to validate soil moisture simulations with biosphere and bucket Models. J. Climate 8:15–35.
624. Roeckner E., Apre K., Bengisson L., Brinkop S., Dumenil L., Esch M., Kirk E. Lunkeit F., Ponater M., Rockel B., Sausen R., Schlege U., Schubert S. and Windelbland M. (1992) Simulation of the present day climate with the ECHAM model: impact of model physics and resolution. Rep. 93. Max Plank Institut fur Meteorologie. Hamburg.
625. Roels J. A. (1983) Energetics and kinetics in biotechnology. Elsevier Biomedical Press, Amsterdam.
626. Roesner L. A., Giguerre P. R. and Evenson D. E. (1977) Computer program documentation for the stream quality model QUAL-II. Prepared for the Southeast Michigan Council of Governments by Water Resources Engineers Inc. Walnut Creek, California.
627. Roesner L. A., Norton W. R. and Orlab G. T. (1974) A mathematical model for simulating the temperature structure of stratified reservoirs and its use in reservoir outlet design. In: Methematical Models in Hydrology, Proceedings of the Warsaw Symposium. 1974. IAHS-UNESCO-WMO. Publish Vol. 2.

628. Ross B. and Walsh J. E. (1986) Synoptic-scale influences of snow cover and sea ice. Mon. Weather Rev. 114:1,795–1,810.
629. Ross P. J. and Bristow K. L. (1990) Simulating water movement in layered and graditional soils using the Kirchhoff Transform. Soil Sci. Soc. Am. J. 54:1,519–1,524.
630. Rowe L. K. (1983) Rainfall interception by an evergreen forest, Nelson, New Zealand. J. Hydrol. 66:143–158.
631. Rui X., Jiang G. and Cheng H. (1998) Study of water level forecasting model with back water effect. Adv. Water Sci. 9(2):124–129. (in Chinese).
632. Rui X., Li Q. and Wang L. (1999) Study of flood routing model based on linear diffusion wave equation. Hydrol. 6:3–7. (in Chinese).
633. Ruimy A., Saugier B. and Dedieu G. (1994) Methodology for the estimation of terrestrial net primary production from remotely sensed data. J. Geophys. Res. 99:5,263–5,283.
634. Runkel R. and Chapra S. C. (1993) An efficient numerical solution of the transient storage equations for solute transport in small stream. Water Resour. Res. 29(1):211–215.
635. Running S. W., Nemani R., Peterson D. L., Band L. E., Potts D., Pierce L. and Spanner M. (1989) Mapping regional forest evapotranspiration and photosynthesis by coupling satellite data with ecosystem simulation. Ecology 70:1,090–1,101.
636. Rusband S. N. (1990) Chaotic Dynamics of Non-linear System. John Wiley and Sons.
637. Saint-Venant B. De (1871) Theory of unsteady water flow, with application to river floods and to propagation of tides in river channels. Computes Rendus Acad. Sci. Paris, 73:148–154, 237–240. (Translated into English by US Corps of Engineers. No. 49-g, Waterways Experiment Station. Vicksbugr, Mississippi, 1949).
638. Saltzman B. (1962) Finite amplitude free convection as an initial value problem. J. Atmos. Sci. 19:329–341.
639. Sarmiento J. L. (1983) A simulation of bomb tritium entry into the Atlantic Ocean. J. Phys. Oceanogr. 13:1,924–1,939.
640. Sarmiento J. L. and Bryan K. (1982) An ocean transport model for the North Atlantic. J. Geophys. Res. 87:395–408.
641. Sarmiento J. L., Hughes T. M. C., Stouffer R. J. and Manabe S. (1998) Simulated response of the ocean carbon cycle to anthropogenic climate warming. Nature 393:245–249.
642. Sasaki Y. (1970) Some basic formalism in numerical variational analysis. Mon. Wea. Rev. 98:875–883.
643. Schmugge T. J., Jakson T, J. and McKim H. L. (1980) Survey of methods for soil moisture determination. Water Resour. Res. 16:961–979.
644. Schopf O. S. and Cane M. A. (1983) On equatorial dynamic mixed layer physics and sea surface temperature. J. Phys. Oceanogr. 13:917–935.
645. Schulze E. D. (1986) Whole-plant responses to drought. Aust. J. Plant Physiol. 13:127–141.
646. Schultz G. A. (1988) Remote sensing in hydrology. J. Hydrol. 100:239–265.
647. Schwarzkopf M. D. and Fels S. D. (1985) Improvements to the algorithm for computing carbon dioxide transmissivitties and cooling rates. J. Geophys. Res. 90:10,541–10,550.
648. Schwarzkopf M. D. and Fels S. D. (1991) The simplified exchange method revisited: an accurate rapid method for computing of infra-red cooling rates and fluxes. J. Geophys. Res. 96:9,073–9,096.

649. Segel I. H. (1975) Enzyme kinetics: Behavior and analysis of rapid equilibrium and steady-state enzyme systems. Wiley-Interscience, New York.
650. Sellers P. J. (1992) Land surface process modeling. In: Trenberth K. E. (Ed.) Climate System Modeling. Cambridge University Press, London.
651. Sellers P. J., Hall F. G., Asrar G., Strebel D. E. and Murphy R. E. (1992) An overview of the First International Satellite Land Surface Climatology Project (ISLSCP) Field Experiment (FIFE). J. Geophys. Res. 97(D17):18,345–18,371.
652. Sellers P. J., Hall F. G., Ransom J., Margolis H., Kelly B., Baldocchi D., J. den artog, Cihlar J., Ryan M., Goodison B., Crill P., Ranson K. J., Lettenmaier D. and Wickland D. E. (1995) The Boreal Ecosystem-Atmosphere Study (BOREAS): An overview and early results from the 1994 field year. Bull. Am. Meteor. Soc. 76:1,549–1,577.
653. Sellers P. J., Mintz Y., Sud Y. C. and Dalcher A. (1986) A simple biosphere model (SiB) for use within general circulation models. J. Atmos. Sci. 43:505–531.
654. Sellers P., Los S. O., Tucker C. J., Justice C. O., Dazlich D. A., Collatz G. J. and Randall D. A. (1996b) A revised land surface parameterization (SIB2) for atmospheric GCMs. part II: the generation of global fields of terrestrial biophysical parameters from satellite data. J. Climate 9:706–737.
655. Sellers P., Randall D. A., Collatz G. J., Berry J. A., Field C. B., Dazlich D. A., Zhang C., Colledo G. D. and Bounoua L. (1996a) A revised land surface parameterization (SIB2) for atmospheric GCMs. part I: Model Formulation. J. Climate 9:676–705.
656. Sellers P. et al. (1996) Comparison of radiative and physiological effects of doubled atmospheric CO_2 on climate. Science 271:1,402–1,406.
657. Sellers P. et al. (1997) Modeling the exchanges of energy, water, and carbon between continents and the atmosphere. Science 275:502–509.
658. Sellers W. D. (1969) A global climate model based on the energy balance of the earth-atmosphere system. J. Appl. Meteor. 8:392–400.
659. Sellers W. D. (1973) A new global climate model. J. Appl. Meteor. 12:241–254.
660. Semtner A. J. (1995) Modeling ocean circulation. Science 269:1,379–1,385.
661. Semtner A. J. and Chervin R. M. (1992) Ocean General Circulation from a Global Eddy-resolving Model. J. Geophys. Res. 97:5,493–5,550.
662. Send U. and Marshall J. (1997) Integral effects of deep convection. J. Phys. Oceanogr. 25:855–872.
663. Seth A. and Giorgi F. (1996) A three-dimensional model study of organized mesoscale circulations induced by vegetation. J. Geophys. Res. 101:7,371–7,391.
664. Seth A., Giorgi F. and Dickinson R. E. (1994) Simulating fluxes from heterogeneous land surface: explicit subgrid method employing the Biosphere–Atmosphere Transfer Scheme (BATS). J. Geophys. Res. 99:18,651–18,667.
665. Shao Y. (2000) Physics and Modelling of Wind Erosion. Kluwer Academic Publishers.
666. Shao Y. and Henderson-Sellers A. (1996) Validation of soil moisture simulation in landsurface parameterization schemes with HAPEX Data. Global and Planetary Change 13:11–46.
667. Shao Y. and Irannejad P. (1999) On the choice of soil hydraulic models in land surface parameterization schemes. Bound. Lay. Meteorol. 90:83–115.
668. Shao Y. and Leslie L. M. (1997) Wind erosion prediction over the Australian continent. J. Geophy. Res. 102:30,091–30,105.
669. Shao Y. and Leslie L. M. (1997) Prediction of soil moisture over the Australian continent. Meteorology and Atmospheric Physics 63:195–215.

670. Shao Y., Leslie L. M., Munro R. K., Irannejad P., Lyons W. F., Morison R., Short D. and Wood M. S. (1997) Soil moisture prediction over the Australian continent. Meteorol. and Atmos. Phys. 63:195–215.
671. Shao Y., Raupach M. R. and Leys J. F. (1996) A model for predicting aeolian sand drift and dust entrainment on scales from paddock to region. Aust. J. Soil Res. 34:309–342.
672. Sherman B. S. and Webster I. T. (1994) A model for the light-limited growth of buoyant pyhtoplankton in a shallow, turbid waterbody. Aust. J. Mar. Freshwater Res. 45:847–862.
673. Shmakin A. B. (1998) The updated version of SPONSOR land surface scheme: PILPS-influenced improvements. Global Planet Change 19:49–62.
674. Showmsky P. A. and Krenke A. N. (1964) Modern cryosphere of the earth and its changes. Geophys. Bulletin, No. 14. (in Russian).
675. Shugart J. H. H. and West D. C. (1977) Development of an Appalachian deciduous forest model and its implication to the assessment of the impacts of the Chestnut Blight. J. Environ. Manage. 5:161–179.
676. Shukla J. (1981) Dynamic predictability of monthly means. J. Atmos. Sci. 38:2,547–2,572.
677. Shukla J. (1991) Short term climate variability and prediction. In: Jager J. and Ferguson H. L. (Eds.) Proc. Second World Climate Conference, Cambridge University Press, 203–210.
678. Shukla J. and Mintz Y. (1982) Influence of land-surface evapotranspiration on the earth's climate. Science 215:1,498–1,501.
679. Shuttleworth W. J. (1988) Evaporation from amazonian rain forest. Proc. Roy. Soc. Lond. 233:321–346.
680. Simmonds I. (1985) Analysis of the "Spinup" of a General Circulation Model. J. Geophys. Res. 90:5,637–5,660.
681. Simmonds I. and Budd W. F. (1991) Sensitivity of the southern hemisphere circulation to leads in the Antarctic pack ice. Q. J. Roy. Meteor. Soc. 117:1,003–1,024.
682. Simmons A. J. and Bengtsson L. (1984) Atmospheric general circulation models, their design and use for climate studies. In: Houghton J. T. (Ed.) The Global Climate. Cambridge University Press, 37–62.
683. Singh V. P. (1989) Hydrological systems. Watershed Modelling. Volume II. Prentice–Hall, a Division of Simon and Schuster Englewood Cliffs, New Jersey.
684. Singh V. P (1995) Computer models of watershed hydrology. Water Resources Publications. Colorado, U.S.A.
685. Slingo J. M. (1987) The development and verification of a cloud prediction scheme for the ECMWF model. Q. J. Roy. Met.Soc. 113:899–927.
686. Slutsky A. H. and Yen B. C. (1997) A macro-scale natural hydrological cycle water availability model. J. Hydrol. 201:329–347.
687. Smagorinsky J. (1963) General circulation experiments with primitive equations, I. The Basic Experiment. Mon. Weather Rev. 91:99–164.
688. Smakhtin V. U. (2001) Low flow hydrology: a review. J. Hydrol. 240:147-186.
689. Smith I. N. (1994) A GCM simulation of global climate trends: 1950-1988 J. Clim. 7:732–744.
690. Smith N. and Lefebvre M. (1998) The global ocean data assimilation experiment. In: International Symposium "Monitoring the Oceans in the 2000s: An Integrated Approach". 15–17 October 1998, Biarritz, France.
691. Smith P. M. and Warr K. (Eds.) (1991) Global Environmental Issues.

692. Smith T. M., Leemans R. and Shugart H. H. (Eds.) (1993) The Application of Patch Models of Vegetation Dynamics to Global Change Issues. Kluwer Academic Publishers. Worshop Summary, Global Change and Terrestrial Ecosystems (GCTE).
693. Smolarkiewicz P. K. and Grabowski W. W. (1990) The multidimensional positive definite advection transport algorithm. J. Comp. Phys. 86:355–375.
694. Soil Conservation Service (1971) Hydrology. SCS National Engineering Handbook, Section 4, U S Department of Agriculture, Washington, D. C.
695. Speer M. S. and Leslie L. M. (1998) Numerical simulation of two heavy rainfall events over coastal southeastern Australia. Meteorol. Appl. 5:239–252.
696. Stammer D., Tokmakian R., Semtner A. J. and Wunsch C. (1996) How well does a 1/4 degree global circulation model simulate large-scale oceanic observations? J. Geophys Res. 101:25,779–25,811.
697. Steer A. (1998) Making development sustainable. In: Blume H. P., Eger H., Fleischhauer E., Hebel A., Reij C. and Steiner K. G. (Eds.) Towards Sustainable Land Use. Advances in Geoecology 31:852–856.
698. Steffen W. L., Walker B. H., Ingram J. S. and Koch G. W. (1992) Global Change and Terrestrial Ecosystems: The Operational Plan. Tech. Rep. 21, IGBP, Stockholm.
699. Stensrud, David J., Harold E. Brooks, Jun Du, Steven Tracton M., Eric Rogers (1999) Using ensembles for short-range forecasting. Monthly Weather Review 127(4):433–446.
700. Stevens D. P. (1990) On open boundary conditions for three dimensional primitive equation ocean general circulation models. Geophys. Astrophys. Fluid Dyn. 51:103–133.
701. Stevens D. P. (1991) The open boundary condition in the United Kingdom Fine-Resolution Antarctic Model. J. Phys. Oceanogr. 21:1,494–1,499.
702. Stommel H. and Arons A. B. (1960) On the abyssal circulation of the world ocean, I. An idealized model of the circulation pattern and amplitude in oceanic basins. Deep Sea Res. 6:217–233.
703. Stramma L. and England M. H. (1999) On the water masses and mean circulation of the South Atlantic Ocean. J. Geophys. Res. 104:20,863–29,883.
704. Street H. W. and Phelps E. B. (1925) A study of the popullation and natural purification of the Ohio River. U. S. Public Health Service Bull. 146, Washington D.C.
705. Stull R. (1988) An Introduction to Boundary Layer Meteorology. Kluwer Acad, Norwell, Mass.
706. Sugawara M. (1995) Tank model. In: Singh V. P. (Ed.) Computer models of watershed hydrology. Water resources Publications, Colorado, USA, 165–214.
707. Sun S. (1997) Compressibility effects in the Miami Isopycnic Coordinate Ocean Model. Ph.D dissertation, University of Miami.
708. Sverdrup H. U., Johnson M. W. and Fleming R. H. (1942) The Oceans: Their Physics, Chemistry and General Biology. Prentice-Hall, Englewoods Cliffs, N. J.
709. Tabios G. Q. and Salas J. D. (1985) A comparative analysis of techniques for spatial interpolation of precipitation. Water Resour. Bull. 21(3):365–380.
710. Tallaksen L. and Erichsen L. (1994) Modelling low flow response to evapotranspiration, Friend: Flow regimes from international experimental and network data. In: Proceeding of the Braunschweig Conference, October, 1993, IAHS Publ. 221:95–102.
711. Tan Y. (1996) Comments on "Modeling of Oxygen transport and pyrite oxidation in acid sulphate soils by Bronswijk et al.". J. Environ. Qual. 25(4):928–930.

712. Tan Y. and Bond W. J. (1994) Modeling subsurface transport of microorganisms. In: Singh V. P. (Ed.) Environmental Hydrology. Kluwer Academic Publishers, New York, 321–355.
713. Tan Y., Wang Z. X., Schneider R. P. and Marshall K. C. (1994) Modelling microbial growth: a statistical thermodynamic approach. J. Biotechnol. 32:97–106.
714. Tan Y., Wang Z. X. and Marshall K. C. (1996) Modelling substrate inhibition of microbial growth. J. Biotechnol. Bioeng. 52:602–608.
715. Tan Y., Wang Z. X. and Marshall K. C. (1998) Modelling pH effects on microbial growth: a statistical thermodynamic approach. J. Biotechnol. Bioeng. 59:724–731.
716. Tan Y. and Wang Z. X. (1997) Derivation of Tsao-Hanson equation using a statistical thermodynamic method. J. Theo. Biol. 185:549–551.
717. Tapper N. J., Garden G., Gill J. and Fernon J. (1993) The climatology and meteorology of high fire danger in the Northern Territory. Rangeland J. 15:339–351.
718. Taylor C. M., Said F. and Lebel T. (1997) Interactions between the land surface and mesoscale rainfall variability during HAPEX-Sahel. Mon. Weather Rev. 125:2,211–2,227.
719. Tegen I. and Fung I. (1994) Modeling of mineral dust in the atmosphere: sources, transport, and optical thickness. J. Geophys. Res. 99:22,897–22,914.
720. Tegen I. and Fung I. (1995) Contribution to the atmospheric mineral aerosol load from land surface modification. J. Geophys. Res. 100:18,707–18,726.
721. Thom A. S. (1975) Momentum, mass and heat exchange of plant communities. In: Montieth J. L. (Ed.) Vegetation and the Atmosphere. Vol. 1, Academic Press, London, 57–109.
722. Thomas E. G. and Crutzen P. J. (1995). Atmosphere, Climate and Change. Scientific American Library, W. H. Freeman Company.
723. Thomas G. and Rowntree P. R. (1992) The boreal forests and climate. Q. J. Roy. Meteorol. Soc. 118:469–497.
724. Thomas W. M. and Wesley P. J. (1994) Predicting sediment yield in stormwater runoff from urban areas. J. Water Resources Planing and Management 120(5):630–650.
725. Thompson Philip Duncan (1977) How to improve accuracy by combining independent forecasts. Monthly Weather Review 105(2):228–229.
726. Thorpe A. J., Miller M. J. and Moncriefe M. W. (1985) Comment on "The dynamical structure of squall line type thunderstorms". J. Atmospheric Sciences 42(2):212–213.
727. Thurman H. V. (1991) Introductory Oceanography. Macmillan.
728. Tian H., Hall C. A. S. and Qi Y. (1998) Modeling primary productivity of the terrestrial biosphere in changing environments: toward a dynamic biosphere model. Crit. Rev. Plant. Sci. 15:541–557.
729. Tiedtke M. (1989) A comprehensive mass flux scheme for cumulus parameterization in large-scale models. Monthly Weather Review 117(8):1,779–1,800.
730. Toggweiler J. R., Dixon K. and Bryan K. (1989) Simulations of radiocarbon in a coarse-resolution world ocean model. I: Steady state prebomb distributions. J. Geophys. Res. 94:8,217–8,242.
731. Toggweiler J. R. and Samuels B. (1992) Is the magnitude of the deep outflow from the Atlantic Ocean actually governed by Southern Hemisphere winds? In: Heimann M. (Ed.) The Global Carbon Cycle. Springer-Verlag, Berlin.
732. Toggweiler J. R. and Samuels B. (1995) Effect of drake passage on the global thermohaline circulation. Deep Sea Res. 42:477–500.

733. Tokmakian R. (1996) Comparisons of time series from two global models with tide-gauge data. Geophys. Res. Lett. 23:3,759–3,762.
734. Tomczak M. and Godfrey J. S. (1994) Regional Oceanography: An Introduction. Elsevier, Oxford.
735. Toole J. M., Schmitt R. W., Polzin K. L. and Kunze E. (1997) Near-boundary mixing above the Flanks of a midlatitude seamount. J. Geophys. Res. 102:947–959.
736. Toth Z. and Kalnay E. (1993) Ensemble forecasting at the NMC: The generation of perturbations. Bull. Amer. Meteorol. Soc. 74:2,317–2,330.
737. Toyota T. and Sato K. (1994) The improvement of the thermodynamical process of JMA operational numerical sea ice model in the southern part of Okhotsk Sea. In: Yu Z. et al. (Eds.) Sea Ice Observation and Modelling. China Ocean Press, Beijing, 147–154.
738. Trenberth K. E., Houghton J. T. and Meira Filho L. G. (1996) The climate system: an overview. In: Houghton J. T., Meira Filho L. G., Callander B. A., Harris N., Kattenberg A. and Maskell K. (Eds.) Climate Change 1995, the Science and Climate Change, 55–64.
739. Tsao G. T. and Hanson T. P. (1975) Extended Monod equation for Batch cultures with multiple exponential phases. J. Biotechnol. Bioeng. 17:1,591–1,598.
740. Tsien H. S. (1954) Engineering Cybernetics. McGraw-Hill, New York.
741. Tubman W. (1996) River ecosystems. Land and Water, Resources Research and Development Corporation, Canberra.
742. Tung Y. K. (1983) Point rainfall estimation for a mountainous region. J. Hydraulic Engineering. America Society of Civil Engineers 109(10):1,386–1,393.
743. Turco R. P. (1992) Atmospheric chemistry. In: Trenberth K. E. (Ed.) Climate System Modeling. Cambridge University Press, 201–240.
744. U. S. EPA (1988) WASP4, A hydrodynamic and water quality model: Model theory. User's manual, and programmers' guide. EPA/600/3–87/039.
745. Van dam J. C., Huygen J., Wesseling J. G., Feddes R. A., Kabat P., van Walsum, P. E. V., Groenendijk P. and van Diepen C. A. (1997) SWAP version 2.0, Theory. Simulation of water flow solute transport and plant growth in the soil–water–atmosphere–plant environment. Technical Document 45, DLO Winnand Staring Centre, Report 71, Department Water Resources, Agriculture University, Wageningen.
746. Van Genuchten M. Th. (1980) A close-form equation for predicting the hydraulic conductivity of unsaturated soils. Soil Sci. Soc. Am. J. 44:892–898.
747. Van Ulden A.P. and Wieringa J. (1996) Atmospheric boundary-layer research at Cabauw. Bound. layer Meteor. 78:39–69.
748. Venkatram A., Karamchandarai P., Parsad P., Sloane C., Saxena P. and Goldstein R. (1997) The development of a model to examine source-receptor relationships for visibility on the Colorado Plateau. J. Air Waste Manage Assoc. 47:286–301.
749. Verseghy D. L. (1991) CLASS – A Canadian land surface scheme for GCMs I. soil model. Int. J. Climatol. 11:111–133.
750. Verseghy D. L., McFarlane N. A. and Lazare M. (1993) CLASS – A Canadian land surface scheme for GCMs. II. vegetation model and coupled runs. Int. J. Climatol. 13:347–370.
751. Verstraete M. M. and Dickinson R. E. (1986) Modeling surface processes in atmospheric general circulation models. Ann. Geophys. 4(B):357–364.
752. Verstraete M., Pinty B. and Myneni R. B. (1996) Potential and limitations of information extraction on the terrestrial biosphere from satellite remote sensing. Remote Sens. Environ. 58:201–214.

753. Vinnikov K. Ya., Groisman P. Ya and Lugina K. M. (1990) Empirical data on contemporary global climate changes (temperature and precipitation). J. Climate 3:662–677.
754. Vinnikov K. Y, Robock A., Speranskaya N. A. and Schlosser C. A. (1996) Scales of temporal and spatial variability of midlatitude soil moisture. J. Geophys. Res. 101:7,163–7,174.
755. Vintzileos A. and Sadnourny R. (1997) A general interface between an atmospheric general circulation model and underlying ocean and land surface models: delocalized physics scheme. Mon. Weather Rev. 125:926–941.
756. Walsh J. E., Tucek D. R. and Peterson M. R. (1982) Seasonal snow cover and short-term climatic fluctuation over the United States. Mon. Weather Rev. 110:1,474–1,485.
757. Walsh J. E. and Johnson C. M. (1979) An analysis of Arctic sea ice fluctuation, 1953–1977. J. Phys. Oceanogr. 9:580–591.
758. Walsh J. E. and Ross B. (1988) Sensitivity of 30-day dynamical forecasts to continental snow cover. J. Climate 1:739–754.
759. Wang Q. J. (1991) The genetic algorithm and its application to calibrated conceptual rain-runoff models. Water Resour. Res. 27(9):2,467–2,471.
760. Wang X., Fan Z., Peng G. and Zhou E. (1990) Statistical study on the spatial-temporal distribution features of the Arctic sea ice extent. Acta Oceanologica Sinica 9:373–387.
761. Wang X., Peng G., Fan Z. and Zhou E. (1992) Connection of the flood season runoff of the upper-middle Yangtze River with Arctic sea-ice indices. Acta Meteorl. Sinica 1:53–58. (in Chinese).
762. Wang Z. and Wu H. (1994) Sea ice thermal processes and simulation of their coupling with the dynamic process. Oceanologia et limnologia Sinica 25:408–415. (in Chinese).
763. Warnant P., Francois L., Strivay D. and Gerard J. C. (1994) CARAIB – A global model of terrestrial biological productivity. Global Biogeochem Cycles 8:255–270.
764. Warnant P., Francois L., Strivay D. and Gerard J. C .(1994) A global model of terrestrial biological productivity. Global Biogeochem. Cy. 8:255–270.
765. Warner, Thomas T., Ralph A. Peterson and Russell E. Treadon (1997) A tutorial on lateral boundary conditions as a basic and potentially serious limitation to regional numerical weather prediction. Bulletin of the American Meteorological Society 78(11):2,599–2,617.
766. Warrilow D. A., Sangster A. B. and Slingo A. (1986) Modelling of land surface processes and their influence on European climate. Dynamical Climatology Tech. Note, No. 38, Metorol. Office, London.
767. Washington W. M., Semtner A. J., Parkinson C. and Morrison L. (1976) On the development of a seasonal change sea ice model. J. Phys. Oceanogr. 6:679–685.
768. Watson R. T., Rodhe H., Oeschger H. and Siegenthaler U. (1990) Greenhouse gases and aerosols. In: Houghton J. T., Jenkins G. J. and Ephraums J. J. (Eds.) Climate Change: The IPCC Scientific Assessment. Cambridge University Press, 1–39.
769. Weaver A. J., Sarachik E. S. and Marotzke J. (1991) Freshwater flux forcing of decadal and interdecadal oceanic variability. Nature 353:836–838.
770. Weaver A. J. and Hughes T. M. C. (1992) Stability and variability of the thermohaline circulation and its link to climate. Trends in Physical Oceanography, Research Trends Series. Council of Scientific Research Integration, Trivandrum, India, 1:1–570.

771. Webb D. J. et al. (17 authors) (1991) An eddy-resolving model of the Southern Ocean. EOS, Transactions of the American Geophysical Union, 72:169–174.
772. Webb D. J. et al. (1998) The First Main Run of the OCCAM Global Ocean Model. Southampton Oceanography Centre, Internal document No.34.
773. Weinberg A. M. (1972) Science and trains-science. In: Civilization and Science, Elsevier, Amsterdam, 105–122.
774. Welander P. (1959) On the vertically integrated mass transport in the oceans. In: Bolin B. (Ed.) The Atmosphere and the Sea in Motion. Rockefeller Inst. Press and Oxford Univ. Press, 95–101.
775. Wen Z. and Liu X. J. (1991) Hydrodynamic model of urban runoff pollution. In: Lee and Cheung (Eds.) Environmental Hydraulics. Balkema, Rotterdam.
776. Westphal D. L., Toon O. B. and Carson T. N. (1988) A case study of mobilisation and transport of Saharan dust. J. Atmos. Sci. 45:2,145–2,175.
777. Wetzel P. J. and Boone A. (1995) A parameterization for land-atmosphere-cloud-exchange (PLACE): documentation and testing of a detailed process model of the partly cloudy boundary layer over heterogeneous land. J. Climate 8:1,810–1,837.
778. Wetzel P. J. and Chang J. T. (1987) Concerning the relationship between evapotranspiration and soil moisture. J. Climate Appl. Meteor. 26:375–391.
779. Whipple Jr. W. (1996) Integration of water resources planning and environment regulation. J. Water Resources Planning and Management 122(3):189–196.
780. White D. H., Howden S. M. and Nix H. A. (1993) Modelling agricultural and pastoral systems. In: Jakeman A. J., Beck M. B. and McAleer M. J. (Eds.). Modeling Change in Environmental Systems. John Wiley and Sons, West Sussex.
781. Wigmosta M. S., Vail L. W. and Lettenmaier D. P. (1994) A distributed hydrology-vegetation model for complex terrain. Water Resour. Res. 30(6):1,665–1,679.
782. Wilson M. F. and Henderson-Sellers A. (1985) A global archive of land cover and soil data for use in general circulation climate models. J. Climate 5:119–143.
783. Wischmeier W. H. and Smith D. D. (1958) Rainfall energy and its relationship to soil loss. Trans. Am. Geog. Un. 39:285–291.
784. World Meteorological Organization (WMO) (1992) Simulated Real-time Intercomparison of Hydrological Models. WMO Oper. Hydrol. Re. 38, WMO 779, Geneva.
785. World Meteorological Organization (WMO) (1998) Bulletin WMO, 47(4):336–343.
786. Wood D. (1998) Legitimizing integrated water management: the Tami Nadu Water resources consoildation project in India. Water Resour. Development 14(1):25–39.
787. Wood E. F. (1991) Global scale hydrology: Advances in land surface modelling. Rev. Geophys. supplement:193–201.
788. Wood E. F., Lettenmaier D., Liang X., Nijssen B. and Wetsel S. W. (1997) Hydrological modelling of continental-scale basins. Annu. Rev. Earth Planet Sc. 25:279–300.
789. Wood E. F., Lettenmaier D. P., Liang X., Lohmann D., Boone A., Chang S., Chen F., Dai Y., Desbourough C., Duan Q., Ek M., Gusev Y., Habets F., Irannejad P., Koster R., Nasanova O., Noilhan J., Schaake J., Schlosser A., Shao Y., Shmakin A., Verseghy D., Wang J., Warrach K., Wetzel P., Xue Y., Yang Z. L. and Zeng Q. (1998) The Project for intercomparison of land-surface parameterization schemes (PILPS) phase-2(c) Red-Arkansas river basin experiment: 1. Experiment description and summary intercomparisons. Global Planet Change 19:115–136.

790. Wood E. F., Sivapalan M. and Beven K. (1990) Similarity and scale in catchment storm response. Rev. Geophys. 28(1):1–18.
791. Woodruff S. D., Slutz R. J., Jenne R. L. and Steurer P. M. (1987) A Comprehensive ocean atmosphere data set. Bull. Amer. Meteor. Soc. 68:1,239–1,250.
792. Woods J. (1985) The world ocean circulation experiment. Nature 314:501–511.
793. Woodward F. (1996) Developing the potential for describing the terrestrial biosphere's response to a changing climate. In: Walker B. and Steffen W. (Eds.) Global Change and Terrestrial Ecosystems. Cambridge University Press, Cambridge. IGBP Book Series, 511–528.
794. Woodward F. I. and Beerling D. (1997) The dynamics of vegetation change: health warnings for equilibrium "dodo" models. Global Ecol. Biogeogr. 6:413–418.
795. Woodward F. I. and Lee S. E. (1995) Global scale function and distribution. Forestry 68:317–325.
796. Woodward F. I., Smith T. M. and Emanuel W. R. (1995) A global land primary productivity and phytogeography model. Global Biogeochem. Cy. 9:471–490.
797. Worthington L. V. (1976) On the North Atlantic Circulation. John Hopkins Oceanographic Studies 6. John Hopkins Univ Press, Baltimore, Maryland.
798. Wu H. (1991) Mathematic representations of sea ice dynamic-thermodynamic processes. Oceanologia et limnologia Sinica 22:221–228. (in Chinese).
799. Wu H., Bai S. and Zhang Z. (1997) Numerical simulation for dynamic processes of sea ice. Acta Oceanologica Sinica, Vol.16, Ocean Press, Beijing, 134–146.
800. Wu H., Bai S. and Zhang Z. (1998) Numerical sea ice prediction in China. Acta Oceanologica Sinica, 17(2), 167–185.
801. Wu H. and Wang Z.(1992) Sea ice thermodynamic process and its simulation. Numerical computation for physical oceanography (in Chinese). In: Computation Series for Science and Engineering, Henan Scientific and Technological Press, 361–428.
802. Wu Y. and Raman S. (1997) Effect of land-use pattern on the development of low-level jets. J. Appl. Meteorol. 36:573–590.
803. Wunsch C. (1996) The Ocean Circulation Inverse Problem. Cambridge University Press, New York.
804. Wyrtki K. (1961) The thermohaline circulation in relation to general circulation in the oceans. Deep-Sea Res. 8:39–64.
805. Wyrtki K. (1985) Water displacements in the Pacific and the genesis of El Nino. J. Geophys. Res. 90:7,129–7,132.
806. Xie Y. (1996) Environment water quality models. Press of Science and technology of China, Beijing. (in Chinese).
807. Xiong L. and Guo S. (1999) A two-parameter monthly water balance model and its application. J Hydrol. 216:111–123.
808. Xue Y. (1996) The impact of desertification in the Mongolian and the Inner Mongolian grassland on the regional climate. J. Climate 9:2,173–2,185.
809. Xue Y., Bastable H., Dirmeyer P. and Sellers P. (1996a) Sensitivitay of simulated surface fluxes to changes in land surface parameterizations – A study using ABRACOS data. J. Appl. Meteorol. 35:386–400.
810. Xue Y., Fennessy M. and Sellers P. (1996b) Impacts of vegetation properties on U.S summer weather prediction. J. Geophys. Res. 101:7,419–7,430.
811. Xue Y. K., Sellers P. J., Kinter J. L. and Shukla J. (1991) A simplified biosphere model for global climate studies. J. Climate 4:345–364.
812. Xue Y. and Shukla J. (1993) The influence of land properties on Sahel climate. Part I: Desertification. J. Climate 5:2,232–2,245.

813. Yang P. C. and Chen N. T. (1990) The predictability of El-Nino/Southern Oscillation. Atmos. Sci. 14:64–71. (in Chinese).
814. Yang J., Zhang R., Wu J. and Allen M. B. (1997) Stochastic analysis of adsorbing solute transport in three-dimensional heteorogeneous unsaturated soils. Water Resour. Res. 33(8):1,947–1,956.
815. Yang X. and Huang S. (1992) The effects of the Arctic sea ice on the variations of the atmospheric circulation and climate. Acta Meteor. Sinica 6:1–14.
816. Yates D. (1997) Approaches to continental scale runoff for integrated assessment models. J. Hydrol. 201:289–310.
817. Ye C. M. (1986) The calculation of water quality in fluvial network without a regular velocity direction. J. Envir. Sci. 6(3):327–333. (in Chinese).
818. Yeh T., Wetherald R. T. and Manabe S. (1983) A model study of the short-term climatic and hydrologic effects of sudden snow removal. Mon. Weather Rev. 111:1,013–1,024.
819. Yi L. and Tao S. (1997) Construction and analysis of a precipitation recycling model. Adv. Water Sci. 8(3):206–211. (in Chinese).
820. Young R. A., Onstad C. A., Bosch D. D. and Anderson W. P. (1989) AGNPS: A nonpoint-source pollution model for evaluating agricultural Watersheds. J. Soil and Water Conservation, March–April.
821. Yu Z., Lakhtakia M. N., Yarnal B., White R. A., Miller D. A., Frakes B., Barron E. J., Duffy C. and Schwartz F. W. (1999) Simulating the river-basin response to atmospheric forcing by linking a meso-scale meteorological model and hydrological model system. J. hydrol. 218:72–91.
822. Yuan X., Liu S. and Chen H. (1999) Application of the artificial neural network method to forecast high sediment flood. Adv. Water Sci. 10(4):398–403. (in Chinese).
823. Zanetti P. (1990) Air Pollution Modelling: Theories, Computational Methods and Available Software. New York, Van Nostrand Reinhold.
824. Zeng Q. C. (1995) Silt sedimentation and relevant engineering problem – an example of natural cybernetics. Math. Res. 87:463–487.
825. Zeng Q. C. (1996) Natural cybernetics. Bulletin of the Chinese Academy of Sciences 10(2):105–113.
826. Zeng Q. C., Yuan C. G., Wang W. Q. and Zhang R. H. (1990) Experiments in numerical extra-seasonal prediction of climate anomalies. Chinese J. of Atmos. Sci. 14(1):1–24.
827. Zeng Q. C. and Zhang X. H. (1982) Perfect energy-conservative time-space finite difference schemes and the consistent split method to solve the dynamic equations of compressible fluid. Scientia Sinica, B., 25, No. 8.
828. Zeng Q. C., Zhang X. H. and Zhang R. H. (1991) A design of an oceanic GCM without the rigidlid approximation and its application to the numerical simulation of the circulation of the Pacific ocean. J. Marine System 1:271–292.
829. Zeng N. (1998) Understanding climate sensitivity to tropical deforestation in a mechanistic model. J. Climate 11:1,969–1,975.
830. Zhang F. and Zhou L. (1994) Maximum ice thickness in China. In: Yu Z. et al. (Eds.) Sea Ice Observation and Modelling. China Ocean Press, Beijing, 118–122.
831. Zhang H., Henderson-Sellers A., Pitman A. J., McGregor J. L., Desborough C. E. and Katzfey J. (2000) Limited-area model sensitivity to the complexity of representation of the land-surface energy balance. J. Climate. (submitted).
832. Zhang J. H. (1999) Global resource and environment in the 20th century. Geo-information Science 1(2):1–5. (in Chinese).

833. Zhang L., Warrick R., Dowes T. and Hatton J. (1996) Modelling hydrological progresses using a biophysically based model – Application of WAVES to FIFE and HAPEX–MOMILHT. J. Hydrol. 185:147–169.
834. Zhang R. H. and Endoh M. (1992) A free surface general circulation model for the tropical Pacific Ocean. J. Geophys. Res. 97:11,237–11,255.
835. Zhang R. H. and Endoh M. (1994) Simulation of the 1986-1987 El-Nino and 1988 La-Nina events with a free surface tropical Pacific Ocean general circulation model. J. Geophys. Res. 99:7,743–7,759.
836. Zhao R. (1980) The Xin'anjiang model. Hydrological Forecasting Proceedings Oxford Symposium, IAHS 129:351–356.
837. Zhao R. (1992) The Xin'anjiang model applied in China. J. Hydrol. 135:371–381.
838. Zhao Y., Ding J. and Deng Y. (1998) Wavelet network model of phase space and its application in hydrological prediction. Adv. Water Sci. 9(3):252–257. (in Chinese).
839. Zheng X., Chu J. and Zhu W. (1997) Unsteady water environmental capacity of river network. Adv. Water Sci. 8(1):25–29. (in Chinese).
840. Zhou G. (1992) Temporal and spatial distribution of the Qinhai-Tibet Plateau and its influence on the precipitation of upper-middle reaches of the Yangtze River. Thesis of a Master of Science. Dept. of Hydrol. Hohai Univ. Nanjing. (in Chinese).
841. Zhou G. Q., Li X. and Zeng Q. C. (1999) An improved coupled ocean-atmosphere general circulation model and its numerical simulation. Prog. in Nature Sci. 9:374–381.
842. Zhu H. (1993) The automatic observing and forecasting system of hydro-information. Beijing: The Press of Water Conservancy and Electric Engineering. (in Chinese).
843. Zhu J., Zeng Q. C., Guo D. J. and Liu Z. (1999) Optimal control of sedimentation in navigation channel. J. Hydraulic Engin. 125(7):750–759.
844. Zou B., Xie S. and Wang Y. (1996) Periodic variation and predictability of Antarctic sea ice. In: Zhou X. and Lu L. (Eds.) The study on the interaction between Antarctic and global climate. China Meteorological Press, 383–392. (in Chinese).

Index

K theory, 372
α and β methods, 190
z-level models, 136

absorption efficiency, 337
accumulated infiltration, 227
acid rain, 6
actual evapotranspiration, 233
adiabatic advection, 139
advection equation, 89, 241
advection-aridity approach, 234
advective processes, 160
aeration, 240
aerodynamic roughness length, 29
aerosols, 411
African savannas, 340
African soils, 340
aggregate, 217
aggregation effects, 203, 206
air chemistry, 361
air pollution, 6, 361
air quality, 371, 385, 418
air quality models, 371
air-quality modelling, 361
air-sea exchanges, 281
air-sea interface, 125, 137, 141, 143
airshed model, 363
airshed modelling, 381
albedo, 28, 29
AMOR model, 252
analysis equations, 167
annual gross flux of carbon, 318
ANSWERS model, 249
Antarctic circumpolar, 126
Antarctic circumpolar current, 126, 146
Antarctic circumpolar wave, 130
Antarctic continent, 279
Antarctic sea-ice cover, 276
antecedent soil moisture condition, 226
anticyclones, 112
Arctic sea-ice cover, 277
artificial neural network, 260

assimilation cycle, 167
atmosphere, 25, 75, 79, 404
atmosphere and land interactions, 27
atmosphere-and-land coupling, 46
atmosphere-and-ocean coupling, 46
atmosphere-biosphere interaction, 29
atmosphere-ice interaction scheme, 48
atmosphere-ocean interactions, 25, 254
atmosphere-ocean-ice interactions, 403, 407
atmospheric boundary layer, 39
atmospheric circulation, 254
atmospheric composition, 3
atmospheric dispersion, 144
atmospheric dispersion equation, 372
atmospheric dispersion models, 372
atmospheric general circulation models (AGCMs), 40
atmospheric global circulation models, 11
atmospheric motion, 25
Australian soils, 340
auto-correlation function, 39
autotropic respiration, 318
available energy, 190, 200

backwater effect, 260
bacteria catalysed reactions, 345
balance of mass and energy, 288
bank flow, 237
baroclinic pressure gradient, 150
baroclinic primitive equations, 387
barotropic boundary flows, 145
barotropic motion, 141
barotropic tides, 140
barotropic velocity field, 141
base flow, 237
baseflow recession, 237
BATS model, 252
below-ground, 241
Bering Sea, 314
BGC model, 322

bifurcation, 19
biochemical model of leaf photosynthesis, 321
biogenic hydrocarbon emission, 382
biogeochemical (BGC) model, 321
biogeochemical exchanges, 318
biogeophysical feedbacks, 329
biological models, 358
biological transformations, 359
biome models, 323
biophysical models, 323
biosphere, 29
biosphere-atmosphere feedbacks, 328
biosphere-atmosphere interactions, 317
biosphere-atmosphere model, 334
Biospheric aspects of the Hydrological Cycle (BAHC), 230
BOD, 239
Boltzmann constant, 348
boreal deforestation, 329
bottom-up approach, 334
boundary conditions, 141, 144, 169, 171
boundary currents, 162
boundary gradients, 147
boundary layer mixing, 147
boundary layers, 135, 140
Boussinesq equation, 238
breeding method, 119
breeding of growing modes, 54
Bryan-Cox achieved global simulations, 131
Bryan-Cox ocean model, 311
Bryan-Cox-Semtner ocean model, 153
bucket model, 179, 180
bulk aerodynamic method, 188
bulk canopy resistance, 192
bulk transfer coefficients, 192, 193, 389
bulk-transfer method, 51, 389, 390
Bénard convection, 57, 58

C_3 pathway, 320
C_3 plant, 320
C_4 pathway, 320
C_4 plants, 320
canopy air resistance, 192
canopy extinction coefficient, 338
canopy resistance, 191
canopy water storage, 195
CARAIB, 323
carbon cycle, 198
carboxylation, 320
Cartesian mixing coefficients, 138
Cartesian system, 139

catchment changes, 250, 251
catchment model, 251
catchment modelling, 257
causal forecasting, 267
cavitating fluid rheology, 309
CE-QUAL-W2 model, 243
cell mapping, 54
CENTURY model, 322
changes in land use, 328
channel flow, 235
channel routing, 235
channel-node-channel model, 244
chaos, 18, 52, 60
chaos system, 42
chaos theory, 57
chemical potential, 350
chemistry of ozone, 362
Chezy equation, 235
chloroplast, 320
Clapp and Hornberger model, 183, 185
climate and seasonal forecasting, 169
climate change, 254, 275
climate processes, 139
climatic efficiency, 337
climatological hydrographic properties, 160
cloud microphysics, 117
CO_2 assimilation pathway, 320
coarse resolution model of Bryan and Lewis, 150
coarse resolution ocean models, 134, 153
coastal ocean currents, 130
coastal ocean dynamics application radar, 163
coastal ocean environment, 163
coastal ocean models, 132, 142, 147
coastal, regional, and global-scale models, 171
cold fronts, 112
complexity, 217
computational environmental models, 9
conceptual models, 253
conceptualisation, 218, 219, 226–228, 233, 244, 245, 249, 255–257, 259
confined aquifer, 238
conservation equations, 31, 33
conservation equations for mass, momentum, heat and moisture, 131
conservation laws of mass, heat and salt, 132
conservation of water vapour, 83
conservation of entropy, 83

conservation of heat, 134
conservation of mass, 82
conservation of mass equation, 235
conservation of momentum, 82
conservation of momentum equation, 235
conservation of salt, 134
constitutive law, 291, 293
constrained least square method, 255
continental biosphere, 173
continental shelf processes, 147
continuity equation, 32, 83, 134, 219, 224, 232, 235–238, 241, 242, 292
continuous methods, 206
control theory, 260
convective boundary layer, 235
convective mixed layers, 151
conversion efficiency, 337
Coriolis effect, 26
Coriolis parameter, 33
cost function, 167, 171
coupled biosphere model, 328
coupled climate models, 127, 131, 150
coupled climate simulations, 145
coupled ice-ocean model, 302
coupled modelling, 171
coupled ocean-atmosphere models, 108, 146, 157, 169
coupled ocean-atmosphere system, 286
coupled sea ice-ocean model, 305
coupled sea-ice-upper-ocean model, 309
coupled SVATs-climate models, 334
coupled systems, 10
Cox ocean model, 307, 308
crop models, 319, 321
cryosphere, 275
cumulative distribution function, 44
cumulative infiltration, 227
curve number, 251
cybernetics, 12

Darcy velocity, 239, 241
Darcy's law, 50, 181, 224, 228, 237, 238
data assimilation, 45, 78, 91, 142, 165, 171, 255, 393, 420
data assimilation systems, 168
decadal time scale, 23
decision-making, 260, 267
deep moist convection, 205
deep water, 126
deforestation, 173
deformation equation, 305
density field, 139

density of sea water, 34
density surfaces, 139
depression storage, 229
desertification, 3, 211, 212
deterministic chaos, 17, 59
deterministic forecasts, 113
deterministic models, 113, 383
deterministic predictions, 45
deterministic systems, 42
deterministic water quality model, 265
DieCast model, 149
diffusion, 134
diffusion coefficient, 239
Digital Earth, 17
digital elevation model (DEM), 230, 250
discharge, 254
discharge forecasting, 254
discrete methods, 207
dispersion flux, 241
dissipative structure, 56–58
distorted-physics method, 307
distributed hydrological model, 248, 250, 252
diurnal time scale, 24
DO, 239
double Ekman spiral, 51
double-mass analysis, 223
downward solar radiation, 337
drag coefficients, 51, 404
drainage, 179, 224
drought, 104
Dupuit-Forchheimer approximation, 238
dust cycle, 412
dust emission, 411, 414
dust storms, 3, 416
dynamic biosphere-climate model, 336
dynamic effects, 204
dynamic global vegetation model, 334
dynamic instability, 11, 40
dynamic method, 11
dynamic-stochastic methods, 383
dynamic-stochastic models, 11, 46
dynamic-thermodynamic model, 292, 299
dynamical balances, 163
dynamical processes, 163
dynamics of coastal ocean flows, 131

earth system, 14
earth systematics, 12, 19
Earth's environment, 1
Earth's environmental system, 75

472   Index

Earth's rotation, 75, 126, 134
Earth's rotation vector, 31
earth-information system, 14
East Australian current, 150
ecological disturbance, 328
ecological modelling, 317, 319, 342
ecological models, 319, 328, 342
ecological systems, 328
ecological water demand, 269
eddies, 127
eddy activity, 162
eddy diffusivity tensor, 139
effective rainfall, 229, 245, 251, 256
effects of biodiversity, 327
Ekman transport, 160
El-Niño, 12, 26, 27, 105, 125, 126, 130, 386
El-Niño and Southern oscillation, 26
empirical method for short-term hydrological forecasting, 255
empirical orthogonal function, 165
energy and water cycles, 173
energy balance, 50, 173
energy balance equation, 281, 288
energy balance model, 288
energy cycle, 407
engineering cybernetics, 12
ensemble average, 39, 53
ensemble forecasting, 117, 255
ensemble mean, 56
ensemble of initial conditions, 117
ensemble prediction, 45, 52, 53, 55, 56
ENSO, 43, 108, 386
ENSO dynamics, 157
ENSO ocean models, 156
entrainment velocity, 407
environmental chamber experiments, 362
environmental changes, 2
environmental chaos, 64
environmental components, 1
environmental modelling, 1, 8, 9, 22, 23, 73, 218, 230
environmental prediction, 10, 11, 383
environmental processes, 11
environmental science, 1
environmental system, 22, 44, 224
enzyme or bacteria catalysed reactions, 345, 346
equation of sea-ice motion, 291
equation of state, 83, 131
equations of motion, 31, 33, 36
estuary, 244

eutrophication, 265
evaporation, 190, 285
evapotranspiration, 180, 203, 224, 233, 256, 318
extended range forecast, 79
external forcing, 9
external hydrological cycle coefficient, 222
extrapolative forecasting, 267

fast processes, 24
Fick's Law, 241
Fickian equation, 137
field capacity, 179
fine resolution Antarctic model, 157
finite-difference method, 85
flood forecasting, 257
flood routing method, 260
flood wave propagation, 235
fluid dynamics, 257
fluvial network model, 244
flux control, 227
flux-gradient relationship, 35, 50
fluxes, 34
fluxes of trace gases, 6
force and restore coefficients, 187
force-restore model, 180, 186
forcing predictability, 52
forecast error, 118, 165
forecast system, 169
forest growth models, 321
forest succession mode, 321
Fourier's law, 242
free surface, 147, 153
free surface condition, 141
free-surface approach, 387
free-surface equations, 141
freshwater fluxes, 143
friction slope, 235
friction velocity, 51, 390
frictional drag, 141
frontal dynamics, 154, 157
frozen, 242
frugality principle, 259
functional ionizable groups, 356
fuzzy theory, 265

gap models, 321
gas constant, 32
gaseous and particle pollutants, 361
Gaussian dispersion method, 372
Gaussian models, 371
Gaussian noise, 120
GBC model, 326

general circulation models (GCMs), 76, 81, 101, 157, 294
generalized inverse method, 167
generic reaction set (GRS) model, 362, 363
genetic algorithm, 259
geochemical tracers, 150
Geographic Information System, 16, 240, 243, 250, 254, 265, 384
geological time scale, 23
geophysical method, 263
geopotential surfaces, 145
geostrophic boundary values, 145
geostrophic flow, 141
geostrophic ocean current, 404
geostrophic wind, 404
GFDL Bryan-Cox primitive equation numerical model, 150
GFDL modular ocean model, 145
giant weighing lysimeters, 239
global carbon cycle, 318
global change, 317
global circulation spectrum model, 297
global climate models, 10
global climate system, 129
global coarse resolution model, 150
global eddy kinetic energy density, 162
global eddy-permitting models, 153
global ocean data assimilation experiment (GODAE), 168
global ocean models, 140, 161
global ocean simulations, 149
global ozone concentration, 5
global positioning system, 384
global radiation balance, 287
global sea-ice model, 309
global simulations, 129
global spectral model, 312
global telecommunication system, 99
global thermohaline transport, 153
global warming, 3, 101, 331
global weather systems, 130
global-scale high resolution models, 157
governing equations, 237
governing equations for ice, 36
governing equations for the atmosphere, 31
graphical user interfaces, 81
gravity, 215
gravity waves, 131, 141
Green-Ampt model, 228
greenhouse effect, 101
grey system theory, 265

grid systems, 132
grid-box resolution, 146
gross primary productivity (GPP), 321
ground-level ozone, 361
groundwater, 215, 237, 247
groundwater flow, 237
groundwater runoff, 237
groundwater/surface water interface, 256

Hadley cell, 282
Hadley circulation, 18
Haldane-Andrews equation, 346, 353
half-Brier score, 116
halocline, 406
HBV, 252
heat balance, 288
heat balance equation, 284
heat conduction, 242
heat energy equation, 286
heat flux, 144
heat flux formulation, 143
heat transport, 138
Heldane equation, 346
heterogeneity, 203, 223, 235, 242, 245
heterotrophic respiration, 318
Hibler ice model, 311
Hibler model, 291
high resolution regional ocean models, 169
high-frequency variability, 160
high-resolution finite element model, 169
higher resolution models, 171
hill slope hydrology, 245
Hooke-Jeeves algorithm, 259
Horton infiltration model, 227, 245
Horton overland flow, 236
human activities, 215, 250
hydraulic conductivity, 35, 50, 183, 224, 228
hydraulic functions, 185
hydraulic head, 36
hydraulic method, 236
hydrograph separation, 259
hydrological cycle, 215, 224, 269, 407
hydrological forecasting, 218, 224, 249, 254
hydrological method, 236
hydrological model, 218, 245, 254, 266
hydrological modelling, 218, 234, 235, 244, 250
hydrological processes, 215, 239, 250, 251

hydrological replacement time, 267, 268
hydrological system, 215, 217, 256
hydrosphere, 215
hydrostatic assumption, 86
hyetal station, 258

ice, 404
ice area forecast, 300
ice boundary, 288
ice content, 242
ice growth/decay, 308
ice thickness, 303, 308
ice-albedo feedback, 28, 29, 44
ice-dynamic process, 293
ice-ocean coupled model, 299, 307
ice-ocean model, 311
ice-ocean-atmosphere interactions, 44
ice-snow cover, 275, 281, 283, 284, 294, 298, 312, 315
ice-snow cover forecast, 299
idealised process-oriented models, 146
impervious strata, 238
implicit method, 141
Indian and East Asian Monsoon, 28
infiltration, 226, 245, 256
infiltration capacity, 227
infiltration rate, 227
infiltration volume, 227
initial conditions, 169
initial perturbations, 53
initial-boundary value problem, 82
Institute of Hydrology Distributed Model (IHDM), 249
integrated empirical rate, 361
integrated empirical rate (IER) model, 364
integrated empirical rate approach, 362
integrated environmental modelling, 384
integrated hydrological modelling system, 250–253
integrated modelling systems, 45
interaction predictability, 43
interactions, 21
interactive biosphere model, 333
interception, 232
interception of precipitation, 318
interflow, 237, 245
intermediate water, 150
internal dynamic predictability, 41, 52
internal hydrological cycle coefficient, 222
internal oceanic variability, 154

International Geosphere-Biosphere Program (IGBP), 230
intrinsic equations, 257
inverse methods, 165, 171
inverse models, 171
ionization, 354
isopycnal layer models, 136
isopycnal mixing scheme, 139
isopycnic ocean model, 149
iterative coupling, 333
IWR-MAIN model, 269

jet streams, 101
judgemental forecasting, 267

Kalman filter model, 261
Kalman filtering, 373
kernel function, 255
kinematic equation, 286
Kirchhoff law, 188
Kirchhoff transform, 183
Kolmogorov entropy, 44
kriging method, 223

La-Nina, 26
lagged average forecasting, 54
Lagrangian atmospheric dispersion model, 377
Lagrangian model, 372
Lagrangian particle models, 372
Lagrangian time scale, 39
Lagrangian velocity, 39
land surface parameterisation, 226
land surface, 27, 34, 173
land surface parameters, 209
land use/cover change, 251
land-atmosphere-ocean interaction, 286
land-surface energy balance, 187
land-surface modelling, 226
land-surface models, 194
land-surface processes, 35, 174, 233
land-surface schemes, 50, 176, 177, 207, 209, 385, 394
large dynamic system, 9
large-scale climate related simulations, 171
large-scale eddy-resolving ocean models, 162
large-scale forecasting, 254
large-scale models, 160
latent heat, 215, 233
law of conservation of mass, 237
leading time, 254
leaf area index, 251, 323, 334

leaf models, 319
leaf-water potential function, 195
leap-frog scheme, 89
lifetime, 216
linear diffusion wave equation, 260
linear input-output model, 261
linear perturbation model, 255
linear storage method, 247
linearised shallow water equations, 169
liquid water, 215
local mass-conserving scheme, 106
long timescale dynamics of vegetation, 333
long-lived systems, 25
long-range predictions, 11, 393
long-term environmental prediction, 62
Lorenz attractor, 62
Lorenz butterfly attractor, 18
Lorenz system, 18, 60
low flow, 237, 252
low-level jets, 174
Lyapunov exponents, 40, 59, 62
Lyapunov vectors, 54

macroscopic stable structure, 57
Manning equation, 236
Manning roughness, 237
Markov process, 39, 115
Marquardt method, 240
mass conservation equation, 36
mass continuity equation, 303
matric head, 242
matrix potential, 186
maximum stomatal conductance, 210
mean areal precipitation, 223
mechanical deformation, 304
mechanical wind forcing, 142
medium range forecast, 105
medium range forecasting, 99
Mellor-Yamada turbulence sub-model, 160
method of adjoint operator, 71
Michaelis-Menten equation, 346
microbial cells, 354
mid- and long-term, 235
mid- and long-term forecasting, 254
mid- and long-term hydrological forecasting, 262
missing data, 223
missing sink, 318
mixed boundary conditions, 143
mixing rates, 139
mixture method, 207

MM5 model, 255
model assessment, 160, 171
model grid systems, 171
model of a snow cover, 288
model of sediment transport, 266
model validations, 163
models of the oceanic carbon cycle, 150
moisture availability factor, 180
moisture storage, 180
MOM grid system, 145
momentum balance, 132, 291, 306
momentum equations, 133, 142, 303
momentum transfer, 257
Monin-Obukhov length, 193
Monin-Obukhov similarity theory, 188, 192, 389
Monod equation, 345, 346
Monod model, 346
monsoon, 298
monsoonal circulation, 101
Monte Carlo approach, 52
Monte Carlo forecast, 53, 119
Monte Carlo procedure, 119
Monteith model, 322
mosaic method, 208
Moser equation, 345
multi-layer convection-diffusion model, 180
multi-layer soil model, 181
multiple Monod equations, 345
multiple substrate limitation, 347
Muskingum routing model, 236, 245

natural cybernetics, 14–16, 19, 20, 66, 73
natural disasters, 6, 77
natural environment, 1
Navier-Stokes equations, 42, 132
nested grid model, 307
net ecosystem production (NEP), 322
net ecosystem productivity, 198
net primary productions, 322
net primary productivity (NPP), 321
net radiation, 174
non-deterministic water quality model, 265
non-eddy-resolving models, 160
non-linear interactions, 11, 38, 40, 43
non-linear Lurmann-Carter-Coyner (LCC) mechanism, 382
non-linearity, 8
non-point pollution model, 266
nonhydrostatic convection, 141
nonpoint source pollution, 250

normalised difference vegetation index, 210, 325, 396
North Atlantic models, 155
NSW shelf model, 158
nudging technique, 395
numerical down-scaling, 110, 111
numerical experiments, 224
numerical forecast of sea ice, 301
numerical schemes, 171
numerical sea-ice prediction, 303
numerical weather prediction, 76, 78, 80, 117, 254
numerical weather prediction models, 10, 142

observation error, 166
observing systems, 168
ocean, 26, 403
ocean circulation, 126, 131, 136, 138, 140, 146, 165
ocean circulation and climate advanced modelling project, 153
ocean circulation models, 171
ocean climate models, 130
ocean currents, 125, 126, 149
ocean dynamics, 157
ocean forecasting, 168
ocean general circulation models, 125, 293
ocean model forcing, 145
ocean modelling, 125, 138
ocean models, 125, 132, 134, 136, 145
ocean system, 125
ocean-surface scheme, 50
oceanic boundary layer, 309
oceanic eddies, 153
oceanic process, 140
oceanic variability, 126
open system, 57
operational forecast process, 81
operational forecasting, 168
optimal interpolations, 420
optimisation, 67, 68
optimisation control theory, 259
optimum engineering, 70
ordinary attractor, 60
ordinary least square method, 255
Oregon continental shelf circulation, 168
organised complexity, 217
Orlanski radiation, 157
osmotic head, 225
overland flow, 235–237

ozone concentration, 361, 366, 369, 422
ozone hole, 5
ozone layer, 5

parallel ocean climate model, 153
parameterisation, 129
parameterisation of subgrid-scale processes, 171
patch model, 224
peculiarity, 217
Penman combination method, 189
people's experience, 261
percolation flow, 237
periodic behavoir, 217
pH effects, 347
pH model, 346
phase space, 60
phase space geometry, 43
Philip two-term model, 229
photo-synthesis, 42
photo-synthetic effect, 29
photochemical processes, 364
photochemical reaction, 363
photochemical smog, 362–364, 366, 369, 373, 381
photochemistry of ozone formation, 361
photolysis of nitrogen dioxide, 362
photolytic rate, 367
photolytic rate coefficient, 365, 367
photosynthesis, 198, 240, 318, 319
photosynthetically active radiation, 337
physical processes, 140
planetary boundary layer, 313
plant growth, 224
plant phenology, 336
plant production and phenology model, 336
plume, 135
polar ice prediction system, 302
poleward heat transport, 140, 156
pollutant transport model, 266
ponding water, 227
population, 2
porous media, 256
positive feedback, 294
positive weight functions, 168
potential evaporation, 189
potential evapotranspiration, 233
potential vorticity, 146
PPP model, 340
practical data assimilation, 168
precipitation, 222, 227, 254, 258
precipitation recycling ratio, 221
precursor emissions, 369, 370

predictability, 11, 38, 40, 44
predictability time scale, 39, 40, 44
prediction indicator, 275
predictive ability of current ecological models, 327
predictive systems, 171
pressure systems, 130
primitive equation model, 299, 311
primitive equation numerical model, 157
primitive equations, 85, 132, 145
primitive euqation ocean circulation model, 131
Princeton ocean model, 147
probability density function, 52, 117, 372
process based BGC model, 322
processed based model, 322
processes based model, 249

QUAL-II model, 240
quantitative precipitation forecasting, 255
quantum state, 348
quasi-dynamic-stochastic method, 19
quasi-geostrophic balance, 157
quasi-geostrophic thermal wind, 150

rainfall-runoff models, 226, 247
rainfall-runoff process, 250
random behavoir, 217
random model, 299
randomness, 217
rate of vapour diffusion, 190
re-aeration, 240
reach forecasting, 264
real time hydrological forecasting, 261
real-time air quality prediction models, 95
regional coastal ocean models, 163
regional models, 129, 141
regional ocean model, 144
regional, primitive equation model, 168
regression based model, 322
remote sensing, 142, 226, 250, 254, 265
remote-sensing based model, 322
representative elementary length, 258
reservoir water quality model, 243
resolution-dependent parameterization, 137
rheological model, 306
Richards equation, 181–183, 224, 227
Richardson number, 140

rigid-lid approach, 387
river, 130
river discharge, 254
river plume models, 171
root length density function, 225
root water uptake, 224, 227, 230, 231
Rosenbrock search method, 255
Rossby radius, 134
roughness lengths, 194, 251
routing time, 254
runoff, 180, 224, 285
runoff formation, 245

Saint-Venant equations, 235
salt content, 33
salt flux, 144
saturated concentration of dissolved oxygen, 240
saturated zone of soil, 227
saturation coefficient, 256
saturation overland flow, 236
savanna ecosystems, 339
savanna fires, 339
scale, 217, 253, 258
scenario simulation, 266
sea level, 386, 387
sea surface heat fluxes, 153
sea-ice, 277, 296, 299
sea-ice area, 314, 315
sea-ice concentration, 295
sea-ice extent, 130
sea-ice forecast, 299
sea-ice model, 293, 305–307, 404
sea-ice-mixed layer-isopycnal ocean model, 299
sea-level, 131
sea-level rising, 3
seasonal time scale, 24
second-order Leapfrog scheme, 88, 89
sediment equation, 237
sediment transport equation, 237
self-organization, 57
semi-Lagrangian scheme, 89, 90, 106
sensible- and latent-heat fluxes, 174, 176
sequential data assimilation, 168
sequential methods, 167
severe weather, 7
shallow convection, 106
shallow surface mixed layer, 163
shallow water equations, 86
shear stress, 51
short and middle-range predictions, 11
short range forecasting, 78, 96

short timescale dynamics of vegetation, 333
short-lived systems, 25
short-term forecasting, 254
short-term hydrological forecasting, 255
sigma-coordinate models, 136
similarity, 217
simple analytic and linear vorticity models, 131
simple linear model, 255
simple linear variable gain factor model, 256
single substrate limitation, 347
single-layer soil model, 176
singular vector decomposition approach, 119
sink and source term, 236, 239, 241, 242
slab ocean model, 386
slaving theory, 58
slow processes, 24
smog, 361
smog producing processes, 362
Smolarkiewicz advection scheme, 160
snow and ice cover, 28
snow cover, 277–279, 289, 312, 313
snow model, 299
soil conservation service (SCS) method, 251
soil erosion, 237, 385, 393
soil heat equation, 242
soil heat transfer, 242
soil moisture, 49, 104, 174, 175, 179, 203, 211, 224, 385, 393–395, 398
soil moisture availability, 318
soil water, 215, 224, 225
soil water availability, 218, 231
soil water content, 224, 226, 252, 256, 259
soil water movement equation, 224, 226, 242, 252
soil-vegetation-atmosphere transfer (SVAT), 251
solar black spots, 42
solar constant, 42, 288
solar radiation, 21, 28, 215
solute movement, 241
sophisticated ocean models, 168
sophistocated assimilation, 168
source and sink term, 244
Southern oscillation, 26
space-time model, 380
space-time scales, 129

spatial and temporal scales, 23
spatial downscaling, 255
specific discharge, 239
specific heat capacity, 215, 242
spectral models, 76, 86
SST anomaly, 105
stability criterion, 89
stand models, 321
standard chemical potential, 350
state, 215, 216, 219, 232
statistic analyses, 314
statistical down-scaling, 110
statistical methods, 11, 254
statistical models, 113, 253
statistical predictions, 45
Stefan-Boltzmann constant, 188
stemflow, 232
stochastic water quality model, 265
stochastic-dynamic modelling, 45
stomata, 320
stomatal resistance, 207
storage-discharge relation, 237, 238
strange attractors, 59–62
stratospheric ozone, 361
stream water quality model, 239, 241
streamflow, 237
streamflow recession, 237
streamfunction field, 147
Street-Phelps model, 240
sub-seasonal time scale, 24
subgrid-scale heterogeneity, 208
subgrid-scale parameterisations, 171
substrate inhibition, 347
substrate inhibition model, 346
substrate molecules, 347
subtropical ridges, 101
sunspot, 300
surface albedo, 188, 196, 197, 329
surface emissivity, 188
surface energy and water budgets, 176
surface energy-balance equation, 49
surface flow, 237
surface forcing, 141, 143
surface heterogeneity, 204, 206
surface information, 168
surface mixed layer, 149
surface net radiation, 188
surface pressure, 141
surface pressure gradients, 141
surface runoff, 179
surface soil hydrological processes, 173
surface velocity data, 168
surface water, 215

surface water solute transport models, 241
surface wind forcing, 157
sustainable flow, 237
symmetry pH equation, 346
synergetics, 12, 58
synoptic method, 262
synoptic-statistic method, 313
synoptic-statistical analyses, 314
system hydrologique European (SHE) model, 249
system noise, 166
systematic analysis, 260
systematical approach for short-term hydrological forecasting, 255

techniques of short-term forecasting, 260
tension water capacity, 245, 246
terrestrial biosphere, 318, 342
terrestrial ecosystem model (TEM), 322
terrestrial radiation, 29
thermal conductivity, 35, 50
thermocline, 406
thermodynamic equilibrium, 131
thermodynamic growth rates, 304
thermohaline circulation, 126, 139
thermohaline forcing, 143, 144
third order upwinding scheme, 91
three-dimensional dispersion model, 95
throughflow, 233, 249
tidal constituents, 171
tidal flows, 131, 171
tidal forcing, 142
tidal models, 125
tides, 125
time scales, 1, 130
time series analysis, 259
top-down approach, 335
TOPMODEL, 249
trade winds, 26
transient storage, 241
transient zone, 241
transition probabilities, 54, 60
transition probability in phase space, 62
transpiration, 29, 191
transport of heat, 129
transportation of moisture and heat, 286
tropical deforestation, 329, 330
tropical deforestation experiments, 330

Tsao-Hanson equation, 345, 346, 351, 353
turbulence sub-model, 147
turbulent exchange drag coefficient, 142
turbulent stratified fluid, 125
typhoons, 80, 97

unconfined aquifer, 238
underlying surface heat status method, 262
unit hydrograph, 245, 251, 255
universal soil loss equation, 237
unsaturated zone of soil, 227
upwelling, 147
upwelling favourable winds, 159, 160, 165
urban airshed models, 361
US Navy sea ice forecasting system, 309

vadose zone, 224
variational data assimilation, 121, 420
vegetation, 318
vegetation dynamics, 333
vegetation response to climate changes, 335
vegetation water, 215, 230
vegetation–atmosphere feedbacks, 330
vegetation-atmosphere interaction, 332
vertical convection, 140
vertical motion, 141
viscosity, 137
viscosity coefficient, 138
viscous-plastic constitutive law, 305
viscous-plastic rheology, 299
von Karman constant, 193

walker cell, 282
walker circulation, 26, 386
WASP4 model, 240
water transfer, 251
water availability, 268
water balance, 222, 256
water balance equation, 256, 258
water capacity, 256
water circulation, 29
water convection, 126
water cycle, 284
water demand, 269
water fluxes, 36
water level forecasting model, 260
water masses, 126
water quality, 239
water quality forecasting, 264
water quality model, 239, 264

water resources, 3
water resources decision support system, 267, 268
water resources management, 269
water retention, 183
water stage, 254
water supply, 244
water supply and water demand, 254, 266, 267
water supply and water demand forecasting, 267
water temperature model, 265
water transfer, 257
water transfer equation, 256–259
water turbidity, 265
water vapour, 215, 219
water-balance model, 49
water-mass, 140
water-mass formation, 160

WATFORE model, 271
wave climate model, 125
weather forecasting systems, 168
weather predictability, 17
weather prediction, 275
western boundary currents, 126, 127, 129, 154, 162
wilting point, 179, 195
wind and thermohaline factors, 130
wind and water erosion, 3
wind erosion, 412
wind stress, 126, 142
world ocean circulation experiment, 154
world ocean model, 131

Xinanjiang model, 233, 245, 246, 253, 259

Yangtze River, 282

Printing (Computer to Film): Saladruck Berlin
Binding: Stürtz AG, Würzburg